VOLUME 3

Surfactants in Solution

VOLUME 3

Surfactants in Solution

Edited by

K. L. Mittal

IBM Corporation, Hopewell Junction, New York

and

B. Lindman

University of Lund, Lund, Sweden

PLENUM PRESS • NEW YORK AND LONDON

7227-3549

CHEMISTRY

Library of Congress Cataloging in Publication Data

Main entry under title:

Surfactants in solution.

"Proceedings of an international symposium on surfactants in solution, held June 27–July 2, 1982, in Lund, Sweden"—T.p. verso.
Includes bibliographical references and indexes.
1. Surface active agents—Congresses. 2. Solution (Chemistry)—Congresses. 3. Micelles—Congresses. I. Mittal, K. L., 1945–   . II. Lindman, Björn, 1943–   .
TP994.S88   1983   668'.1   83-19170
ISBN 0-306-41483-X (v. 1)
ISBN 0-306-41484-8 (v. 2)
ISBN 0-306-41485-6 (v. 3)

Proceedings of an international symposium on Surfactants in Solution,
held June 27–July 2, 1982, in Lund, Sweden

© 1984 Plenum Press, New York
A Division of Plenum Publishing Corporation
233 Spring Street, New York, N.Y. 10013

Printed in the United States of America

PREFACE

This and its companion Volumes 1 and 2 document the proceedings of the 4th International Symposium on Surfactants in Solution held in Lund, Sweden, June 27–July 2, 1982. This biennial event was christened as the 4th Symposium as this was a continuation of earlier conferences dealing with surfactants held in 1976 (Albany) under the title "Micellization, Solubilization, and Microemulsions"; in 1978 (Knoxville) under the title "Solution Chemistry of Surfactants"; and in 1980 (Potsdam) where it was dubbed as "Solution Behavior of Surfactants: Theoretical and Applied Aspects."The proceedings of all these symposia have been properly chronicled.[1,2,3] The Lund Symposium was billed as "Surfactants in Solution" as both the aggregation and adsorption aspects of surfactants were covered, and furthermore we were interested in a general title which could be used for future conferences in this series. As these biennial events have become a well recognized forum for bringing together researchers with varied interests in the arena of surfactants, so it is amply vindicated to continue these, and the next meeting is planned for July 9–13, 1984 in Bordeaux, France under the cochairmanship of K. L. Mittal and P. Bothorel. The venue for 1986 is still open, although India, inter alia, is a good possibility. Apropos, we would be delighted to entertain suggestions regarding where and when these biennial symposia should be held in the future and you may direct your response to KLM.

The response to these biennial events has been growing and as a matter of fact we had to limit the number of presentations in Lund. Even with this restriction, the Lund Symposium program had 140 papers from 31 countries by more than 300 authors. So it is quite patent that this meeting was a veritable international symposium both in spirit and contents. It should be added that the program contained a number of overviews by prominent researchers, as it is imperative to include some overviews to cover the state-of-knowledge of the topic under discussion.

As for these proceedings [containing 126 papers (2156 pages) by 324 authors from 29 countries], these are arranged in nine parts. Parts I and II constitute Volume 1; Parts III–VI comprise Volume 2; and Parts VII–IX are the subject of Volume III. Apropos,

the papers in the proceedings have been rearranged (from the order
they were presented) in a more logical manner. Among the topics
covered include: Phase behavior and phase equilibria in surfactants
in solution; structure, dynamics and characterization of micelles;
thermodynamic and kinetic aspects of micellization; mixed micelles;
solubilization; micellar catalysis and reactions in micelles;
reverse micelles; microemulsions and reactions in microemulsions;
application of surfactants in analytical chemistry; adsorption and
binding of surfactants; HLB; polymerization of organized surfactant
assemblies; light scattering by liquid surfaces; and vesicles.

    A few salient aspects of these proceedings should be recorded
for posterity. All papers were reviewed by qualified reviewers so
as to maintain the highest standard. As a result of this, most
papers were returned to respective authors for major/minor revi-
sions and some did not pass the review. In other words, these
proceedings are not simply a collection of unreviewed papers,
rather the peer review was an integral part of the total editing
process. It should be added that we had earnestly hoped to include
discussions at the end of each paper or group of allied papers, but
in spite of constant exhortation, the number of written questions
received did not warrant undertaking such endeavor. However, it
must be recorded that there were many spontaneous and brisk discussions both formally in the auditorium and informally in other
more suitable (more relaxed) places. Most often the discussions
were enlightening, but on occasions one could feel some enthalpy as
these tended to be exothermic.

    Also a general concern was expressed about the possibility of
correlating research done in different laboratories. In particular
in the microemulsion field it was felt that a few selected, stable
and well-defined systems should be chosen for collaborative work
between a number of active groups using a variety of techniques.
The response to such discussion (initiated by Prof. M. Kahlweit,
Göttingen) was very heartening and culminated in the so-called Lund
Project (coordinator Prof. P. Stenius, Stockholm), which is a coor-
dinated collaboration between a number of research groups in different countries. A report meeting was hosted by M. Kahlweit in
Göttingen in the spring 1983 and further results will be presented
in Bordeaux in 1984. In addition, throughout the meeting, small
groups of people were seen to be leisurely discussing more specific
topics of mutual interest. In other words, there were ample and live-
ly discussions in various forms during the span of this symposium.

    Coming back to the proceedings, even a cursory look at the
Table of Contents will convince even the most skeptic that the
field of surfactants in solutions has come a long way, and all
signals indicate that the accelerated tempo of interest and
research in this area is going to continue. Also it is quite clear
that as we learn more about the amphiphilic molecules, more excit-

ing research areas and pleasant applications will emerge. It should
be added that these proceedings cover a wide spectrum of topics by
a legion of prominent researchers and provide an up-to-date coverage of the field. The coverage is inter- and multidisciplinary and
both overviews and original unpublished research reports are included. Also it should be pointed out that both the aggregation and
adsorption of surfactants are accorded due coverage. These proceedings volumes along with the earlier ones in this vein (total $\sim$ 5000
pages) should serve as a repository of current thinking and research dealing with the exciting field of surfactants in solution.
Also these volumes should appeal to both veteran and neophyte researchers. The seasoned researchers should find these as the source
for latest research results, and these should be a fountainhead of
new research ideas to the tyro.

Acknowledgements: One of us (KLM) is thankful to the appropriate management of IBM Corporation for permitting him to participate in this symposium and to edit these proceedings. His special
thanks are due to Steve Milkovich for his cooperation and understanding during the tenure of editing. Also KLM would like to
acknowledge the assistance and cooperation of his wife, Usha, in
more ways than one, and his darling children (Anita, Rajesh, Nisha
and Seema) for creating only low decibel noise so that Daddy could
concentrate without frequent shoutings. The time and effort of the
reviewers is sincerely appreciated, as the comments from the peers
are a desideratum to maintain standard of publications. We are
appreciative of Phil Alvarez, Plenum Publishing Corp., for his
continued interest in this project. Also we would like to express
our appreciation to Barbara Mutino for providing excellent and
prompt typing service. Our thanks are due to the members of the
local Organizing Committee (Thomas Ahlnäs, Thomas Andersson, Gunnar
Karlström, Ali Khan, Mary Molund, Gerd Olofsson, Nancy Simonsson
and Marianne Swärd) who unflinchingly carried out the various
chores demanded by a symposium of this magnitude. The financial
support of the Swedish Board for Technical Development, the Swedish
National Science Research Council, and the University of Lund is
gratefully acknowledged.

K. L. Mittal   B. Lindman
IBM Corporation   University of Lund
Hopewell Junction, NY 12533   Lund, Sweden

1.   K. L. Mittal, Editor, Micellization, Solubilization and
     Microemulsions, Vols. 1 & 2, Plenum Press, New York, 1977.
2.   K. L. Mittal, Editor, Solution Chemistry of Surfactants,
     Vols. 1 & 2, Plenum Press, New York, 1979.
3.   K. L. Mittal and E. J. Fendler, Editors, Solution Behavior
     of Surfactants: Theoretical and Applied Aspects, Vols. 1 & 2,
     Plenum Press, New York 1982.

CONTENTS

PART VII. REVERSE MICELLES

PART VIII. MICROEMULSIONS AND REACTIONS IN MICROEMULSIONS

CONTENTS OF VOLUME 1

PART II.   STRUCTURE, DYNAMICS AND CHARACTERIZATION
OF MICELLES

CONTENTS OF VOLUME 2

PART III. THERMODYNAMIC AND KINETIC ASPECTS
OF MICELLIZATION

PART IV.   SOLUBILIZATION

PART V. MICELLAR CATALYSIS AND REACTIONS IN MICELLES

Part VII
Reverse Micelles

KINETIC CONSEQUENCES OF THE SELF ASSOCIATION MODEL IN REVERSED MICELLES

Charmian J. O'Connor and Terence D. Lomax

Department of Chemistry
University of Auckland
Private Bag, Auckland, New Zealand

The aggregation behaviour of surfactants in non-polar solvents has been reviewed in terms of the phase separation model and the sequential indefinite self association model. It is considered that the existence of $(AB)_h$ type aggregates probably holds the answer to most of the apparent anomalies in these systems.

The decomposition of p-nitrophenylacetate has been measured, at 341 K, in benzene solutions of a series of α,ω-alkyldiamine bis(dodecanoates) and in a solution of dodecylammonium propionate. The non-linear dependence of the observed pseudo first order rate constant, k_ψ, on salt concentration may be analysed by application of the pseudophase model to these "reversed micellar" systems. However, a more valid interpretation to reactivity in these systems may lie in application of the sequential indefinite self association model. The observed rate data may be mimicked by a rate equation which includes contributions due to the square of the monomer concentration and to the concentrations of dimer and larger oligomers. At high concentrations of detergent the increase in dielectric constant causes an increase in the effective concentration of monomer and this, in turn, is reflected in an increased dependence of k_ψ upon detergent concentration. It is suggested that aggregation may have an inhibitory effect upon the kinetics of the reaction.

INTRODUCTION

Theories and models for aggregation[1] of reversed micellar systems, in which the aggregates in the nonpolar solvent are held together by electrostatic forces and are size-limited by steric effects,[2-4] were until recently based upon those derived from aqueous systems[2,5,6] The phase separation model postulates *ab initio* the micellization to be a phase transition. Use of such a model was advocated by Shinoda and Hutchinson,[7] and successfully applied by Singleterry[2] and Fowkes,[8] to describe the aggregation of dinonylnaphthalene sulphonates in benzene.

The narrow distribution of aggregate sizes centering around an average micellar size ('monodispersity') seen in these studies has also been observed by Kon-no and Kitahara.[9] The phase separation model follows the description of a two-phase equilibrium, and thus overemphasizes cooperativity with respect to aggregation. Assuming ideality, the chemical potential of the surfactant in each phase must be equal, and thus is given by Equation (1).

$$\mu^{\ominus}_{micelle} = \mu^{\ominus}_{solution} + kTln(CMC) \qquad (1)$$

In its simplest form the model does not contain a size-limiting step and therefore is of little value in understanding the formation of the small aggregates seen in nonpolar media.

The mass action model develops from the application of the mass action law to the overall aggregation process, Equation (2), where

$$nM \underset{}{\overset{K_n}{\rightleftharpoons}} M_n \qquad (2)$$

K_n is the association constant of the "all or nothing" process. Thus, with conservation of mass, one obtains Equation (3), where

$$\frac{[m]}{[D]} + n(\frac{[m]}{[D]})^n [D]^{n-1} K_n = 1 \qquad (3)$$

[D] is the total molal concentration of detergent. From Equation (3), assuming that K_n is the product of n-1 individual and equal mass action constants, Equation (4) is obtained, and CMC is then obtained from Equation (5).

$$\frac{[m]}{[D]} + n\frac{[M_n]}{[D]} = 1 \qquad (4)$$

$$CMC = \frac{1}{K} = [D] \qquad (5)$$

where [D] . K = 1.

This model stresses the monomer as one aggregation state and allows for only one other aggregational state, (M_n); it therefore cannot account for a distribution of molecular weights. Further it does not address the existence of a size limit of the aggregate, or why the aggregation process should be cooperative.

Many experimental observations in these nonpolar solvent systems show relatively smooth transitions from monomer $\rightleftharpoons$ dimer $\rightleftharpoons$ trimer *etc.* with concentration-dependent growth of the aggregates. [1]H n.m.r. work on alkylamine carboxylates[10-14] suggests that the aggregation of surfactant molecules in organic solvents occurs in a stepwise manner, rather than in a monomer $\rightleftharpoons$ n-mer equilibrium, and that the best fit to the data occurs when all the association equilibrium constants are equal, *i.e.* in the set of equilibria given in Equation (6) the identities of Equation (7) can be assumed to hold.

$$\text{monomer} \underset{}{\overset{K_{12}}{\rightleftharpoons}} \text{dimer} \underset{}{\overset{K_{23}}{\rightleftharpoons}} \text{trimer} \rightleftharpoons \dots \rightleftharpoons \text{n-mer} \tag{6}$$

$$K_{12} = K_{23} = \dots = K_{ij} \tag{7}$$

The multiple equilibrium model, which corresponds to the type of aggregation shown in Equation (6), does not provide for any critical concentration. On the basis of the experimental data for aggregation, Kertes and Gutmann[15] have argued that the concept of a critical micelle concentration (as used for the aqueous system) is not applicable. They say that the search for a CMC is likely to be the result, at least in some cases, of preconceived concepts visualized in terms of imitation of the behaviour of surfactants in aqueous media.[16]

Eicke and Denss[17] have analysed several surfactant systems according to the multiple equilibrium model and the pseudo-phase model, and have suggested that the search for evidence of a CMC in a nonpolar medium is not meaningful unless the aggregation data can be fitted by the pseudo-phase model. They studied three alkyl-ammonium surfactants and of these only dodecylamine benzoate (DABz) showed aggregation behaviour which corresponded clearly to the pseudo-phase model. The behaviour of dodecylamine propionate (DAP) and dodecylamine formate (DAF) could not be clearly attributed to either model. With the notable exception of dodecylamine benzoate, the alkylamine carboxylates so far studied do not seem to form aggregates to which the term "reversed micelle" can be applied. Analysis of vapour pressure osmometry[10,18-23] and [1]H n.m.r. data gives rise to small aggregation numbers and a sequential indefinite type of association, Equation (6).

The positron annihilation technique,[18-20] which utilises the dependence of the formation and reactions of the positronium ion upon its environment, shows abrupt changes in I_2 (the intensity of the long-lived component in the positron lifetime spectrum) over a small concentration range. In the DAP cyclohexane system this concentration range corresponds to the "operational CMC" as determined by [1]H n.m.r.[11] The change in I_2 does not correspond to the smooth continuity of physical properties predicted by Kertes and Gutmann.[15] Jean and Ache[19] considered that the nature of their results, the coincidence of the abrupt breaks in the I_2 *versus* concentration profiles with Fendler's "operational CMS's",[11] coupled with the correlation of these with chain length of the alkylamine carboxylates, solvent polarities and rate measurements, and Eicke's observation of abrupt changes in electric field measurements,[24] were not reconcilable with a multiple equilibrium model. They suggested that the observed breaks in the I_2-concentration curves may signal phase transitions, similar to the formation of closed aggregates postulated in Eicke's pseudo-phase model.[24] However, Jean and Ache did not feel justified in concluding that this process could be considered as the formation of reversed micelles or that the surfactant concentration at which the phase transition occurs could be called the CMC in these systems.

Kertes[16] emphasizes that the stepwise aggregation process in water is different from the stepwise aggregation of ionic surfactants occurring in hydrocarbon solvents. In the latter case the dipole-dipole interactions which promote aggregation give rise to the characteristic stability of aggregates with low numbers of monomers and aggregates at very low concentrations of detergent (typified, for example, by alkylamine carboxylate systems).

Levy[25] has proposed a model for this type of aggregation by treating the aggregates as rigid spheres containing point dipoles. The average interaction energy between the solute molecules was regarded as a composition of a) a repulsive energy component due to the molar volume of the solute particles, and b) an attractive energy component which includes the dipole-dipole interaction. Eicke[3] has suggested that surfactant monomers can be regarded as both dipoles and dissociating species, continuously distributed around a polar core of volume V. The aggregation process can then be treated in terms of the contribution of the dispersion and coulombic interactions to the Gibbs free energy change.

Eicke[24] used the dielectric field effect technique to investigate the aggregation behaviour of Aerosol OT, (AOT, sodium di-2-ethyl-hexylsulfosuccinate) in cyclohexane, and suggested that the formation of trimers and hexamers is a fundamental process in the aggregation of this surfactant. This approach had been previously applied to solutions of alkylamine salts in benzene, xylene and dioxane,[26] with the similar conclusion that $(AB)_3$ and $(AB)_6$ type aggregates were present in significant proportions. (However, see reference 14 which suggests that trimers will disproportionate to dimers and tetramers for alkylamine carboxylates.) This latter study was supported by conductivity[27] and vapour pressure osmometry[28] data for these salt solutions.

The $(AB)_n$ type aggregate probably holds the answer to most of the apparent anomalies seen in these systems. A recent study[29] has investigated, by cryoscopic and calorimetric methods, the non-ideal behaviour of tri-n-octylamine halides in benzene solution and has analysed the results by use of an association model in which the first stepwise association reaction occurs less readily than the remainder. Figure 1 shows the distribution of monomeric, dimeric, trimeric and higher oligomeric species of tri-n-octylamine bromide in benzene solution at the freezing point of benzene. The experimental and calculated values agreed to within 10^{-3} mol kg^{-1}.

Recent studies[30,31] have indicated that there may be preferred complexes (besides the obvious 1:1 complex) between amines and carboxylic acids, that have three carboxylic acid groups to one amine molecule. These complexes, which were noted in studies on carboxylic acid/amine mixtures with no solvent present, were described as "orientationally ill-defined 3:1 aggregates with extreme influence on thermodynamic and structural properties". The 3:1 complex was interpreted as being the result of an interaction between a highly polar form of the 1:1 complex and the carboxylic acid dimer.[30,32]

De Tar and Nowak,[33] in an i.r. investigation of various carboxylic acids with triethylamine, consider not only the formation of 1:1 salts and 2:1 salts, but also state that it is necessary to allow for the dimerisation of the acid. They note that ion pair formation is the expected behaviour for primary and secondary amines, but triethylamine with acetic acid or benzoic acid in CCl_4 undergoes hydrogen bonding without proton transfer.

Bruckenstein and Saito,[34] in a study by i.r. and v.p.o. techniques, showed that it is possible to establish the stoichiometry of the acid-base reactions. They postulated three types of ion pairs: BH^+X^-, $BH^+HX_2^-$ and $BH^+X(HX)_2^-$ and noted that the tendency of these ion pairs to form oligomers was dependent upon

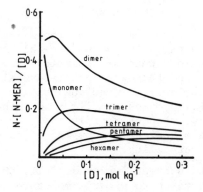

Figure 1. The distribution of monomeric, dimeric and higher oligomeric species of tri-n-octylamine bromide in benzene solution at the freezing point of benzene. From reference 29.

the strength of the acid or base. (Jasinski *et al.*[35] have found
that the reaction of nitrophenols with aliphatic amines in benzene
produces only the 1:1 complex.) Bruckenstein and Saito believed, in
contrast to Yerger and Barrow's results[36] in CCl_4, that the absence
of a marked effect of oligomer formation upon special features was
the result of long range electrostatic forces rather than hydrogen
bonding stabilizing the oligomers. Values for the aggregation size
of these oligomers noted in Bruckenstein and Saito's study (1 for
tertiary amines to 6 for primary amines, but these values are
structure and concentration dependent) agree with the trends found
by Lo *et al.*[10] The reaction of the ion pair or higher oligomer
with additional acid (BHX + HX) was shown to be non-quantitative.
(A similar non-quantitative reaction[36] of tertiary amines with
carboxylic acids in formation of the 1:1 complex gives rise to
average aggregation numbers < 1).

Whilst not implying that these types of complexes do necessarily
exist in solutions of alkylamine carboxylate systems in nonpolar
solutions, their existence, as well as that of the simple $(AB)_n$
type aggregate, must be considered in formulation of a possible
answer to some of the phenomena that occur in these surfactant
solutions.

Analysis of the data in such systems should also consider the
possibility of ionic species. A theoretical model,[37] which
assumed the possible existence of "some or all of the following
species: HA, H^+, A^-, AHA^-, HAH^+, $(HA)_2$, $A(HA)_2^-$, $H(HA)_2^+$, $(HA)_3$,
$A(HA)_3^-$, and $H(HA)_3^+$", was able to explain experimental conductance,
potentiometric and photometric data from the literature in terms of
appropriate acid-base equilibrium systems. The existence of these
complex ions (reference 27, p 2387), together with a simple
aggregation model, and the non-applicability of concepts derived
from aqueous systems indicate the problems encountered in interpreting physical and kinetic measurements in these systems.

EXPERIMENTAL

Dodecylammonium propionate (DAP) and the α,ω-alkyldiamine
bis(dodecanoates) were prepared by the general method of Kitahara.[38]
The C_4, C_5 and C_7 diamines were precipitated only after cooling
the solution to 253 K and they were not recrystallised because re-
crystallisation successively lowered the melting point 1-2 K. A
study[30] of the properties of carboxylic acid/amine mixtures
indicates that the m.p. changes rapidly in the region of stoichiometric equivalence, and that a 3:1 complex of carboxylic acid:
amine also has marked stability.

All surfactants and solvents were carefully dried and, as far
as possible, care was taken to exclude moisture in all experimental
procedures.

p-Nitrophenylacetate (PNPA) decomposes in dry benzene in the
presence of alkylamine carboxylates to yield *p*-nitrophenol (PNP) as

one of the products. The formation of this yellow product can be followed at kinetic concentrations ($\sim 10^{-5}$ mol dm^{-3}) by the appearance of an absorption maximum in the u.v. spectrum at *ca* 314 nm.

Solutions of the alkylamine carboxylates in benzene solvent were made up by weight (±0.00005 g) directly into stoppered 10 mm quartz cells, and the total weight of cells and solution was determined after the kinetic run was completed. The reaction was initiated by addition of the appropriate µℓ volume of the ester, from a stock benzene solution, by microsyringe to give, generally, an absorbance (at infinite time) of between 0.5 and 1.0.

The spectrophotometric measurements for the kinetics of decomposition of PNPA in the various surfactant solutions were made using a Cary 14 spectrophotometer. The cell compartment was maintained at the required temperature (±0.1 K for runs up to $\sim$4 hours, ±0.4 K for overnight runs) by water circulation from a Grant SC10 or a Grant LE8 water circulator. The temperature was monitored before and after each kinetic run by thermometer (Zeal 0-100°C) using a glycerol sample in the sample chamber.

RESULTS AND DISCUSSION

We have studied the reactivity of PNPA in benzene solutions of α,ω alkyldiamine bis(dodecanoates) and of DAP at 341 K. The former reactions are generally faster than that with DAP, but the plots of the pseudo first order rate constant, k_ψ, against detergent concentration, [D], are similar - an initial sigmoidal increase in k_ψ followed by a region of linear increase with increasing [D]. Figure 2 shows a typical rate profile for the system, that for tetramethylenediamine bis(dodecanoate), 4DBD, and also analysis of the rate data according to Equation (8) which is derived[5] from application of the pseudophase model (Equation 2) to reversed micellar systems.

k_O and k_M are the rate constants in the bulk and micellar phase, respectively, K is the micelle-substrate binding constant, and N is the aggregation number.

$$\frac{k_\psi - k_o}{k_M - k_\psi} = \frac{K}{N}([D] - CMC) \qquad (8)$$

Equation (8) has been previously applied to alkylamine carboxylate systems.[22,39,40]

Although alkyldiamine salts have two cationic headgroups per monomer, it is expected that they will undergo an aggregation pattern similar to that of alkylamine carboxylates. However, they do exhibit a much greater range of pK$_a$ values for the ammonium headgroup and the dependence of rate on pK$_a$ has been determined.[41,42]

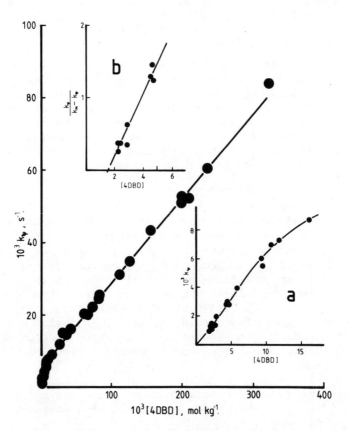

Figure 2. Rate profile for decomposition of PNPA in tetramethylenediamine bis(dodecanoate), 4DBD, in benzene at 341 K; Inset a, the
same plot expanded to show sigmoidal dependence of k_ψ at low [4DBD];
Inset b, analysis of the rate data by use of Equation (8), leading
to values of K = 432 mol^{-1} dm^3 and CMC = 1.6 x 10^{-3} mol dm^{-3}.

The changing dependence of rate on detergent concentration is more easily seen in plots of k_ψ/[D] against [D]. Figure 3 shows such a plot for 4DBD and also a plot of log k_ψ against log [4DBD].

Plots of k_ψ/[D] against [D] illustrate the critical dependence of the shape, height and position of the maximum in the profile, which occurs at small [D], upon the length of the alkyldiamine chain. Figures 4 and 6 illustrate the variations seen as the number of carbon atoms is increased. Figure 4 refers to tri-methylenediamine bis(dodecanoate), 3DBD, and Figure 6 to dodecyldi-amine bis(dodecanoate), 12DBD. Figure 6 also shows that a similar behaviour pattern is probably true for DAP. Figure 5 expands the low concentration region of Figure 4 and indicates that the apparent scatter may in fact be a real phenomenon, as seen for the reactivity of tris(oxalato)chromate(III) in DAP,[43] and for the reactivity of methyl orthobenzoate in bile salt solutions.[44]

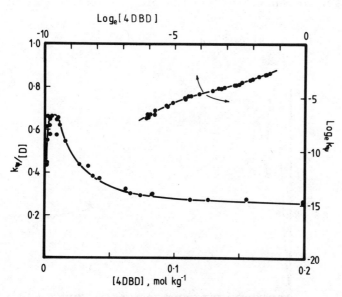

Figure 3. Plot of k_ψ/[4DBD] against [4DBD], showing the distorted bell shaped profile followed by a plateau and (on the upper and right hand axes) the rate data plotted in the form log k_ψ against log [4DBD]. (4DBD represents tetramethylenediamine bis(dodeca-noate).)

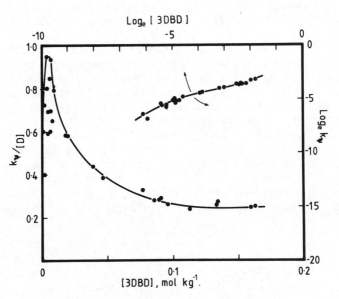

Figure 4. Plot of k_ψ /[D] against trimethylenediamine bis(dodecanoate), 3DBD, concentration and the plot of log k_ψ against
log [3DBD].

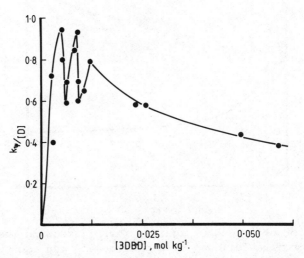

Figure 5. Plot of k_ψ /[D] against [3DBD] on an expanded concentration scale to illustrate non random scatter at low concentrations.

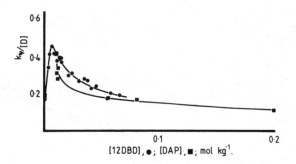

Figure 6. Plot of $k_\psi/$[D] against dodecamethylenediamine bis-(dodecanoate), 12DBD, concentration, ● , and the same plot for DAP, ■

Similar profiles were obtained for each dodecanoate salt in the series ethylenediamine to decamethylenediamine and for 12DBD; examination of literature data on alkylamine based systems[22,39,40,45] suggests that a similar treatment would also yield distorted bell shaped plots.

In Figure 7 it is seen that the data for DAP appear to lie on an extension of that for 12DBD.

The shape of the plots of $k_\psi/$[D] against [D], Figures 3, 4 and 6, is very similar to that of [dimer]/[D] against [D], Figure 8, calculated for the sequential indefinite self aggregation model.[10] Figure 8 shows the dependence of the fraction of various aggregates in solution upon [D] for a value of the association constant K_{ij} = 100. (This arbitrary value of K_{ij} was chosen because it lies within the range 50-300 obtained for alkylamine salts in non polar solvents.[10,29] Profiles for K_{ij} = 50 or 300 are similar to those shown in Figure 8, with higher values of K_{ij} shifting the peak towards the ordinate and pulling the start of the plateau region towards the origin.)

This similarity and the apparent second-order dependence of k_ψ upon detergent concentration at low [D] suggest that the dimer (and perhaps higher oligomers), as well as the monomeric detergent, contribute to the overall reaction with the ester.

The most noticeable difference in the shape of the plots of $k_\psi/$[D] against [D] compared with that of fraction of dimer against [D] is the higher value of their plateaux, thus suggesting a

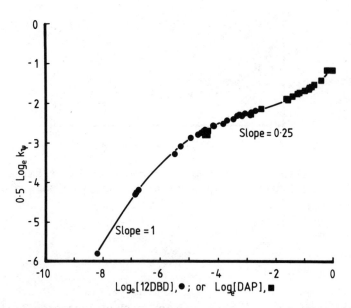

Figure 7. Plots of 0.5 log k_ψ against log [12DBD], ●, and against log [DAP], ■, indicating that the rate data for DAP appear to lie on an extension of those for 12DBD. Note also the upwards curvature as the mole fraction of detergent (χ_D) approaches unity.

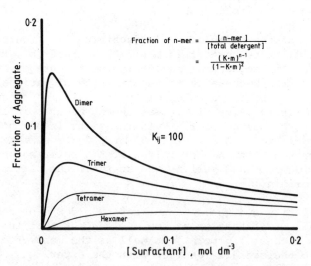

Figure 8. Plots of fraction of dimer, trimer, tetramer and hexamer (profiles from top to bottom of graph), against increasing detergent concentration using a value of $K_{ij} = 100$.

contribution to the rate profiles by larger aggregates.

If we now postulate that the observed rate constant includes equal contributions from all the aggregates present in solution, as well as from the monomeric detergent, then the kinetic results may be represented by Equation (9).

$$\frac{k_\psi}{[D]} = \frac{k_1 [monomer]^2}{[D]} + \frac{k_2[dimer]}{[D]} + \frac{k_3[trimer]}{[D]} + --- + \frac{k_n[n\text{-}mer]}{[D]} \quad (9)$$

Substitution of Equation (10) (an arbitrary but useful allowance

$$k_1 [monomer]^2 = k_2 [dimer] \quad (10)$$

for monomer contribution) into Equation (9) leads to the conclusion that $k_\psi/[D]$ should be proportional to ([total aggregate] + k_2[dimer]).

Plots of fraction of total aggregate, ([total aggregate]/[D]), against [D] and of ([total aggregate] + k_2[dimer])/[D] against [D] are shown in Figures 9 and 10 for values of K_{ij} = 100 and 300 respectively and $k_1 = k_2 = ---- = k_n = 1$.

It is seen that the upper profiles in Figures 9 and 10, in which allowance is made for contributions from the monomer and all oligomers now more closely resemble those obtained from the kinetic results, Figures 3, 4 and 6, suggesting that the self-aggregation model, Equation (6), may better explain the observed kinetics in these systems than does the pseudophase model, Equation (2).

Cox and Jencks[46] have postulated a preassociation mechanism for acid catalysis in aqueous media. Application of their reasoning to catalysis by these surfactant systems in apolar solvents suggests a forced preassociation. Data obtained from the Bronsted plot for this alkyldiamine series of detergents support[42] the concept of such a preassociation for the slope of the plot changes from 0.085 to 0.17 with increasing [D]. Amine attack from within a substrate-oligomer complex is possible, as also is attack by the free amine generated by dissociation of n-mers. (For example, an A_3B_3 type aggregate could dissociate to yield a stable A_3B aggregate and two free amine moieties.) Hydrogen bonded amine is likely to be less reactive.

As the concentration of detergent is increased, we have shown (Figure 7) that the linear dependence of rate upon [D], which is observed at moderate [D], no longer exists and that the rate profile bends upwards at high [D]. Similar non-linear dependence at high [DAP] was observed for decomposition of p-nitrophenyl-carbonate.[47] Prigogine and Defay[48] have noted that the fraction of monomer, f, which is normally equal to $\gamma_{alcohol}/\gamma_{solvent}$, in

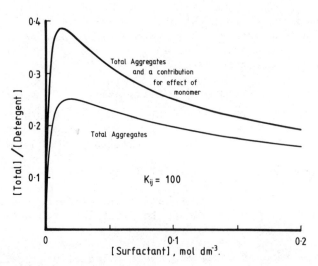

Figure 9. Plot of fraction of total aggregate, lower plot, and of
([total aggregate] + k_2[dimer])/[D], upper plot, against [D] for
K_{ij} = 100.

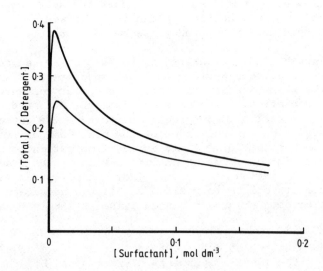

Figure 10. Plot of fraction of total aggregate, lower plot, and of
([total aggregate] + k_2[dimer])/[D], upper plot, against [D] for
K_{ij} = 300.

associating alcohol/CCl_4 systems tends to become larger than the value expected from activity coefficient considerations alone. In the range of detergent concentrations that have been studied, f is related to the stoichiometric concentration of detergent by Equation (11).

$$\frac{1 - f^{\frac{1}{2}}}{f} = K_{ij}[D] \tag{11}$$

K_{ij} is inversely proportional to the dielectric constant of the medium, which in turn is a function of both the dielectric constant of the solvent and of the detergent. As χ_D increases, the dielectric constant of the medium and therefore f increases. If one postulates that the contribution of the monomer to the overall kinetics is equal to or greater than that of the oligomeric species, then at high χ_D, when f increases, the dependence of k_ψ upon [D] once again approaches second order.[47] However, at moderate [D], formation of oligomers tends to remove monomer from the solution and values of k_ψ, which become more nearly dependent on [D], reflect this association.

Thus, while we accept that aggregation affects the rate of reaction, we hesitate to postulate that it causes catalysis. We believe the catalysis to be due to a preassociation mechanism in the apolar solvent and we suggest that aggregation has an inhibitory effect upon the kinetics of the reaction.

ACKNOWLEDGEMENTS

Financial assistance from the Research Committees of the New Zealand Universities' Grants Committee and the University of Auckland Research Committee (to C.J.O'C.) and the Maori Education Foundation (to T.D.L.) is gratefully acknowledged.

REFERENCES

1.   H.F. Eicke, Topics in Current Chemistry, 87, 85 (1980).
2.   C.R. Singleterry, J. Am. Oil Chemists Soc., 32, 446 (1955).
3.   H.F. Eicke, in "Micellization, Solubilization and Microemulsions", K.L. Mittal, Editor, Vol. 1, p429, Plenum Press, New York, 1977.
4.   A. Kitahara and K. Kon-no, in "Colloidal Dispersions and Micellar Behaviour", K.L. Mittal, Editor, A.C.S. Symposium Series, No.9, p225, American Chemical Society, Washington, D.C., 1975.
5.   J.H. Fendler and E.J. Fendler, "Catalysis in Micellar and Macromolecular Systems", Chapter 10, Academic Press, New York, 1975.

6.   E.J.R. Sudhölter, G.B. van de Langkruis, and J.B.F.N. Engberts, Recueil Review, J. Netherlands Chem. Soc., 99, 73 (1980).

7.   K. Shinoda and E. Hutchinson, J. Phys. Chem., 66, 577 (1962).

8.   F.M. Fowkes, J. Phys. Chem., 66, 1843 (1962).

9.   K. Kon-no and A. Kitahara, J. Colloid Interface Sci., 35, 409, 636 (1971).

10.   F.Y.-F. Lo, B.M. Escott, E.J. Fendler, E.T. Adams, R.D. Larson, and P.W. Smith, J. Phys. Chem., 79, 2609 (1975).

11.   J.H. Fendler, E.J. Fendler, R.T. Medary, and O.A. El Seoud, Trans. Faraday Soc., 69, 280 (1973).

12.   E.J. Fendler, J.H. Fendler, R.T. Medary, and O.A. El Seoud, J. Phys. Chem., 77, 1432 (1973).

13.   O.A. El Seoud, E.J. Fendler, J.H. Fendler, and R.T. Medary, J. Phys. Chem., 77, 1876 (1973).

14.   N. Muller, J. Phys. Chem., 79, 287 (1975).

15.   A.S. Kertes and H. Gutmann, in "Surface and Colloid Science", E. Matijevic, Editor, Vol. 8,pp193-295, Wiley-Interscience, New York, 1975.

16.   A.S. Kertes, in "Micellization, Solubilization and Micro-emulsions", K.L. Mittal, Editor, Vol. 1,pp445-454, Plenum Press, New York, 1977.

17.   H.F. Eicke and A. Denss, J. Colloid Interface Sci., 64, 386 (1978).

18.   Y.C. Jean and H.F. Ache, J. Am. Chem. Soc., 100, 984 (1978).

19.   Y.C. Jean and H.F. Ache, J. Am. Chem. Soc., 100, 6320 (1978).

20.   L.A. Fuagauchi, B. Djermouni, E.D. Handel, and H.J. Ache, J. Am. Chem. Soc., 101, 2841 (1979).

21.   A. Kitahara, Bull. Chem. Soc. Japan, 31, 288 (1958).

22.   K. Kon-no, T. Matsuyama, M. Mizuno, and A. Kitahara, Nippon Kagaku Kaishi, 11, 1857 (1975).

23.   U. Hermann and Z.A. Schelly, J. Am. Chem. Soc., 101, 2665 (1979).

24.   H.F. Eicke, R.F.W. Hopmann, and H. Christen, Ber. Bunsenges. Phys. Chem., 79, 667 (1975).

25.   O. Levy, J. Phys. Chem., 76 1752 (1972).

26.   J.A. Rutgers and M. de Smet, Trans. Faraday Soc., 48, 635 (1952).

27.   R.M. Fuoss and C.A. Kraus, J. Am. Chem. Soc., 55, 21, 476, 1019, 2387, 3614 (1933).

28.   F.M. Batson and C.A. Kraus, J. Am. Chem. Soc., 56, 2017 (1934).

29.   C. Klofutar and S. Paljk, Trans. Faraday Soc., 77, 2705 (1981).

30.   F. Kohler, H. Atrops, H. Kalali, E. Liebermann, E. Wilhelm, F. Ratkovics, and T. Salamon, J. Phys. Chem., 85, 2520 (1981).

31.   F. Kohler, R. Gopel, G. Götze, H. Atrops, M.A. Demiriz, E. Liebermann, E. Wilhelm, F. Ratkovics, and B. Palagyi, J. Phys. Chem., 85, 2524 (1981).

32. F. Kohler, L. Liebermann, G. Miksch, and C. Kainzu, J. Phys.
     Chem., 76, 2764 (1972).

33. D.-L.F. De Tar and R.W. Nowak, J. Am. Chem. Soc., 92, 1361
     (1970).

34. G.S. Bruckenstein and A. Saito, J. Am. Chem. Soc., 87, 698
     (1965).

35. T. Jasinski, A.A. El-Harakany, A.A. Taha, and H. Sadek,
     Croatica Chem. Acta, 49, 767 (1977).

36. E.A. Yerger and G.M. Barrow, J. Am. Chem. Soc., 77, 4474,
     6204 (1955).

37. N.G. Sellers, P.M.P. Eller, and J.A. Caruso, J. Phys. Chem.,
     76, 3618 (1972).

38. A. Kitahara, Bull. Chem. Soc. Japan, 28, 234 (1955).

39. C.J. O'Connor and R.E. Ramage, Aust. J. Chem., 33, 757, 771,
     779 (1980).

40. O.A. El Seoud, A. Martins, L.P. Barbur, M.J. Da Silva, and
     V. Aldrigue, J. Chem. Soc. Perkin Trans. II, 1674 (1977).

41. C.J. O'Connor, T.D. Lomax, and R.E. Ramage, in "Solution
     Behavior of Surfactants: Theoretical and Applied Aspects",
     K.L. Mittal and E.J. Fendler, Editors, Vol. 2, pp803-831,
     Plenum Press, 1982.

42. C.J. O'Connor and T.D. Lomax, (1982), unpublished data.

43. C.J. O'Connor and R.E. Ramage, Aust. J. Chem., 33, 695 (1980).

44. C.J. O'Connor, R.G. Wallace, and B.T. Ch'ng, these proceed-
     ings.

45. C.J. O'Connor and K.J. Mollett, J. Chem. Soc. Perkin Trans.
     II, 369 (1976).

46. M.M. Cox and W.P. Jencks, J. Am. Chem. Soc., 103, 572 (1981).

47. H. Kondo, K. Fujiki and J. Sunamoto, J. Org. Chem., 43, 3584
     (1978).

48. I. Prigogine and R. Defay, "Chemical Thermodynamics", D.H.
     Everett, Translator, pp414-415, Longmans and Green, 1954.

DYNAMICS OF REVERSED MICELLES

Z.A. Schelly

Department of Chemistry
The University of Texas at Arlington
Arlington, Texas   76019-0065

The possible dynamic processes in simple reversed
micelle systems (surfactant-nonpolar solvent-trace
amount of water) are summarized. Major differences
between aqueous and reversed micelles are pointed out.
It is emphasized that the equilibrium description
of a system is essential for the interpretation of
the kinetic data on the formation of, and solubilization by, reversed micelles. Results on these two
processes, available for a few systems, are described.

INTRODUCTION

The abrupt change in the colligative and other solution-properties of surface active molecules in dilute aqueous solution at a certain concentration, known as the critical micelle concentration (cmc), is attributed to the formation of molecular aggregates called micelles. The micelles are in chemical equilibrium with the monomers. Usually, the name micelle is used for any soluble aggregate (excluding vesicles) which is spontaneously and reversibly formed from amphiphilic molecules or ions.

It is generally accepted that in aqueous solution the non-polar tails of the associating molecules comprise mainly the interior of the micelle, while the polar head groups, charged or uncharged, are located on the exterior, maintaining contact with the solvent and keeping the micelle in solution.

In nonpolar solvents the structure of the micelle is inverted. Here, the polar head groups form the interior and the hydrophobic parts of the monomers the exterior of the aggregate. Hence the name "reversed (or inverted) micelle."

The physical and chemical equilibrium properties of most aqueous micelle systems are well understood.[1] On the other hand, our knowledge of reversed micelles is still in its infancy.[2] Even the basic concepts such as the existence of a cmc, as well as the kinds of aggregates present in solution have been the subject of controversy.[3-8] The typical features of aqueous micelle systems, viz., the existence of a cmc, and the often found concentration-independence of the aggregation number, are not generally characteristic of reversed micelles. In apolar solvents, it is more appropriate to talk about an operational cmc, which is essentially the lowest surfactant concentration range where aggregation can be detected by a particular method used.

Although, in principle, reversed micellar solutions could be two-component systems (surfactant and the apolar solvent), in reality, trace amounts of a third component (water or another polar substance) are always present, comprising the core of the aggregates. If the third component is water, it is usually termed the "pool"[9] of the micellar core.

Dynamic Processes in Reversed Micellar Systems

The most important of the possible rate processes that may occur in the simplest 3-component (surfactant, nonpolar solvent, and trace amount of water) reversed micelle system are the

i)   association-dissociation dynamic equilibria

$$A + A \rightleftharpoons A_2 \tag{1}$$

$$\vdots$$

$$A_i + A \rightleftharpoons A_{i+1} \tag{2}$$

$$A_n + A_m \rightleftharpoons A_{n+m} \tag{3}$$

between the surfactant monomer A and the micellar aggregates A_{i+1} or A_{n+m}, since these reactions represent the formation of micelles. Our knowledge of these processes is very limited.

Much more is known about the

ii)   molecular dynamics (translation, rotation, vibration, and conformational changes) of all the species present in a reversed micellar solution.

Diffusion methods,[10] conductance,[11] NMR,[12] and other spectroscopic techniques have been successfully used in the explorations.

At high surfactant concentration, there is the possibility of

iii)   phase transitions $P_1 \leftrightarrow P_2$ and/or   (4)

phase separations $P_1 \leftrightarrow P_1 + P_2$ (5)

between several different phases P_i present. These processes may involve different micellar and liquid crystalline phases. Again, very little is known about the rate and mechanism of the transitions.

If one introduces a substrate B that is (usually) more soluble in the pool than in the bulk solvent, then the

iv)   solubilization of the substrate[13-15]

$$\underset{M}{\bigcirc} + B \rightleftharpoons \textcircled{B} \tag{6}$$

also has to be considered as an additional rate process. Equation (6) represents the overall reaction in which the substrate penetrates the reversed micelle M. Such 4-component systems, and the process of solubilization, have great industrial and biological importance.[16,17]

If two reactive substrates B and C are introduced, one deals with a 5-component system. Such systems (containing large pools) offer the possibility for the investigation of the kinetics of

v) reactions between compartmentalized (solubilized)
    species[18]

$$\text{(B)} + \text{(C)} \rightleftharpoons \text{(BC)} + \bigcirc \qquad (7)$$

The rate of the overall reaction (7) may also provide information
about some of the elementary processes involved, such as collision frequency between reversed micelles, their fusion-fission
(Equation 3), and penetration reactions (Equation 6).

DISCUSSION

Our own work has focused on the investigation of points i)
and iv), the dynamics of reversed micelle formation, and their
penetration by probe molecules. All our experiments have been
carried out on carefully dried systems, containing only trace
amounts of water.

Equilibrium Description of Reversed Micellar Solutions

The ultimate objective of the equilibrium investigation of
such complex systems is, of course, the establishment of the
population distribution of the species present, as a function of
the total surfactant concentration and temperature. Needless to
say, the knowledge of the equilibrium description of a given
system is essential for the interpretation of the kinetic data
obtained on the formation of, and/or the solubilization by,
the aggregates. Unfortunately, however, the experimental
equilibrium methods available afford only the conjecture of
possible distributions. An equilibrium model leading to the
optimal distribution must be selected in such a way that the
number of adjustable parameters of the model is minimal, and the
distribution obtained is consistent with the observed equilibrium
as well as dynamic behavior of the system (such as rates of
solubilization and micelle formation). Thus, the results of
rate studies can also serve as additional boundary conditions
for selecting the best aggregation model for a given system;
and vice versa, the equilibrium model determines the selection
of the reaction mechanism involved. Nevertheless, the measurement of a colligative property of the solution (e.g. using
vapor pressure osmometry VPO) which yields the total solute
concentration C as a function of the surfactant concentration C_s,
provides the primary experimental check on any equilibrium model
considered.[19]

Our investigation of 17 alkylammonium carboxylates in
benzene and cyclohexane, using VPO, optical absorption and
fluorescence probes[20] has confirmed the usually low mean

aggregation number ($\bar{n}$ = 3), and established the typical population distribution of these systems (Figure 1).

Our VPO, spectroscopic and kinetic studies of Aerosol-OT in benzene[13,14] have shown that of 11 aggregation models tested, the bimodal monomer-m-mer-n-mer (m = 6, n = 14) association fits all the data the best. This association model assumes that only the monomer A, sixmer A_6 and fourteenmer A_{14} are present at significant concentration (Figure 2). The actual distribution in bimodal systems, however, is probably not discrete but rather clustered around the two favored aggregation numbers. A possible explanation for the bimodality of the Aerosol-OT distribution may be the relative stability of an octahedral-like geometrical arrangement in the 6-mer, and a face-centered-cubic-like arrangement in the 14-mer, of the polar head groups in the tightly packed aggregates.

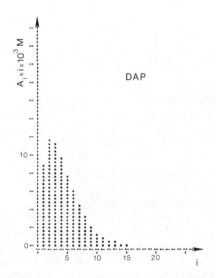

Figure 1. Example for the concentration distribution of aggregates of alkylammonium carboxylates in benzene at 37°C. The distribution given is for dodecylammonium propionate (DAP) at a concentration of 7.3 x 10^{-2} M. The concentration of the ith aggregate is given in millimoles of surfactant monomer per liter. The largest fraction of the surfactant ($\sim$19%) is in the dimer (i = 2) form, and less than 0.5% is in the form of micelle with aggregation number i = 15 (Reprinted with permission - from Ref. 20.)

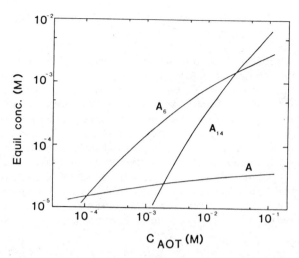

Figure 2. Equilibrium concentrations of the monomer, 6-mer and
14-mer of Aerosol-OT in benzene at 25°C, computed by using the
bimodal association model (Reprinted with permission from Ref. 13).

For dodecylpyridinium iodide in benzene, in the concentration
range of 1-5 mM, we found that aggregation can be adequately
described by the simple $7A \rightleftharpoons A_7$ equilibrium.[15]

Kinetics of Reversed Micelle Formation

In contrast to aqueous micelles,[21,22] very little is known
about the kinetics of the formation of reversed micelles. The
reason for this lies in the experimental difficulties one faces
in the investigation of very rapid processes in transparent,
nonconducting solutions. The few studies that have been reported
utilized stationary relaxation methods:[23] dielectric relaxation
and ultrasonic absorption. Experiments that were carried out
on Aerosol-OT in cyclohexane suggest[24] that micelle formation
may be described by a mechanism where aggregation advances only
to the trimer stage (in essentially diffusion controlled steps)
at concentrations much smaller than the operational cmc. At
higher concentrations, additional monomers are bound according
to reaction in Equation (2). The oligomers formed can undergo
a topological transformation which modifies the structure of
the larger aggregates in such a way that further acceptance of
monomers becomes too difficult and the growth process stops.[2]

Our experiments on Aerosol-OT in benzene,[14] using the con-
centration- and solvent-jump relaxation methods,[25] have revealed
that the aggregates are too stable to be significantly perturbed
by a twofold dilution. The experiments also provided indirect
evidence that reactions (1) and (2) are much faster than the process
in Equation (6). Clearly, a faster and more effective perturbation is needed in conjunction with sensitive detection. Presently,
experiments are in progress in our laboratory using laser pulse
induced electric birefringence, to elucidate the kinetics of
the formation of reversed micelles.

Kinetics of Solubilization by Reversed Micelles

Results on a few but extensively studied systems have revealed that the rate and mechanism of solubilization of (at least
partially) polar substrates are functions of the surfactant,
the nature of the substrate,[26] and the size of the water pool.[27]
Typically, in such studies, it is convenient to choose a substrate that can act as an optical probe for the progress of
its own solubilization, i.e., the absorption spectrum of the
probe in the bulk solvent should be different from that in the
solubilized state.

We investigated the dynamics of two systems in detail: the
solubilization of picric acid (HP) by Aerosol-OT,[14] and TCNQ
(7,7,8,8-tetracyanoquinodimethane) by DPI[15] (dodecylpyridinium
iodide) reversed micelles, both in dried benzene, using the
concentration-jump and pressure-jump relaxation methods,
respectively.

In the Aerosol-OT system,[14] the penetration of HP into the
pool of the aggregates involves at least four distinct rate
processes: one in the submillisecond and three in the milli-
second time range. The first rapid step may be associated with
the adsorption of the probe at a hydrated carboxyl group of
the chain. In a second step, HP moves further into the interior
of the aggregate and adsorbs at a hydrated sulfonate group of
the micelle in the pool. This desorption-adsorption process is
assumed to be slow, and it corresponds to two relaxations (one
for the reaction in the 6-mer and the other for that in the 14-
mer present). With increasing surfactant concentration, the
pool size is shrinking to such an extent, that above 4×10^{-3} M
Aerosol-OT there is not enough water present to hydrate all the
polar groups of the micelle. Consequently, part of the carboxyl
groups are dehydrated. When this is the case, an additional
relaxation appears. This relaxation may be viewed as a slowed-
down version of the first rapid step. Slowed-down either by the
tighter packing of the surfactant molecules in the aggregates,
or by the necessity of replenishment of the water pools prior
to successful penetration of the micelle by the probe.

In the TCNQ-DPI system,[15] the probe is solubilized as TCNQ$^-$
anion radical. Only one relaxation can be observed by the
pressure-jump method. The solubilization process can be subdivided into several steps, where TCNQ first penetrates the
hydrophobic coat of the micelle in a rapid, diffusion-controlled
step. For the completion of the electron transfer to TCNQ at
the boundary of the pool, apparently the condition for the
formation of the water-soluble I_3^- in the pool must be created
by the entrance of a second electron-acceptor TCNQ molecule
into the micelle. The entrance of the second TCNQ and the
transfer of electrons seem to be the rate-determining reactions.
A micelle with two TCNQ molecules at the interface of the pool
is crowded and unstable. We assumed that this species is in a
steady state. In the final steps, the TCNQ$^-$ anion radicals and
I_3^- are redistributed among the micelles.

Based on the limited amount of information available, there
are only a few generalizations that can be made. i) Polar
probes can penetrate the hydrophobic shell of the aggregates
at a diffusion-controlled rate. ii) The rate and mechanism of
penetration of the polar regions of the aggregates are system-
specific. The polar regions are presented by possible polar
groups in the surfactant chain (as in Aerosol-OT, for example),
and, more importantly, by the interfacial region of the pool.
iii) It seems that probe-induced nucleation of the surfactant
is insignificant as a solubilization mechanism. iv) Once the
probe reaches the water-solvent interface, only the polar
group of the probe dips into the pool. v) The rate of the
dipping step may be controlled by reactions (such as hydration,
dissociation, charge transfer, etc.) taking place at the interface.

ACKNOWLEDGMENT

Our work was partially supported by the R.A. Welch Foundation and the Organized Research Fund of UTA. Acknowledgment is
made to the donors of the Petroleum Research Fund, administered
by the American Chemical Society, for additional partial support.

REFERENCES

1. C. Tanford, "The Hydrophobic Effect", 2nd ed., Wiley,
    New York, 1980.
2. H.-F. Eicke, in "Topics in Current Chemistry", Vol. 87,
    Springer Verlag, Berlin, 1980.
3. H.-F. Eicke and V. Arnold, J. Colloid Interface Sci., 46,
    101 (1974).

4. H.F. Eicke and H. Christen, J. Colloid Interface Sci., 48, 281 (1974).
5. N.J. Muller, J. Phys. Chem., 79, 287 (1975).
6. F.J.F. Lo, B.M. Escott, E.J. Fendler, E.T. Adams, Jr., R.D. Larsen, and P.W. Smith, J. Phys. Chem., 79, 2609 (1975), and references cited therein.
7. A.S. Kertes and H. Gutmann, in "Surface and Colloid Science," Vol. 8, E. Matijevic, Editor, pp. 193–295, New York, 1976.
8. Y. Jean and H.J. Ache, J. Am. Chem. Soc., 100, 6320 (1978), and references cited therein.
9. F.M. Menger, J.A. Donohue, and R.F. Williams, J. Am. Chem. Soc., 95, 286 (1973).
10. B. Lindman, N. Kamenka, B. Brun, and P.G. Nilsson, in "Microemulsions," I.D. Robb, Editor, Plenum Press, New York, 1982, and references therein.
11. T.A. Bostock, M.P. McDonald, and G.J.T. Tiddy, these proceedings.
12. B. Lindman and P. Stilbs, these proceedings.
13. K. Tamura and Z.A. Schelly, J. Am. Chem. Soc., 103, 1013 (1981).
14. K. Tamura and Z.A. Schelly, J. Am. Chem. Soc., 103, 1018 (1981).
15. S. Harada and Z.A. Schelly, J. Phys. Chem., 86, 2098 (1982).
16. K. Shinoda, "Colloidal Surfactants," Academic Press, New York, 1963.
17. J.H. Fendler and E.J. Fendler, "Catalysis by Micellar and Macromolecular Systems," Academic Press, New York, 1975.
18. B.H. Robinson, these proceedings.
19. Z. A. Schelly, in "Aggregation Processes in Solution", J. Gormally and E. Wyn-Jones, Editors, Elsevier, Amsterdam, in press.
20. U. Herrmann and Z.A. Schelly, J. Am. Chem. Soc., 101, 2665 (1979).
21. E.A.G. Aniansson, S.N. Wall, M. Almgren, H. Hoffmann, I. Kielmann, W. Ulbricht, R. Zana, J. Lang, and C. Tondre, J. Phys. Chem., 80, 905 (1976).
22. M. Kahlweit and M. Teubner, Adv. Colloid Interface Sci., 13, 1 (1980).
23. M. Eigen and L. DeMaeyer, in "Technique of Organic Chemistry," Vol. 8, Part 2, S.L. Friess, E.S. Lewis, and A. Weissberger, Editors, Wiley-Interscience, New York, 1963.
24. H.F. Eicke, R.F.W. Hopmann, and H. Christen, Ber. Bunsenges, Phys. Chem., 79, 667 (1975).
25. D.Y. Chao and Z.A. Schelly, J. Phys. Chem., 79, 2734 (1975); Z.A. Schelly and D.Y. Chao, Adv. Molec. Relaxation Processes, 14, 191 (1979).
26. Z.A. Schelly, G. Sumdani, S. Harada, and K. Tamura, to be published.
27. S. Harada and Z.A. Schelly, to be published.

REVERSE STRUCTURES IN A P-NONYLPHENOLPOLYETHYLENEGLYCOL (9.6 MOLE ETHYLENE OXIDE) - WATER SYSTEM

A. Derzhanski and A. Zheliaskova

Institute of Solid State Physics
Bulgarian Academy of Sciences
72 Lenin Blvd., 1184 Sofia, Bulgaria

The system polyoxyethylene with 9.6 mole ethylene oxide-water was investigated by NMR spectroscopy for surfactant concentration between 70 and 100 weight per cent at intervals of 5%. The behavior of the absorption spectrum was explained in terms of the formation of supermolecular reverse aggregates.

INTRODUCTION

The phase structure and molecular mobility of different phases of liquid crystals (including lyotropic systems as well) have been the subject of much research. Especially effective has been the use of broad line NMR spectroscopy method[1]. The line width and form of the absorption spectrum give information about the mobility of a mesogenic molecule as a whole, and also about the mobility of its constituent atomic groups[2].

EXPERIMENT

Broadline NMR spectra of optically isotropic solutions of heavy water with p-nonylphenolpolyethyleneglycol with 9.6 mole ethylene oxide,abbreviated by us as Arkopal 9.6,were recorded for weight concentrations of the surfactant between 70% and 100% in intervals of 5%. These concentrations are well above the range of the existence of the lamellar phase[3,4]. The spectra were recorded by means of a homemade NMR spectrometer with operating frequency 14 MHz. The experiment was carried out at a temperature of 25°C.

1463

EXPERIMENTAL RESULTS

The experimental curves obtained with widened wings are of the so-called "super-Lorentzian" kind. This shows that the atoms of the surfactant molecule take part in different movements. Because of the hindered reorientation of the molecular axes our system resembles a solid lattice whose spectrum has a Gaussian shape[10,11]. The different degrees of freedom of the two ends of the alkyl chain give as a result a more complicated shape of the spectrum. That is why we consider the spectrum as a superposition of two Gaussian curves. A computer program was prepared which determines the parameters of the two Gaussian curves A and B so that the coincidence with the experimental curve is the best. The curve A has a narrow width (about 1 gauss); it practically does not depend on the concentration. It is probably associated with the fast intramolecular reorientation of the CH_2 groups. A similar spectral line of shape close to Gaussian one and a halfwidth about 0.5 gauss was observed by Bloom et al.[11] The other curve B is wider (several gauss) and in all probability its width expresses the molecular mobility of Arkopal 9.6 as a whole. If similar component existed in the spectrum of the substances examined by Bloom et al.[11], it could not be registered by them due to the short time (several microseconds) of free induction decay. The width $\Delta\omega$ of the broad line of the pure surfactant is considerable -- about 9.1×10^4 s^{-1} (3.4×10^{-4} T; see Table I). This indicates hindered molecular rotation. Supermolecular aggregates probably appear in the sample. They are constructed like linear two-dimensional or three-dimensional polymer formations. Possibly these structures arise due to hydrogen bonds; protons from the OH groups which belong to the end of the polyoxyethylene chains as well as protons from the residue water molecules which exist in our sample can take part. Chou and Carr[5,6] have assumed that analogous supermolecular aggregates exist in some thermotropic liquid crystals whose molecules are able to form hydrogen bonds. The addition of 5% water leads to a narrowing of the spectrum, thus showing that the molecular mobility increases. At 90% concentration of Arkopal 9.6 (10% D_2O) the spectrum widens strongly and its halfwidth increases to about 9.6×10^4 s^{-1} (3.6×10^{-4} T). A further increase in the water content leads to a monotonic narrowing of the spectrum to about 4.3×10^4 s^{-1} (1.6×10^{-4} T) (at 20% D_2O, cf. Table I). The behavior of the NMR spectrum can be explained for this concentration interval in terms of the formation of compact molecular aggregates (for example reverse micelles or maybe reverse vesicles) in the sample which practically fills up the whole volume at 10% D_2O. Increasing the water content in the sample increases the number of these aggregates while at the same time it decreases their sizes.

DISCUSSION

Let us consider the case of reverse micelles. For the surfactant used by us, it can be expected that the micellar borders
are not sharp at all because of the low hydrophilicity of the
polyoxyethylene chain. Because of this, the sizes of the micelles
can be considerably larger than the sizes of the molecules which
form the micelles. This was established[7] for the micelles of
Triton X100 which is structurally similar to the surfactant we
have used. The micelles examined possess a hydrophilic core
(sperical or cylindrical) with a comparatively small radius. The
molecules near the core are radially oriented. The molecular
orientation changes gradually from radial, with respect to the
first core, to radial with respect to the second one, when passing
from one core to another. We assume that the diameter of the
micelle core does not in practice depend on the water content of
the sample.

We assume that the volume of the hydrophilic core of
spherically micelles, V_o is constant. The volume V_s of the whole
micelle will be $V_s = (4/3) \pi R_s^3$, where R_s is the radius of the
micelle. Then the specific water volume in one micelle is V_o/V_s
and the water content per unit volume of the solution $C_{w,s}$ is
inversely proportional to the third power of the radius R_s of the
micelle:

$$C_{w,s} \propto 1/R_s^3 \tag{1}$$

The volume V_o of the hydrophilic core of cylindrical micelles
is $V_o = S_o \ell$, where S_o is the area of the cross-section of the
core, and ℓ is the length of the micelle. In this case the volume
of the micelle V_c is $V_c = \pi R_c^2 \ell$, R_c being the radius of the micelle.
The ratio of these volumes V_o/V_c gives the water content per unit
volume of the solution. Thus the water concentration $C_{w,c}$ is
inversely proportional to the second power of the radius of the
micelle:

$$C_{w,c} \propto 1/R_c^2 \tag{2}$$

It is possible in our case for aggregates to exist which are
similar to those observed in the lipid-water systems[13,14]. In the
area near the lamellar phase the aggregates can appear as tubelike
or spherical vesicles. They are constructed of a lipid bilayer
whose hydrophobic core forms a spherical or cylindrical shell. The
hydrophilic lecithin heads and water molecules are positioned
inside and outside this shell.

The aggregates which exist in our sample would probably have a reverse structure. The spherical and cylindrical shells would consist of structured water and the hydrophilic polyoxyethylene parts of the Arkopal. The molecules whose hydrophobic parts form the shell as well as neighboring molecules of the environment are oriented perpendicular to it.

Here we assume that the hydrophilic shell of the aggregates has a fixed thickness d. A change in the water content changes the number and size of the aggregates including the radius of the shell. The volume of the water shell for spherical (V_{ovs}) and cylindrical (V_{ovc}) vesicles is then $V_{ovs} = 4\pi R_{vs}^2 d$ and $V_{ovc} = 2\pi R_{vc}$ xℓd respectively. R_{vs} and R_{vc} are the radii of the spherical and cylindrical vesicles respectively and ℓ is the length of the tube-like vesicles. The volume of the spherical and tubelike vesicles can be expressed as $V_{vs} = (4/3)\pi R_{vs}^3$ and $V_{vc} = \pi R_{vc}^2 \ell$, respectively. It is clear that the radius of the hydrophilic core and the corresponding radius of the whole vesicle do not differ substantially By calculating the corresponding ratio, volume of the hydrophobic shell to volume of the whole vesicle V_o/V_v, we find that in both cases the water concentration $C_{w,v}$ is inversely proportional to the radius of the vesicle:

$$C_{w,v} \propto 1/R_v \qquad\qquad\qquad (3)$$

It should be remarked that the concept of the radius of the aggregate here is not precisely defined as in the micelles of ionic surfactants and lecithin vesicles. In our case, the radius of the spherical aggregates is half the average distance between the geometric centers of two neighboring aggregates; for cylindrical aggregates with long enough axes, this is half the distance between two neighboring axes if the aggregates were parallel to each other.

We suppose that the cylindrical aggregates are more probable than the spherical ones. Moreover in the concentration interval under investigation, which is close to the lamellar phase, many surfactants manifest a reverse hexagonal phase constructed of closely packed cylindrical molecular aggregates which are practically infinitely long. We did not observe optical anisotropy in our sample. This indicates that the cylindrical micelles which determine the hindered molecular reorientation are not long enough to be packed parallel to one another.

The molecules of the surfactant change their own orientation in the sample by diffusional motion, either by turning around the hydrophilic core or by passing from one aggregate to another. In both cases it is necessary for the orientation change to cover a correlation distance of the order of R, the radius of the aggregate.

According to the diffusion laws, the correlation time τ_c for the change of molecular orientation is $\tau_c = R^2/6D$, where D is the diffusion coefficient of the surfactant molecules. In our case, the diffusion coefficient has a tensor character because of the nonspherical molecular shape and because of the microscopical anisotropy of the medium. The scalar D used by us is in fact some average value.

The viscosity of the sample was also determined. It depends slightly on the concentration in the range studied and has a value of about $\eta = 0.3$ Kg $m^{-1}s^{-1}$. The calculated diffusion coefficient is $D = 1 \times 10^{-14}$ $m^2 s^{-1}$.

We explain the observed narrowing of the component B of the NMR spectrum in terms of molecular reorientation which is characterized by the correlation time τ_c. According to the NMR theory[8] we obtain for the dependence of the spectrum width from the molecular reorientation

$$\delta\omega^2 = \delta\omega_o^2 \frac{2}{\pi} arctg(\alpha \ \delta\omega \ \tau_c) \qquad (4)$$

where $\delta\omega^2$ is the second moment of the observed sample, and $\delta\omega_o^2$ is the second moment of the corresponding solid lattice. The magnitude α is of the order of one, and that is why we do not take it into consideration in our calculations. According to Ref. 9, the second moment for immovable CH_2 groups is $\delta\omega_o^2 = 1.79 \times 10^{10}$ s^{-1}. For three water concentrations of 10%, 15%, 20%, we obtained corresponding values (Table I) for correlation time τ_c and aggregate radius R by means of experimental data for the broad line $\delta\omega$ and expression (4). In Figure 1, the water concentration was plotted on the x-axis and the corresponding value of R on the y-axis both in logarithmic scale. The three points lie well on a straight line slope 1.3. This gives $C_w \propto 1/R^{1.3}$.

Table I. Values of the Correlation Time τ_c and the Aggregate Radius R for Different Water Concentrations.

Concentration of surfactant	$\delta\omega$ (s^{-1})	ΔB (T)	τ_c (s)	R (m)
80%	4.3×10^4	1.6×10^{-4}	3.8×10^{-6}	48×10^{-10}
85%	5.9×10^4	2.2×10^{-4}	5.3×10^{-6}	57×10^{-10}
90%	9.6×10^4	3.6×10^{-4}	8.5×10^{-6}	81×10^{-10}

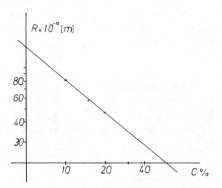

Figure 1. Dependence of the aggregate radius R on water concentration in the solution.

The fact that this value does not coincide with the expected values for spherical (1) and cylindrical (2) micelles and vesicles (3) could be due to the approximations we have used or to the existence of several kinds of aggregates at the same time, mainly vesicles and cylindrical micelles.

REFERENCES

1. G. J. T. Tiddy, Phys. Reports, 57, 1 (1980).
2. Å. Johansson and B. Lindman, in "Liquid Crystals and Plastic Crystals", G. W. Gray and P. A. Winsor, Editors, Vol. 2, p. 192, Halsted Press, New York, 1974.
3. V. Luzzati and F. Husson, J. Cell. Biol., 12, 207 (1962).

4. A. Derzhanski and A. Zheliaskova, Mol. Cryst. and Liq. Cryst., in press.
5. L. S. Chou and E. F. Carr, in "Liquid Crystals and Ordered Fluids", J. F. Johnson and R. S. Porter, Editors, Vol. 2, p. 39, Plenum Press, New York, 1973.
6. L. S. Chou and E. F. Carr, Phys. Rev. A, 7, 1639 (1973).
7. R. J. Robson and E. A. Dennis, J. Phys. Chem., 81, 1075 (1977).
8. A. Abragam, "The Principles of Nuclear Magnetism", p. 456, Pergamon Press, Oxford, 1961.
9. I. Ya. Slonim, Vysokomol. Soed., 6, 1371 (1964).
10. J. Charvolin and P. Rigny, J. Chem. Phys., 58, 3999 (1973).
11. M. Bloom, E. E. Burnell, S. B. W. Roeder and M. I. Valic, J. Chem. Phys., 66, 3012 (1977).
12. N. Bloenbergen, E. M. Pursell and R. V. Pound, Phys. Rev., 73, 679 (1948).
13. R. M. Servuss, W. Harbich and W. Helfrich, Biochim. Biophys. Acta, 436, 900 (1976).
14. H. H. Hub, U. Zimmermann and H. Ringsdorf, Febs. Letters, 140, 254 (1982).

REACTIVITY STUDIES IN A.O.T. REVERSE MICELLES

M. P. Pileni, J. M. Furois and B. Hickel

Laboratoire de Chimie-Physique
Université P et M Curie, 57231 Paris cedex 05 France;
and CEN-Saclay, DPC/SCM, 91191 Gif sur Yvette, France

It is shown that the hydrated electron formed by pulse radiolysis can be used as a probe to provide information on the water pool in reverse micelles.

The study of the photoelectron transfer from magnesium tetraphenylporphyrin to viologen has been studied in reverse micelles. The rate constant of electron transfer and the intramicellar exchange have been determined.

The photoreduction of water soluble protein such as cytochrome C by N methylphenothiazine has been observed in reverse micelles.

INTRODUCTION

Water can be readily dispersed in an organic medium such as
isooctane using sodium 2 ethyl hexyl sulfosuccinate (Aerosol OT,
AOT) as a dispersant. AOT possesses a hydrophobic head group which,
in non polar solvents in the presence of water, forms aggregates of
colloidal dimensions known as reverse micelles. A relatively large
amount of water can easily be incorporated into the micellar core
and the state and the properties of such water pools have been
previously investigated [1]. The number of water molecules in a
pool can vary considerably and depends principally on w, the
molar ratio H_2O/AOT in the system. For a given value of w, the
size of the pool does not vary significantly with the total
concentration of AOT and experimental evidence favours the formation
of spherical particles with a fairly monodisperse size distribution.
As in the case of normal micelles, the presence of two regions of
distinctly different polarity in reverse micelles provides
a unique environment in which chemical reactions may be carried
out. Indeed many reactions are dramatically catalyzed in such
systems. During the last few years, several authors have investigated proteins dissolved in such systems [2]. According to these
authors, enzymes dissolved in the aqueous polar core of the
reverse micelles are protected against denaturation by the
fact that the surface of the interfacial "phase" between the
protein globule and the organic solvent phase is stabilized by the
surfactant molecules.

As is illustrated by photosysnthesis, electron transport systems
play an important role in biological energy conversion. It may
therefore be useful to arrange a biomimetic electron transport
system as a first step in energy storage[3] and micellar systems
are thought to be suitable vehicles for promotion of photochemical
studies for the conversion of light into chemical energy.

In this paper we report the dependence of the hydrated electron
formation with the water content of the micelles and the photoelectron transfers from magnesium tetraphenylporphyrin to various
viologens which are thought to differ from each other by their
location in the aggregate. We show for the first time that proteins
can be used as an electron acceptor in the photoelectron transfer.

EXPERIMENTAL

Isooctane and methanol were obtained from Fluka and were of
"puriss" quality. They were used without further purification.
The incorporated water was deionized and twice distilled. Aerosol

OT from Merck was purified by dissolving it in methanol and adding
active charcoal. This solution was stirred for several hours,
concentrated by evaporation, and the AOT extracted with isooctane.
Methanol and water were eliminated from the AOT isooctane solution
by distillation taking into account the azeotropic point of
isooctane-water.

The structure of viologen, MVS, used is :

$$R-N \overset{+}{\underset{}{}} \begin{matrix} CH-CH \\ CH \ CH \end{matrix} \ \ C-C \ \ \begin{matrix} CH-CH \\ CH=CH \end{matrix} \underset{+}{} N-R$$

with R = $-(CH_2)_3-SO_3^-$; MVS, was synthetized by the Willner méthod.

Magnesium tetraphenylporphyrin, MgTPP, was Sigma product and
cytochrome C was obtained from Fluka. N methylphenothiazine, MPTH,
was synthetized by Dr A.M. Braun, N benzynicotinamide, BNA, was
purchassed from K & K.

Absorption measurements were made on Beckman spectrophotometer.
Continuous irradiation experiments were carried out using an Osram
1000 watts lamp as a light source. A monochromator was placed in
the beam to select a wavelength band. The spectrofluorimeter used
for all quenching studies was a Perkin Elmer MPF - 2A recording
spectrofluorimeter equipped with a Hamamatsu type R 446 photomultiplier tube. Quenching of magnesium tetra phenylporphyrin, MgTPP,
was examined in undegassed aqueous solution. It was established
that quenching of florescence by oxygen was negligible. The micro-
second flash photolysis apparatus employed has been proviously
described [4]. Flash photolysis experiments used a 20 ns. 530nm light
pulse from a Q switched flash photolysis apparatus (Quantel Lasers),
the transients being monitored by fast kinetic spectroscopy. The
pulse radiolysis experiments were carried out with Febetron 707
delivering single pulses of electrons in the energy range 1.6 -
1.8 Mev. The pulse shape is triangular, the total base width being
less than 20ns. Absorption spectra were recorded using a fast spectrophotometric detection system. Two detectors were used : a Hamamatsu R 446 photomultiplier and, for the infrared range, a silicon
diode coupled with a fast amplifier. The Suprasil irradiation cell
has an optical path length of 2.5 cm. Relative doses were monitored
by measuring the total pulse charge with a charge integrating cir-
cuit. The induced voltage was a linear function of the dose absorbed in the irradiation cell.

The concentration of photolytic products were monitored at
395 nm where the porphyrin cation and the reduced viologen absorb.

RESULTS AND DISCUSSION

I Hydration of Electron :

Two papers have been previously published on the hydrated electron in AOT reverse micelles [5,6]. The results obtained by the authors are not in total agreement. In this paper, we report the dependence of the hydrated electron formation on the water content of the micelles.

Figure 1A shows the transient absorption spectra of AOT/H_2O/isooctane obtained by pulse radiolysis at various w values. At high water content, w=60, the absorption spectrum is close to that of the hydrated electron in homogeneous aqueous media, Figure 1B. Such absorption spectra are attributed to electrons hydrated in micellar water pools. By decreasing the water content, a decrease in the hydrated electron optical density extrapolated to zero time and a blue shift with a broadening of the absorption spectra are observed Figure 1. Figure 2. illustrates the variation of the optical density, O.D, with the water content, w. At w values above 15 the O.D. of the hydrated electron is independent of the water content in the water pool; at lower values, the hydrated electron O.D. falls with decreasing w. To explain these results we assume that excess electrons formed in hydrocarbons are captured by the water pool. Hence the efficiency of electron capture by the micellar particles diminishes with the size of the water pool when the water is largely immobilized.

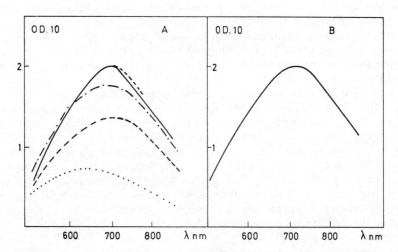

Figure 1. Absorption spectra extrapolated to zero time.
A - in reverse micelles at various water content : w=5 (....),
w=10 (— — —),w=15 (— . . —), w=30 (———) and w=60 (---);
B - in aqueous solution.

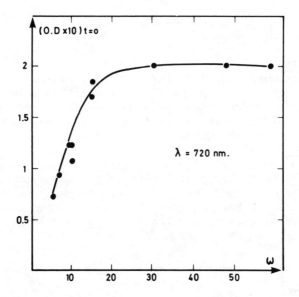

Figure 2. Evolution of the optical densities, O.D, observed at 720nm with water content, w.

The width of the absorption spectra of hydrated electron in reverse micelles depends on the water content . Insert in figure 3 shows the variation of the width of the absorption band with w. At low water content the absorption spectrum is broader than that observed in aqueous solution or at w values above 15 . This broadening could be attributed to changes in the properties of water in reverse micelles at low water content. The corresponding pseudo first order rate constants are shown as a function of w in figure 3. In aqueous solution, the hydrated electron reacts with AOT with a rate constant, k_1 , equal to $4.7.10^7$ $M^{-1}.s^{-1}$. Hence the half-life of the hydrated electron shorter by an order of magnitude than that in water under similar conditions, is not due to the decrease in the local pH in the reverse micelle but to its reaction with a surfactant molecule. At high water content, w > 15, the electron lifetime is equal to 230ns. On diminishing w, we observe a decrease in the electron lifetime. This effect could be attributed to the close proximity of the surfactant to the hydrated electron formed in the pool.

Figure 2 and 3 show a change in the shape of the curve around a w value equal to 15 . Similar changes in the variation of the physical properties with water content have been observed by NMR relaxation [7] and by fluorescence [8] . However these changes occured at w = 6 corresponding to the solvation shell of the counter ion and to rigid micellar core . At w values lower than 6 no solvation of the electron has been observed in reverse micelles[9]. To explain

our results and those obtained by different authors [5,6,9] we
propose the following model:
- at low w values the water in the pool is highly immobilized and
not able to solvate electrons .
- when w increases the water at the center of the pool remains free
from the interface and the probability of electron solvation also
increases. The fact that the absorption spectra of hydrated electrons
inside the micelle are shifted toward short wavelengths compared to
bulk water shows that the water in the pool is different from the
different from the bulk .
- when w increases further, the probability of electron capture
and solvation by the water pool increases and the spectra of the
hydrated electron become closer and closer to that observed in
the bulk water. These results are in agreement with those already
obtained by fluorescence measurements [10] .

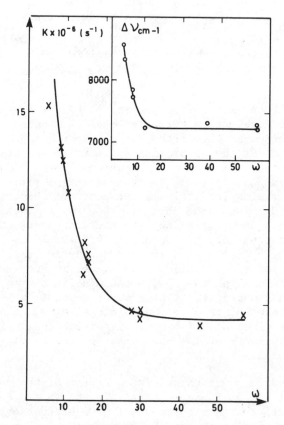

Figure 3 . Variation of the hydrated electron rate constant with
the water content , w .
insert: Variation of the width of the absorption spectra of the
hydrated electron with w .

II Photoreduction of Viologen :

   Magnesium tetraphenylporphyrin, MgTPP, is insoluble in water
and sparsely soluble in isooctane. Addition of AOT up to 3.10^{-2}M
increases the MgTPP solubility considerably indicating that the
MgTPP is probably located in the hydrocarbon tail or at the
interface of the reverse micelle. The absorption spectrum of MgTPP
in AOT reverse micelles is characterized by a Soret band centered
at 420 nm and two α and β absorption bands centered respectively at
560nm and 600nm. No difference in the absorption bands is observed
on increasing the water content in the pool from w = 2.5 to w = 60.
Also no differences have been observed in MgTPP absorption spectrum
on adding MVS to these micelles.

   The fluorescence of 5.10^{-6}M MgTPP exhibits maxima centered
at 602nm and 653nm. The fluoresecence lifetime is equal to 9 ns.
In the absence of quencher or in the presence of MVS no change in
the fluorescence quantum yield was observed on increasing w.

   By laser photolysis of MgTPP in AOT reverse micelles a triplet-
triplet absorption spectrum of MgTPP, characterized by a maximum
centered at 460nm, is observed. Its extinction coefficient, ε_{TT},
determined at 460nm and its triplet quantum yield, ϕ_T, are
$1.7.10^5$ $M^{-1}.cm^{-1}$ and 0.88 respectively. The triplet quantum yield
and the shape of the transient absorption spectrum are independent
of water content.

   In the presence of $2.5 \cdot 10^{-4}$M MVS with w > 10, photoelectron
transfer from MgTPP to viologen is observed. Figure 4 shows the
transient spectra obtained in the presence of viologen. At the
end of the laser pulse, the triplet-triplet absorption spectrum
of MgTPP is observed. A few microseconds later, the triplet spectrum
decreases and a new species appear characterized by an absorption
band around 395-400nm. By comparison with the absorption spectra
given in the literature [13] for the porphyrin cation and the reduced
viologen, the quenching of MgTPP triplet by viologen is attributed
to an electron transfer reaction of the form:

$$^3\text{MgTPP} + \text{MVS} \rightarrow \text{MgTPP}^+ + \text{MVS}^- \qquad (1)$$

The optical densities of the photolytic products, observed at 395nm,
increase as a function of the water content, w. This increase in
the efficiency is due to the increase in the size of the micelle
and the greater probability of finding one viologen per micelle.
Reaction (1) occurs with a kinetic rate constant, k_q. To determine
k_q, it is assumed that the reactants are distributed in the water
pool according to a Poisson distribution[14]. Assuming the value
of the intermicellar quenching rate constant, k_e, the kinetic
model leads to the following time dependence for MgTPP triplet:

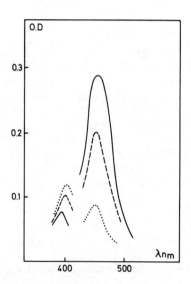

Figure 4. Transient absorption spectra obtained by laser photolysis :
(——) at the end of the laser pulse, (- - -) 5μ s after the laser pulse
and (·····) 10μ s after the laser pulse.

$$f(t) = \mathrm{Ln}\, \frac{^3\mathrm{MgTPP^*}_t}{^3\mathrm{MgTPP^*}_{t=0}} = \mathrm{Ln}\, \frac{O\, D_t}{O\, D_{t=0}} = $$
$$-(k_o + k_e . [\mathrm{MVS}])t - \bar{n}.(1 - \exp(-k_q t)) \tag{1}$$

where $^3\mathrm{MgTPP^*}_t$ and $^3\mathrm{MgTPP^*}_{t=0}$ are the MgTPP triplet concentrations at
time t and time zero respectively, $O.D_t$ and $O.D_{t=0}$ are the corresponding optical densities of the MgTPP triplet at these times, k_o
is the first order rate constant governing the MgTPP triplet decay
in the absence of viologen, k_e is the bimolecular rate constant for
the exchange process involving water pool collision, and $\bar{n}$ is the
average number of viologen molecule per water pool, $\bar{n} = \mathrm{MVS/WP}$ and
WP is the water pool concentration. At long times the previous
equation reduces to :

$$\phi(t) = \mathrm{Ln}\, \frac{OD_t}{OD_{t=0}} = - p.t - \bar{n} \quad \text{with} \quad p = k_o + k_e . [\mathrm{MVS}] \tag{2}$$

Figure 5 shows the variation of the logarithm of $OD_t/OD_{t=0}$ versus
time. The slope of the long time decay gives p. By plotting p versus
MVS, k_e is obtained, see insert in figure 5. From Table I we
deduced that the intermicellar exchange rate constant depends on
the water content in the pool but is independent of the viologen
used. Using these p values in equation (2) to fit the short time
experimental data enables k_q to be determined. (At w = 15 and
w = 25 the k_q values are equal to $7.10^5 s^{-1}$).

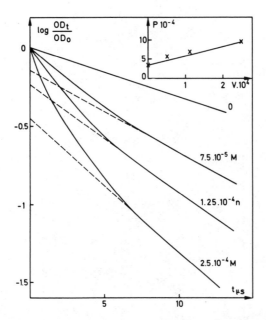

Figure 5. Decay of MgTPP triplet at 460nm at different viologen
concentrations. Insert. Variation of p versus viologen concentration

Table I. Exchange Rate Constants Obtained from Equation (3) Using
MVS as a Quencher of MgTPP Triplet :

$$k_e . 10^{-8} (M^{-1}.s^{-1}) \qquad\qquad 1.2^5 \qquad 1.9^{15} \qquad 2.3^{25}$$

III - Photoreduction of Protein such as Cytochrome C.

In this paper we will analyze a photoredox reaction which involves the reduction of a cytochrome C, $Fe^{3+}C$, by an excited N methylphenothiazine, MPTH[13] :

$$MPTH^* + Fe^{3+}C \rightarrow MPTH^+ + Fe^{2+}C$$

The absorption spectrum of MPTH exhibits a maximum centered at 300nm.
MPTH is hydrophobic and thus, in reverse micellar solution, will be
preferantially associated with the surfactant aggregate and with
the solvent. The reduced cytochrome C, $Fe^{2+}C$, is characterized by
two bands centered at 520nm and 550 nm, respectively. A reverse micellar solution containing : $10^{-4}M$ MPTH, $5.10^{-4}M$ benzylnicotinamide,

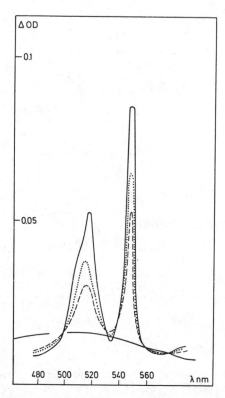

Figure 6. Difference spectra before and after various irradiation
times between the irradiated solution and non irradiated solution
(---) t = 60nm, (•••) t = 120nm, (——) t = 180nm.

BNA, and, 4.10^{-6}M cytochrome C, was deaerated and irradiated continuously with 300nm light. Under these conditions, the concentration
of MPTH remained unchanged throughout the irradiation period. The
absorption spectra of the photolytic product (Figure 6) are in good
agreement with the absorption spectrum of reduced cytochrome C
obtained by chemical reduction. Fe^{2+}C formed by photolysis disappears on exposure to air and is not formed in the dark. The role of
BNA is to donate an electron to MPTH cation before the reoxidation
of Fe^{2+}C can occur :

$$MPTH^+ + BNA \rightarrow MPTH + BNA^+$$

CONCLUSION

From the data obtained by pulse radiolysis and according to those obtained previously by different authors[5,6,9], we can conclude that the observed yield of hydrated electrons depends on the water content of the micelles. The change in the water content affects the properties of the hydrated electron. The water aggregate in reverse micelles is different from the bulk water at low w, and the properties of bulk water only appears in this system at low values above 20. Hence the hydrated electron can be used as a probe to provide information on the water pool in reverse micelles.

From the photochemical experiments, we can conclude that a photoelectron transfer from the sensitizer excited in the triplet state to viologen can occur. From the kinetic studies, assuming the reactants are distributed in the micelle with a Poisson distribution, the electron transfer rate constant for viologen used is determined, and the intermicellar exchange rate constants at various contents are deduced.

For the first time in such system, it has been possible to use a protein as an electron acceptor in the photoelectron transfer study. The mechanism of such electron transfer seems to be different from that observed with small molecules such as viologen.

ACKNOWLEDGEMENT

The authors would like to thank Dr J.Sutton and Dr J.Tabony for a critical reading of the manuscript.

REFERENCES

1. J.H.Fendler Acc.Chem.Res., $\underline{9}$, 1953 (1976)
2. P.J.Curtis Biochem.Biophys.Acta, $\underline{255}$, 833 (1972)
3. A.A.Krasnovskii and A.N.Cuganshaya Dokl.Acad.Nauk.SSSR $\underline{223}$, 229 (1979)
4. M.P.Pileni J.Chim.Phys., $\underline{75}$, 32 (1978)
5. M.Wong, M.Gratzel and J.K.Thomas Chem.Phys.Letters $\underline{30}$, 329 (1975)
6. V.Calvo-Perez, G.S.Beddard and J.H.Fendler J.Phys.Chem., $\underline{85}$, 863 (1981)
7. M.Wong, J.K.Thomas and T.Novak J.Am. Chem. Soc., $\underline{99}$, 4730 (1977)
8. E.Keh and B.Valeur J.Phys.Chem., $\underline{85}$, 1062 (1981)
9. G.Bakale, G.Beck and J.K.Thomas J.Phys.Chem., $\underline{85}$, 1062 (1981)
10. P.E.Zinsli, J.Phys.Chem., $\underline{83}$, 3223 (1979)
11. P.A.Trudinger Analyt.Biochemistry, $\underline{36}$, 222 (1970)
12. S.S.Atik and J.K.Thomas Chem.Phys Letters, $\underline{79}$, 351 (1981)

PREPARATION OF COLLOIDAL IRON BORIDE PARTICLES IN THE CTAB-n-HEXANOL-WATER REVERSED MICELLAR SYSTEM

N. Lufimpadio, J. B.Nagy and E.G. Derouane

Laboratoire de Catalyse, Département de Chimie
Facultés Universitaires Notre-Dame de la Paix
61, rue de Bruxelles, B-5000 Namur, Belgium

The reduction by $NaBH_4$ of Fe(III) ions solubilized in the water core of reversed micelles provides a convenient way to produce iron boride particles of predetermined size and size distribution. The reduction is achieved under nitrogen to avoid back-oxidation of the iron boride particles. Those are then deposited on silicagel and examined by electron microscopy. The particles have a narrow size distribution ($\pm$ 2-5 Å) with average size varying from about 30 to 80 Å. It is greatly influenced by the size of the micelles (studied by ^{19}F-NMR), the lability of the interface, and the absolute concentrations of Fe(III) ions and $NaBH_4$ reactants.

INTRODUCTION

Catalysis plays an important role in modern industrial processes. Nearly all industrial chemicals come into contact with a
catalyst before they reach the consumer [1,2]. Metals constitute an
important class among the usual heterogeneous catalysts.

Iron catalysts with high specific surface area are of great
interest in catalysis, in particular for the ammonia and Fischer-
Tropsch syntheses. A fine dispersion increases the activity of the
catalyst due either to the increased number of active sites and/or
to the change in the nature of these active sites. It is however
difficult to obtain, by classical methods, small and nearly monodisperse iron particles deposited on a support, particularly for
sizes lower than 100 Å. This difficulty stems from the fact that
metallic iron particles sinter easily during the reduction by hydrogen of the precursor oxidic species [3].

In order to prepare small and nearly monodisperse iron boride
particles, we developed a method which consists of reducing by
sodium borohydride (NaBH4) Fe(III) ions solubilized in the water
core of reversed micelles in the ternary system CTAB-n-Hexanol-
water. The existence domain of reversed micelles is shown in
Figure 1 [4]. Reversed micelles are formed by self-association of
surfactant molecules in non-aqueous media. The core of the micelles is formed of water and constitutes a "cage" where chemical

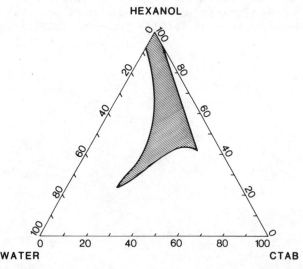

Figure 1. Domain of existence of reversed micelles (following ref.4)

reactions may occur. We show in this paper that the size of the water core, the lability of the interface and the concentration of metallic ions are the parameters which are critical in the preparation of small iron boride particles of controlled size.

EXPERIMENTAL

1. N.m.r. measurements

a) ^{13}C-n.m.r. of n-Hexanol and CTAB

The interactions between Fe(III) ions and the carbon atoms of the micellar components were studied by ^{13}C-n.m.r. The spectra were recorded on a Bruker CXP-200 spectrometer at ambient temperature (ca. 22°C) or at 34°C. To avoid the problem of volume susceptibility corrections, we used as internal reference the methyl carbon of the n-Hexanol molecule. The stabilization of the magnetic field was carried out using a deuterium lock of D_2O contained in the external part of two concentric tubes, the inner tube being filled with the micellar solution.

b) ^{19}F-n.m.r. of 6-fluorohexanol

The approximate sizes of the micelles were determined by ^{19}F-n.m.r. of 6-fluorohexanol, which was used as a label compound in the micellar solution (ca. 1-5 wt.%) [5]. 6-fluorohexanol was synthesized according to the Pattison method [6]. The ^{19}F-n.m.r. spectra were recorded on a Bruker CXP-200 spectrometer with C_6F_6 as external reference.

2. E.P.R. measurements

The interface dynamics were studied by E.P.R. using spin labels. The E.P.R. spectra were recorded on a Bruker BER-420 spectrometer operating in the X-band with 100-kHz field modulation. The samples were placed in capillary tubes of ca. 1 mm diameter. The concentration of spin labels was kept low (ca. 5×10^{-4} molal) in order to avoid strong spin-spin interactions and perturbation of the micelles [7]. The following spin labels were used :

I : $CH_3-CH_2-\underset{\substack{O \quad N-O^{\cdot} \\ \rule{1em}{0.4pt}\underset{}{}}}{C}-(CH_2)_{14}-COOH$

II : $CH_3-(CH_2)_{12}-\underset{\substack{O \quad N-O^{\cdot} \\ \rule{1em}{0.4pt}\underset{}{}}}{C}-(CH_2)_3-COOH$

3. Preparation of small iron boride particles

The preparation of small iron boride particles is carried out by reducing with sodium borohydride the Fe(III) ions dissolved in the inner water core of the reversed micelles. After degassing with nitrogen, the micellar solution is transferred into a glove box. An excess of sodium borohydride is then progressively added at 0°C. The black solution which is obtained is vigorously stirred. Finally the particles are deposited on silicagel. All operations (filtration, washing, drying, ...) are achieved under nitrogen atmosphere to avoid a back-oxidation of iron particles.

4. Electron microscopy

The determination of average size of particles deposited on silicagel was made by electron microscopy, using a Philips-301 electron microscope working in the transmission mode.

RESULTS AND DISCUSSION

1. Fe(III)-Micellar interactions and site of solubilization

The nature of the interactions between the Fe(III) ions and the carbon atoms of the micellar components, and the location of the solubilized Fe(III) ions, were studied by ^{13}C-n.m.r. of the carbon atoms of CTAB and n-Hexanol in the presence of the Fe(III) ions. First, the chemical shifts of n-Hexanol carbon atoms were determined as a function of the concentration of the Fe(III) ions dissolved in pure n-Hexanol (Table I, Figure 2).

Table I. ^{13}C-n.m.r. chemical shifts (δ in Hz) of n-Hexanol carbon atoms vs concentration of Fe(III) ions.

[Fe(III)] x 10^2 (molal)	$\delta^{a,b}$ (Hz)				
	C_1	C_2	C_3	C_4	C_5
2.5	−15	81	15	0	0
5	−26	152	30	−4	−4
7.5	−19	178	33	−7	−7
10	−26	222	48	−11	−11

a. Methyl carbon of n-Hexanol taken as reference ; average error ; ±3 Hz
b. Spectrometer frequency : 50.3 MHz

These results show clearly that Fe(III) ions interact strongly with n-Hexanol and that this interaction affects mostly carbon-2. This strong influence on carbon-2 indicates that the interactions between the Fe(III) ions and the carbon atoms of n–Hexanol include an important contribution of the contact type, stemming from the direct interaction of the oxygen atom of n–Hexanol and the Fe(III) ions [8].

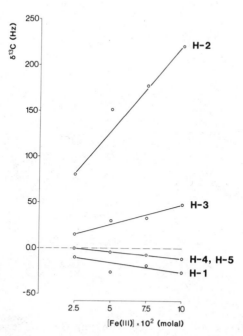

Figure 2. ^{13}C chemical shifts of n–Hexanol vs Fe(III) concentration.

    When dissolved in a micellar solution, Fe(III) ions have a quite different influence on the carbon atoms of n–Hexanol. Table II and Figure 3 show the great variation in the chemical shifts of CTAB carbon atoms due to the presence of Fe(III) ions, while the carbon atoms of n–Hexanol are only slightly influenced (from –30 to +20 Hz).

Table II. The chemical shifts (δ in Hz) of CTAB carbon atoms vs concentration of Fe(III) ions. Micellar composition : CTAB 40% wt.%, n–Hexanol 50 wt.%, water 10 wt.%.

[Fe(III)] x 10^2	δ a,b (Hz)		
(molal)	C–O	C–1	C–m
2.5	96	41	–4
5	218	96	0
7.5	337	155	15
10	–	203	–

a,b : see Table I and Figure 3.

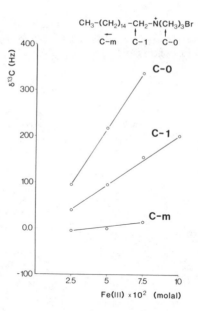

Figure 3. ^{13}C chemical shifts of CTAB vs Fe(III) concentration.

For CTAB molecules, the chemical shifts of carbons C-0, C-1 and C-m (see Figure 3) increase with increasing Fe(III) ion concentration but they decrease rapidly within the hydrocarbon chain, the more remote carbon atoms (C-m) being less influenced by the presence of Fe(III) ions. This observation indicates that the interaction between the Fe(III) ions and the CTAB molecules is mostly of dipolar type.

From the ^{13}C-n.m.r. chemical shifts of CTAB and n-Hexanol carbon atoms, we therefore conclude that the Fe(III) ions dissolved in the micellar solution are essentially solubilized in the water core of the micelles where they are largely solvated by water molecules. There is no direct contact between the n-Hexanol oxygen atom and the Fe(III) ion, and in contrast to the direct Ni(II) ions-n-Hexanol interaction [9], the n-Hexanol molecules are excluded from the first coordination shell of the Fe(III) ions.

2. Determination of micellar sizes

The determination of the approximate sizes of micelles by ^{19}F-n.m.r. is based on the distribution of 6-fluorohexanol in the water core, the interface and the bulk organic phase formed by n-Hexanol [5,9]. Because of fast exchange between those phases, the

^{19}F-n.m.r. spectra consist only of a single component. The relationship between the observed chemical shift (δ_{obs}) and the quantity of n-Hexanol in each phase is given by the following equation:

$$\delta_{obs} = X\delta_W + Y\delta_I + Z\delta_H$$

in which δ_W, δ_I and δ_H are the ^{19}F-chemical shifts of 6-fluorohexanol in water, interface and n-Hexanol, respectively ; X, Y and Z are the corresponding mole fractions in water, interface and organic phase. Due to the very low solubility of n-Hexanol in water [5], X $\simeq$ 0 and the relationship becomes :

$$\delta_{obs} = Y\delta_I + Z\delta_H \quad \text{with} \quad Z + Y = 1$$

If δ_H and δ_I are determined independently, one can calculate Y and Z. δ_H is determined experimentally by taking the average of ^{19}F-chemical shifts of 6-fluorohexanol dissolved in n-Hexanol with different concentrations : δ_H = -9571 Hz.

δ_I is determined from normal micelles formed by CTAB in aqueous solutions in which 1-2 wt.% of 6-fluorohexanol is dissolved. The observed ^{19}F chemical shifts obtained in solution with different concentrations of CTAB are related, in the fast-exchange approximation, to δ_I, δ_W and the CMC of the surfactant [10] according to :

$$\delta_{obs} = \delta_I + \frac{CMC}{[CTAB]} (\delta_W - \delta_I)$$

From the variation of δ_{obs} with respect to $1/[CTAB]$, the intercept of the straight line yield δ_I. Table III shows the ^{19}F-n.m.r. data in normal micellar solutions.

Table III. ^{19}F chemical shifts in normal micellar solutions.

sample	water (wt.%)	6-fluorohexanol (wt.%)	CTAB (wt.%)	1/[CTAB] (m^{-1})	$\delta^{19}F$ (Hz)
1	58.8	1.2	40	0.54	-9273
2	71.8	1.5	26.7	0.98	-9261
3	76.3	1.6	22.1	1.26	-9256
4	81.6	1.7	16.7	1.79	-9241

Extrapolation to infinite CTAB concentration yields δ_I = -9287 Hz (Figure 4). The mole fraction of n-Hexanol molecules in the interface (Y) is thus easily computed (Table IV).

The size of the inner water core can therefore be calculated assuming that all CTAB molecules are located in the interface [11] and assuming the cross section of the CTAB and n-Hexanol molecules to be equal to 34 Å^2 (according to the model of the CTAB molecule)

and 21 Å^2 [5] respectively. The number of CTAB and n-Hexanol molecules in the interface yield the total area (S_T) of the interface
($S_T = S_{Hex} + S_{CTAB}$).

$$ST = N_{CTAB} \times 34 + N_{Hex}^{int} \times 21 \ (\text{Å}^2)$$

$$\text{where } N_{CTAB} = \frac{(\text{wt.\% CTAB})/100}{(\text{M.W. CTAB} = 364)} \times 6.02 \times 10^{23}$$

$$\text{and } \quad N_{Hex}^{int} = Y \ \frac{(\text{wt.\% Hex})/100}{(\text{M.W. Hex} = 102)} \times 6.02 \times 10^{23}$$

$$V = (\text{wt.\% } H_2O/100) \times 10^{24} \ (\text{Å}^3)$$

Finally, the average radius $\langle r \rangle$ of the inner water core, assuming
a monodisperse distribution of spherical inverted micelles is evaluated as :

$$\langle r \rangle = \frac{3V}{S_T}$$

where V is the total volume of water in the micellar solution.

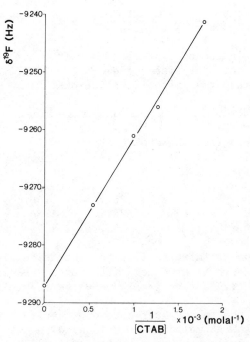

Figure 4.   Determination of δ_I.

Table IV. ^{19}F chemical shifts in reversed micellar solutions and average radii of micelles.

Sample	Micellar composition (wt.%)			$\delta^{19}F(Hz)$	Y	<r>(Å)
	CTAB	n-Hexanol + 6-F-hexanol	Water			
1	40	50	10	-9477	0.331	7.0
2	39	49	12	-9474	0.342	8.4
3	38.5	47.5	14	-9470	0.356	9.9
4	38	47	15	-9468	0.363	10.6
5	38	46	16	-9467	0.366	11.4
6	37.5	45.5	17	-9466	0.370	12.2
7	37	45	18	-9466	0.370	13.2
8	36.5	44.5	19	-9466	0.370	14.0
9	36	44	20	-9464	0.377	14.7

The average radii of the micellar water cores are reported in Table IV for various micellar solutions with a constant [CTAB]/[n-Hexanol] ratio.

Figure 5 shows the average size of the micellar cores as a function of water concentration. The average size increases with increasing water concentration, when the [CTAB]/[n-Hexanol] ratio remains constant. Under these conditions, the slight change in the n-Hexanol mole fraction composition (Y) of the interface corresponds to an almost constant surface (S_T) and the average radius of the inner water core is directly proportional to the volume occupied by water in the reversed micellar solutions.

3. The dynamics of the interface

The lability of the interface influences the diffusion rate of the reducing agent (NaBH$_4$) across the interface, hence also the rate of formation of the iron boride particles. When the interface is more rigid, the diffusion of NaBH$_4$ across it is slower. In these conditions, the water cores are also better isolated and limit, therefore, the aggregation of iron boride particles which should eventually yield smaller iron boride particles.

The dynamics of the interface were studied by the technique of EPR using spin labels I and II. Spin label I, in which the nitroxyl radical is quite distant from the polar head group, probes the non-polar region of the interface ; while spin label II with the nitroxyl radical close to the polar head group explores the polar region of the interface. Figure 6 shows a typical EPR spectrum of spin label II in a micellar solution. From the heights (h) and the linewidths (ΔH) of the different multiplets (due to the electron-nitrogen hyperfine coupling), one can calculate the rotational correlation-times (τ_c) of these spin labels [12-15] from the relationship

$$\tau_c = A \cdot H_{(+1)} \left[(h_{(+1)}/h_{(-1)})^{1/2} - 1 \right]$$

where $H_{(+1)}$ is the peak-to-peak width (in gauss) of the low-field resonance line and $h_{(+1)}$ and $h_{(-1)}$ are the peak-to-peak heights for the low and high-field lines, respectively. A is a constant assumed equal to 6.6×10^{-10} s.gauss^{-1} [12]. Table V lists the τ_c values obtained from different micellar solutions.

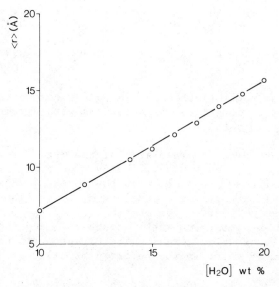

Figure 5. Average radius of micelle vs water concentration.

Table V. τ_c values for spin labels I and II in micellar solutions.

Micellar composition (wt.%)			$\tau_{c_I} \times 10^{10}$ (s)	$\tau_{c_{II}} \times 10^{10}$ (s)
CTAB	n-Hexanol	Water		
40	50	10	3.7	16
30	40	30	3.3	12.9
19	65	16	2.7	9.7
15	75	10	2.8	9.1
–	100	–	1.9	7.1

Table VI gives the correlation times for spin label II for micellar solutions with constant [CTAB]/[water] ratios.

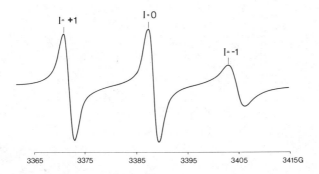

Figure 6.   Typical EPR spectrum of the spin label II in a micellar solution.

Table VI. τ_{cII} values in micellar solutions with constant [CTAB]/[water] ratios.

Micellar composition (wt.%)			τ_{cII} x 10^{10}(s)	[CTAB]/[n-Hexanol] (interface)
CTAB	n-Hexanol	Water		
31.6	52.6	15.8	11.7	0.49
18.5	72.1	9.4	9.9	0.29
6.1	90.8	3.1	7.3	–
–	100	–	7.1	0.00

We observe that the rotational correlation times (τ_c) depend strongly on the composition of the micellar solution and that $\tau_{cII} > \tau_{cI}$ for all the investigated micellar solutions. This shows clearly that the polar region of the interface is more rigid than the non-polar part of the micelle. In addition, it is obser- ved that the mobility of the interface (τ_c) is related linearly to the [CTAB]/[n-Hexanol] ratio in the interface (see Figure 7) at constant [CTAB]/[water] ratio. When the former ratio decreases the rotational correlation time also decreases, indicating a "dilution" of the interface. This, in turn, leads to a decreased interaction between the CTAB molecules and results in a more labile interface.

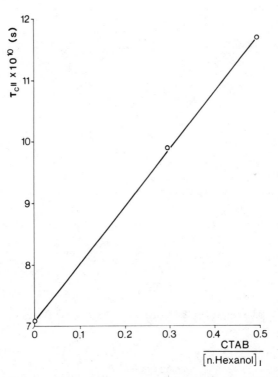

Figure 7. $\tau_{c_{II}}$ vs [CTAB]/[n-Hexanol] ratio in the interface.

4. Preparation of small iron boride particles

The average size of iron boride particles obtained by reduction by sodium borohydride of Fe(III) ions dissolved in a micellar solution, depends on the concentrations of the Fe(III) ions and of the $NaBH_4$ reagent in addition to the other micellar parameters. Table VII shows the variation of particle size with different Fe(III) ion concentrations for micellar solutions of the same composition (CTAB 30 wt.%, n-Hexanol 60 wt.%, water 10 wt.%).

Table VII. Particles size vs Fe(III) ion concentration.

Fe(III) molal	$\overline{d}$(nm)
5×10^{-3}	3
3.4×10^{-2}	5
5×10^{-2}	8

We observe that the average size of iron boride particles increases with increasing Fe(III) ion concentration. Table VIII gives the variation of the particle size with the composition of

various micellar solutions containing the same total Fe(III)
concentration (5 x 10^{-3} molal).

   It is seen that the average particle size increases with decreasing water content in the micelle. The comparison is even
more striking between the variation of the average particle size
and the local Fe(III) concentration in the aqueous micellar core :
higher Fe(III) concentration leads to larger particles. In addition,for all these iron boride particles, the size distribution
is rather small (σ varies from 2 to 5 Å).

Table VIII. Particle size as a function of micellar composition
containing the same total concentration of Fe(III) ions
(5 x 10^{-3} molal).

Micellar composition (wt.%)			$\bar{d}$(nm)	$[Fe(III)]_1^a$ x 10^2 (M)	N_n^b/N_M^c x 10^5
CTAB	n-Hexanol	Water			
40	50	10	4.7	5	1.5
38	47	15	4.1	3	4.8
36	44	20	3.3	2	18.9
31.5	38.5	30	3.2	1	-

a. local Fe(III) concentration : water only taken as solvent.
b. number of nuclei determined from the average size of particles
   and the total weight of Fe(III) ions assuming the formation of
   FeB.
c. number of micellar water cores calculated from ^{19}F-n.m.r data.

   The average number of iron boride nuclei (N_n) per micellar
water core (N_M), formed in the early stage of the reduction increases with increasing micellar water content (Table VIII) and varies
inversely with the local Fe(III) ion concentration. The higher is
the Fe(III) ion local concentration, the lower the number of iron
boride nuclei, therefore yielding larger particles. The opposite
behaviour was found in the reduction of Ni(II) ions by $NaBH_4$ in a
similar reversed micellar system [16] where the nucleation process
was essentially governed by the initial local concentration of the
reducing agent $NaBH_4$ and/or the average number of Ni(II) ions per
micelle. The higher the $NaBH_4$ local concentration, the higher the
number of nickel boride nuclei, yielding smaller particles.

   It seems that the nucleation process for the iron boride particles is much easier than for the nickel boride particles. The
average size of the particles is essentially controlled by the
local Fe(III) ion concentrations : it increases with increasing
Fe(III) concentration as well as with decreasing water content.
In addition, the greater mobility of the interface at low water
content also favors the crystal growth process as the rapid

rearrangement of the inner water cores is accompanied by a transport of Fe(III) ions to the nucleation centers.

The sizes of the particles are much greater than those of the
inner water cores of the reversed micelles. This difference cannot
stem from the approximations inherent to the ^{19}F-n.m.r. indirect
method for the determination of micellar sizes (neglect of the presence of CTAB monomers in n-Hexanol and in water or errors in the
determinations of chemical shifts ; this latter would contribute
about 5% to the mole fraction of hexanol at the interface [5]).
Therefore a micellar reorganization must be included during the
formation of the iron boride particles. The micellar reorganization is supposed to be a "homogeneous phenomenon" throughout the
micellar solution, leading therefore to "almost monodispersed"
particles, the number of which is determined in the early stage of
the nucleation process.

CONCLUSION

The reduction of Fe(III) ions by $NaBH_4$ in the CTAB-n-Hexanol-
water reversed micellar system leads to small iron boride particles (30-80 Å) with a narrow particle size distribution. The micellar water cores act as reaction cages for the reduction. The
particle size is controlled by the concentration of the reactants,
the size of the inner water core and the lability of the interface.

REFERENCES

1. D.L. Trimm, Chimia, 33, 415 (1979).
2. G.C. Bond "Principles of Catalysis". The Chemical Society
   Monographs for Teachers n°7, London (1972).
3. H. Topsøe, J.A. Dumesic, E.G. Derouane, B.S. Clausen, S. Mørup,
   J. Villadsen, and N. Topsøe, Stud. Surf. Sci. Catal., 3, 365
   (1979).
4. S.I. Ahmad and S. Friberg, J. Am. Chem. Soc., 94, 5196 (1972).
5. T. Nguyen and H.H. Ghaffarie, J. Chim. Phys., 76, 513 (1979).
6. F.L.M. Pattison and W.C. Howell, J. Org. Chem., 21, 739 (1956).
7. H.W. Offin, D.R. Dawson, and D.F. Nicoli, J. Colloid Interface
   Sci., 80, 118 (1981).
8. A.A. Obynochnyi, O.I. Bel'chenko, P.V. Schastnev, R.Z. Sagdeev,
   A.V. Dushkin, Yu. N. Molin, and A.I. Rezvukhin, Zhur. Strukt.
   Khim., 17, 620 (1976).
9. A. Gourgue, Dr. Sc. Thesis, Facultés Universitaires Notre-Dame
   de la Paix, Namur (1981).
10. B. Lindman and H. Wennerström, Topics in Current Chemistry,
    87, 3 (1980).
11. H.L. Rosano, J. Soc. Cosmet. Chem., 25, 609 (1974).
12. H. Yoshioka, J. Am. Chem. Soc., 101, 28 (1979).
13. A.S. Waggoner, O.H. Griffith, and C.R. Christensen, Proc. Nat.

   Acad. Sci., $\underline{57}$, 1198 (1967).
14. Y. Murakami, A. Nakano, K. Iwamoto, and A. Yoshioka, Chem.
   Lett., 951 (1979).
15. G.I. Likhtenshtein "Spin Labelling Methods in Molecular Biology", Wiley Interscience, pp. 1-21, New York, 1976.
16. J. B.Nagy, A. Gourgue, and E.G. Derouane, Stud. Surf. Sci.
   Catal., Elsevier, Amsterdam, in press.

Part VIII
Microemulsions and Reactions in
Microemuslions

MICROEMULSIONS - AN OVERVIEW

Th. F. Tadros

I.C.I. Plant Protection Division
Jealotts Hill Research Station
Bracknell, Berkshire   RG12 6EY

The theories of microemulsion formation and stability
have been reviewed. Three main approaches, namely
mixed film, solubilisation and thermodynamic theories,
have been discussed. Of these, the most recent
thermodynamic theories consider the free energy of
formation of microemulsion ΔG_m to consist of three
main contributions: (i) A free energy term ΔG_1 due to
mixing of surfactant and water and cosurfactant and
oil; (ii) a free energy term, ΔG_2, due to the
increase in interfacial area; and (iii) a free energy
term, ΔG_3 due to mixing of the droplets into the
continuous phase. From a consideration of the
relative magnitude of each term, one arrives at the
conclusion that for ΔG_m to become zero or negative,
the interfacial tension including the electrical term
must have very low but slightly positive value.

Application of scattering methods for characterisation
of microemulsions is not straightforward since dilution
techniques, normally used to reduce interparticle
interactions, cannot be applied to microemulsions. Of
the scattering techniques, time average light
scattering can be used for estimating the droplet size
providing an allowance is made for the structure factor.
The latter can be calculated using the hard sphere
fluid theory which may be applied to concentrated
dispersions of very small particles. Small angle
neutron scattering (S.A.N.S.) gives valuable
information regarding the "structure" in the system,
since the range of scattering vector that can be

1501

applied is relatively wide when compared with light
scattering. The results of S.A.N.S can be used to
determine the radius of the droplets either by
application of the Bragg's equation or by fitting the
scattering curve, using the hard sphere model. The
results obtained from S.A.N.S are consistent with those
obtained from light scattering, thus confirming the
validity of the hard sphere model. However, dynamic
measurements using quasielastic light scattering are
relatively more difficult to interpret, due to the lack
of adequate theoretical treatments that take into
account all possible variables (such as polydispersity,
dynamic structure factor, etc).

Conductivity measurements also give valuable
information on the structure of microemulsions. In
particular, the trends of conductivity versus volume
fraction of droplets can be used to distinguish between
the two classes of "transparent" systems obtained which
depend to a large extent on the chain length of the
alcohol (cosurfactant). Systems with short chain
alcohols (<C6) show percolative behaviour indicating
that the transparent system is best represented by
quaternary molecular solutions. On the other hand,
with longer chain alcohols ($\geqslant$ C6) the systems show low
conductivity and are therefore non-percolating. Such
systems represent "true" microemulsions with definite
water cores.

Further information on the structure of microemulsions
can be obtained from measurements of the self diffusion
coefficient of all constituents, eg, using NMR
techniques. These results, show that with a short chain
alcohol (<C6) the self diffusion coefficient of all
constituents is high and roughly the same, indicating
either the presence of a bicontinuous structure or an
easily deformable and flexible interface with very
short life time. On the other hand, with longer chain
alcohols (>C6) the self diffusion coefficient of the
component constituting the internal phase (eg, water in
w/0 microemulsion) is relatively small compared with
that of other constituents, indicating the presence of
well defined "cores" with a more pronounced separation
into hydrolphilic and hydrophobic regions.

INTRODUCTION

The term microemulsion was first introduced by Hoar and
Schulman[1] to describe the "transparent" or "translucent" systems

obtained by titration of a milky emulsion with a medium chain
alcohol such as pentanol or hexanol, which was later referred to
as cosurfactant. There has been much debate about the use of the
term microemulsion to describe such systems and many other terms
such as swollen micellar solutions or solubilised micellar
solutions[2] have been introduced. From thermodynamic treatments
of microemulsion formation (see later) one usually arrives at the
conclusion that the origin of stability of microemulsions is
associated with the fact that the free energy of their formation
is zero or negative, explaining their spontaneous formation. From
that point of view, it is perhaps convenient to describe such
systems as thermodynamically stable liquid in liquid dispersions.
This definition, however, falls short of describing the isotropic
nature of the system and the fact that in many cases well defined
droplets are not indeed present (see section on structural
aspects). For that reason, Danielsson and Lindman[3] have adopted
the following definition for microemulsions, "a system of water,
oil and amphiphile which is a single optically isotropic and
thermodynamically stable liquid solution".

     The literature on microemulsions has been growing at a very
fast rate in the last few years. This is not surprising in view
of the advances in modern techniques of scattering, spectroscopic
methods, etc, which now make it possible to obtain quantitative
information on the "structure" of microemulsions. Moreover, such
systems are ideal as models for studying interaction between
particles in concentrated systems. As we will see later, a number
of theoretical treatments have also emerged in the last five years
or so to describe the reasons for stability of microemulsions. On
the applied front, microemulsions have attracted considerable
attention in view of their potential applications in enhanced and
tertiary oil recovery.

     Due to this rapid progress, available reviews[2,4,5] fall
short of some of the recent developments. However, such reviews
should form the basis for any reader seeking detailed information.
In this overview (which is by no means exhaustive in view of the
vast literature available) I will attempt to highlight the most
recent viewpoints on the theories of microemulsion formation and
stability. The earlier theories which have been adequately
reviewed before[2,4,5] will only be briefly mentioned for
comparison. This will then be followed by a section on the recent
developments in scattering techniques for characterisation of
microemulsion droplets and studying their structure. The less
rigorous methods such as conductivity, dielectric constant and
viscosity measurements will be described only briefly. This will
then be followed by a section on structural and dynamic aspects of
microemulsions which have been obtained from applications of
spectroscopic techniques of nuclear magnetic resonance.

THEORIES OF MICROEMULSION FORMATION AND STABILITY

It is perhaps convenient to classify these theories into three main categories. The first is the so called interfacial or mixed film theories which were originally introduced by Schulman and Coworkers[4,6,7] and Prince[8,9]. The second class of theories are the so called solubilisation theories which were proposed by Shinoda, Friberg and collaborators[2,10,16]. The third category of theories, which are the most recent, are the thermodynamic treatments of Ruckenstein and collaborators[17-23] and Overbeek[24]. The mixed film and solubilisation theories have been adequately reviewed[2,4] and therefore will only be briefly described. On the other hand, the thermodynamic treatments will be discussed in relatively more detail.

Mixed Film Theories[4,6-9]

The essential features of the mixed film theories is to consider the film as a liquid, two dimensional third phase in equilibrium with both oil and water, implying that such a monolayer could be a duplex film, i.e., one having different properties on the water side than the oil side[4]. Schulman and Coworkers[6,7] originally considered that the interactions in the interphase among surfactant, cosurfactant and oil produce a reduction of the original oil/water, interfacial tension, $\gamma_{o/w}$, to zero. In other words, the two dimensional spreading pressure of the mixed film, π, reaches a value that either becomes equal to $\gamma_{o/w}$ or even exceeds it. Thus, the total interfacial tension can be represented by,

$$\gamma_T = \gamma_{o/w} - \pi \qquad (1)$$

Contributions to π were considered to be the crowding of the surfactant and cosurfactant molecules and penetration of the oil phase into the hydrocarbon part of the interphase. According to equation (1) if $\pi > \gamma_{o/w}$, γ_T becomes negative leading to expansion of the interface until γ_T reaches zero.

The main drawback of the above concept is the high values of π that need to be postulated to reach the condition $\gamma_T < 0$, since many hydrocarbon oils give a $\gamma_{o/w}$ of the order of 50mN m^{-1}, and such high π values required would tend to eject the hydrocarbon molecules from the mixed surfactant film. However, Prince[4,8,9] pointed out that the presence of the cosurfactant in the oil phase reduces $\gamma_{o/w}$ considerably (to about 15mN m^{-1}) and therefore $\gamma_{o/w}$ should indeed be replaced by the value reached in the presence of the cosurfactant, e.g., alcohol, ie, $(\gamma_{o/w})a$. Thus, equation (1) should be written as,

$$\gamma_T = (\gamma_{o/w})a - \pi \qquad\qquad (2)$$

which implies that reasonable values of π (of the order of 15mN m^{-1}) need to be postulated to reach the condition $\gamma_T \lessgtr 0$.

The above concept of duplex film can be used to explain both the stability of microemulsions and the bending of the interface. Considering that initially the flat duplex film has different tensions on either side of it, then the driving force for film curvature is the stress of the tension gradient which tends to make the pressure or tension on both sides of the curved film the same. This can be illustrated from the surface pressure area curves, schematically illustrated in Figure 1.

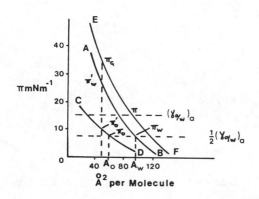

Figure 1. Schematic representation of π-A curves.

A-B represents the π-A curve for the side of the duplex film against water and C-D that for the side of the duplex film against oil, whereas EF is the actual π-A curve for the mixed duplex film (measured on a Langmuir trough) which is the sum of curves AB and CD. Assuming an area/molecule of ~50A^2 (reasonable value for surfactant molecules including that of the alcohol), the value of π'_o and π'_w for the flat film are 10 and 30 mN m^{-1}, whereas π_G for the duplex film derives from the relative magnitude of π'_w and π'_o. Under the stress due to the difference in surface pressures π'_w and π'_o and penetration of oil, expansion of the interface on both sides takes

place spontaneously until the two pressures become equal and π_G
drops to a value that is equal to $(\pi_{o/w})a$ (which as mentioned
above is about 15mN m^{-1}). Since $\pi = \pi_o + \pi_w$ then expansion
occurs until $\pi_o = \pi_w = 7.5$mN m^{-1}. This is illustrated in
Figure 1. This means that π'_w which originally for a flat film
was 30mN m^{-1} now drops to 7.5mN m^{-1}, whereas π'_o also drops
from 10 to 7.5mN m^{-1}. Thus, the driving force for curvature is
$\pi_G - \pi$ or $\pi_G - (\gamma_{o/w})a$ which according to equation (2) is
equal to $\pi - \gamma_T$, i.e., the negative interfacial tension before
curvature.

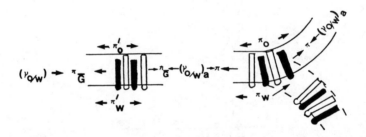

Figure 2.   Schematic representation of film bending.

The above case is one where $\pi'_w > \pi'_o$ and in this case
the film has to expand much more at the water side than at the oil
side which means that an o/w microemulsion will result (see Figure 2
On the other hand, if $\pi'_o > \pi'_w$ then the film has to expand
more at the oil side of the interface than at the water side (ie,
A_o would be higher than A_w) resulting in the formation of a
w/o microemulsion (see Figure 2). Since it is more difficult to
fill the water side of an o/w microemulsion with molecules than
to fill the oil side of an w/o microemulsion with oil molecules,
it becomes clear why it is easier to find w/o microemulsions than
o/w ones.

Solubilization Theories[2,10-16]

The idea of treating microemulsions as swollen micellar systems i.e., with oil or water solubilised in normal or reverse micelles stemmed from the studies of the three and four component phase diagrams on the one hand and the solubilisation studies of water and hydrocarbons by nonionic surfactants on the other[10-16]. For example, the phase diagram of a three component system of water, ionic surfactant and alcohol (or carboxylic acid) usually displays one isotropic aqueous liquid region L_1 and one isotropic liquid region from the alcohol (or acid) corner containing reversed micelles L_2 and several liquid crystalline structures. The alcohol solution L_2 contains a large amount of water and surfactant in the form of reversed micelles. This solution can dissolve a large amount of a hydrocarbon oil. Alternatively, such inverse micelles may be produced if the alcohol is dissolved in the hydrocarbon followed by addition of water and surfactant. Since the final solution is isotropic and no phase separation takes place when going from the pure hydrocarbon state to the microemulsion state, there does not seem to be any justification of describing these systems as two phase microemulsions.

Further support for consideration of microemulsions as solubilised systems came from Shinoda's[10-16] work using nonionic surfactants. Of particular significance in this respect is the solubilisation observed with nonionic surfactants with poly-oxyethylene oxide head groups[10]. At low temperatures the ethoxylated surfactant is soluble in water. At a given surfactant concentration, such a solution solubilises a given amount of oil which rapidly increases with increase of temperature (near the cloud point of the surfactant). This is schematically shown in Figure 3a which shows the influence of temperature on solubilisation, as well as the cloud point curve of the surfactant. As the amount of oil added exceeds that determined by the solubilisation curve, the excess oil separates as a separate phase. Moreover, as the temperature is increased above the cloud point, then separation into oil, water and surfactant takes place. However, an isotropic region exists between the solubilisation and cloud point curves; this is the o/w solubilised systems.

If, on the other hand, water is added to an oil solution of the surfactant solubilisation of the water takes place, which decreases with increase of temperature. This is represented by the solubilisation curve of Figure 3b. Any increase in the water weight fraction beyond the solubilisation curve results in its separation as a separate phase. Figure 3b also shows the haze-point curve; as the temperature is reduced below the haze point curve, separation into oil, water and surfactant takes place. Again an isotropic region of w/o solubilised system exists between the solubilisation and haze point curve.

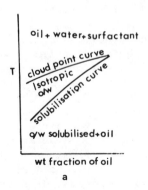 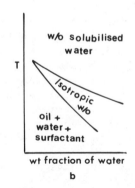

(a) oil solubilised in a   (b) water solubilised in
    nonionic surfactant   a nonionic surfactant
    solution   solution

Figure 3.   Schematic representation of solubilisation

Thermodynamic Theories[17-24]

Two main treatments have been considered, namely by
Ruckenstein et al[17-23] and by Overbeek[24]. Although the
treatments follow roughly the same procedure, yet they vary
somewhat in detail. Ruckenstein et al[17] treatments consider
the free energy of formation of microemulsions ΔG_m to consist of
three main contributions: ΔG_1 an interfacial free energy term,
ΔG_2 an energy of interaction between the droplets term, and ΔG_3
an entropy term accounting for the dispersion of the droplets into
the continuous medium. The interfacial free energy term ΔG_1 was
considered to consist of two contributions due to the creation of
an uncharged surface (given by the product of area created and
specific surface free energy of the interface) and a contribution
due to the formation of electrical double layers (which is given
by the product of the interfacial area and the specific surface
free energy due to creation of an electrical double layer). The
double layer contribution was calculated using the Debye-Hückel
approximation.

For calculation of ΔG_2 a pairwise additivity of interaction potentials was assumed using two approaches: a continuous approach which replaces the sum of interaction potentials by an integral and a non-continuous approach whereby the sum was divided into a discrete part representing the interaction with nearest neighbours (assumed to be 12) and an integral part accounting for the interaction with the rest of the droplets. The latter approach is similar to that proposed by Albers and Overbeek[25]. The pair-potential U_r was considered to consist of two parts, namely the van der Waals attraction potential $U_1(r)$ and the double layer repulsive potential $U_2(v)$. The former was calculated using Lifshitz theory as established by Ninham and coworkers[26] whereas the double layer repulsion was calculated using the expression given by Verwey and Overbeek[27].

For the calculation of the entropy contribution term ΔG_3, a lattice model was used to calculate the number of configurations Ω of a liquid mixture formed from N_1 molecules of continuous phase and N_2 molecules of the dispersed phase, the latter being in the form of m equal-sized droplets. The entropy term can be calculated from the Boltzmann relationship, ie, $\Delta S_M = k \ln \Omega$ where k is the Boltzmann constant. However, the exact computation of Ω taking into account all possible configurations was not possible and only a lower and upper bound of Ω was derived, thus making it possible to provide only a lower and upper limit of ΔS_M and hence ΔG_3.

The variation of ΔG_M with droplet radius R, and constant volume fraction ϕ_2 of the droplets was then determined using the relationship

$$\Delta G_M(R) = \Delta G_1 + \Delta G_2 - T\Delta S_M \qquad (3)$$

From equation (3), the condition for spontaneous micro-emulsion formation, with the most stable droplet size (R*) for a given volume fraction may be obtained,

$$\left. \frac{d\,G_M}{dR} \right|_{R = R^*} = 0 \qquad (4)$$

and,

$$\left. \frac{d\,G_M}{dR^2} \right|_{R = R^*} > 0 \qquad (5)$$

The variation of ΔG_M with R was found to follow four curves (see Figure 4), which illustrate the transition from instability $\rightarrow$ kinetic stability $\rightarrow$ thermodynamic stability. Thus, curve D represents the case of instability since ΔG_M is positive for all values of R. With curve C, kinetic stability is possible providing the height of the energy maximum is significant. On the

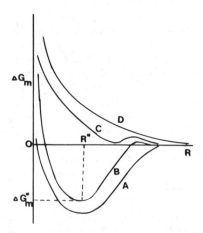

Figure 4.   Variation of ΔG_M with R.

other hand, the situation represented by curves A and B show
the case where a negative ΔG_M within a certain range of R
exists.   This means that dispersions with droplet radii within
that range are thermodynamically stable and should show no phase
separation.   The radius corresponding to the minimum in the ΔG_M-
R curve, namely R* represent the most stable radius.   Based on
equation (3), the transition from $D \rightarrow C \rightarrow B \rightarrow A$ may be obtained by
reducing the value of ΔG_1, i.e., reducing the specific surface
energy f_s.   This shows the need for producing an ultra-low
interfacial tension.   Using the condition ΔG_M to denote
thermodynamic stability, plots of f_s versus R for various values
of ϕ were constructed, which showed the limit of thermodynamic
stability of both o/w and w/o microemulsions.   From a comparison
of the values of ΔG_M* for o/w and w/o, the most stable type
(with lowest value of ΔG_M*) could be predicted.   Moreover, it
was shown that phase inversion occurred when the ΔG_M* values
were the same.   Thus, using this treatment, it was possible to
predict phase inversion as well as which type of microemulsion
would form for a given composition.

The reduction of f_s to sufficiently small values was later
accounted for by Ruckenstein[18] in terms of the so called
'dilution effect'.   Accumulation of surfactant and cosurfactant at
the interface not only causes signficant reduction in the

interfacial tension, but also results in reduction of the chemical
potential of surfactant and cosurfactant in bulk solution. The
latter reduction may exceed the positive free energy caused by the
total interfacial tension and hence the overall ΔG of the system
may become negative. Further analysis by Ruckenstein and
Krishnan[20-23] have shown that micelle formation encountered with
water soluble surfactants reduce the dilution effect as a result
of the association of the surfactant molecules. However, if a
cosurfactant is added, it can reduce the interfacial tension by
further adsorption and introduce a dilution effect. The treatments
by Ruckenstein and Krishnan[20-23] also highlighted the role of
interfacial tension in the formation of microemulsions. A
negative dynamic interfacial tension may be possible, but at
equilibrium $\gamma_{o/w}$ should have a small but positive value.

As mentioned above, the treatment by Overbeek[24] was rather
different in detail from that of Ruckenstein and Chi[17]. Before
considering the free energy contributions, it is perhaps useful to
consider the effect of addition of cosurfactant on the interfacial
tension. For that purpose, Overbeek[24] used the Gibbs adsorption
equation in this form.

$$\left(\frac{\delta\gamma}{\delta\mu i}\right)_{T,p,\ \text{all }n_j\ \text{except }n_i,\ A} = -\left(\frac{\delta n i}{\delta A}\right)_{T,p,\mu_i,\ \text{all }n_j\ \text{except }n_i} = \Gamma^i \qquad (6)$$

By using equation (6) twice for the cosurfactant (i=co) and
for the surfactant (i=sa), the following equation can be derived.

$$\gamma=\gamma_o-[\int_{c(co)=o}^{c(co)} \Gamma_{co}\ \mu_{co}]_{T,p,c(sa)=o}-[\int_{c(sa)=o}^{c(sa)} \Gamma_{sa}\ d\mu_{sa}]_{T,p,n(co),n_{oil},w} \qquad (7)$$

It is clear from equation (7) that the addition of a second
surfactant results in further decrease in γ; the essential
requirements being a not too small adsorption of the second
surfactant. Whether it replaces the first surfactant or is
adsorbed in addition to it is immaterial, just as it is not
essential for the two surfactants to form a complex. If the two
surfactants are of the same type, e.g., both water soluble anionic
surfactants, they will form mixed micelles and this will lower the
activity of the second surfactant added and decreases both its Γ
and $d\mu$. However, if the two surfactants are different in nature,
e.g., one predominantly water soluble and the other oil soluble,
they will only slightly affect each other's activity and their
combined effect on the interfacial tension may be large enough to
bring γ to zero at finite concentrations.

The free energy of microemulsion formation ΔG_m was also
subdivided by Overbeek[24] into three main contributions: ΔG_1
due to the Gibbs energy of mixing of surfactant and water and of
cosurfactant and oil; ΔG_2 due to the interfacial area, and ΔG_3

due to the mixing of droplets into the continuous phase. To
simplify the term ΔG_1, Overbeek[24] assumed that the mutual
solubility of the water and oil to be negligible and also the
solubility of surfactant into oil and cosurfactant into water to
be negligible. Thus ΔG_1 is simply given by the sum of the
product of the number of moles of each component and its chemical
potential with reference to that of the standard state. The
interfacial area term ΔG_2 is given by the product of the final
interfacial tension and area of the interface plus a chemical
potential term due to adsorption of surfactant and cosurfactant.
Finally, the free energy term due to mixing of droplets into the
continuous medium has been obtained using the hard sphere model of
Percus-Yevick[28] and Carnahan and Starling[29] which was
originally used by Agterhof et al[30] to describe the nonideal
behaviour of concentrated microemulsions.

Overbeek[24] derived an expression for dG_m taking the above
three contributions into account, the resulting equation reads:

$$dG_m = dA \left[\gamma_{uncharged} + \int \psi_o d\sigma + \frac{A^2 \, kT}{12\pi (n_w v_w)^2} \left(\ln\phi - 1 + \phi \frac{4-3\phi}{(1-\phi)^2} + \ln \frac{V_o}{V_{hs}} \right) \right] \quad (8$$

where dA is the change in interfacial area. The first term
between the square brackets is the interfacial tension term,
$\gamma_{uncharged}$ that is obtained if no double layers had been formed.
$\int \psi_o d\sigma$ is the electrical contribution to the interfacial tension
as a result of formation of an electrical double layer, with ψ_o
being the surface potential and σ the surface charge density. The
third term is the osmotic contribution due to mixing of
microemulsion droplets (treated as hard spheres) with the
continuous medium. Here ϕ is the volume fraction of hard spheres
and V_{hs} is their molar volume. The number of hard sphere
droplets is given by,

$$n_{hs} = \frac{A^3}{36\pi (n_w V_w)^2} N_{av} \qquad (9)$$

where A is the area, n_w is the number of water molecules and
V_w is their molar volume.

The three terms of equation (8) are unequal in magnitude.
The last term is always negative. For example, for a high ratio
of surfactant to water $n_s/n_w = 0.04$ (corresponding to a mass
ratio of about 0.7) and $\phi = 0.5$, this term is ~0.2mN m^{-1}
(taking $A \sim n_{sa}/\Gamma_{sa}$). For $\phi = 0.1$, the value would be -0.5mN m^{-1}
and to make this term -1.0mN m^{-1}, ϕ has to be below 10^{-5},
i.e., a very dilute microemulsion.

On the other hand, the electrical free energy per unit area

of double layer (second term) is high and positive even for
relatively low surface potential. The contribution of this term
could be tens of mN m^{-1}. This requires $\gamma_{uncharged}$ to have a
high negative value to reach the condition dG = 0. The conclusion
so far reached from this analysis is that in a microemulsion the
interfacial tension, including the electrical term, must have very
low but slightly positive value. The small variations in the
total interfacial tension required to balance the variation in the
free energy of mixing (osmotic term) can be easily obtained by
small variations in the amount of surfactant and cosurfactant at
the interface leading to variations in $\gamma_{uncharged}$.

Several factors play a role in determining whether an o/w or
w/o microemulsions is formed. Generally speaking since the osmotic
term increases with increasing ϕ, the most stable microemulsion
would be that in which the phase with the smaller volume fraction
forms the droplets. For w/o microemulsion, the hard sphere volume
is only slightly larger than the water volume, since the
hydrocarbon tails of the surfactant may interpenetrate to a
certain extent, when two droplets come close together. For an oil
in water emulsion, on the other hand, the double layer may extend
to a considerable extent, depending on the electrolyte
concentration (the thickness of the double layer 1/κ is of the order
of 100nm for 10^{-5}mol dm^{-3} and 10nm for 10^{-3} mol dm^{-3} 1:1
electrolyte). Thus, the hard sphere radius can be increased by 5nm
or more unless the electrolyte concentration is very high (say 10^{-1}
mol dm^{-3} where 1/κ ~1nm). Thus this factor works in favour of
the formation of w/o microemulsions especially for small droplets.
Furthermore, establishing a curvature of the adsorbed layer at a
given adsorption is easier with water as the disperse phase, since
the hydrocarbon chains of the surfactant will have more freedom
around than if they were inside the droplet. Thus, other things
being equal, the interfacial tension tends to be lower for a w/o
microemulsion than for an o/w microemulsion.

SCATTERING METHODS FOR CHARACTERISATION OF MICROEMULSIONS

Scattering techniques provide the most obvious methods for
obtaining quantitative information on the size, shape and
structure of microemulsions. The scattering of radiation, eg,
light, neutrons, etc., by particles have been successfully applied
for the investigation of many systems such as polymer solutions,
micelles, colloidal dispersions, etc. In most of these cases
measurements could be made at sufficiently low concentrations, to
avoid complications produced as a result of particle-particle
interactions. The results obtained are then extrapolated to
infinite dilution to obtain the desired property such as radius of
gyration, size, shape etc. However, application of scattering
methods to microemulsions are not straightforward since dilution
is not possible, as this results in the disappearance of the

microemulsion droplets. Therefore, measurements can only be made at finite concentrations where particle-particle interactions must be taken into account. Below, a brief account of how this interaction can be incorporated into the basic equations is given and this is followed by a description of the three most commonly used scattering techniques, namely time average light scattering, neutron scattering and quasi-elastic light scattering, together with some illustrations of their application.

In any scattering technique one usually measures the intensity of scattered radiation I(Q) as a function of the scattering vector (Q) given by the equation,

$$Q = \frac{4\pi}{\lambda} \sin(\theta/2) \tag{10}$$

where λ is the wave length of the radiation and θ is the angle at which the measurement is made.

For a fairly dilute system the intensity of scattered radiation I(Q) is proportional to the number of particles, N, the square of the volume of an individual scattering unit, V_p, and some property of the system (i.e., the refractive indices of the components when dealing with light scattering and, scattering length density when dealing with neutron scattering). For more concentrated systems, the scattering intensity also depends on the interference effects arising from particle-particle interaction[31,32]. Thus, the following general expression gives the dependence of I(Q) on the above-mentioned parameters[31],

$$I(Q) = [\text{Instrument Constant}] \times [\text{Material Constant}] \times NV_p^2 \times P(Q)$$
$$\times S(Q) \tag{11}$$

where P(Q) is the particle scattering form factor which allows the scattering from a single particle of known shape and size to be predicted as a function of the scattering vector Q. S(Q) is the so called "structure factor" which takes into account the particle-particle interaction.

Thus, for example, to estimate the particle radius from equation (11), it is necessary to estimate P(Q) and S(Q). For spherical particles of radius R, the particle scattering form factor is given by,

$$P(Q) = [3 \sin QR - QR \cos QR / (QR)^3]^2 \tag{12}$$

The structure factor S(Q) can be estimated if a model for particle interaction in a concentrated dispersion is adopted. As pointed out by Agterhof et al[30] the hard sphere fluid theory may be applied to concentrated dispersions of very small particles, providing the interaction is of short-range nature, i.e., determined by the thickness of the surfactant layer.

By analogy with the liquid state theory, the structure factor can be given by the following expression,

$$S(Q) = 1 + \frac{4\pi N}{Q} \int_0^\infty [[g(r) - 1]\, r \sin Qr\, dr \qquad (13)$$

where $g(r)$ is the particle-pair radial distribution function and r is the centre-to-centre particle separation. For hard sphere interaction, a useful expression can be obtained from the Percus-Yevick approximation[28] in the form given by Ashcroft and Lekner[33],

$$S(Q) = \frac{1}{[1-NC(2QR_{HS})]} \qquad (14)$$

with

$$C(2QR_{HS}) = -32\pi\, R_{HS}^3 \int_0^\infty \frac{\sin(2s Q\, R_{HS})}{2s\, QR_{HS}}\, (\alpha + \beta s + \gamma s^3)\, s^2 ds \qquad (15)$$

where R_{HS} is the hard sphere radius given by,

$$R_{HS} = R_c + t \qquad (16)$$

where R_c is the radius of the droplet core and t is the thickness of the surfactant and cosurfactant layer which determines the hard sphere interaction. The coefficients α, β and γ are defined as,

$$\alpha = (1+2\, \phi_{HS})^2 / (1-\phi_{HS})^4 \qquad (17)$$

$$\beta = -6\phi_{HS}\, (1+0.5\phi_{HS})^2 / (1-\phi_{HS})^4 \qquad (18)$$

$$\gamma = 0.5\phi_{HS}\, (1+2\phi_{HS})^2 / (1-\phi_{HS})^4 \qquad (19)$$

where ϕ_{HS} is the hard sphere volume fraction that is given by,

$$\phi_{HS} = \frac{4\pi}{3} R_{HS}^3 \cdot N \qquad (20)$$

Time Average Light Scattering

In light scattering one usually measures the intensity of scattered light from the sample at various angles applying some correction factors (for depolarisation, dissymmetry, etc.) and then plots the intensity at some chosen angle (usually $90°$ to the direction of propagation of the incident beam) as a function of concentration of microemulsion droplets (usually expressed as volume fraction). For example the intensity i_{90} of the scattered light beam is related to that of the incident beam I_o by,

$$i_{90} = \frac{Io}{r_s^2} K_o \ MC \ P(90) \ S(90) \tag{21}$$

where M is the molecular mass of the particle scattering light, c
its concentration, r_s is the distance of scattering volume from
the detector and K_o is an optical constant given by the
equation,

$$K_o = \frac{9\pi^2 n_m^4}{2N_A \lambda_o^4} \left[\frac{n_o^2 - n_m^2}{n_o^2 + 2n_m^2} \right]^2 \tag{22}$$

where n_m is the refractive index of the medium, n_o that of the
core of the microemulsion droplet and N_A is the Avogadro's
number.

$(i_{90}/Io)r_s^2$ is usually referred to as the Rayleigh
ratio R_{90}, i.e.,

$$R_{90} = K_o \ MC \ P(90) \ S(90) \tag{23}$$

For small particles (as is the case with microemulsions) $P(90) \sim 1$
and since $M = (4/3)\pi R_c^3 \rho_c N_A$, and $C = \phi_c \rho_c$ equation (23)
becomes,

$$R_{90} = K_o \ \frac{4}{3} \pi R_c^3 \rho_c N_A \cdot \phi_c \rho_c \ S(90) \tag{24}$$

or $\quad R_{90} = K_1 \phi_c R_c^3 \ S(90) \tag{25}$

with $\quad\quad K_1 = K_o \ \frac{4}{3} \pi N_A \rho_c^2 \tag{26}$

As an illustration, $R_{90} - \phi H_2 0$ for water in xylene
microemulsions obtained using sodium dodecylbenzene sulphonate
(NaDBS) and hexanol[34] is shown in Figure 5 at various NaDBS
concentrations. To calculate R_c from equation (24) one needs to
calculate S(90) as a function of ϕ. Calculation of S(90) was
carried out using equations (14-19) in a similar manner as
performed by Cebula et al[31,32] assuming two values for t,
namely 5 and 8 A respectively. The results for t = 5A are shown
in Figure 6, which clearly demonstrates the linear increase of
R_c with $\phi_{H_2 0}$ on the one hand and its decrease at any given
$\phi_{H_2 0}$ with increase in NaDBS, on the other.

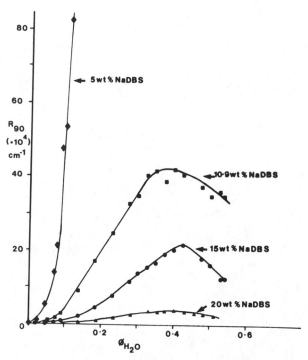

Figure 5. Scattering from water in xylene microemulsions as a
function of volume fraction of water at 25°C

Neutron Scattering

The intensity of scattered neutrons $I(Q)$ at a scattering
vector Q can also be given from equation (11) by[32]

$$I(Q) = [\text{Instrument Constant}]\ (\bar{\rho} - \rho_o)^2 \times NV_p^2 \times P(Q) \times S(Q) \qquad (27)$$

where $\bar{\rho}$ is the mean scattering length density of the
particle and ρ_o that of the solvent. Thus equation (26) is
equivalent to that used for light scattering except that $(\rho - \rho_o)$
is used for the material constant in place of the refractive index
difference used in light scattering.

One of the main advantages of neutron scattering over light
scattering is the Q range at which one can operate[31]. With
light scattering the range of Q is small (~0.0005 - 0.005 $\overset{o}{A}^{-1}$)
which for small angle neutron scattering (S.A.N.S.) is relatively
large (0.02 -0.18 $\overset{o}{A}^{-1}$). The small Q values for light scattering
means that $P(Q) \rightarrow 1$; whereas for S.A.N.S. a wide variation of $P(Q)$
with Q becomes available thus making fitting procedures applicable
and increasing the certainty of the size analysis in the range of
small particle sizes.

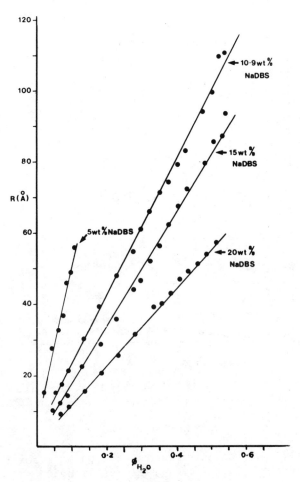

Figure 6.   Variation of water core radius R_C with ϕ_{H2O} for
            water in xylene microemulsions at four difference
            NaDBS concentrations

    As an illustration of S.A.N.S., the results obtained by
Cebula <u>et al.</u>[31] , for water in xylene microemulsions at various
volume fractions of water are shown in Figure 7. For $\phi_c > 0.147$
the microemulsions show distinct peaks in the values of $I(Q)$. As
ϕ_c increases, the peaks become more defined and the values of
$I(Q)$ at the peak also increases with increase of ϕ_c up to $\phi_c =$
0.436. Moreover, the position of the peak shifts to lower Q with
increasing ϕ_c. However, as ϕ_c is increased from 0.436 to
0.533, $I(Q)$ decreases. This is the same volume fraction at which
the intensity of light scattered also decreases (see Figure 5).
The peaks in the S.A.N.S. data are taken as indication of the
structural features of the system, i.e., the degree of order
introduced by the interactions between the particles, particle
size and polydispersity[31]. For very dilute colloidal

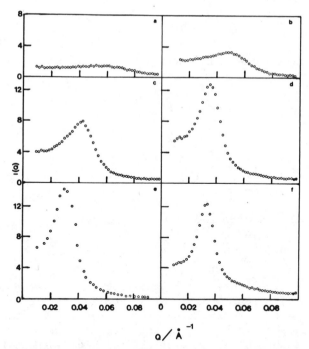

Figure 7. S.A.N.S. data for water in xylene microemulsions[31] at various volume fractions of water (a) $\phi_C = 0.147$; (b) $\phi_C = 0.198$; (c) $\phi_C = 0.301$; (d) $\phi_C = 0.380$; (e) $\phi_C = 0.436$; (f) $\phi_C = 0.533$

dispersions where particle-particle correlations in space are essentially absent $S(Q)\sim 1$ and $I(Q)\sim P(Q)$. For concentrated colloidal dispersions, if $P(Q)$ is known, then one can obtain $S(Q)$ from $I(Q)$ using equation (27).

However, analysis of neutron scattering data for microemulsions is not as simple as that for model colloids such as monodisperse polystyrene latices[35]. Two main problems are encountered with microemulsions: the first deals with the particle shape for which there is no certainty that they remain spherical (particularly at high ϕ_C); the second and more serious problem is that of polydispersity which is not taken into account in equation (27).

In spite of the above uncertainties, Cebula et al.[32] used the S.A.N.S. to calculate R_C as a function of ϕ_C. Two procedures were used. In the first procedure the Q_m values (at the peak) were used to calculate the lattice spacing $d\sqrt{2/3}$, assuming these to be face centred cubic (f.c.c), using the Bragg's equation. From a comparison of the volume fraction for f.c.c close packing (0.74) with the actual volume fraction of the microemulsion,

$$\phi = \phi_c + \phi_s = \frac{4\pi}{3} (R_c + t_s)^3 N \tag{28}$$

where ϕ_c is the volume fraction of the water core and ϕ_s is that of the surfactant active agent shell, R_c can be calculated, since,

$$\frac{\phi}{0.74} = (\frac{R_c + t_s}{(d/2)})^3 \tag{29}$$

and

$$\phi_s = \phi_c [(1 + \frac{t_s}{R_c})^3 - 1] \tag{30}$$

which when combined with equation (29) gives

$$\frac{\phi_c \bar{d}^3}{5.92} = R_c^3 \tag{31}$$

The second procedure for calculating R_c consisted of fitting the I(Q) versus Q data, using models for P(Q) and S(Q). In this case, a Percus-Yevick model was used for calculating S(Q) and the polydispersity was taken into account (using an arbitary distribution function). The best fit for the S.A.N.S. results was obtained when t was taken to be equal to 3Å. The results are shown in Figure 8 and for comparison the results obtained from light scattering data are also shown for two t values of 3 and 8Å. These data give a good correlation with the S.A.N.S. data, confirming the validity of simple hard-sphere model for analysing the scattering data.

Quasi-Elastic Light Scattering

In quasi-elastic light scattering (some times referred to as photon correlation spectroscopy (p.c.s)) one measures the fluctuation in intensity that occurs over very short time intervals due to the Brownian motion of the particles in solution. In this sense the measurements are time dependent and should give information on the dynamics of the system. This technique has been successfully applied to determine the z-average translational diffusion coefficient and the corresponding hydrodynamic radius of macromolecules, micelles and colloidal suspensions[36]. The application of p.c.s. to microemulsions is difficult since measurements can only be made at finite concentrations (see below). In p.c.s. measurements one usually plots ln $[C^{1/2} g^{(1)}(Q,\tau)]$ versus $Q^2\tau$, where $g^{(1)}(Q,\tau)$ is the scattered-field correlation function, the square root of the time dependent part of the measured intensity correlation function; C is an apparatus constant dependent on geometry, signal to background count ratio and τ is the correlation delay time. If such plots are linear,

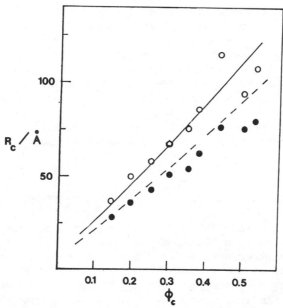

Figure 8.   Comparison of R_c versus ϕ_c values obtained from
S.A.N.S and light scattering (l.s) results:(——) l.s.
data for t = 8Å; (---) l.s. data for t = 3Å; (o) S.A.N.S
data from Q; (●) S.A.N.S data from fit to the full
curve of I(Q) against Q

then it is possible to calculate the diffusion coefficient from the
slope of these plots. However, with microemulsions linear plots
are scarcely obtained indicating that the observed correlation
functions are not single exponential curves. This has been
recently demonstrated by Cebula et al[32] for xylene-water
microemulsions. The extent of departure from exponential behaviour
was found to increase with increase of volume fraction of the water
droplets.

The observed deviation was attributed to the interaction of
water droplets in the concentrated microemulsion system. Under
these conditions, the scattered field correlation function is
given by[37]

$$g^{(1)} (Q, \tau) = F^M (Q, \tau) / S^M(Q) \qquad\qquad (32)$$

where $S^M(Q)$ is the measured static structure factor and
$F^M(Q, \tau)$ the dynamic one.

Thus, analysis of the p.c.s. results require estimation of
these structure factors and introduction of the polydispersity
effect. Using this procedure, Cebula et al[32] calculated the

average diffusion coefficient D of the microemulsion droplets as a
function of ϕ_c. The results clearly showed the reduction of D
with increase in ϕ_c as expected. Calculation of the
hydrodynamic radius R_h from D is only valid for dilute systems
($\phi_c < 0.1$) and this shows the limitation of the p.c.s.
measurements for determining the droplet size of microemulsions.
Moreover, the theoretical analysis of the data is far from being
simple and requires a number of assumptions to be made. Thus,
although p.c.s. measurements are fast and relatively easy to
obtain, yet their value in characterisation of microemulsions is
questionable.

CONDUCTIVITY AND DIELECTRIC MEASUREMENTS

Conductivity and dielectric measurements, combined with other
investigations or used alone, may provide valuable information
concerning the structural behaviour of condensed systems such as
micellar solutions and microemulsions. In the early applications
of conductivity measurements, the technique was used to determine
the nature of the continuous phase and detecting phase inversion
phenomena[38-39]. It was expected that o/w microemulsions should
give fairly high electrical conductivity in contrast with w/o ones
which should be poorly conducting. However, later careful
analysis of the conductivity measurements enabled some authors to
use conductivity and dielectric measurements more quantitatively
to study the structure of microemulsions.

For w/o microemulsions, two main trends in the variation of
conductivity κ with volume fraction of water ϕ_w have been
observed in the literature. In the first case, the conductivity
was initially very low and then it showed a sharp increase as ϕ_w
passes a critical volume fraction. This is illustrated in
Figure 9 for water in toluene microemulsions prepared using
potassium oleate and butanol[40]. Similar results were obtained
for water in cyclohexane using sodium dodecyl sulphate and 1-
pentanol.   For low ϕ_w, κ is approximately proportional to ϕ_w
but within a narrow concentration range $\phi_w{}^c$, κ increases
steeply over five orders of magnitude when the concentration of
droplets increases.

The second trend observed is illustrated in Figure 10 for the
system water/hexadecane/potassium oleate/hexanol[41]. In this
case κ increases with increase of p_w, the water weight fraction,
reaches a maximum at a critical p'_w and then decreases again.
The conductivity then increases sharply at another critical volume
fraction p''_w. Similar results were obtained by Baker and Tadros[42]
for the system water/xylene/ sodium dodecyl benzene sulphonate/
hexanol.

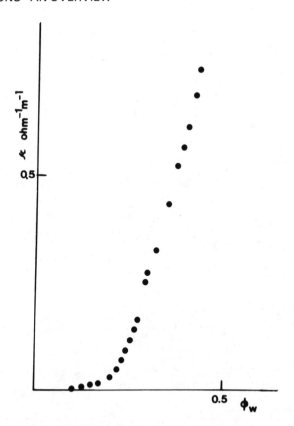

Figure 9.   Variation of κ with ϕ_w for water toluene/potassium
             oleate/butanol microemulsions.

The first trend in $\kappa - \phi_w$ curves can be accounted for in
terms of percolative conduction originally proposed to explain the
behaviour of conductance of conductor - insulator composite
materials[43]. In the latter model, the effective conductivity is
practically zero as long as the conductor volume fraction is
smaller than a critical value ϕ^P, called the percolation
threshold, beyond which ϕ^P suddenly takes non-zero value and
rapidly increases with increase of ϕ_w. Under these conditions,

$$\kappa \sim (\phi_w - \phi_w^P)^{8/5} \quad \text{when} \quad \phi_w > \phi_w^P \qquad\qquad (33)$$

$$\text{and} \quad \kappa \sim (\phi_w^c - \phi_w)^{-0.7} \quad \text{when} \quad \phi_w < \phi_w^P \qquad\qquad (34)$$

By fitting the conductivity data to equation (33), ϕ^P was found
to be 0.176 ± 0.005, which when compared with the theoretical
value of 0.29 gave a ratio of R_{HS}/R_C of 1.18 ± 0.01. This
value is in excellent agreement with the value of 1.17 obtained by
Lagues et al[44] from neutron scattering data. This illustrates
the validity of the percolative model on the one hand, and the

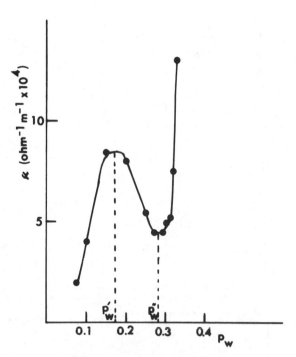

Figure 10.   Variation of κ with p_w for water/hexadecane/
potassium oleate/hexanol microemulsion.

value of conductivity measurements in characterisation of microemulsions on the other.

The second trend observed (see Figure 10) cannot be accounted
for in terms of percolation theory.   Indeed plots of $κ^{5/8}$ versus
$φ_w$ did not show the linear behaviour expected in the region of
the percolation threshold[42].   Thus, the transition observed must
be due to more subtle changes in the conduction units as the water
concentration is increased.   The initial increase in conductivity as
$φ_w$ is increased can be ascribed to the enhancement of
surfactant solubilisation with added water[41].   Alternatively the
increase in conductivity in this region is likely to be due to the
increase in the dissociation of the surfactant on the addition of
water[42].   Beyond the maximum, addition of water mainly causes
micelle swelling, ie, a definite water core (microemulsion
droplets) begins to form, which may be considered as a dilution
process leading to a decrease in conductivity.   In other words,
the decrease in conductivity beyond the maximum may be due to the
replacement of the hydrated surfactant - alcohol aggregates with
microemulsion droplets.   The sharp increase in κ beyond the
minimum observed in Figure 10 must be associated with a

"facilitated" path for ion transport. This can be ascribed to the
formation of non spherical particles resulting from swollen
micelle clustering and subsequent cluster interlinking, a process
that is indicative of system stability breakdown and of
crystalline structure organisation[41].

In a recent publication, Clausse et al[45] demonstrated the
value of conductivity measurements in establishing the type of
system produced which depends on the chain length of the alcohol
(cosurfactant). Four quarternary systems involving water, dodecane,
sodium dodecyl suplhate (SDS) and four saturated primary
alcohol with carbon numbers of 4,5,6 and 7 were investigated.
For both 1-butanol and 1-pentanol, the transparent region is best
represented by quarternary molecular solutions, since no
separation into zones when moving from the water poor to the water
rich system is observed. With 1-hexanol and 1-heptanol, on the
other hand, the region is split into two disjoined sub-regions
which are separated by a zone over which viscous turbid
birefringent media are encountered. This behaviour is clearly
reflected in the conductivity-water volume fraction curves, which
are shown to depend to a large extent on the chain length of the
alcohol. For 1-butanol and 1-pentanol, the conductivity is fairly
high and varies with p_w, the water mass fraction according to
the equations derived from the percolation and effective medium
theories. Thus, such systems can be classified as percolative
systems. In contrast, 1-hexanol and 1-pentanol exhibited low
conductivity where variations with p_w were smooth. Thus, these
systems present those of "true" microemulsion (i.e with definite
water cores) and are therefore non-percolating. Thus conductivity
measurements can be used to distinguish between the two types of
systems.

Further information on the structural aspects of
microemulsions may be obtained from dielectric measurements.
Hanai[46], derived the following relationship between the
relative complex permittivity of an emulsion ε^* and that of the
dispersed phase ε_1^* and medium ε_2^*,

$$\frac{\varepsilon_1^* - \varepsilon_2^*}{\varepsilon_1^* - \varepsilon^*} \frac{\varepsilon^*}{\varepsilon_2^*} = \frac{1}{(1-\phi)^3} \qquad (35)$$

where ϕ is the volume fraction of the dispersed phase. In
equation (35) the complex permittivity ε^* is made of a real part
ε' and an imaginary (or loss) part ε'' ($\varepsilon^* = \varepsilon' - J\varepsilon''$). The real
part ε' is frequency independent, whereas ε'' is inversely
proportional to the frequency. Thus, by carrying out dielectric
measurements as a function of frequency, it is possible to isolate
ε' from ε''. Using this procedure, Senatra and Giubilaro[48,49]
investigated water in dodecane microemulsions (using potassium

oleate and hexanol) as a function of the volume fraction of water. These microemulsions showed a transition from optically clear to translucent at a critical water volume fraction of 0.31 followed by another transition from milky to clear at another critical ϕ_w of 0.58. At both transitions, there is a sharp increase in ε' and ε''. However, the dielectric permittivity was found to be strongly frequency dependent at the first transition, but frequency independent at the second transition (corresponding to phase inversion). The frequency dependence observed at the first transition was attributed to the presence of nematic-like structure (nematic mesophase). In the second transition ($\phi_w \sim 0.58$) hysteresis-like phenomenon and metastable states with relaxation times of the order of several days were observed. The behaviour of the relative complex permittivity at this critical concentration value was ascribed by the authors to a phase transition of the system which develops following a critical phenomenon pattern.

VISCOSITY MEASUREMENTS

Few systematic measurements of viscosity of microemulsions that extend over a wide range of volume fractions have been carried out. As an illustration, the results obtained recently by Baker and Tadros[42] for water in xylene microemulsions are shown in Figure 11 at four different sodium dodecylbenzene sulphonate concentrations. The η_r-ϕ_{H_2O} plots show the typical behaviour usually observed with concentrated dispersions. Analysis of viscosity data for concentrated systems is not simple, since the particle interactions must be taken into account. Various empirical equations are available to treat the viscosity volume fraction curves, of these the Mooney equation[50]

$$\eta_r = \exp \left(\frac{a\phi}{1-k\phi} \right)$$

(36)

where a and k are constant, may be used. However, plots of the viscosity data using equation (36) in the form,

$$\frac{\phi}{\log\eta_r} = \frac{2.303}{a} - \frac{2.303}{a} k\phi$$

(37)

did not give the expected straight lines between $\phi/\log\eta_r$ versus ϕ. This points to the fact that the parameters a and k may not be constant, perhaps due to the fact that microemulsion droplets do not behave as rigid spheres. For that reason, the viscosity data were analysed using computational analysis. The best fit was obtained when a was set between the limits 1-10 and k between 1.35 and 1.91, the theoretical limit estimated by Mooney[50]. The value of k was fairly constant indicating that hydrodynamic interactions play a major role, whereas a increases from 3.0 at

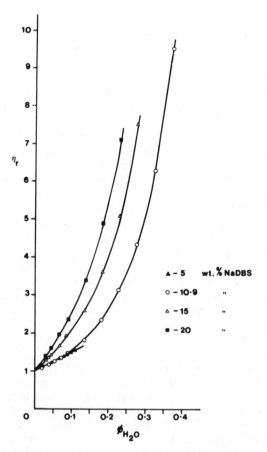

Figure 11.   Relative viscosity versus volume fraction curves for
             water-in-xylene microemulsions.

10.9 wt% NaDBS to 6.0 at 20% NaDBS. This suggests that
increasing the surfactant concentration results in a signficant
increase in the degree of solvation of the microemulsion droplets.

STRUCTURE AND DYNAMICS OF MICROEMULSIONS -
NMR INVESTIGATIONS

     One of the most important questions regarding microemulsions
is their possible organisation into polar and non-polar domains
and the rigidity of the barriers between them. For micelles and
liquid crystalline phases this separation is quite distinct.
Lindman and collaborators[51-57] have clearly demonstrated that
the question of organisation of microemulsions can be elucidated
from measurement of self diffusion coefficient of all the
constituents of the microemulsion. Such studies can be made either

by radio tracer method[54,55] or by NMR methods[54,56,57]. The
basic principles underlying the use of self diffusion coefficients
in elucidating the structural domains are as follows[57]. Within a
micelle, the molecular motion (translational, reorientation chain
flexibility) is almost as rapid as in a liquid hydrocarbon.
Similarly in a reverse micelle, water molecules and counterions are
highly mobile[53]. On the other hand, in a lamellar liquid
crystal, all motions appear to be very rapid in the direction of
the lamella while in the perpendicular direction translational
motion is relatively slow[58]. It is characteristic for many
surfactant-water systems that there is a rather distinct spatial
separation between hydrophilic and hydrophobic domains[51,52].
Under these conditions, the passage of species between different
regions is an improbable event and thus it occurs slowly. Thus the
self diffusion, if studied over macroscopic distances, should
reveal whether the process is rapid or slow depending on the
geometrical properties of the inner structure. For example, a
phase that is water continuous but oil discontinuous should exhibit
rapid diffusion of hydrophilic components, while hydrophobic ones
should be diffusing slowly. Conversely, an oil continuous but
water discontinuous system should exhibit rapid diffusion of
hydrophobic components. A bicontinuous phase should give rapid
diffusion of all components.

The concentration dependence of the self diffusion
coefficient should also have characteristic features. Thus,
increasing the concentration of oil droplets in a water continuous
phase results in a decrease in the diffusion coefficient of the
continuous medium due to the obstruction effect from the
aggregates. Moreover, the diffusion coefficient of the
discontinuous phase also decreases as a result of increasing
interaggregate interaction. Thus, with an oil in water
microemulsion, the diffusion coefficients of hydrophilic and
hydrophobic species should decrease with increase in the oil
concentration. For a water in oil microemulsion, this coefficient
should decrease with increasing water concentration. For a
bicontinuous system (or one which lacks distinct internal
interface) the diffusion coefficient of hydrophiles should decrease
with increasing oil while that of a hydrophobe increases.

In order to acquire a more general picture of the microdynamic
behaviour with compositional changes, Lindman et al[57] have
undertaken an investigation on several microemulsion systems using
a novel Fourier transform pulsed spin echo ^{1}H and ^{13}C NMR to
obtain the self diffusion coefficient for the components. For
microemulsions consisting of water,hydrocarbon, an anionic
surfactant and a short chain alcohol (C_4 and C_5) the self
diffusion coefficients of both water, cosurfactant and hydrocarbon
were quite high (of the order of 10^{-9} m^2 s^{-1}) being orders of
magnitude higher than the value to be expected for a discontinuous

medium (10^{-11} m^2 s^{-1} and below). This can be attributed to
three alternative effects or a combination of all of them: (i)
bicontinuous solutions, ie, both oil and water continuous; (ii)
easily deformable and flexible interface, or; (iii) absence of
large aggregates. Lang et al[59] have shown that addition of
short chain alcohol to a micellar solution results in considerable
reduction of life-time of surfactant monomer in the micelle. This
result is consistent with the picture of disorganised and flexible
interfaces. Thus, the picture of microemulsions based on an
anionic surfactant and short chain alcohols is that of very small
aggregates with any internal interface with very limited spatial
extension or very dynamic and flexible structure which break up and
reform at very high speed, or both.

On the other hand, with microemulsions based on an anionic
surfactant and a long chain alcohol, D_w was fairly low for
certain concentrations, indicating that distinct water droplets in
a hydrophobic medium may form. The system investigated by
Lindman et al[57] was based on sodium   octonoate - decanol -
octane - water. This means that the anionic "surfactant" used
contains only seven carbons in the alkyl chain which is fairly
short. With longer chain surfactants, one would expect well
defined "water cores" provided the alcohol is also long-chain.
Such well defined "water cores" have also been confirmed by
Lindman et al[57] for the Aerosol OT - hydrocarbon system. In
this case the self diffusion coefficient concentration curve shows
a behaviour distinctly different from the cosurfactant
microemulsions. D_w has a quite low value throughout the
extension of the isotropic solution phase up to the highest water
contents. This implies that a model with closed droplets
surrounded by surfactant anions in a hydrocarbon medium gives an
adequate description of these solutions. However, since D_w was
found to be significantly higher than D_s, the authors came to
the conclusion that a non-negligible amount of water must exist
between the emulsion droplets.

Thus, in summary, self diffusion coefficient measurements
using NMR by Lindman et al[51-57] have clearly indicated that the
structure of microemulsions depends to a large extent on the chain
length of the cosurfactant (alcohol), the surfactant, and the type
of system. With short chain alcohols (<C6) and/or surfactants,
there is no marked separation into hydrophobic and hydrophilic
domains and the structure is best described by a bicontinuous
solution with easily deformable and flexible interfaces.

With long chain alcohols (>C6), on the other hand, well defined
"cores" may be distinguished with a more pronounced separation
into hydrophobic and hydrophilic regions. The same is true for
other specific systems such as that based on Aerosol OT without
the presence of a cosurfactant.

ACKNOWLEDGEMENTS

Some of the results described in this overview paper were taken
from the thesis of Dr R C Baker (to be published later). Dr Baker
has carried out his research work under my supervision, thus
giving me the opportunity to pursue my research interest on
microemulsions. Many valueable discussions have been held with him
during the course of his research. I am aslo grateful to
Prof. R.H. Ottewill for establishing the basis of interpretation
of scattering results.

REFERENCES

1.   T.P. Hoar and J.H. Schulman, Nature (London) $\underline{152}$, 102 (1943).
2.   K. Shinoda and S. Friberg, Adv. Colloid Interface Sci. $\underline{4}$, 281
     (1975).
3.   I. Danielsson and B. Lindman, Colloids and Surfaces $\underline{3}$, 391
     (1981).
4.   L.M. Prince, Editor, "Microemulsions Theory and Practice",
     Academic Press, New York (1977).
5.   L. Rosano, J. Soc. Cosmet. Chem. $\underline{25}$, 609 (1974).
6.   J.E. Bowcott and J.H. Schulman, J. Electrochem. $\underline{54}$, 283,
     (1955).
7.   J.H. Schulman, W. Stoeckenius and L.M. Prince, J. Phys. Chem.
     $\underline{63}$, 1677 (1959).
8.   L.M. Prince, J. Colloid Interface Sci. $\underline{23}$, 165 (1967).
9.   L.M. Prince, J. Soc. Cosmet. Chem. $\underline{21}$, 193 (1970).
10. H. Saito and K. Shinoda, J. Colloid Interface Sci. $\underline{24}$, 10
     (1967).
11. K. Shinoda and H. Saite, J. Colloid Interface Sci. $\underline{26}$, 70
     (1968).
12. K. Schinoda and H. Kunieda, J. Colloid Interface Sci. $\underline{42}$, 381
     (1973).
13. K. Shinoda and T. Ogawa, J. Colloid Interface Sci. $\underline{24}$, 56
     (1967).
14. S.I. Ahmad, K. Shinoda and S. Friberg, J. Colloid Interface
     Sci. $\underline{47}$, 32 (1974).
15. S. Friberg and I. Burasczenska, Prog. Colloid Polymer Sci.
     $\underline{63}$, 10 (1978).
16. S. Friberg and I. Burasczenska, in "Micellization,
     Solubilization and Microemulsions" K L Mittal, Editor,
     Vol. 2, p. 791, Plenum Pressm, New York, 1977.
17. E. Ruckenstein and J.C. Chi, J. Chem. Soc. Faraday Trans $\underline{71}$,
     1690 (1975).
18. E. Ruckenstein, J. Colloid Interface Sci. $\underline{66}$, 369 (1978).
19. E. Ruckenstein, Chem. Phys. Letters $\underline{57}$, 517 (1978).

20. E. Ruckenstein and R. Krishnan, J. Colloid Interface Sci. 71, 321 (1979).
21. E. Ruckenstein and R. Krishnan, J. Colloid Interface Sci. 75, 476 (1980).
22. E. Ruckenstein and R. Krishnan, J. Colloid Interface Sci. 76, 188 (1980).
23. E. Ruckenstein and R. Krishnan, J. Colloid Interface Sci. 76, 201 (1980).
24. J.Th.G. Overbeek, Faraday Disc Chem. Soc. 65, 7 (1978).
25. W. Albers and J.Th.G. Overbeek, J Colloid Sci 14, 510 (1959)
26. B.W. Ninham, V.A. Parsegian and G.H. Weiss, J. Stat. Phys. 2, 323 (1970).
27. E.W.J. Verwey and J.Th.G. Overbeek "Theory of Stability of Lyophobic Colloids" Elsevier, Amsterdam, (1948).
28. J.K. Percus and G.J. Yevick, Phys Rev. 110, 1 (1958).
29. N.F. Carnahan and K.E. Starling, J. Chem. Phys. 51, 635 (1969).
30. W.G.M Agterof, J.A.J. van Zomeren and A. Vrij, Chem. Phys. Letters. 43, 363 (1976).
31. D.J. Cebula, R.H. Ottewill, J. Ralston and P.N. Pusey, J. Chem. Soc. Faraday Trans. I, 77, 2585 (1981).
32. D.J. Cebula, D.Y. Myers and R.H. Ottewill, Colloid Polymer Sci. 260, 96 (1982).
33. N.W. Ashcroft and J. Lekner, Phys. Rev. 45, 33 (1966).
34. R.C. Baker and Th.F. Tadros, to be published.
35. R.H. Ottewill, Colloid Polymer Sci. 59, 14 (1976).
36. B. Chu "Laser Light Scattering" Academic Press, New York, (1974).
37. J.C. Brown, P.N. Pusey, J.W. Goodwin and R.H. Ottewill, J. Phys. A 8, 664 (1975).
38. J.H. Schulman, T.S. Roberts, Trans Faraday Soc. 42, 165 (1946).
39. J.H. Schulman and D.P. Riley, J. Colloid Sci. 3, 383 (1948).
40. B. Lagourette, J. Peyrelasse, C. Boned and M. Clausse, Nature, 281, 60 (1979).
41. C. Boned, M. Clausse, B. Lagourette, J. Peyrelasse, V.E.R. McClean and R.J. Sheppard, J. Phys. Chem. 84, 1520 (1980).
42. R.C. Baker and Th.F. Tadros, to be published.
43. S. Kilpatrick, Rev. Mod. Phys. 45, 574 (1973).
44. M. Lagues, R. Ober and C. Taupin, J. Phys. Lett. Fr. 31, L487 (1978).
45. M. Clausse, J Heil, J Peyrelasse and C. Boned, J Colloid Interface Sci. 87, 584 (1982)
46. T. Hanai, Kolloid Z. 171, 23 (1960).
47. T. Hanai, Kolloid Z. 175, 61 (1961).
48. D. Senatra and G. Giubilaro, J. Colloid Interface 67, 448 (1978).
49. D. Senatra and G. Giubilaro, J. Colloid Interface 67, 457 (1978).
50. M. Mooney, J. Colloid Sci. 6, 162 (1950).

51. B. Lindman and H. Wennerstron, Top. Current Chem., <u>87</u>,
     1 (1980).

52. H. Wennerstrom and B. Lindman, Phys Reports <u>52</u>, 1 (1979).

53. B. Lindman and H. Wennerstrom in "Solution Behaviour of
     Surfactant - Theoretical and Applied Aspects" , K.L. Mittal
     and E.J. Fendler, Editors, Vol. 1, pp. 3-25, Plenum Press,
     New York, 1982.

54. B. Lindman, N. Kamenka, T.M. Kalhopoulis, B. Brumand, P.G.
     Nilsson, J. Phys. Chem. <u>84</u>, 1485 (1980).

55. H. Fabre, N. Kamenka and B. Lindman, J. Phys. Chem. <u>85</u>, 3493
     (1981).

56. P. Stilbs, M.E. Nosely and B. Lindman, J. Magn. Reson. Sci.
     401 (1981).

57. B. Lindman, P. Stilbs and M.E. Moseley, J. Colloid Interface
     Sci. <u>83</u>, 569 (1981).

58. G. Lindblom and H. Wennerstrom, Biophys. Chem. <u>6</u>, 167 (1977).

59. J. Lang, A. Djavanbakht and R. Zana, in "Microemulsions"
     I.D. Robb, Editor, Plenum Press, New York (1982) p. 233-255.

THE WATER-IN-OIL MICROEMULSION PHENOMENON: ITS UNDERSTANDING AND PREDICTABILITY FROM BASIC CONCEPTS

Hans-Friedrich Eicke[1], Rudolph Kubik[1], René Hasse[2]
and Iris Zschokke[2]

[1]Institute for Physical Chemistry
[2]Institute for Applied Physics
University of Basel
Klingelbergstr. 80 and 82, CH-4056 Basel, Switzerland

This review considers the thermodynamically stable three component system water, Aerosol OT and iso-octane. It tries to correlate molecular details with macroscopic thermodynamic properties of the system. In particular, the stability of the aqueous microphases is essentially based on the existence of the semi-diffuse electrical double layer. This concept also permits one to estimate the (very small) distribution of microphase sizes. Time-domain-spectroscopic studies yield the dielectric loss function which could be successfully correlated with the penetration of oil molecules into the monomolecular surfactant layers of the aqueous microphases.

INTRODUCTION

The ability of the anionic surfactant Aerosol OT (sodium di-2-ethylhexylsulfosuccinate) to form thermodynamically stable, three component microemulsions with water and oil has been known for about a decade. The realization of thermodynamically stable microemulsions, with the lowest possible number of components, has distinguished this system as a sort of reference for fundamental studies by an increasing number of laboratories.

1533

The various investigations are concerned with
(i)   thermodynamically stable, unconstrained interfaces
(ii)   physical properties of aqueous microphases and their ensemble
        (W/O-microemulsions); and
(iii)   the final goal of these investigations is seen to attain a
        self-consistent picture of more involved systems with more
        than three components and to classify these microemulsions
        within the frame of the thermodynamically stable, multi-
        component mixtures.

The Hard Sphere Concept

     To start with let us consider Figure 1 which displays the
existence region of the so-called W/O-microemulsions, an apparent
one-phase domain which, however, comprises an oil external phase
and an aqueous phase dispersed into micro-droplets, typically of
10 - 50 nm diameter, and a monomolecular layer of surfactants (AOT).
It is also seen from the figure that there exists a non-linear
transition from micelles to microphases which will become more
evident from the following figures.

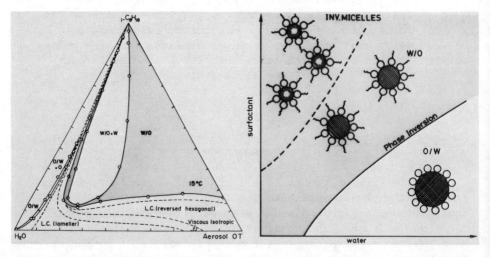

Figure 1.   Phase diagram of Water, AOT, iso-octane; right hand side;
sketch of the W/O-microemulsion domain with phase inversion region [23]

The particular suitability of AOT to form water-in-oil microemulsions is also apparent from Figure 1.

Two aspects regarding microphases were recently discussed:
microphases with well-defined interfaces which were reasonably
described by "hard-sphere" models, [1,2,3,4] and which apparently
conform to the original view of microemulsions. There is evidence,
however, that such microphase domains may be less well-defined, [5,6]
indicating possible transitions towards bi-continuous or partly molecularly dispersed states of oil and water in the presence of
surfactant.

The system discussed in this paper corresponds primarily to
the first case, which is evidenced from recent experiments and theoretical considerations, [7], regarding interfacial tension of micro-

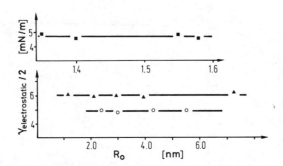

Figure 2. Calculated parameter $\gamma_{electrostatic}/2$ for different
microphase radii R_o.
Systems:▲H_2O/AOT/isooctane [17]; T = 298 K
         ○H_2O/0.05M AOT/n-heptane [15]; T = 298 K
         ■H_2O/0.1M AOT/toluene [18]; T = 393 K

phases. Schulman [8,9] postulated the main feature of microemulsions to be the zero interfacial tension. As Overbeek [10] pointed out this physical state generally necessitates the application of two rather different surfactants (i.e. surfactant and co-surfactant). Thus, it is remarkable that in the present case Aerosol OT already fulfills the condition of zero interfacial tension (γ). Since AOT is an ionic surfactant, one may split γ, the total equilibrium interfacial free energy, into two terms, $\gamma_{uncharged}$ and $\gamma_{electrost}$. The electrostatic contribution of the interfacial free energy has recently been calculated [11] by solving the Poisson-Boltzmann equation and considering the boundary conditions valid for reversed aggregates (microphases). Inserting our experimental data into the theoretical results yields $\gamma_{electrost}$= constant, i.e. independent of the curvature of the spherically assumed microphases (see Figure 2). This result very nicely confirms the prediction that $\gamma_{uncharged}$ = const., since this contribution should depend only on the nearest neighbour (short range) interactions, [10] and should accordingly be independent of curvature. The quantitative values found for $\gamma_{electrost}$ were used to calculate the water activity a_{H_2O} as a function of the aliquots of water (w_o= $[H_2O]/[AOT]$) added to the system. Figure 3 exhibits the very satisfactory agreement between experimental and theoretical plots.

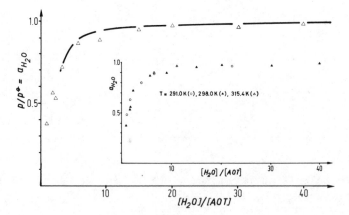

Figure 3. Activity of water a_{H_2O} against amount of water in the ternary H_2O, AOT, i-C_8H_{18} system, [7]. Calculated Plot (——)

Also another important parameter which characterizes the
system can be derived in the frame of this model: the polydispersity of the microphases. The calculations (see Appendix) predict a
very small polydispersity, i.e. deviations from a mean microphase
radius of a few percent (see Figure 4). This result confirms Vrij's
assumption, [4] of an isodispersed suspension of hard spheres. The
surprisingly small polydispersity of particles with an interfacial
tension of almost zero is a direct manifestation of the Gibbs-
Marangoni effect [10]. It also explains the fact that these microphases experience elastic repulsive forces after thermal
collisions, [12].

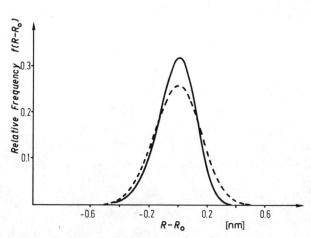

Figure 4. Polydispersity of aqueous microphases: relative frequency, $f(R-R_o)$, of deviation $R-R_o$. R_o, radius of the microphase
which corresponds to the minimum Gibbs free energy.
R_o: ——— : 2.0 nm, ------: 10 nm.

An additional set of experiments concerning optical and
dielectric properties of microphases serve to elucidate some of
their individual properties. The optical anisotropy ($\Delta n = n_{||} - n_{\perp}$)
which is observed with the help of an electro-optical Kerr
equipment is due

to purely induced dipole moments as could be demonstrated by a reversal of the electric field pulse (Figure 5). This finding conforms to the assumption of a spherical symmetry of the microphases.

An interesting observation is made if Δn is plotted as a function of the microphase concentration (Figure 6) for a particular field strength. There are clearly two regions to be distinguished. A linear one which conforms to Kerr's law and a parabolic (quadratic) increase of Δn which suggests that dimers of such microphases are formed. We assume that the development of the semi-diffuse double layer will influence the mobility of the ions and increase the concentration of dimers, probably due to a w_o-dependent charge exchange with a concomitant formation of charged microphase pairs [13]; this question has been discussed again quite recently (Figure 7) [14]. This also conforms to ideas developed earlier, [12] regarding the exchange of

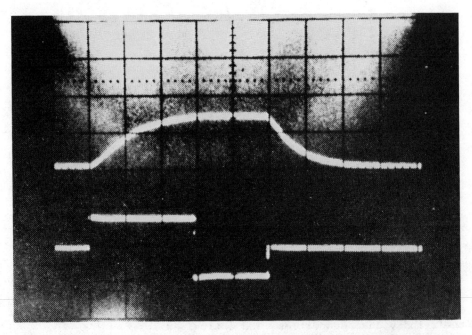

Figure 5. Photocurrent signal (above) and electric field pulse (250 kV m^{-1}) (below) in electro-optical Kerr effect measurements [24].

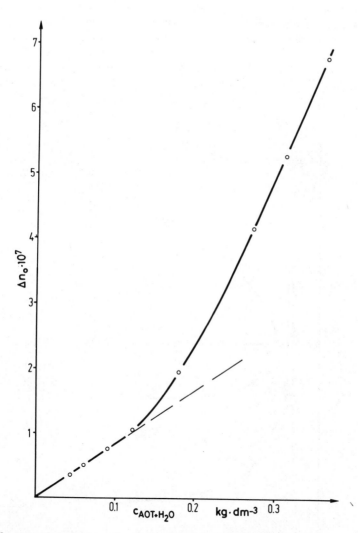

Figure 6. Steady-state electric birefringence $\Delta n_0 = n_{\|} - n_{\perp}$ against concentration of AOT and H_2O in the ternary system: H_2O, AOT, $i\text{-}C_8H_{18}$. Parameter: $E^2 = 10^5 (kV\ m^{-1})^2$, $W_0 = 74$, $T = 292$ K[24].

solubilizate between microphases. These concepts were successfully
applied to investigations on the kinetics of such exchanges, [15, 16].

In relation to the properties of microphases and their ensemble
(microemulsions) of ternary systems the effect of non-amphiphilic
organic additives has to be considered. This aspect is relevant to
the question of the interaction between the continuous oil phase
and the microphases. A frequently discussed matter with respect to
these interactions is the possible penetration, i.e. solvation, of
the palisade layer of the surfactant molecules in the oil/water

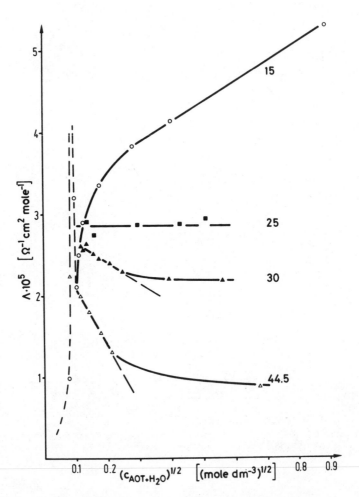

Figure 7 . Molar conductivities Λ of the ternary system H_2O, AOT,
$i\text{-}C_8H_{18}$ against the square-root of the water and AOT concentration.
Parameter: $[H_2O]/[AOT]$, T = 293 K, [14].

interface. Recently, time-domain spectroscopic data became available
which proved to be particularly suitable to follow up this problem.
The principle of this technique is to apply a short electric pulse
of about 20 ps to the sample; the answer function, i.e. the reflected
part of the original signal, then covers a broad frequency spectrum
in the radio- and lower microwave region. The result of the measurement is the spectrum of the complex dielectric function $\varepsilon^*(f)$ as
function of the frequency (f), [18]. Figure 8 shows a plot of the
dielectric loss $\varepsilon''(f)$ against the frequency for the system water/
AOT/oil where each curve corresponds to a particular "oil" (external hydrocarbon phase). On the right side of the diagram the
molar volumes of the different solvents are plotted. From a qualitative comparison of both ordinates it is seen that the dielectric
loss follows approximately the molecular volume of the solvents
used. Independent of the model applied to describe the dielectric
loss function, it is apparent from the plot that the addition of
cyclohexane makes the loss function to almost vanish within the
displayed frequency region. In other words the (to be extrapolated)
maximum of the loss function is shifted towards considerably lower

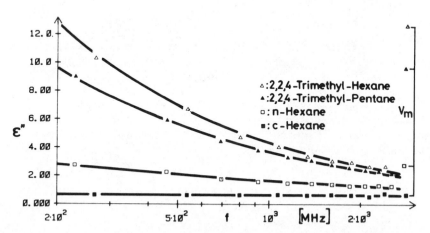

Figure 8. Time-domain-spectroscopic dielectric loss measurements
in H_2O, (0.4 mol^{-3}) AOT, oil systems. $[H_2O]/[AOT]= 40$, T= 298 K.
Dielectric loss function $\varepsilon''(f)$ against frequency (f). Right hand
side: molecular volumes of oil molecules [25].

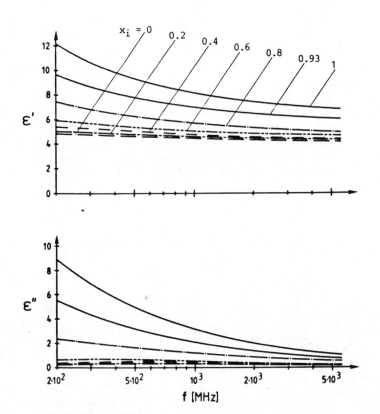

Figure 9. Time-domain spectroscopic measurements in H_2O, (0.3 mol dm^{-3}) AOT, $i-C_8H_{18}/c-C_6H_{12}$. $[H_2O]/[AOT]= 40$, T= 298 K. Dielectric functions ε' and ε'' against frequency (f). Parameter: mole-fractions x_i of $i-C_8H_{18}$.[25]

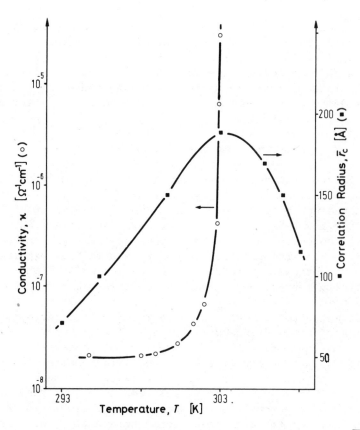

Figure 10. Temperature-dependent mean correlation radii; $\overline{r}_c$ and specific conductivity κ of the ternary system: H_2O, AOT, $i\text{-}C_8H_{18}$; $[H_2O]/[AOT]= 65$, [26].

frequencies. With increasing molecular volumes of the external oil-phase molecules, the amplitude of $\varepsilon''(f)$ increases. The results of the time-domain spectroscopy might be interpreted in terms of the relaxation of the Aerosol OT molecule within the oil/water interface, as could also be inferred from variations of w_o and the AOT concentration. Hence the dependence of $\varepsilon''(f)$ on the molecular volume of organic additives is then due to their penetration into the surfactant monolayer.

The trend of the different oil molecules to accumulate in the interface is revealed by Figure 9 where the real part of the complex dielectric function $\varepsilon'(f)$ and the dielectric loss function $\varepsilon''(f)$ are plotted against the frequency. The curves differ in the mole fractions of cyclohexane in iso-octane and exhibit the particular tendency of cyclohexane to be enriched in the interfacial layer.

Deviation from the Hard Sphere Concept?

Considering now the possibility of less well-defined interfaces, as mentioned in the first part of this paper, one has to approach the upper coexistence curve (the so-called solubilization curve) (see right hand side of Figure 1) within the transparent, apparently homogeneous, microemulsion phase. Within the temperature range of the reversible coagulation process various experiments like the electro-optical Kerr-effect, the quasi-elastic light-scattering, and viscosity measurements, demonstrate the apparent approach to a critical point [19,20]. This process is due to a segregation of microphases from the continuous oil phase leading to a macroscopic phase separation of two continuous oil phases. The frequently observed preceding maximum of the plot describing the phase transition is apparently due to an equilibrium between microphases of different aggregative states. The formation of network structures corresponding to the latter phenomenon is believed to be verified by the observation of a percolation phenomenon (Figure 10). In this state part (or all?) of the microphase forms apparently a gel-like network to allow an increase of the specific electric conductivity by approximately four orders of magnitude (Figure 11). It can be shown that the quasi-sudden increase of the conductivity can be produced by adding typical cosurfactants, i.e. $C_1 - C_6$ straight-chain alcohols, to the ternary system (Figure 12) instead of varying the temperature. Depending on the distribution coefficient of the particular alcohol between water and oil, the onset of the percolation is shifted towards lower or higher temperatures with respect to the "ternary reference system" (Figure 13). This is another example of the predicted dependence of the critical temperature on the additive concentration in ternary [21] and pseudo-ternary systems [22].

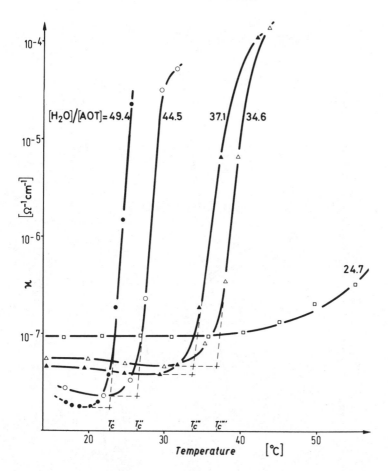

Figure 11.   Temperature-dependent specific conductivities κ of ternary system: H_2O, AOT, $i-C_8H_{18}$. Parameter: added amount of water, [23].

CONCLUSION

In retrospect it appears that the ternary water, Aerosol OT and oil system indeed represents an exceptional example of thermodynamically stable emulsions with oil-continuous phases. In view of the many experimental details regarding this system which are already known, this particular ternary solution may actually be recommended as a reference system, especially with respect to the extensive studies with more than three-component oil external systems.

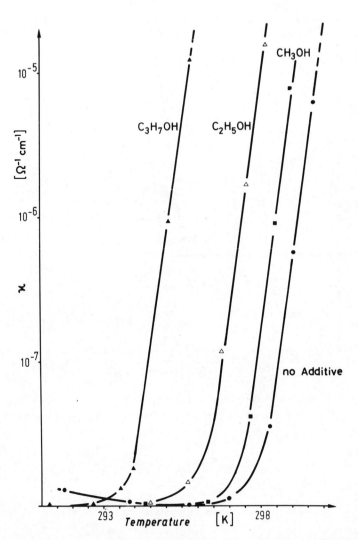

Figure 12. Temperature-dependent specific conductivities κ of H_2O, (0.1 mol dm^{-3}) AOT, $i\text{-}C_8H_{18}$, alcohol-systems. $[H_2O]/[AOT] = 48.9$; $[ALCOHOL]/[AOT] = 0.34$

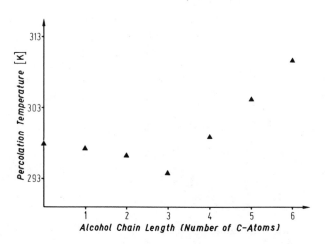

Figure 13. Shift of percolation temperature against chain lengths
of added alkylalcohols in quaternary system: H_2O, AOT (0.1 mol dm^{-3}),
$i-C_8H_{18}$, alcohol: $[H_2O]/[AOT]$ = 48.9, $[ALCOHOL]/[AOT]$ = 0.34

APPENDIX

Microphase size distribution of the ternary system:
H_2O, AOT, Oil

The Gibbs free energy G of a single microphase can be written
as [10,11]

$$G = G^\sigma + G_f + G_{mix} + G_s \qquad (1)$$

where G^σ is the free energy of the standard state, G_f that due to
electric field of the semi-diffuse double layer, G_{mix} the free
energy of mixing of the counter-ions belonging to the diffuse part
of the double layer with the water of the aqueous core, and G_s the
free interfacial energy of the uncharged system. The transition
from a single microphase to an ensemble, i.e. a (W/O)-microemulsion,
would require calculation of an additional free energy of mixing
of the microphases with the continuous oil phase. This energy con-

tribution has been neglected since it is at least an order of magnitude smaller than the other terms of Equation (1).

If the above defined Gibbs free energy of a single microphase is known, the size distribution of the microphases characterized by their radii R at a constant amount of water is accessible via Boltzmann's distribution law. The relative probability (frequency), $f(R-R_o)$, for the occurrence of a deviation $(R-R_o)$ from the radius R_o which corresponds to the minimum of the free energy G_o can be expressed by

$$f(R - R_o) = const \cdot exp\{-(G - G_o)/kT\} \qquad (2)$$

where the value of the constant is obtained by the normalization condition

$$\int_{-\infty}^{\infty} f(R - R_o) \, dR = 1 \qquad (3)$$

The calculation demonstrates that the microemulsion is almost monodispersed. The half-width of the distribution curve is about 0.2 nm and increases slightly with the amount of water. A small asymmetry can be seen which is more pronounced for smaller w_o-values.

ACKNOWLEDGEMENT

This work is part of several projects of the Swiss National Science Foundation. R.H. and I.Z. are grateful to the Stiftung Hasler Werke for financial support.

REFERENCES

1. J. K. Percus and G.J. Yevick, Phys. Rev., 110, 1 (1958).
2. E. Thiele, J. Chem. Phys., 39, 474 (1963).
3. N.F. Carnahan and K.E. Starling, J. Chem. Phys., 51, 635 (1969).
4. W.G.M. Agterof, J.A.J. van Zomeren and A. Vrij, Chem. Phys. Letters, 43, 363 (1976).
5. B. Lindman, N. Kamenka, Th.-M. Kathopoulis, B. Brun and P.-G. Nilsson, J. Phys. Chem., 84, 2485 (1980).
6. H. Fabre, N. Kamenka and B. Lindman, J. Phys. Chem., 85, 3493 (1981).
7. R. Kubik, H.-F. Eicke and B. Jönsson, Helv. Chim. Acta, 65, 170 (1982).
8. T.P. Hoar and J.H. Schulman, Nature, 152, 102 (1943).
9. J.H. Schulman and J.B. Montagne, Ann. N.Y. Acad. Sci., 92, 366 (1961).
10. J.Th.G. Overbeek, Faraday Disc. Chem. Soc., 65, 7 (1978).

11. B. Jönsson and H. Wennerström, J. Colloid Interface Sci., <u>80</u>,
     482 (1981).
12. H.-F. Eicke, J.C.W. Shepherd and A. Steinemann, J. Colloid
     Interface Sci., <u>56</u>, 168 (1976).
13. H.-F. Eicke and A. Denss, in "Solution Chemistry of Surfactants,"
     K.L. Mittal, Editor, Vol. 2, p. 699, Plenum Press, New York, 1979.
14. H.-F. Eicke, H. Hammerich and G. Vasta, J. Colloid Interface
     Sci., in Press (1982).
15. P.D.I. Fletcher and B.H. Robinson, Ber. Bunsenges. Phys. Chem.,
     <u>85</u>, 863 (1981); also in "Microemulsions," I.D. Robb, Editor,
     p. 221, Plenum Press, New York, 1982.
16. B.H. Robinson, D.C. Steytler and R.D. Tack, J. Chem. Soc.
     Faraday Trans., <u>1</u>, 75, 481 (1979).
17. H.-F. Eicke and J. Rehak, Helv. Chim. Acta, <u>59</u>, 2883 (1976).
18. R.A. Day, B.H. Robinson, J.H.R. Clark and J.V. Doherty, J.C.S.
     Faraday I, <u>75</u>, 132 (1979).
19. D.F. Nicoli, F. de Buzzaccarini, L.S. Romsted and C.A. Bunton,
     Chem. Phys. Letters, <u>80</u>, 422 (1981).
20. A.M. Cazabat, D. Langevin, J.Meunier and A. Ponchelon, Adv.
     Colloid Interf. Sci., <u>16</u>, 175 (1982); J. Physique-Lettres <u>43</u>,
     89 (1982).
21. C. Wagner, Z. phys. Chem., <u>132</u>, 273 (1928).
22. H.-F. Eicke, J. Colloid Interface Sci., <u>59</u>, 308 (1977).
23. H.-F. Eicke, Chimia, <u>36</u>, 241 (1982).
24. Z. Marković, PhD thesis, Univ. Basel, 1980.
25. R. Hasse, PhD thesis, Univ. Basel, 1983.
26. H.-F. Eicke and Z. Marković, J. Colloid Interface Sci. <u>79</u>,
     151 (1981).

PHASE BEHAVIOR OF MICROEMULSIONS: THE ORIGIN OF THE MIDDLE PHASE, OF ITS CHAOTIC STRUCTURE AND OF THE LOW INTERFACIAL TENSION

Eli Ruckenstein

State University of New York at Buffalo
Department of Chemical Engineering
Amherst, New York 14260

A thermodynamic approach to microemulsions is developed using a Helmholtz free energy which contains, in addition to the usual quantities, a free energy term, Δf, due to the entropy of disperson of one phase in the other. This dispersion is considered to occur in the form of spherical globules of a single size. The van der Waals and double layer interactions between the globules are included in their interfacial tension. The treatment leads to two basic equations: one of them relates the above interfacial tension to the entropy of dispersion (and can be used to establish a relation between the radius r of the globules and the volume fraction ϕ of the dispersed phase); the other, a generalized Laplace equation, relates the difference between the pressures p_2 inside the globules and p_1 near the globules in the continuous phase to the entropy of dispersion and the bending stress and plays a role in phase equilibria. The thermodynamic formalism uses either the surface of tension or an arbitrary surface as the (Gibbs) dividing surface. Conditions of phase equilibrium are derived by assuming that the concentrations of the components in the globules are the same as those in the excess dispersed phase and that the compositions in the continuous and excess continuous phases are also equal. On this basis, one is led to conclude that an excess dispersed phase appears when $p_2 = p$ (where p is the external pressure) and that a third phase, the excess continuous phase, forms, when in addition, $p_1 = p$. While in the two phase region the microemulsion contains spherical globules, its struc-

ture changes with the appearance of the third, excess
continuous, phase. This happens because in the latter
case $p_2 = p_1$ and this equality is unlikely to be compatible with a stable finite radius of curvature. Arguments are adduced to demonstrate that in this case
extended interfaces are generated (between the dispersed and continuous phases) which fluctuate chaotically.
For this reason we suggest this structure be called
chaotic.

Equations are derived to identify the transition
to the two phase region as well as for the transition
to the three phase region. Various expressions are used
for the entropy of dispersion to derive relations linking the interfacial tension at the surface of the globules to their radius. Since intuition suggests that
the environment near the surface of the globules is
similar to that near the interface between the microemulsion and the excess dispersed phase, these expressions are extended, under some restrictions, to the interfacial tension of the latter interface. These equations are in agreement with recent neutron diffraction
experiments; they are valid as long as the microemulsion contains spherical globules and show that the
interfacial tension between the microemulsion and the
excess dispersed phase is essentially caused by the
adsorption of surfactant and cosurfactant on the interface. In the three phase region there are no longer
spherical globules and the interface between the two
media of the microemulsion fluctuates chaotically. This
behavior, which is similar to that occurring near a
critical point, explains why the interfacial tensions
between microemulsion and each of the excess phases
could be represented by expressions valid near a critical point and why the thicknesses of the corresponding interfaces are large. An important conclusion is
that while Δf is small, it nevertheless plays a major
role as regards both the interfacial tension and phase
behavior.

I. INTRODUCTION

Two pathways can be used to prepare microemulsions. In one, the starting point is an emulsion stabilized by the adsorption of surfactant molecules on the surface of the globules. The addition of a medium length alkyl alcohol - a cosurfactant - leads spontaneously to a microemulsion which has globules of almost the same size, lying between 10^2 and 10^3 Å, and which, in contrast to emulsions, can be optically transparent[1-3]. Their spontaneous formation suggests, again in contrast to emulsions, that their stability is thermodynamic in nature. Another pathway, which emphasizes even better their thermodynamic stability, starts from micellar solutions. Surfactants dissolved in water form, above the critical micelle concentration, a substantial number of large aggregates. Hydrocarbon molecules, though almost insoluble in water, can be solubilized by micellar solutions[4-8]. In general the solubilized molecules are located among the hydrocarbon chains of the micelles. For some surfactants or, more generally, when a cosurfactant is present, the solubilized molecules can form a core surrounded by a layer of surfactant, or surfactant and cosurfactant. Micellar aggregates or micellar aggregates containing solubilized molecules belong to a class of systems which are thermodynamically stable. One can therefore expect, by extrapolation, that "swollen micelles" containing a core of solubilized molecules (microemulsions) are also, at least under some conditions, thermodynamically stable. At least two kinds of theoretical treatments could be used regarding microemulsions, each associated with one of the pathways described above. One can, for instance, extrapolate to microemulsions the procedures employed for surfactant aggregation and solubilization[8]. Such a procedure is now being examined and will be published elsewhere. The other procedure, which is used in the present paper, starts from an emulsion, assumes that the globules are of uniform size and tries to identify the conditions under which such a structure is thermodynamically stable.

In order to formulate the main questions which require answers, it is appropriate to summarize some experimental results. Experiment shows that microemulsions can exist as single phases, or can coexist in equilibrium with excess oil and/or excess water. For illustrative purposes we consider the system composed of oil, water, surfactant, cosurfactant, and salt which exhibits all these possibilities[9-13]. For sufficiently large concentrations of surfactant, a single phase may exist, either as a microemulsion or as a liquid crystal, while at moderate surfactant concentrations, two or three phases can coexist. For moderate amounts of NaCl, an oil phase is in equilibrium with a water-continuous microemulsion, whereas for high salinity, a water phase coexists with an oil-continuous microemulsion. At intermediate salinities, a middle phase composed of oil, water, surfactants and salt forms between excess

water and oil phases. Extremely low interfacial tensions, of the
order of 10^{-2} to 10^{-3} dynes/cm, are found between different pairs
of coexisting phases, with the lowest occurring in the three phase
region. These systems have attracted attention because of their
application to tertiary oil recovery, since the displacement of
oil is most effective at very low interfacial tensions[14].

The structure of microemulsions has been investigated with
various experimental techniques, such as low-angle X-ray diffraction[3], light scattering[2], ultra-centrifugation[15], electron microscopy[16], viscosity measurements[17] and very recently by the Fourier
transform NMR self-diffusion method[18]. These seem to indicate
that, when the volume fraction ϕ of the dispersed phase is not too
large, the single phase microemulsion contains spherical droplets
almost uniform in size, whereas when ϕ is sufficiently large, the
microemulsions have very flexible internal interfaces which open
up and reform on a short time scale. While, in general, the microemulsions coexisting with excess oil or water also contain spherical globules, the structure of the middle phase is more complex,
since no stable globules appear to exist in this case.

The experimental facts enumerated above suggest the following
obvious questions:

   (i)   What is the origin of the thermodynamic stability of
         microemulsions?
  (ii)   Under what conditions does an excess phase coexist with
         a microemulsion?
 (iii)   Under what conditions can both excess phases coexist
         with a (middle phase) microemulsion?
  (iv)   What is the origin of the change in structure at the
         transition between a microemulsion in equilibrium with
         excess dispersed phase to one coexisting with both
         excess phases?
   (v)   What is the origin of the low interfacial tension
         between a microemulsion and either of the excess
         phases?

A qualitative answer to these questions will be given in the
next section. This will be followed by a formulation of the basic
thermodynamic equations. The fourth section examines phase equilibria involving microemulsions and contains a discussion concerning
the structure of microemulsions. Expressions for the entropy of
dispersion of the globules in the continuous phase are presented
in the fifth section and are used to establish a relation between
the interfacial tension at the surface of the globules and their
radius. The sixth section considers the problem of the interfacial
tension between a microemulsion and the excess dispersed medium
and in the seventh we deal with the problem of phase inversion.
Finally, the main results of the paper are summarized.

II. QUALITATIVE DISCUSSION

II.1   The Origin of the Thermodynamic Stability of Microemulsions

The origin of the thermodynamic stability of microemulsions
can be understood more easily if a reference state is defined in
which water and oil are not dispersed and the surfactant and cosurfactant are distributed at equilibrium between the two. Let us
now disperse one phase in the other in the form of globules. This
increases the entropy of the system and leads to the adsorption of
surfactant and cosurfactant on the large interfacial area thus
created. This adsorption decreases the interfacial tension from
about 50 dynes/cm, characteristic of a water-oil interface devoid
of surfactants, to some very low value. In addition, the concentrations of surfactant and cosurfactant are decreased as a result
of adsorption, and their chemical potentials thereby reduced below
the values in the reference state, generating a free energy change
which is negative. I call, for obvious reasons, this change in
free energy, the dilution effect. Dispersions that are thermodynamically stable can be generated when the negative free energy
change due to the entropy of dispersion of the globules in the
continuous phase combined with the dilution effect overcomes the
positive product of the low interfacial tension and the large
interfacial area produced[19].

The existence of an "optimum" radius for which the free
energy of the system is a minimum, which is the expected globule
radius, can be easily understood observing that: (a) If the radius
is large, the surface area generated is relatively small, the
amount adsorbed per unit interfacial area is large (but the total
amount adsorbed is small) and the bulk concentrations of surfactant and cosurfactant stay relatively large. As a consequence,
while the free energy associated with the interfacial tension is
low, the chemical potentials of surfactant and cosurfactant remain
relatively large. (b) If the radius is small, the surface area
generated is large, the bulk concentrations of surfactant and
cosurfactant and hence their chemical potentials are low. Consequently, the interfacial tension is relatively large and thus
so is the associated free energy (interfacial tension multiplied
by the surface area). At intermediate values of the radius both
contributions can be moderate and thus a minimum in the free
energy can occur.

II.2   Roles of the Cosurfactant and Salt

The cosurfactant provides a "dilution effect" in addition to
that of the surfactant and a further decrease of the interfacial
tension. In addition, a water soluble surfactant forms large
aggregates (micelles) above the critical micelle concentration

(cmc) and thus the largest chemical potential available for the surfactant in the reference state is that corresponding to the critical micelle concentration. The dilution effect will not exist if the final concentration remains at cmc, and the decrease in the interfacial tension might not be sufficiently large for a microemulsion to form without the cooperation of a cosurfactant. A low solubility of the surfactant also provides a low chemical potential in the reference state, a low dilution effect and a relatively small decrease of the interfacial tension. Actually, the role of the cosurfactant is probably more complex since it may affect the structure of water and the distribution of the surfactant between phases and interface.

The addition of salt has the following effects: At low concentrations, in systems containing ionic surfactants, it shields the electric field produced by the adsorption of the charged surfactant molecules and thus facilitates an even greater adsorption of surfactant. This reduces the interfacial tension and increases the dilution effect. At large ionic strengths, the electrical field is completely shielded and double layer effects can be ignored. However, at large concentrations, salt has another effect, on both ionic and nonionic surfactants, because of the extensive organization of water molecules induced by ions with relatively small ionic radius, such as Na^+. As a consequence, the interactions between water and the hydrocarbon chains of the surfactant molecules become even less favorable and the favorable interactions between the polar head groups and water are diminished. Therefore, the surfactant molecules are salted out as micelles in the water phase and/or as individual molecules in oil (globules and excess oil phase, if present), and at the interface. The salting out to the oil phase will be dominant, when the favorable interactions of the head groups with water are decreased to the point where they are outweighed by the larger entropy available in the oil phase (where, since there is no critical micelle concentration, the surfactant molecules essentially do not form many large aggregates) as compared to that in water (where they form many large aggregates). The salting out increases the amount of surfactant adsorbed at the interface and thus decreases the interfacial tension between water and oil.

II.3 Phase Behavior of Microemulsions

As indicated in the Introduction, if the salinity is increased at constant amounts of oil, water, surfactant and cosurfactant, three regions can be identified: at relatively low salinities, a water continuous microemulsion coexists with excess oil; at intermediate salinities, a three phase system involving a microemulsion and both excess phases forms; and at still higher salinities, an oil continuous microemulsion coexists with excess water. Since it

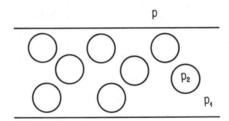

Figure 1.   The pressures in a microemulsion.

is reasonable to assume that the compositions of the dispersed
phase and of the excess dispersed phase are the same, it is clear
that the occurrence of the excess phase cannot be associated with
the composition dependence of the electrochemical potential. This
suggests at once that with such an assumption mechanical equilibrium is responsible for the phase separation. In what follows
arguments will be adduced to demonstrate that this can indeed be
the case.

The pressure inside the globules, p_2, is different from the
pressure p_1 near the globules in the continuous phase (Fig. 1).
We call p_1 and p_2 micropressures because they are defined on the
scale of the globule size and also to contrast them to the macro
(thermodynamic) pressure which is defined on the scale of the
entire microemulsion. The macropressure is defined on the basis
of a free energy which takes into account, via the entropy of dispersion in the continuous phase, the collective behavior of the
globules, while the micropressures characterize the behavior of
individual globules. Each of the micropressures augmented by
contributions from the collective behavior leads to the thermodynamic pressure. Because of mechanical equilibrium, the global
(thermodynamic) pressure equals the external pressure, p. Now,
let us consider a globule at the interface between the microemulsion and the external world. As long as the pressure p_2 is
smaller than p, the globule will stay in the microemulsion; the
mechanical equilibrium is still satisfied because of the
additional entropic contribution (which together with the
micropressure constitute the macropressure). When $p_2 > p$ the
above equilibrium is no longer satisfied and the globules will
separate from the continuous phase, i.e., the microemulsion gives
way to separate oil and water phases. An excess dispersed phase
in equilibrium with a microemulsion will appear when $p_2 = p$.

Similarly, we expect that an excess continuous phase will be
generated when p_1 = p. A three phase system composed of both
excess phases and a microemulsion will form when p_2 = p_1 = p.
However, the equality of the two micropressures is unlikely to be
compatible with the existence of a stable, finite radius of curvature. For this reason a change in structure is expected to occur
in the vicinity of the transition from two to three phases, since
the microemulsion can no longer consist of spherical globules.
Arguments are offered later in the paper that the interface between the two media of the microemulsion is extended and fluctuates chaotically.

At least in principle, it is possible for the pressures p_2
and p_1 to become equal without any phase separation (i.e., without
their common value being equal to p). In such cases a change in
structure accompanied by chaotic fluctuations of the interface is
also expected to occur. The spherical shape may also become unstable because of the instability of the low interfacial tension
interfaces to external and thermal perturbations. This may explain the large fluctuations of the interface detected experimentally[18], for values of the volume fraction that are not too
large.

II.4 The Origin of the Low Interfacial Tension Between a
Microemulsion and Excess Dispersed Phase.

The above qualitative discussion clearly indicates that a
microemulsion forms only if the interfacial tension at the surface
of the globules is sufficiently low. Because of this, the interfacial tension between a microemulsion and excess dispersed phase
is also low. Indeed, the concentrations of various components in
the interior of the globules and in the excess dispersed phase are
likely to be the same or at least very near to one another. Thus,
the surface of the globules and the interface between the microemulsion and excess dispersed phase separate essentially the same
two media. If, in addition, the globule size is sufficiently
large compared with the thickness of its interface, so that the
effect of curvature of the latter on its interfacial tension can
be neglected, then this interfacial tension should be nearly equal
to that between the microemulsion and excess dispersed phase (a
more detailed discussion is provided later in the paper). Consequently, the origin of the low interfacial tension at the latter
interface, also, must be sought in the adsorption of the surfactant and cosurfactant there. The explanation sometimes offered
that the low interfacial tension is due to nearness to a critical
point ignores the features specific to microemulsion systems. Another current explanation, namely that the low interfacial tension
between microemulsion and excess dispersed medium is due to a
thick layer of globules, ignores the fact that both the existence

of globules and the low interfacial tension are caused by the
adsorption of surfactant and cosurfactant on the interfaces. In
fact, it has been shown experimentally that the interfacial tension between the excess dispersed phase and the continuous medium
of microemulsion remains the same as that between microemulsion
and excess dispersed medium[20]. Of course, the above considerations involve microemulsions which contain spherical globules.
The internal interface of a middle phase microemulsion displays,
however, a dynamics which is chaotic. Since this behavior is
similar to that occurring near a critical point, it is reasonable
to consider that the interfaces between the middle phase and each
of the excess phases have characteristics similar to those encountered near a critical point.

III. BASIC THERMODYNAMIC EQUATIONS

III.1 The Free Energy of a Microemulsion.

For computational purposes, we assume in what follows that
the microemulsion contains spherical globules of a single size.
In the light of the experimental observations already mentioned,
such an assumption holds in numerous cases. The equations are
valid or can be easily extended to other structures, for instance
parallel layers of oil and water.

The first thermodynamic treatment of microemulsions was developed by Ruckenstein and Chi[21] who emphasized the role of the
entropy of dispersion and derived expressions for it. A variant
was published by Ruckenstein[19] and Overbeek[22]. The treatments in
Ref. 19 and 22 are largely equivalent. The present treatment is
an improvement of the previous ones since additional equations,
such as Eq. (7), are derived which yield some insight into the
thermodynamics and structure of the middle phase microemulsion.

For given numbers of molecules of each species, temperature
and external pressure, we consider an ensemble of systems in which
the radii and volume fractions of the globules can take arbitrary
values. Because the equilibrium state of the system is completely
determined by the number of molecules of each species, the temperature and the external pressure, the actual values of the radius
and volume fraction will emerge from the condition that the free
energy of the system be a minimum.

We begin with the observation that the dispersion of the
globules in the continuous phase is accompanied by an increase in
the entropy of the system and denote by Δf the free energy change
per unit volume of microemulsion due to this effect. However, let
us consider a "frozen" system of globules inside the continuous

medium, ignoring for the time being their entropy of dispersion. For each species i, we define an electrochemical potential μ_i for each of the two media constituting the "frozen" microemulsion. In other words, we consider the derivatives of the free energy of each of the media with respect to the number of molecules of various species they contain, when the appropriate quantities are kept constant. Of course, for each component the two electrochemical potentials are equal.

If the surface of tension is chosen as the (Gibbs) dividing surface, the variation at constant temperature T of the Helmholtz free energy f_o per unit volume of the "frozen" microemulsion is given by:

$$df_o = \gamma dA + \Sigma_i \mu_i dn_i - p_2 d\phi - p_1 d(1 - \phi), \qquad (1)$$

where γ is the interfacial tension, A the interfacial area per unit volume of microemulsion, n_i the number of molecules of species i per unit volume of microemulsion, ϕ the volume fraction of the dispersed phase, p_2 the pressure inside the globules and p_1 that in the continuous medium. The area A is related to ϕ and the radius of the globules r via:

$$A = 3\phi/r. \qquad (2)$$

The interfacial tension γ is defined by a variation in the frozen free energy with area at constant ϕ, n_i and T. We remark that the interfacial tension γ includes also those interactions between globules, such as the van der Waals and double layer interactions, which change with the surface area A. Now a change in A at constant ϕ also changes the shortest distance, 2h, between the surfaces of two neighboring globules. Thus, the virtual change used to define γ changes the spatial distribution of the globules not just their surface area. Therefore, γ includes, in addition to the van der Waals and double layer interactions at a given distance 2h, the effect of the reversible work done against these forces as h changes. Denoting by F the force per unit interfacial area of the globules and considering this force to be positive for repulsion, the variation 2dh of the distance produces, per unit volume of microemulsion, the work $AFdh$. An interfacial tension, γ_h, from which the effect of the variation of h is eliminated, can be defined by:

$$\gamma dA = \gamma_h dA - AF(\frac{\partial h}{\partial A})_{\phi,n_i,T} dA \qquad (3)$$

This new interfacial tension is a more familiar quantity. However, since the equations written in terms of γ are more compact than those in terms of γ_h, the thermodynamic treatment which follows uses Eq. (1) as the starting point.

As emphasized in the previous section, the quantities p_2 and p_1 represent local pressures inside the globules and in their vicinity in the continuous medium. In contrast to these micropressures, which are defined on the scale of the individual globules, one can define the usual thermodynamic pressures characteristic of the ensemble of globules, by taking into account in the expression of the free energy their collective behavior reflected in the entropy of dispersion. Appendix A contains a thermodynamic formulation of the problem using pressures and an interfacial tension which include the effect of the entropy of dispersion. As a result, the latter quantity does not appear explicitly in the expression of the free energy considered there.

In the representation used here, the free energy per unit volume of microemulsion is given by the sum

$$f = f_o + \Delta f, \tag{4}$$

which, together with Eq. (1), leads to

$$df = \gamma dA + \Sigma_i \mu_i dn_i - p_2 d\phi - p_1 d(1 - \phi) + d\Delta f. \tag{5}$$

Because the radius of the globules is relatively large, Δf is a small quantity. However, as shown in what follows, it plays a major role in relation to both the interfacial tension and the phase behavior.

III.2 An Equation for the Equilibrium Radius and a Generalized Laplace Equation

While all the variables r, n_i, ϕ, T and p are necessary to specify an arbitrary virtual state, the equilibrium state of a microemulsion is completely determined by n, T and p. Values of r and ϕ will therefore result from the condition that the microemulsion be in internal equilibrium. This requires that the free energy f be a minimum with respect to r and ϕ, which leads to the equations:

$$\gamma = - \left(\frac{\partial \Delta f}{\partial A}\right)_\phi \qquad\qquad \text{(constant } n_i \text{ and } T) \tag{6}$$

and

$$P_2 - P_1 = (\frac{\partial \Delta f}{\partial \phi})_A \qquad \text{(constant } n_i \text{ and T)} \qquad (7)$$

For spherical globules $A = \frac{3\phi}{r}$ and Eq. (6) becomes:

$$\gamma = \frac{r^2}{3\phi} (\frac{\partial \Delta f}{\partial r})_\phi . \qquad (8)$$

Eq. (8) was already established in Ref. 19.

Since,

$$(\frac{\partial \Delta f}{\partial \phi})_A = (\frac{\partial \Delta f}{\partial \phi})_r + \frac{r}{\phi} (\frac{\partial \Delta f}{\partial r})_\phi ,$$

Eq. (7) can be rewritten for spherical globules as:

$$P_2 - P_1 = \frac{r}{\phi} (\frac{\partial \Delta f}{\partial r})_\phi + (\frac{\partial \Delta f}{\partial \phi})_r . \qquad (9)$$

It is instructive (see below) to write Eq. (9) in a different form. To do this we observe that at constant m, where m is the number of globules per unit volume of microemulsion $(m = \phi/\frac{4}{3}\pi r^3)$,

$$(\frac{\partial \Delta f}{\partial r})_m = \frac{3\phi}{r} (\frac{\partial \Delta f}{\partial \phi})_r + (\frac{\partial \Delta f}{\partial r})_\phi .$$

Combining the above equation with Eqs. (9) and (8), one obtains:

$$P_2 - P_1 = \frac{2\gamma}{r} + \frac{r}{3\phi} (\frac{\partial \Delta f}{\partial r})_m . \qquad (10)$$

Eq. (8) yields the equilibrium radius. For, the interfacial tension γ depends upon the concentrations of surfactant and cosurfactant in the bulk, which in turn depend on the amounts adsorbed on the interfacial area of the globules, because n_i are fixed (the system is closed). Hence, mass balances introduce a dependence of the bulk concentrations on the radius r. However, before mass balances can be applied, explicit expressions relating the surface excesses Γ_i of species i to the concentrations in the bulk are needed as well as the distribution coefficients of surfactant and cosurfactant between phases. An additional dependence of γ on r occurs when the thickness of the adsorbed layer is not small compared with r. This effect is similar to that of the curvature on the interfacial tension of small droplets, an effect originally discussed by Defay et al.[23] and Tolman[24]. It will be shown later in the paper that Δf is also a function of r and ϕ. However, while r and ϕ used in Eq. (8) correspond to the surface of tension, those on which Δf depends are the real ones (corresponding to the actual boundary of the globules, to the extent to which these can be defined). Therefore a model providing relations between the two is needed. Only in the limiting case in which the thickness of the adsorbed layer is small compared with the radius can one assume that the "real" r and ϕ and those corresponding to the surface of tension coincide.

As concerns the second basic equation established above, Eq. (9), it is clear that it is a consequence of the condition of mechanical equilibrium between the continuous and dispersed media of the microemulsion, being, as Eq. (10) more obviously reveals, a generalized Laplace equation.

III.3   The Derivation of Basic Equations Using an Arbitrary Choice for the Dividing Surface

In the previous subsection, the surface of tension was chosen as the (Gibbs) dividing surface for the interface. If the dividing surface is chosen arbitrarily, additional terms which account for bending stresses must be included. Following Gibbs phenomenological theory[25,26], we can write, for curved surfaces, characterized by the principal curvatures c_1 and c_2, the following expression for the frozen free energy f_o:

$$df_o = \gamma dA + C_1 dc_1 + C_2 dc_2 + \Sigma_i \mu_i dn_i - p_2 d\phi - p_1 d(1 - \phi). \quad (11)$$

Here γ is a generalized interfacial tension defined for the particular dividing surface chosen and C_1 and C_2 are bending stresses. When the surface of tension is chosen as the dividing surface, $C_1 = C_2 = 0$. For spherical globules $c_1 = c_2 = 1/r$ and $C_1 = C_2 = C/2$, and the above equation becomes:

$$df_o = \gamma dA + Cd(1/r) + \Sigma_i \mu_i dn_i - p_2 d\phi - p_1 d(1 - \phi). \quad (12)$$

Now, the equilibrium radius satisfies the equation

$$-\frac{3\phi}{r^2} \gamma - \frac{C}{r^2} + \left(\frac{\partial \Delta f}{\partial r}\right)_\phi = 0 \quad (13)$$

and the generalized Laplace equation has the form:

$$P_2 - P_1 = \frac{2\gamma}{r} - \frac{C}{3\phi r} + \frac{r}{3\phi}\left(\frac{\partial \Delta f}{\partial r}\right)_m. \quad (14)$$

An equivalent form of the latter equation, obtained by eliminating γ between Eqs. (13) and (14), is:

$$P_2 - P_1 = -\frac{C}{\phi r} + \frac{r}{\phi} \left(\frac{\partial \Delta f}{\partial r}\right)_\phi + \left(\frac{\partial \Delta f}{\partial \phi}\right)_r . \tag{15}$$

It is important to emphasize that the values of r, ϕ and γ depend upon the particular choice made for the dividing surface and that in the above equations Δf is supposedly expressed in terms of the r and ϕ which characterize the dividing surface. We also observe that γ and C are not independent quantities, being related via the generalized Gibbs adsorption equation:

$$d\gamma + \Sigma \Gamma_i d\mu_i = \frac{rC}{3\phi} d(1/r), \tag{16}$$

where Γ_i is the surface excess of species i in molecules per unit surface area. This equation can be derived by employing the procedure used to derive the usual form of the Gibbs adsorption equation. Integrating Eq. (12) at constant $1/r$, γ, μ_i, P_1 and P_2, one obtains the expression:

$$f_o = A\gamma + \Sigma n_i \mu_i - P_2 \phi - P_1 (1 - \phi),$$

which when differentiated and combined with the Gibbs-Duhem equations for the two media and with Eq. (12) leads to Eq. (16). We do not provide more details concerning this derivation because Eq. (16), which is the same as that for a single droplet, is already available for the latter case[26].

Eq. (13) can be used to relate the equilibrium radius of the globules to the volume fraction ϕ. To achieve this goal, explicit expressions would be needed for: the generalized surface tension and bending stress; Γ_i; the distribution coefficient of the surfactant and cosurfactant between the continuous and dispersed media of the microemulsion; and Δf. These expressions together with mass balances would yield the desired relation between the equilibrium radius and ϕ, but clearly a great deal of calculation would be involved. We will not enter into further details here, since the goal of this paper is not so much to make quantitative predictions as to gain insight.

Concerning Eq. (15), it will be shown in the next section that it plays an important part in phase equilibria involving microemulsions, a conclusion already suggested by the qualitative discussion of II.3.

IV.   PHASE EQUILIBRIA IN MICROEMULSIONS

IV.1   The Free Energy of the Entire System

As already mentioned, experiment shows that a microemulsion can be in equilibrium with excess oil and/or excess water. In what follows, the conditions under which two or three phases occur are identified. The equations will be written for an arbitrary dividing surface, since those obtained when the surface of tension is chosen may be regarded as a particular case.

The variation at constant temperature of the Helmholtz free energy F of the entire system can be written as:

$$dF = \gamma d(AV) + \Sigma \mu_i dN_i + CVd\ (1/r) - p_2 d(V\phi) - p_1 d(V(1 - \phi)) +$$

$$d(V\Delta f) + \Sigma \mu_i' dN_i' - p'dV' + \Sigma \mu_i'' dN_i'' - p''dV'', \tag{17}$$

where the primes and double primes denote excess oil and water phases, respectively, whereas the quantities without superscript refer to the microemulsion. As usual V is the volume, p the pressure, N_i the number of molecules of species i and μ_i their electrochemical potential. The pressures p' and p" are obviously equal to the external pressure p.

IV.2   The Mechanical Equilibrium Condition

Let us consider a variation dV of the volume of microemulsion at constant T, N_i, N_i' , N_i'' , ϕ and r. Since dV is equal to either -dV' or to -dV", Eq. (17) leads, for mechanical equilibrium, to

$$\frac{3\phi}{r} \gamma - (p_2 - p_1)\phi + (p - p_1) + \Delta f = 0. \tag{18}$$

Combining Eqs. (13), (15) and (18), one obtains

$$p_2 - p = \Delta f + (1 - \phi) \left(\frac{\partial \Delta f}{\partial \phi}\right)_r - \frac{C}{\phi r} + \frac{r}{\phi} \left(\frac{\partial \Delta f}{\partial r}\right)_\phi \tag{19}$$

and

$$p_1 - p = \Delta f - \phi \left(\frac{\partial \Delta f}{\partial \phi}\right)_r \tag{20}$$

Eqs. (19) and (20) relate the micropressures p_2 and p_1 to the external pressure p, to the free energy Δf due to the entropy of dispersion of the globules in the continuous phase and to the bending stress C. Obviously, Eqs. (19) and (20) remain valid for a single phase microemulsion subject to the external pressure p.

IV.3   The Chemical Equilibrium

Now, let us consider a variation dN_i at constant T, N_i, N_i', N_i'' (with i $\neq$ j) V, V', V", ϕ and r. Since dN_i is equal to either $-dN_i'$ or to $-dN_i''$, one obtains, (if Δf depends only on r and ϕ and does not depend also separately on n_j) as the chemical equilibrium condition

$$\mu_i = \mu_i' = \mu_i'' . \tag{21}$$

Of course, if only two phases coexist then only the corresponding electrochemical potentials are equal.

Let us assume that the composition of the globules is the same as that of the excess dispersed phase. Because the concentration dependent part of the chemical potentials of a given component is the same in the two phases,

$$\mu_j - \mu_j' = v_j (p_2 - p), \tag{22}$$

where v_j is the molecular volume of species j. The chemical equilibrium condition   leads to

$$p_2 - p = 0, \tag{23}$$

which, because of Eq. (19), becomes

$$\Delta f + (1 - \phi) \left(\frac{\partial \Delta f}{\partial \phi}\right)_r - \frac{C}{\phi r} + \frac{r}{\phi} \left(\frac{\partial \Delta f}{\partial r}\right)_\phi = 0. \tag{24}$$

If the composition in the globules and in the excess dispersed phase are different, then Eq. (24) has to be replaced by the system of equations:

$$v_j \left\{ \Delta f + (1 - \phi) \left(\frac{\partial \Delta f}{\partial \phi}\right)_r - \frac{C}{\phi r} + \frac{r}{\phi} \left(\frac{\partial \Delta f}{\partial r}\right)_\phi \right\} = kT\ln (a_j/a_j'),$$
$$\tag{24a}$$

where a_j is the activity of component j in the globules and a_j' that in the excess dispersed phase.

Eq. (24) provides a relation between r and ϕ when excess oil (water) coexists with a water (oil) continuous microemulsion. Eqs. (13) and (24), supplemented by expression for Δf and for other quantities already listed in the previous section, provide values for both ϕ and r when the above phases coexist. At a given temperature and for given amounts of surfactant, cosurfactant and continuous phase, Eqs. (13) and (24) provide the maximum amount of dispersed phase allowed if there is to be no excess dispersed phase as well as the corresponding radius of the globules. Any additional amount of dispersed phase (having the concentrations of surfactant and cosurfactant equal to those of the globules) will appear as an excess phase. Consequently, any addition to the system of excess dispersed phase having the same composition as the globules will change neither ϕ nor r in the microemulsion, as soon as ϕ becomes equal to the above mentioned maximum value.

Eq. (24) was already established in a previous paper using the surface of tension as the dividing surface[27]. A variant of it together with Eq. (13) was used there to interpret, qualitatively, the temperature dependence of the solubilization of water in oil by polyoxyethylene surfactants[27].

Consider now that an excess continuous phase forms and again that its composition is equal to that of the continuous medium of the microemulsion. Since, in this case

$$\mu_j - \mu_j'' = v_j \, (p_1 - p),$$
(25)

the equilibrium condition and Eq. (20) lead to

$$\Delta f - \phi \, (\frac{\partial \Delta f}{\partial \phi})_r = 0.$$
(26)

When the concentrations in the continuous medium of microemulsion and in the excess continuous medium are different, Eq. (26) must be replaced by the system of equations:

$$v_j \, (\Delta f - \phi \, (\frac{\partial \Delta f}{\partial \phi})_r) = kT \ln \, (a_{jc}/a_j''),$$
(26a)

where a_{jc} is the activity of component j in the continuous medium of microemulsion and a_j'' is that in the excess dispersed medium.

When a microemulsion is in equilibrium with both excess phases and the equality of concentrations of the components is assumed between globules and excess dispersed phase and between the continuous phase and the excess continuous phase, then

$$p_1 = p_2 = p.$$
(27)

IV.4 The Origin of the Change in Structure with the Occurrence
      of the Middle Phase

As anticipated in the qualitative discussion, an excess dispersed phase appears when the micropressure p_2 inside the globules
equals the external pressure p, whereas a third phase forms when,
in addition, the micropressure p_1 becomes equal to p. The equality $p_2 = p_1$ is, however, unlikely to be compatible with the occurrence of a stable, finite radius of curvature and a change in the
structure of the microemulsion is expected to occur. A regular
structure consisting of alternating planar water and oil layers of
constant thicknesses will have zero curvature, but is unlikely for
the following reasons: For such an ordered structure the entropy
of dispersion is expected to be much smaller than for a globular
dispersion. In fact, if this entropy is evaluated on the basis of
the number of distinguishable configurations by using Boltzmann's
formula, one obtains that for the regular planar structure $\Delta f = 0$,
since only one such configuration exists. Because Eqs. (6) and (7)
are of general validity, they can be applied to this regular
planar configuration to obtain $\gamma = 0$. Under such conditions the
planar interfaces are unstable to external and thermal
perturbations and a more disordered structure results.

However, the deformation and possible "fracture" of the regular structure, produces "defects" in the system and as a result
the entropy of dispersion increases. Simultaneously, the area of
the interface increases. Equation (6), if it were extended to the
instantaneous behavior of the system, would therefore suggest that
the interfacial tension increases. Such an increase opposes further growth of the perturbation and leads to a stable oscillatory
behavior. (Alternatively, the increase of this dynamic interfacial
tension could be explained by the larger amount of surfactant and
cosurfactant adsorbed upon the increased interface and the corresponding decrease of their concentrations in the bulk and thus per
unit area of the interface). The detailed (hydro)dynamics of these
fluctuations in structure is extremely complex, and is affected by
the viscosities of the bulk phases, the surface viscosity and
other characteristics of the interfacial layer. (Some studies have
been undertaken of the fluctuations which occur in thin films[29-33]
as well as of the roughness of an interface produced by thermal
perturbations[34]. The present case is however much more complicated.)

The above dynamics is expected to increase in complexity as
the volume fraction ϕ increases because the probability of interlinking among various parts of the dispersed phase becomes greater. The interlinking is particularly important near the phase
inversion, where it probably leads to a bicontinuous structure.
This explains why the phase inversion occurs in a region around
$\phi \approx 0.5$ rather than at a fixed value.

Consequently, it is likely that as soon as the micropressures p_2 and p_1 are near to one another, extended internal interfaces form which are subject to fluctuations on short time scales. Because of the nature of the perturbations and the complexity of the motions involved, their dynamics cannot be characterized by single time (frequency) and length scales, in other words the oscillations in structure are not periodic but disorderly. For this reason, we suggest this structure be called chaotic.

We close this subsection by observing that the above thermodynamic equations are no longer valid in their initial form for the chaotic structure, because we do not have spherical globules of only one size, but much more complex structures, and because the area, the interfacial tension, the micropressures p_1 and p_2 and the concentrations are now fluctuating quantities. In addition, probably chaotic breakup and coalescence occur.

IV.5   Two Microemulsions in Equilibriuim

Assuming that the dispersed phase is the same in both microemulsions, considerations similar to those of the previous sections lead to the equalities:

$$p_1' = p_1'' \quad \text{and} \quad p_2' = p_2'' \ ,$$

where the primes and double primes refer to the two microemulsions. The equilibrium radius in each of the microemulsions is given by Eq. (13).

In this case, there is a critical point at which the two microemulsions become identical. Of course, near this critical point strong fluctuations occur in both microemulsions and at the interface between the two.

We also note that, at least in principle, the two microemulsions can have different dispersed and continuous media (in which case $p_1' = p_2''$ and $p_2' = p_1''$) and that the two microemulsions can coexist with one or both of the excess phases.

V.   EXPRESSIONS FOR THE ENTROPY OF DISPERSION

In a previous paper[21], using a lattice model, upper and lower bounds for the entropy of dispersion of spherical globules of the same size in the continuous medium of a microemulsion have been established.

The free energy corresponding to the upper bound on the entropy has the form:

$$\Delta f = - \frac{3\phi kT}{4\pi r^3} \{- \frac{1}{\phi} \ln (1 - \phi) + \ln [\frac{4\pi r^3}{3v_c} \frac{(1 - \phi)}{\phi}]\}$$ (28)

and that corresponding to the lower bound the form:

$$\Delta f = - \frac{3\phi kT}{4\pi r^3} \ln \{\frac{4\pi r^3}{3v_c} [(\frac{0.74}{\phi})^{1/3} - 1]^3 \}$$ (29)

Here v_c is the molecular volume of the continuous medium. It is important to note that in Eqs. (28) and (29), ϕ and r represent the "real volume fraction and radius" of the globules as far as these can be clearly defined.

The expression based on the Carnahan-Starling approximation for hard spheres, suggested in reference 22,

$$\Delta f = - \frac{3\phi kT}{4\pi r^3} [1' - \ln \phi - \phi \frac{4 - 3\phi}{(1 - \phi)^2} + \ln \frac{4\pi r^3}{3v_c}],$$ (30)

leads to results which practically coincide with those of Eq. (29).

These equations show that Δf is extremely small, being of the order of -10 kT per globule.

By combining the thermodynamic equation (13) with any of the expressions (28) to (30), relations can be established between the interfacial tension and r. However, r and ϕ in Eqs. (28) to (30) are "real" radius and volume fraction, while in Eq. (13) they correspond to the particular dividing surface; so, strictly speaking, we need additional equations relating the two pairs of variables. However, when the thickness L of the surfactant layer is small compared with the radius, the surface of tension almost coincides with the "actual surface" and one can take the r and ϕ in Eq. (13) as the actual values and C = 0. Consequently when L is small compared with r,

$$\gamma(r) = \frac{r^2}{3\phi} (\frac{\partial \Delta f}{\partial r})_\phi,$$ (31)

where r and ϕ are the radius and volume fraction involved in the entropy of dispersion. Combining (31) with each of the expressions (28) to (30), one obtains in the same order[28]:

$$\gamma = \frac{3kT}{4\pi r^2} \{ \ell n \, [\frac{1 - \phi}{\phi} \frac{4\pi r^3}{3v_c}] - \frac{\ell n(1 - \phi)}{\phi} - 1 \}, \tag{32}$$

$$\gamma = \frac{3kT}{4\pi r^2} (\ell n \, \{\frac{4\pi r^3}{3v_c} \, [(\frac{0.74}{\phi})^{1/3} - 1]^3 \} - 1) \tag{33}$$

and

$$\gamma = \frac{3kT}{4\pi r^2} (\ell n\{\frac{4}{3} \frac{\pi r^3}{v_c}\} - \ell n \, \phi - \frac{\phi(4 - 3\phi)}{(1 - \phi)^2}). \tag{34}$$

It is important to observe that these equations have been established from thermodynamics, namely from the equation which determines the equilibrium radius, and do not involve any specific model of the interface. They show that γ is almost inversely proportional to r^2, with a proportionality constant of about 4kT which is weakly dependent on r and ϕ.

Of course, the above expressions for Δf could be used together with Eq. (24) to obtain some information concerning phase separation in the two phase region. Such calculations have been carried out in the limiting case in which r >> L. In this case r and ϕ in the thermodynamic equations (24) are the same as those in Eqs. (28) to (30) and C = 0. However, the values obtained for r are only a few Angstrom or less, for ϕ < 0.5. This seems to indicate that no phase separation can occur under the conditions stated above. One possibilitiy is that the expression for the entropy of dispersion is not accurate enough and Eq. (24), containing small terms, is very sensitive to slight errors. Another possible explanation is that the concentrations of the components which we have assumed to be the same in the globules and in the excess dispersed phase are, in reality, somewhat different. If this is true, it is also possible that even the activity coefficients of water and oil which are near to unity, or their small mutual solubilities, are significant. Since Δf is very small, such small differences could matter. One may also note that the expressions established for the entropy of dispersion do not take into account the interactions between the globules.

The above discussion involves the assumption that r is sufficiently large and hence that the surface of tension coincides with the actual interface. When, however, the inequality between r and the thickness L of the interfacial layer is not so strong and the surface of tension is still used as the dividing surface, specific models are needed to relate the ϕ and r in the thermodynamic expression (24) to those in any of the expressions derived for Δf. For other choices of the dividing surface, expressions for C are also needed. They can be, however, provided by molecular theories only. Eq. (24) could predict phase separation if such models were developed, but we will not undertake this task here. Instead, in the next section we use the main results obtained here to interpret some experimental results concerning the dependence of the interfacial tension (between a microemulsion and excess dispersed phase) on the radius as well as on the salinity, and to explain the origin of the low interfacial tension between microemulsion and excess dispersed phase.

VI.   INTERFACIAL TENSION BETWEEN A MICROEMULSION AND EXCESS PHASES

As already pointed out, if one assumes that the interior of the globules and the excess dispersed phase have nearly the same composition, the surface of the globules and the interface between the microemulsion and excess dispersed phase separate essentially the same two media. If, in addition, the globule size is sufficiently large compared with the thickness of its interface, so that the effect of the curvature of the latter on its interfacial tension can be neglected, then this interfacial tension should be nearly equal to that between the microemulsion and the excess dispersed phase. Now it is true that double layer and van der Waals interactions have an effect on γ and that such interactions among globules in the bulk are expected to be somewhat different from those between the globules and the interface between the two phases. However, at the high salinities which are of interest here (see below) the effect of the double layer forces is negligible, and for values of ϕ not too large, the effect of the van der Waals interaction forces is small in comparison with that due to the entropy of dispersion[28]. So the interfacial tension is largely determined by the adsorption of surfactant and cosurfactant at the interfaces.

The thermodynamic considerations of the previous sections provide a relation between γ at the interface of the globules and r. Under the above conditions, this relation can be applied to the interfacial tension between the microemulsion and the excess dispersed phase. The dependence of the radius on the independent

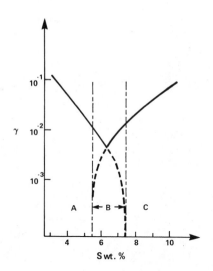

Figure 2.   Interfacial tensions γ (dyne/cm) against salinity for
46.8 wt% water, 46.25 wt% toluene, 1.99 wt% sodium dodecylsulfate
and 3.96 wt% butanol (T = 293°K).   See text for additional details.   The full line is the interfacial tension between microemulsion and excess dispersed medium, while the broken line is the
interfacial tension between microemulsion and excess continuous
medium.

variables, such as temperature, salinity, and the amounts of surfactant, cosurfactant, oil and water can be obtained only from a
detailed analysis of the phase equilibria.   This has not yet been
attempted.   However, the above equations are still useful for a
qualitative comparison with experiment.

Figure 2 shows how the interfacial tensions can change with
salinity, S, for a particular case[20].   The full line is the interfacial tension between microemulsion and excess dispersed medium
and the broken line is that between microemulsion and excess continuous medium.   At relatively low salinities, an O/W microemulsion coexists with excess oil (region A); at high salinities, a
W/O microemulsion coexists with excess water (region C); whereas
in the intermediate range of salinities, a three-phase region
(region B) exists.

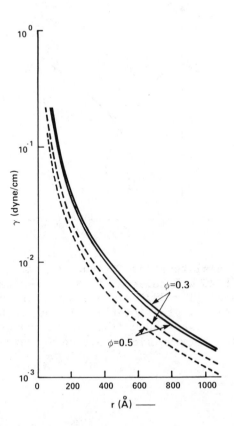

Figure 3. Interfacial tension γ (microemulsion–excess dispersed medium) against the equilibrium radius for hexane in water microemulsion. The continuous lines are based on Eq. (32), and the broken lines on Eq. (33).

The structure of the middle phase microemulsion was discussed earlier in the paper and related to the fluctuations which occur when $p_2 = p_1$. Their effect is important and a comment will be made in this respect later. For the time being however let us compare the above equations (which ignore fluctuations) with experiment. These equations are certainly valid for a microemulsion coexisting with an excess dispersed phase as long as the microemulsion contains spherical globules.

To do this we plot (in Figure 3) Υ against r, as given by Eqs. (32) and (33). (Eq. (34) provides almost the same results as Eq. (33)).

Experiment shows that the radius of the globules varies between 10^2 and 10^3Å, and that at relatively low concentrations of salt it increases with the concentration, while the opposite occurs at high salt concentrations[35]. The theoretical curves indicate that, for the above range of sizes,the interfacial tension varies between 10^{-2} and 10^{-3} dynes/cm, as do the experimental values, and that Υ decreases with increasing radii, again as shown by experiment. It is important to emphasize that Eqs. (32) to (34) can be applied to the experiments of Fig. 2, since the ionic strength is high (double layer forces negligible) and the volume fraction is below 0.5 (van der Waals interactions negligible).

It is noteworthy that a direct verification of Eqs. (32) to (34) is now available. Preliminary neutron diffraction experiments accompanied by interfacial tension measurements between microemulsion and excess dispersed phase have shown that $\Upsilon \sim \frac{kT}{r^2}$ with a proportionality constant between 3 and 4, weakly dependent on the radius[36], as predicted by theory.

The above considerations have to be modified when the internal interface of a microemulsion fluctuates chaotically. One of the effects of these fluctuations is that even at the interface between the microemulsion and the excess dispersed phase two similar media may be brought into contact. Similarly, at the interface with the excess continuous phase, there is some contact between two dissimilar media. These fluctuations will therefore cause the interfacial tensions for the two pair of phases in contact to be equal at the inversion point. This happens because $\phi \sim$ 1/2 at this point and therefore the probability of having water or oil at any point of either interface, on the microemulsion side, is also 1/2.

Let us note that the dynamics of the internal interface of the middle phase microemulsion is similar (without being the same) to that near a critical point. It is therefore not surprising that the interfacial tensions between the microemulsion and each of the excess phases could be represented by expressions valid for the latter case[37,38]. The large thicknesses of the interfaces between the middle phase and each of the excess phases[38] can be similarly explained. They are caused by the strong fluctuations which lead to large correlation lengths.

To explain the occurrence of a minimum in the interfacial tension (Fig. 2), it is necessary to examine the effect of salinity on the interfacial tension between water and oil, as well as on the distribution of surfactant and cosurfactant between phases.

At low salt content, the surfactant molecules accumulate at the interface where their hydrocarbon chains can be immersed in oil and the polar head groups still enjoy their favorable interactions with water. At higher salt concentrations, the organization of water molecules induced by ions (for structure making ions) leads to the salting out of the surfactant molecules either as micelles and/or into oil (globules and excess dispersed phase) and on the interface. At very high salt concentrations, the water might be so strongly organized that not only is the surfactant salted out completely to the oil phase, but also the interactions of head groups are more favorable with oil than water. Under such circumstances, negative adsorption of the surfactant may occur. Because the chemical potential of the surfactant increases with the concentration in the oil phase, the Gibbs adsorption equation shows that γ increases with increasing salinity, as soon as the negative adsorption starts. Thus, a minimum in γ can occur. An alternative explanation may be provided by the negative adsorption of salt. Since oil is nonpolar, or only slightly polar, the distribution of ions near the surface will tend to be such that the ions enjoy the full strength of the favorable hydration interaction. While it is known that the corresponding increase in interfacial tension is small, it may well be comparable to the magnitude of the interfacial tension itself.

VII. PHASE INVERSION

As regards the phase inversion from an oil in water to a water in oil microemulsion one may observe that before and after the inversion point the free energy of the actual system is smaller than that of the (virtual) inverted one, but that they are equal at the inversion point. To allow further analysis, the microemulsion will be assumed to have a globular structure so that the following expression for the free energy of a microemulsion per unit volume may be used:

$$f = \frac{3\phi}{r} \gamma + \Sigma \, n_i \mu_i + \Delta f - p_1 \, (1 - \phi) - p_2 \phi.$$

This expression was obtained by integrating Eq. (12) at constant p, r, μ_i, p_2 and p_1 and by using Eq. (4).

When water changes from being continuous to the dispersed medium, its chemical potential changes by $(p_2' - p_1)v_w$, where v_w is the molecular volume of water and p_2' is the pressure in the globules after inversion; similarly, the change for oil is $(p_1' - p_2)v_o$,

where v_o is the molecular volume of oil and p_1' is the pressure in
the continuous medium after inversion. There are, of course,
changes in the chemical potentials of the surfactant and cosurfactant molecules as well, but because their numbers are much smaller
than those of oil and water, the effect on the free energy is
negligible. Any change in the composition of the phases, which
occurs during inversion, is likely to be small and is therefore
also neglected.

At the inversion point the free energies of the two kinds of
systems are equal:

$$\frac{3\phi\gamma}{r} + \Delta f - \phi p_2 - (1 - \phi)p_1 \approx \frac{3\phi'\gamma'}{r'} + \Delta f' + n_w v_w(p_2' - p_1)$$

$$+ n_o v_o (p_1' - p_2) - \phi' p_2' - (1 - \phi') p_1', \qquad (35)$$

where ϕ and ϕ' are subject to the constraint

$$\phi + \phi' \approx 1. \qquad (36)$$

Here the prime refers to the state after inversion and the
subscripts w and o to water and oil. The interfacial tension γ
and Δf are related via Eq. (13), which contains also the bending
stress C. However, let us consider the case in which near the
inversion point the size of the globules is so large that their
surface can be considered as the surface of tension. Consequently
C can be taken as zero and r, r', ϕ and ϕ' in the expressions of γ
and γ' coincide with the actual values. Thus, Eq. (13) (with
C = 0) provides two additional equations relating r to ϕ and r' to
ϕ'. The latter two equations together with Eqs. (35) and (36)
form a system of four equations from which one can obtain the
values of r, ϕ, r' and ϕ' at the inversion point.

Should Δf be independent of any characteristics of the continuous medium (such as v_c), but dependent only on ϕ and r, γ will
also depend only on ϕ and r. Because the effect of curvature on
the interfacial tension is assumed negligible near the inversion
point, the relation between ϕ and r provided by Eq. (13) is also
likely to be independent of which phase is dispersed. (The latter
assertion involves the approximation that only minor changes occur
in the composition of oil and water during inversion.) Similarly,
Eqs. (19) and (20) indicate that the pressures p_2 and p_1 depend only
on ϕ and r, and are independent of which phase is dispersed. Under
such conditions one can easily verify that $\phi = \phi' = 1/2$ and r =
r'. Indeed, for these values of ϕ and r, we have $\Delta f = \Delta f'$, $\gamma = \gamma'$,
$n_w v_w(p_2' - p_1) = \phi' (p_2 - p_1) = - n_o v_o(p_1' - p_2) = - \phi(p_1 - p_2)$ and
Eq. (35) and the constraint (36) are satisfied. The qualitative
arguments brought above indicate that these values satisfy also
the other two equations resulting from Eq. (13).

In reality [see Eqs. (28) to (30)], Δf has a weak dependence on v_c and, in addition, some approximations are involved. For this reason, phase inversion occurs near $\phi = 1/2$ and not exactly at 1/2. The presence of fluctuations, which was ignored in the above considerations, makes the transition more diffuse because it favors interlinking in the dispersed phase and hence the formation of a bicontinuous structure. These conclusions are in agreement with experiment[10,11].

VIII.   SUMMARY AND CONCLUSIONS

A thermodynamics of microemulsions is developed starting from a Helmholtz free energy which is composed of the free energy of a frozen system, which ignores the entropy of dispersion of one phase in the other, and the free energy Δf associated with the latter quantity. The free energy of the frozen system contains an interfacial tension which includes the effect of the interaction forces between the globules, and the micropressures p_2 in the globules and p_1 in the continuous phase near the globules. These micropressures represent the real pressures on the scale of the size of the globules. They are contrasted to the thermodynamic pressure which is calculated by including in the expression of the free energy the collective behavior of the globules reflected in the entropy of dispersion. Two basic equations have been derived. One relates the interfacial tension at the surface of the globules to the entropy of dispersion and to the bending stress, while the other relates the pressure difference $p_2 - p_1$ to the entropy of dispersion and to the bending stress. The first equation provides the equilibrium radius and the second constitutes a generalized Laplace equation and plays a part in phase equilibria. The thermodynamic formalism uses either the surface of tension or an arbitrary surface as the (Gibbs) dividing surface. A comment that should be made is that the various choices for the dividing surface are different but equivalent ways of describing the same phenomenon. Of course, for each choice, the physical quantities, such as Δf, have to be expressed in terms of the corresponding radius r and volume fraction ϕ. Assuming that the globules and the excess dispersed phase have the same composition, it is shown that an excess dispersed phase appears when $p_2 = p$, where p is the external pressure. A similar assumption shows that a third phase, the excess continuous phase, forms when, in addition, $p_1 = p$. Expressions have been established in terms of Δf and bending stress which can, in principle, provide the radius r and the volume fraction in the two phase region and which can identify the transition to the three phase system.

The change in the structure of the microemulsion which occurs at or near the transition to a three phase system is associated with the equality $p_2 = p_1$, which is unlikely to be compatible with

a stable, finite radius of curvature. Arguments are adduced which
suggest that a planar regular structure cannot replace the spherical globules since the former is unstable to external and thermal
perturbations, and that chaotic oscillations between extended and
curved structures occur. For this reason, we call this structure
"chaotic". Using an explicit expression for the entropy of dispersion, a relation is derived between the interfacial tension at
the surface of the globules and their radius. Since the environment near the interface between a microemulsion and excess dispersed phase is likely to be similar to that near the surface of
the globules, the above equation is used to relate the interfacial
tension between the microemulsion and excess dispersed phase to
the radius of the globules. The equation is in excellent agreement with recent neutron diffraction experiments. One concludes
that the latter interfacial tension, like the former, is essentially caused by the adsorption of surfactant and cosurfactant on
the interface. Because of the chaotic fluctuations of the internal interface of the middle phase, the dynamics of this microemulsion is similar to that near a critical point. This explains why
the interfacial tensions between the middle phase and each of the
excess phases could be represented by expressions valid near a
critical point and why the thicknesses of the corresponding interfaces are large. The occurrence of a minimum in the interfacial
tension (see Fig. 2) may be associated with a possible negative
adsorption of surfactant on the interface. This may occur at high
salinities because of the strong effect which ions of small ionic
radius have on the organization of water.

Finally, it is explained that the phase inversion from an O/W
to a W/O microemulsion is driven by the entropy of dispersion and
occurs near $\phi = 1/2$.

An important conclusion to emerge from the discussion is that
while Δf is small it plays a major part regarding both the interfacial tension and phase behavior.

APPENDIX A

The variation at constant temperature T of the Helmholtz free
energy F, defined for the volume V of the microemulsion, can be
written in a number of equivalent forms. One of them, considered
in the paper, is the following:

$$dF = \gamma d(AV) + CV\, d\,(1/r) + \Sigma\, \mu_i dN_i - p_1 d(V(1 - \phi)) - p_2 d(V\phi) + d\Delta F$$

$$(A-1)$$

where $\Delta F = V\Delta f$. Here the free energy ΔF associated with the
entropy of dispersion is added to the free energy of a "frozen"

system and p_1 and p_2 are micropressures (see text for their physical meaning).

Another form is

$$dF = \gamma^* d(AV) + \sum_i \mu_i^* dN_i - p_1^* dV_1 - p_2^* dV_2, \tag{A-2}$$

where $V_1 = V(1 - \phi)$ and $V_2 = V\phi$. In this expression, ΔF and $CVd(1/r)$ do not appear explicitly, but are distributed among γ^*, μ_i^*, p_1^* and p_2^*. Indeed, the interfacial tension γ^* is defined by a variation in the free energy with area at constant N_i, V_1 and V_2. Now, a change in A at constant V_1 and V_2 changes the radius of the globules and their number and hence the entropy of dispersion. Similarly, a variation of V_2 at constant AV, N_i and V_1, by which the pressure p_2^* is defined, changes the entropy of dispersion. One can therefore conclude that, indeed γ^*, p_1^* and p_2^* contain the effect of the entropy of dispersion. In contrast to the micro-pressures p_1 and p_2, the pressures p_1^* and p_2^* are defined on the basis of a free energy which includes the collective behavior of the globules, via the free energy ΔF. For this reason we call them macropressures or thermodynamic pressures.

Because the two formulations have to lead to the same results, one obtains

$$\gamma^* = \gamma + \frac{C}{3\phi} - \frac{r^2}{3\phi} \left(\frac{\partial \Delta f}{\partial r}\right)_\phi, \tag{A-3}$$

$$p_1^* = p_1 - \Delta f + \phi \left(\frac{\partial \Delta f}{\partial \phi}\right)_r, \tag{A-4}$$

$$p_2^* = p_2 - \Delta f - (1 - \phi) \left(\frac{\partial \Delta f}{\partial \phi}\right)_r + \frac{C}{\phi r} - \frac{r}{\phi} \left(\frac{\partial \Delta f}{\partial r}\right)_\phi \tag{A-5}$$

and

$$\mu_i^* = \mu_i + \left(\frac{\partial \Delta f}{\partial n_i}\right)_{r, \phi, n_j(j \neq i)}. \tag{A-6}$$

Finally, one may observe that at equilibrium $\gamma^* = 0$. This equation is equivalent to Eq. (13) of the text.

Because the system is subject to the external pressure, the mechanical equilibrium condition leads to $p_1^* = p_2^* = p$ and expressions (A-4) and (A-5) become identical to Eqs. (20) and (19), respectively. If one assumes (as in the paper) that $\Delta f = \Delta f(\phi, r)$, Eq. (A-6) becomes $\mu_i^* = \mu_i$.

REFERENCES

1. T. P. Hoar and J. H. Schulman, Nature, 152, 102 (1943).
2. J. H. Schulman and D. R. Riley, J. Colloid Sci., 3, 383 (1948).
3. W. Stoeckenius, J. H. Schulman and L. M. Prince, Kolloid Z., 169, 170 (1960).
4. P. H. Elworthy, A. T. Florence and C. B. MacFarlane, Solubilization by Surface Active Agents and Its Applications in Chemistry and Biology, Chapman and Hall, London, 1968.
5. K. Shinoda, J. Colloid Interface Sci., 24, 4 (1967).
6. K. Shinoda and T. Ogawa, J. Colloid Interface Sci., 24, 56 (1967).
7. K. Kon-No and A. J. Kitahara, J. Colloid Interface Sci., 34, 22 (1970); 37, 469 (1971).
8. R. Nagarajan and E. Ruckenstein, Separation Sci. Technol., 16, 1429 (1981).
9. P. A. Winsor, "Solvent Properties of Amphiphilic Compounds", Butterworth's Scientific Publications, London, 1954.
10. R. N. Healy, R. L. Reed and D. G. Stenmark, Soc. Pet. Eng. J. Trans., AIME, 261, 147 (1976).
11. R. N. Healy and R. L. Reed, Soc. Pet. Eng. J. Trans., AIME, 263, 129 (1977).
12. R. C. Nelson and G. A. Pope, Soc. Pet. Eng. J. Trans., AIME, 265, 325 (1978).
13. M. Bourrel, C. Koukounis, R. Schechter and W. Wade, J. Dispersion Sci. Tech., 1, 13 (1980).
14. W. R. Foster, J. Pet. Tech. Trans., AIME, 255, 205 (1973).
15. R. Hwan, C. A. Miller and T. Fort, Jr., J. Colloid Interface Sci., 68, 221 (1979).
16. J.E.L.Bowcott and J. H. Schulman, Z. Electrochem., 59, 283 (1955).
17. J. W. Falco, R. D. Walker, Jr. and D. O. Shah, AIChE J., 20, 510 (1974).
18. B. Lindman, P. Stilbs and E. Moseley, J. Colloid Interface Sci., 83, 569 (1981).
19. E. Ruckenstein, Chem. Phys. Lett., 56, 518 (1978).
20. A. Pouchelon, J. Meunier, D. Langevin, D. Chatenay and A. M. Cazabat, Chem. Phys. Lett., 76, 277 (1980).
21. E. Ruckenstein and J. C. Chi, J. Chem. Soc. Faraday Trans. II, 71, 1690 (1975).
22. J. Th. G. Overbeek, Faraday Discussions of the Chem. Soc. No. 65, 7 (1978).
23. R. Defay, I. Prigogine and A. Bellemans, "Surface Tension and Adsorption", Longmans, London, 1966.
24. R. C. Tolman, J. Chem. Phys., 17, 333 (1949).
25. J. W. Gibbs, Collected Works, Vol. 1, Yale University Press, New Haven, CT, 1948.

26. F. R. Buff, J. Chem. Phys., 19, 1591 (1951).
27. E. Ruckenstein and R. Krishnan, J. Colloid Interface Sci., 71, 321 (1979).
28. E. Ruckenstein, Soc. Pet. Eng. J. Trans., AIME, 21, 593 (1981).
29. A. Vrij, F. Hesselink, J. Lucassen and M. van den Tempel, Proc. Kon. Ned. Akad, Wet., B73, 124 (1970).
30. E. Ruckenstein and R. K. Jain, J. Chem. Soc. Faraday Trans. II, 70, 132 (1974).
31. C. Maldarelli, R. K. Jain, I. B. Ivanov and E. Ruckenstein, J. Colloid Interface Sci., 78, 118 (1980).
32. G. I. Sivaskinsky and D. M. Michelson, Prog. Theor. Phys., 63, 2112 (1980).
33. T. Shlang and G. I. Sivaskinsky, J. Physique, 43, 459 (1982).
34. L. Mandelstam, Ann. d. Phys., 41, 609 (1913).
35. J. W. Benton, Personal Communication, Sept., 1980.
36. J. S. S. Huang, Personal Communication, April, 1982.
37. P. D. Fleming, J. E. Vinatieri and G. R. Glinsman, J. Phys. Chem., 84, 1526 (1980).
38. A. M. Cazabat, D. Langevin, J. Meunier and A. Pouchelon, Adv. Colloid Interface Sci., 16, 175 (1982).

ACKNOWLEDGEMENT

This work was supported by the National Science Foundation.

INFLUENCE OF THE COSURFACTANT CHEMICAL STRUCTURE UPON THE PHASE
DIAGRAM FEATURES AND ELECTRICAL CONDUCTIVE BEHAVIOR OF WINSOR IV
TYPE MEDIA   (SO-CALLED MICROEMULSIONS)

M. Clausse[*], J. Peyrelasse, C. Boned, J. Heil,
L. Nicolas-Morgantini and A. Zradba

Université de Pau et des Pays de l'Adour
Institut Universitaire de Recherche Scientifique
Avenue Philippon, F-64000, Pau, France

     Series of so-called microemulsion samples were made
up of water and benzene, using surfactant/alcohol combinations. The surfactant used was sodium dodecylsulfate, the
cosurfactants being various straight or branched alkanols
with carbon number ranging from 2 to 7. Irrespective of
the alcohol used, the surfactant/alcohol weight ratio was
kept at 1/2. It was found that the configuration of the
so-called microemulsion realm of existence (monophasic,
fluid and isotropic domain) was strikingly influenced by
the alcohol chemical structure. Two types of systems were
defined. For the longer straight chain alkanols, 1-pen-
tanol, 1-hexanol and 1-heptanol, the monophasic domain is
split into two disjointed areas. The oil-rich area is
separated from the water-rich one by a composition zone
over which viscous and turbid phases with long range
ordered structures are encountered, (Type S systems).
The existence of such structured media is to be considered
as reflecting a phase inversion mechanism that affects, at
the macroscopic level, the mechanical and optical properties of the medium. For the shorter straight alkanols,
ethanol, 1-propanol and 1-butanol, the monophasic domain
forms a unique area whose shape is the classical triangular one, (Type U systems). In this case, there is
no macroscopically detectable boundaries between the
oil-rich and water-rich regions, the media of intermediate
composition being monophasic, fluid and isotropic as well.

*Present address: Universite de Technologie de Compiègne,
              B.P. 233,60206 COMPIEGNE Cedex, France.

The same behavior was observed with 2-propanol and the
1-pentanol tertiary isomer 2-methyl-2-butanol. With the
1-pentanol secondary isomers 3-methyl-2-butanol, 3-pent-
anol and 2-pentanol and the primary isomers 2-methyl-1-
butanol and 3-methyl-1-butanol, hybrid configurations
were put into evidence, the unique monophasic area pre-
senting an indentation located between the oil-rich and
water-rich regions, (Type U* systems). An even more
peculiar situation was evidenced in the case of 2,2-dimethyl-
1-propanol, (Type S* systems). It was observed that each
type of system can be characterized by a typical electrical conductive behavior. In the case of Type S systems,
over the oil-rich area, the conductivity values are low
(around $10^{-3} Sm^{-1}$) and it undergoes non-monotonous variations as the water content increases. In the phase diagram,
the composition points corresponding to conductivity maxima
define a curve Γ_1 that can be considered as a structural
transition locus separating, in the oil-rich area, a region
of pre-micellar hydrated surfactant aggregates from a
region of water-swollen spherical micelles. This result
is consistent with data reported by other scientists. In
contrast, in the oil-rich region of the monophasic domain
of Type U systems, the conductivity takes rapidly high
values (around $0.5 Sm^{-1}$) and varies linearly as the water
content increases. This behavior can be analyzed in terms
of the Percolation and Effective Medium theories, which
suggests the existence of strong attractive interactions
between the disperse inverted micelles. Upon increasing
the water content further, the linear conductivity regime
is followed by a non-linear increase up to a maximum and
then by a non-linear decrease down to the value of the
conductivity of the water used. A correlated analysis
of this conductive behavior and of the phase diagram led
us to define in the monophasic domain two curves C_b and
C_m that could be internal boundaries separating, respectively, the W/O and O/W regions from IZ, the intermediate composition zone over which the w/o and o/w
transformation (phase inversion) would take place through
a progressive and diffuse mechanism.

INTRODUCTION

It is well known that by mixing water and a hydrocarbon component (oil), in the presence of adequate amounts of surface active
agents (amphiphiles), it is possible to form over broad composition
ranges, and with no or little mechanical agitation, macroscopically
monophasic, fluid, transparent and isotropic media that, according
to the classification proposed by Winsor[1,2], can be labeled Type
IV phases. The so-called microemulsions studied by Schulman and
his co-workers[3-10] are Winsor IV phases that incorporate as surface active agents combinations of an ionic surfactant (an alkaline
metal soap-type compound) with a so-called cosurfactant (an alcohol
or an amine, for instance). Similar media can be formulated as
well without any addition of a cosurfactant, by using double chain
ionic surfactants (sulfosuccinic esters such as Aerosol OT, AOT)[11-17]
or non-ionic surfactants[18-23].

During the past ten years or so, intensive investigations have
been devoted to these different types of Winsor IV phases which
have important applications in many industries[24-30]. Moreover,
these media have attracted the attention of fundamentalist physico-
chemists interested in understanding the properties and structures
of these intriguing and versatile systems which, in spite of their
chemical diversity[31], belong to a special field of colloidal systems located at the border of molecular solutions, micellar solutions, liquid crystals, gels and macroemulsions. All these investigations have provided quite significant data concerning the
phase behavior, physico-chemical properties and structure of Winsor
IV phases[32-38]. However, the structure of these media is a problem that is far from being fully solved in all its aspects. On
the basis of results obtained using several complementary techniques, Schulman and his co-workers[3-10] depicted their so-called
microemulsions as stable and monodisperse suspensions of spherical
microdroplets either of the w/o type (oil-external or "inverted"
microemulsions) or of the o/w type (water-external or "direct"
microemulsions). The microdroplets were viewed by these authors as
composite globules consisting of a central aqueous (w/o), or organic
(o/w), core surrounded by a monomolecular film of interspersed surfactant and cosurfactant molecules with polar heads oriented towards the aqueous medium. Recent studies[38-44] carried out on
different systems involving ionic surfactant/alcohol combinations
by means of ultracentrifugation, small angle neutron scattering
and dynamic light scattering techniques have proved that, as far
as w/o type systems are concerned and under certain composition
conditions, the composite globule model is applicable and can be
further refined significantly. The microemulsions studied appear
to be depictable as fairly monodisperse populations of dynamic
spherical microglobules made up of an inner spherical aqueous core
(radius r_w) surrounded by a monomolecular shell of mixed surfactant
and cosurfactant. The microglobules are suspended in an organic

phase (hydrocarbon in which part of the cosurfactant and a small amount of water are dissolved) and are submitted to interactions that derive from a hard-sphere potential corrected with a small attraction term. Thus, from the viewpoint of their interactions, the microglobules can be considered as consisting of a spherical hard core, defined by the "chemical" radius r_c, and of an outer compressible region that is partially penetrated by molecules of the surrounding organic medium. The value of the "chemical" radius r_c is smaller than that of the hydrodynamic radius r_h, determined either from ultracentrifugation or light scattering experiments, and is greater than r_w, the aqueous core radius. The "chemical" thickness (r_c-r_w) appears to be compatible with the molecular length of the alcohol used as cosurfactant in the different systems investigated. As for the hydrodynamic radius r_h, its value may differ from that of the microglobule theoretical radius r_m (r_w plus the length of the extended surfactant molecule) by ± 5Å or so, owing to the fact that, on the one hand, surfactant molecule lipophilic moieties retain some flexibility and that, on the other hand, each microglobule carries along during its motion a layer of molecules of the suspending medium. In light of the results obtained by different groups[15-17,45-47], it is most likely that the model of stable and fairly monodisperse populations of composite globules (they may be called, liquid particles, micro-droplets, micelles or swollen micelles of the inverted type) can be extended to the case of w/o systems incorporating sulfosuccinic esters or non-ionic surfactants as surface active agents. As concerns o/w type systems, definite structural data are very scarce in the literature[47-50], but it can be reasonably assumed that the composite globule model may still be applicable, although some authors have reported that some o/w media appeared to be thermo-dynamically unstable[51,52].

It has been pointed out that, because of molecular and geo-metrical constraints[53-55], the model of monodisperse populations of composite globules cannot be valid over the entire realm of existence of Winsor IV phases. These theoretical predictions are supported by various experimental results that show the structural diversity that characterizes Winsor IV media[56]. It has been proved by different groups[14,15,57-67] that, at low water contents, ternary or quaternary Winsor IV media incorporating ionic surfactants contain most probably premicellar entities consisting of hydrated surfactant aggregates. Several authors have stressed that, in the central part of the phase diagram, it is difficult to characterize Winsor IV phases as typical w/o or o/w systems[56,68-71]. Other experiments have suggested the existence of micelle aggregation phenomena[41,56,62-64]. In addition, it has been experimentally proven[42,63-65,68,72-76] that slight variations in component chem-ical structure induce drastic changes in several physicochemical properties of Winsor IV media of analogous compositions. In this connection, it appears that, other things being equal, the chemical structure of the alcohol plays a key role.

This rapid overview of data available in the literature shows that the problem of the structure of Winsor IV media is a complex one that calls for thorough investigations using complementary methods and techniques. In this context, an extensive study[64] was undertaken to show the influence of composition factors upon both Winsor IV domain configuration and Winsor IV phases electrical properties that reflect the medium structure[77]. The results presented in the following sections were obtained with quaternary systems incorporating sodium dodecylsulfate as the surfactant and various alcohols as cosurfactants. These results complement and support the data published partially in previous articles[66,76,78,79].

EXPERIMENTAL

The microemulsion samples were made up of distilled water and pure benzene (99.5% from Fluka A.G.), the surfactant being high grade sodium dodecylsulfate, (99% from Touzart and Matignon, France). The alcohols incorporated as cosurfactants were the C_2 to C_7 straight chain alkanols, (ethanol, 1-propanol, 1-butanol, 1-pentanol, 1-hexanol and 1-heptanol), the C_3 isomer 2-propanol, and the seven isomers of 1-pentanol (primary isomers 3-methyl-1-butanol, 2-methyl-1-butanol and 2,2-dimethyl-1-propanol; secondary isomers 2-pentanol, 3-pentanol and 3-methyl-2-butanol; tertiary isomer 2-methyl-2-butanol). All these compounds were used as supplied (Prolabo, France, or Fluka A.G.) with labeled grades higher than 98% (except for 3-methyl-2-butanol, 97%). Table I gives the solubilities in water of the eight isomeric pentanols, as determined at 25°C by Ginnings and Baum[80].

Table I. Solubilities in Water of the Eight Isomeric Pentanols at 25°C (From Ginnings and Baum[80]).

Alcohol	$s_{a/w}$	Alcohol	$s_{a/w}$
1-pentanol	2.19	2-pentanol	4.46
3-methyl-1-butanol	2.67	3-pentanol	5.15
2-methyl-1-butanol	2.97	3-methyl-2-butanol	5.55
2,2-dimethyl-1-propanol	3.50	2-methyl-2-butanol	11.00

($s_{a/w}$ is expressed in weight percent of alcohol in solution).

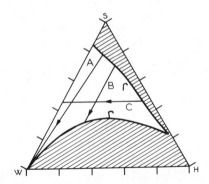

Figure 1. Mass pseudo-ternary phase diagram of the quaternary
system water/sodium dodecylsulfate/2-methyl-2-butanol/benzene.
Weight ratio of surfactant to alcohol: k = 1/2. T = 25°C. W:100%
water; H:100% benzene; S:100% surfactant/alcohol combination. For
A, B, and C, see text.

The samples were prepared by stirring together the required amounts
of the different constituents whose weight fractions were predetermined to within 0.002. Irrespective of the alcohol used, the weight
ratio k of surfactant to alcohol was kept equal to 1/2. Consequently
the configuration of the Winsor IV domains (i.e., the composition
ranges over which stable, monophasic, fluid, transparent and isotropic media are formed) was investigated using the mass pseudo-
ternary phase diagram representation, as illustrated in Figure 1.
It has been stressed by Friberg[81] that this representation is "a
very practical tool for understanding the association phenomena of
importance to microemulsions". In such a diagram, W designates the
100% water apex, H the 100% hydrocarbon (benzene) apex, and S the
apex corresponding to a theoretical 100% concentration of the surface active agent mixture, (surfactant and alcohol combined with
a weight ratio k equal to 1/2). The SW side of the phase diagram
represents ternary systems involving only water, alcohol and surfactant, no hydrocarbon being present. Similarly, the HS side
corresponds to ternary systems involving only hydrocarbon, alcohol
and surfactant, with no added water. Owing to the very low mutual

solubility of water and benzene, the WH side can be considered as representing almost exclusively trivial binary,macroscopically diphasic, systems in which a water phase coexists with a hydrocarbon one. Inside the pseudo-ternary phase diagram, every point represents a quaternary system whose overall composition is entirely defined by any pair among the three weight fractions p_w (water); p_h (hydrocarbon); and p_s (surfactant/alcohol combination). In the phase diagram of Figure 1, the blank area represents the Winsor IV domain, delineated by the boundary Γ. As it will be explained further on,the Winsor IV domain may consist either of a unique area, as shown in Figure 1, or of two disjointed areas. The boundaries of the Winsor IV areas were determined at 25°C by visual observations, by varying system composition and recording the composition points at which turbid-clear or clear-turbid transitions occurred. Samples of the different systems studied were stored and subjected to regular observations over a few weeks to check their stability, especially for composition points close to the boundary Γ of the Winsor IV domain.

Step-by-step low frequency conductivity measurements were performed over the Winsor IV domains along experimental paths such as A, B, or C in Figure 1. Thus, the conductivity variations with water content were recorded for sample series defined by either a constant value of r, the weight ratio of benzene to surfactant/ alcohol combination (A, r-type paths); or a constant value of p_h, the weight fraction of benzene (B, p_h-type paths); or a constant value of p_s, the surfactant/alcohol combination mass fraction (C, p_s-type paths). These procedures allow monitoring of trends in conductive behavior as the system composition varies all over the Winsor IV domain. The low frequency conductivity σ was measured to within 1% by means of a semi-automatic precision Wayne-Kerr B331 conductance bridge working at the angular frequency $\omega = 10^4$ rad. s^{-1}. The cells used were of the Mullard type (Philips) with silver plate electrodes. They were immersed in a thermostatically controlled bath and all measurements being taken at 25.0 ± 0.1°C.

PHASE DIAGRAMS

The diagrams in Figure 2 show the configuration of the Winsor IV domains determined for systems incorporating any of the straight chain alkanols with carbon number n_a ranging from 2 to 7 as cosurfactants. For the sake of comparison with the 1-propanol, the diagram obtained with 2-propanol is displayed. It is readily seen that the alkanol chain length influences strikingly the features of the Winsor IV domain. Up to the threshold value $n_c = 4$ (1-butanol) of the carbon number n_a, the Winsor IV domain consists of a unique area that has the shape of a curvilinear triangle leaning against a large portion of the SW side of the phase diagram. The systems for which this phenomenon is observed will be referred

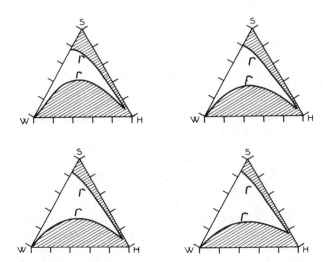

Figure 2.   Mass pseudo-ternary phase diagrams of quaternary systems
water/sodium dodecylsulfate/alkanols/benzene. Weight ratio of
surfactant to alcohol: k = 1/2. T = 25°C. The blank areas are
the Winsor IV domains.

    (a)   ethanol   (b) 2-propanol
    (c)   1-propanol   (d) 1-butanol

to as Type U systems. For n_a greater than the threshold value n_c,
the Winsor IV domain is split into two disjointed areas that are
separated by a composition zone over which viscous, turbid and
birefringent media are encountered. This second class of systems
will be referred to as Type S systems. Moreover, it can be observed
that the Winsor IV domain is the largest for $n_a = n_c$. Below n_c,
the unique monophasic area shrinks as n_a decreases. Above n_c the
reunion of the two disjointed monophasic areas is, in all cases,

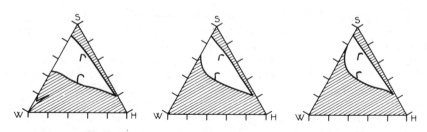

Figure 2 (continued).   (e) 1-pentanol   (f)1-hexanol   (g)1-heptanol.

smaller than the unique area obtained at n_c (i.e. with 1-butanol).
In addition, each of both of the monophasic areas becomes individually smaller as n_a increases. In particular the monophasic area
located near the W apex of the phase diagram vanishes drastically
upon substituting 1-hexanol for 1-pentanol and is eventually reduced to a tiny spot in the 1-heptanol case.

    All these findings are analogous to previous ones obtained on
several quaternary systems incorporating either a different surfactant (potassium oleate) or a different hydrocarbon (1-dodecane)
or both (potassium oleate, n-hexadecane)[64,65,76]. In each case,
it was possible as well to show a straight chain alkanol critical
carbon number n_c marking a sharp transition from Type U to Type S
systems. With systems involving 1-dodecane and sodium dodecylsulfate, n_c was found to be 5 (1-pentanol)[76]. In agreement with a
suggestion made by Bansal et al.[74], this result can be tentatively
expressed in terms of molecular length compatibility, since the
molecular length of 1-pentanol (~7Å) is the one which when added
to that of 1-dodecane (~14Å) fits best the length of the extended
sodium dodecylsulfate molecule (~21Å). Although this molecular
length compatibility concept cannot be transposed straightforwardly to the present case where an aromatic compound is involved, it

remains that 1-butanol appears to be the straight chain alkanol
that optimizes the composition range over which Winsor IV media
can be formed with the formulation retained (benzene as the hydrocarbon component, sodium dodecylsulfate as the surfactant, and
weight ratio of surfactant to alcohol equal to 1/2).

In the case of Type S systems (Figures 2e to 2g) it can be
taken for granted that the small monophasic area located in the
vicinity of the W apex represents the realm of existence of water-
external media that are most probably o/w microemulsions of the
Schulman's type. This assertion is based on the fact that, over
this area, water is by far the major component, with a weight
fraction in the range of 0.8 - 1; whereas the mass fraction of
benzene does not exceed a maximum value of 0.1 or so in the best
case, i.e., for 1-pentanol. The situation appears much less
straightforward as concerns the monophasic area located in the
upper part of the phase diagram. This area spans over a fairly
broad range of compositions, which implies most probably structural
diversity. Inside the corner directed towards the H apex of the
phase diagram, the benzene is the major component, and it can be
reasonably assumed that the media are of the oil-external type.
In the vicinity of the SW side of the phase diagram, the media
contain only small amounts of benzene and are made up essentially
of water ($0.2 < p_W < 0.5$ or so), surfactant and alcohol. They
appear so to be very akin to the ternary water-surfactant-alcohol
Winsor IV type media whose realm of existence is represented by
the "monophasic" portion of SW. According to Friberg and co-
workers[60,81,82,83], such ternary monophasic compounds are solutions
of spherical inverted micelles in alcohol and the quaternary systems issued from them upon addition of a hydrocarbon retain the
same structure. On this basis, it may be considered that, in the
vicinity of SW, the media are of the oil-external type as well,
the suspending medium consisting of an alcohol-hydrocarbon mixture. However, it must be pointed out that, at the lowest water
contents, i.e., in the corner formed by SW and the upper branch of
the boundary Γ of the Winsor IV area, the systems are poor in both
water and benzene, the major component then being the surfactant/
alcohol combination. For such special compositions, it can be
questioned whether the medium can still retain a microscopic
spherical geometry. Results reported by Bellocq et al.[56] tend to
indicate that, in this small surface active agent rich region, the
structure could be of the lamellar type. In the vicinity of the
upper branch of the boundary Γ, owing to the low water-to-surfactant ratios, the medium could be, at least partially, made up of
pre-micellar entities consisting of surfactant aggregates with
highly bonded water molecules. This assumption is strongly supported by results obtained by different groups on various systems
containing ionic surfactants[57-67]. In contrast, in the vicinity
of the lower branch of the boundary Γ, the water-to-surfactant
ratios are medium or high ones. As indicated by structural data

obtained by means of ultracentrifugation, small angle neutron
scattering and light scattering techniques[39-44], the media consist
of populations of aqueous composite globules (inverted swollen
micelles) and can be properly labeled w/o type microemulsions, as
Schulman and his co-workers did. However, it must be pointed out
that this picture is based mainly on structure studies performed
on dilute systems obtained from initial saturated Winsor IV phases
whose compositions are represented by points lying on Γ. It is
most probable that, owing to aggregation phenomena, clusters of
micelles are present in the initial saturated Winsor IV phases[41,
56,62-64,84,85]. Finally over the central region of the upper
Winsor IV area, it can be reasonably assumed that the w/o composite
globule model depicts correctly the medium structure, since the
water-to-surfactant ratio is sufficiently high and the water-to-
hydrocarbon ratio sufficiently low to allow the formation of discrete water pools. So, with the support of available structural
data, this qualitative analysis points out the structural diversity
existing within the upper Winsor IV area that, nevertheless, can be
considered, on the whole, as the realm of existence of oil-external
(w/o) media. As illustrated by the diagrams in Figures 2e to 2g,
the lower o/w Winsor IV area is separated from the upper w/o one by
a fairly broad composition region over which, as mentioned previously, viscous, turbid, sometimes birefringent, and unstable media are
formed. This means that, in conformity with the first of the two
alternative predictions of Winsor's intermicellar equilibrium
model[86], the w/o to o/w transformation or vice-versa (phase inversion) is driven by a mechanism characterized by the build-up of
long range ordered phases, which entails modifications of the structure of the medium at both the microscopic and macroscopic levels.
On the basis of results obtained using various methods on the
quaternary system water/potassium oleate/1-hexanol/n-hexadecane,
Shah and co-workers[87-89] proposed the following sequence of structural changes: fluid, transparent, isotropic dispersions of water
spheres in oil - highly viscous, turbid, birefringent dispersions
of water cylinders in oil - highly viscous, turbid birefringent
lamellar structures of interspersed layers of water, surface active
agents and oil - fluid transparent isotropic dispersions of oil
spheres in water. Similar conclusions were arrived at by Clausse
et al.[90] who showed that a divergent behavior of both the low frequency condictivity and permittivity is associated with the first
transparent/turbid transition. So, in the case of Type S systems,
the phenomenon of phase inversion in microemulsions can be easily
detected through the striking alterations that affect suddenly and
simultaneously the mechanical, optical and electrical behavior of
the medium, when the latter is progressively enriched with added
water, for instance along r-type composition paths (see Figure 1).

   In the case of Type U systems, whose phase diagram displays a
unique monophasic domain (Figures 2a to 2d) the same qualitative
analysis as above can be carried out as concerns structural

diversity, which leads to similar conclusions, consistent with experimental data obtained by several authors on systems of the same kind[56,91]. But there is an additional complication since the phase diagrams of these Type U systems do not exhibit any "solubilization gap" composition zone separating the w/o and o/w areas, unlike what is observed for Type S systems. Upon increasing the water content along experimental paths of the r-type, the medium remains of the Winsor IV type, i.e., macroscopically monophasic, stable, fluid, transparent and isotropic; although it can be taken for granted that it undergoes phase inversion in a certain composition range. The phase inversion phenomenon appears to be in that case a diffuse and progressive process whose mechanism does not imply the build-up of long range ordered structures inducing drastic and macroscopically detectable changes in the mechanical, optical and electrical properties of the medium. This observation is consistent with the second term of Winsor's intermicellar equilibrium model alternative that predicts the possibility of a "continuous passage" from Winsor IV phases of the S_1 type (o/w) to Winsor IV phases of the S_2 type (w/o)[86]. Winsor suggested that, over the composition region corresponding to the "continuous passage", the medium could consist of coexisting w/o and o/w micelles. Although this situation can be predicted also from the geometrical model analyzed by Biais et al.[55], it is most probable that it can be rejected as unrealistic because the suspending medium should exhibit contradictory properties to accommodate simultaneously the presence of direct and inverted micelles. Other suggestions may be made that seem more suitable, the medium undergoing phase inversion being considered either as a cosolubilized system[72] or as a dynamic equilibrium bicontinuous phase[92-96]. Whichever model is retained to depict the medium structure during the phase inversion process, the exact location and delineation of the phase inversion region in the phase diagram is a problem which is still unsolved. Investigations performed in this regard by several groups have provided no definite answer, while indicating nevertheless the absence of well defined structures at intermediate compositions[56,68-70,91,97,98].

The phase diagrams in Figures 2b and 2c show that the substitution of 2-propanol for 1-propanol has only little effect on the configuration of the Winsor IV domain. In contrast, as illustrated by the set of phase diagrams in Figure 3, drastic alterations are observed when using by turns the eight isomeric pentanols. As already mentioned, with 1-pentanol, (Figure 3h), the system is of the S type, its monophasic domain being split into two disjointed areas. In contrast, with the 1-pentanol tertiary isomer 2-methyl-2-butanol, (Figure 3a), the system is of the U type, its monophasic domain consisting of a unique triangular area leaning against a large portion of the SW side of the phase diagram. With the secondary isomers 3-methyl-2-butanol, 3-pentanol and 2-pentanol, (Figures 3b to 3d), the monophasic domain consists also of a unique

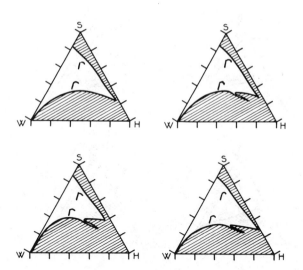

Figure 3. Mass pseudo-ternary phase diagrams of quaternary systems water/sodium dodecylsulfate/isomeric pentanols/benzene. Weight ratio of surfactant to alcohol: k = 1/2. T = 25°C. The blank areas are the Winsor IV domains.

    (a)   2-methyl-2-butanol   (b)   3-methyl-2-butanol
    (c)   3-pentanol   (d)   2-pentanol

area but exhibits an indentation. The same phenomenon is even more evident in the case of the primary isomers 2-methyl-1-butanol, (Figure 3f), and 3-methyl-1-butanol (Figure 3g). These systems, whose supposed w/o and o/w regions are connected by a monophasic channel and form a contorted unique Winsor IV area, will be referred to as Type U* systems. A very special situation is that illustrated by Figure 3e, which corresponds to the 1-pentanol primary isomer 2,2-dimethyl-1-propanol, (Type S* systems). This system is of the S-type, but its supposed w/o area exhibits an indentation similar to the one existing in the case of Type U systems. Several interesting conclusions can be derived from the analysis of the phase diagrams of Figure 3. By substituting an isomeric pentanol for another one, starting with 2-methyl-2-butanol and ending with 1-pentanol, a progressive transition from Type U

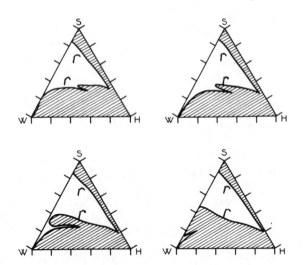

Figure 3 (continued).
        (e)   2,2-dimethyl-1-propanol   (f)   2-methyl-1-butanol
        (g)   3-methyl-1-butanol   (g)   1-pentanol

systems to Type S ones is evidenced, in contrast with the sharp
transition observed upon substituting a longer straight alkanol
for a shorter one, (Figure 2). This progressive transition is
characterized by a progressive moderate shrinking of the Winsor
IV domain surface, which shrinking follows the decrease of the
solubility in water of the isomeric pentanols (see Table I) with
the exception of 3-pentanol that displays a slightly anomalous
behavior. The incorporation of the least water-soluble isomer
1-pentanol gives the least extended Winsor IV domain (split into
two separated areas, Type S), and that of the most water-soluble
isomer 2-methyl-2-butanol the most extended Winsor IV domain
(forming a unique area, Type U) that is though smaller than the
one obtained with 1-butanol. The most significant issue of this
phase diagram study of systems containing any of the eight isomeric
pentanols is the attainment of Type U* systems for which the w/o
and o/w areas are connected through a channel that widens when the
alcohol solubility in water increases (Figures 3b to 3d and 3f to
3g). As illustrated more particularly by Figure 3g, which is the

most demonstrative one, depending upon the value of r, (hydrocarbon to surfactant/alcohol combination weight ratio), the w/o
to o/w transformation (phase inversion) may occur either through
the build-up of long range organized structures, as in the case of
Type S systems (higher values of r) or in a diffuse way that does
not alter at the macroscopic level the Winsor IV character of the
medium, as in the case of Type U systems (lower values of r). So,
for these peculiar systems, both terms of Winsor's intermicellar
equilibrium model, i.e., "gel stage" or "continuous passage", must
be retained to characterize the phase inversion process. Since
the channel and the non-Winsor IV protuberance are adjacent composition regions, it may be considered, for proximity and continuity reasons, that the channel is part of the composition zone over
which the diffuse phase inversion phenomenon takes place, within
the Winsor IV domain; whereas the non-Winsor IV protuberance defines
the composition region over which the phase inversion phenomenon
occurs through the formation of long range ordered media.

From the preceding developments, it appears clearly that the
configuration of the Winsor IV domain of the quaternary systems
water/sodium dodecylsulfate/alkanol/benzene, (surfactant/alcohol
weight ratio equal to 1/2) is strongly influenced by the chemical
structure of the alcohol used as the cosurfactant, which allows to
define two main types of systems depending on whether their Winsor
IV domain consists of two disjointed areas (Type S) or of a unique
area (Type U and U*). In the following it is shown that the electrical properties of Winsor IV media are strikingly influenced as
well by the alcohol chemical structure, and that each type of
systems is characterized by a typical electrical conductive behavior.

CONDUCTIVITY MEASUREMENTS

General Remarks

Figure 4 shows the comparative plots of the conductivity
variations with Φ_w, the water volume fraction, as recorded along
the same p_s-type path (p_s = 0.45), for the different normal alkanols used. It is readily seen that the conductive behavior is
strongly influenced by the nature of the alcohol, other things
being equal. For the shorter ethanol, 1-propanol and 1-butanol,
that give Type U systems, the conductivity rapidly takes high
values as the water content increases. Its variations are characterized by an initial smooth non-linear increase, followed by
a steep linear increase whose onset occurs at lower values of Φ_w,
as the alcohol is the shorter (Figure 4). In contrast, with the
longer 1-pentanol, 1-hexanol and 1-heptanol, that give Type S
systems, the conductivity does not attain high values and its

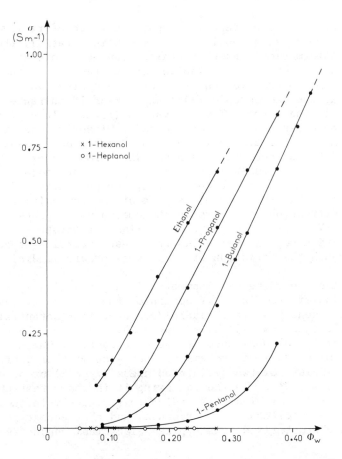

Figure 4. Variations of the electrical conductivity σ (in Sm^{-1}) versus Φ_w, the water volume fraction, as recorded at T = 25°C along a p_s-type composition path (p_s = 0.45). Influence of the alcohol chemical structure.

variations with Φ_w are smooth, as illustrated by the 1-pentanol curve; even, in the case of 1-hexanol and 1-heptanol, very flat variations are observed compared to those recorded for the shorter alkanols. Similar observations were made for other values of p_s. These results are consistent with the previous data obtained on systems containing different surfactants and hydrocarbons[63-65,72,76,99].

The discrepancies existing between the electrical conductive behavior of Type U Winsor IV media and that of Type S ones can be appreciated as well by comparing the conductivity plots of Figures 5 and 10. These plots show the conductivity variations as the system composition varies along p_h-type or r-type experimental paths.

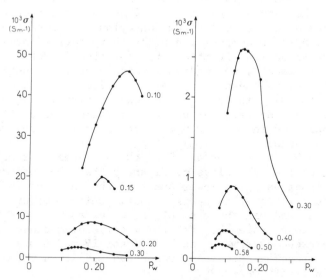

Figure 5.   Variations of the electrical conductivity σ (in Sm^{-1}) versus p_w (water weight fraction) as recorded at T = 25°C over the w/o Winsor IV area of the system water/sodium dodecylsulfate/ 1-hexanol/benzene, along different p_h-type compositions paths.

In the case of Type S systems (Figure 5), all over the Winsor IV w/o area, the conductivity has low values (smaller than $10^{-2}Sm^{-1}$) and undergoes non-monotonous variation as the water content increases up to the value corresponding to the clear-to-turbid transition. On the contrary, in the case of Type U systems (Figure 10), over the same water content range (p_w < 0.35 or so), the conductivity takes high values (10^2 times greater than for Type S systems) and displays a steep linear increase. As illustrated by the conductivity plots of Figures 7 and 8, the same behavior is observed in the case of Type U* systems.

It thus appears clearly that irrespective of the type of the composition path followed, the conductive behaviors for Type S systems, on the one hand, and for Type U and U* systems, on the other hand, are strikingly dissimilar, which suggests the existence of structural discrepancies between the Winsor IV media issued from the two kinds of systems. In the following, the electrical conductive behavior will be analyzed for the different types of systems.

Type S Systems

Figure 5 shows the variation of the electrical conductivity σ versus p_w, the water weight fraction, as recorded over the w/o Winsor IV area of the water/sodium dodecylsulfate/1-hexanol/benzene system, along p_h-type composition paths characterized by values of p_h, the hydrocarbon mass fraction. ranging from 0.10 to 0.58. As pointed out previously, the medium remains poorly conducting all over its realm of existence, σ retaining values lower than 0.05 Sm^{-1}. Whatever the value of p_w, the conductivity is the higher as p_h is lower. For any value of p_h, the variations of σ with p_w are non-monotonous. On increasing p_w, σ begins to increase, reaches a maximum for a certain value $(p_w)_c$ of p_w, and then decreases until the lower branch of the boundary Γ of the w/o area is reached. It is readily seen from the plots of Figure 5 that the conductivity maximum is higher as p_h is lower and that the value of $(p_w)_c$ at which it occurs is lower as p_h is higher. Quite similar observations have been made in the case of the water/sodium dodecylsulfate/1-heptanol/benzene system[64]. For the water/sodium dodecylsulfate/1-pentanol benzene system, whose w/o Winsor IV media are more conductive, the phenomena are less pronounced and it is more difficult to show conductivity maxima[64].

In Table II, the values of $(p_w)_c$ are reported versus those of p_h. From these values, it is possible to determine in the phase diagram the composition points for which the conductivity maxima are observed. It appears that these points are not distributed at random but define within the Winsor IV area a curve labeled Γ_1 in Figure 6. At higher values of p_h, Γ_1 displays a linear portion that is directed towards the H apex of the phase diagram. This feature indicates that, along this linear portion, the water to surface active agent combination weight ratio remains constant, which implies that the water to surfactant ratio remains constant as well. This is clearly shown by the values given in the fourth column of Table II showing that, for values of p_h ranging from 0.30 to 0.58, the water to sodium dodecylsulfate molar ratio w_o remains equal to 12 or so. At lower values of p_h, Γ_1 exhibits a curvature which is reflected in Table II by the increase of w_o as p_h increases. The same analysis is valid in the case of the water/sodium dodecylsulfate/1-heptanol/benzene system, for wich a constant w_o value of 11 was found along the linear portion of the curve Γ_1[64].

All these results are consistent with previous findings reported on similar Type S systems[62-66] and with data reported in the literature by several groups who used different techniques[14-17, 57-61,67].

Table II. Determination of the Line Γ_1 (see Text) and Evaluation of the Micellar Radius Values along Γ_1, for the Water/Sodium Dodecylsulfate/1-Hexanol/Benzene System, at T = 25°C.

P_h	$(P_w)_c$	P_{SDS}	w_o*	r_w(Å)	r_m(Å)**
0.10	0.30	0.20	23	103	124
0.15	0.22	0.21	17	76	97
0.20	0.19	0.20	15	67	88
0.30	0.14	0.19	12	54	75
0.40	0.12	0.16	12	54	75
0.50	0.10	0.14	11	49	70
0.58	0.08	0.11	12	54	75

* w_o: water/sodium dodecylsulfate molar ratio.

** $r_m = r_w + \ell_s$, with $\ell_s = 21$ Å length of the extended sodium dodecylsulfate molecule.

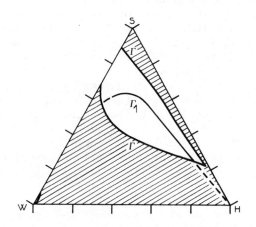

Figure 6. Partition of the w/o Winsor IV area of the water/sodium dodecylsulfate/1-hexanol/benzene system.

The fact that the conductivity increases up to a maximum can be correctly depicted using a simple model proposed by Eicke and Denss[100]. These authors explained experimental conductivity data they obtained for the ternary system water/Aerosol OT/benzene, from which w/o Winsor IV phases can be formed without addition of a cosurfactant, by means of the following model for electrical charge production and exchange. Two processes were considered. The first one is dissociation of the AOT molecule inside the disperse micellar entities which can be assumed to be independent of the surrounding medium. The second one is the transfer of the anionic charge thus produced to another micellar aggregate which was originally electroneutral. This latter process was considered to represent a phase transfer between the pseudo-phase made up of all the disperse neutral micellar entities (Phase I) and the pseudo-phase consisting of the dilute solution of charged micellar entities in benzene (Phase II). On this basis, the overall medium conductivity is determined, on the one hand, by two dissociation processes (dissociation of single surfactant molecules in Phase I, and dissociation of ion pairs in Phase II) and, on the other hand, by a viscosity effect linked to the micellar size increase with the addition of water. A good agreement was found between the theoretical conductivity variations with the water to surfactant molar ratio w_o, as derived from the model, and the experimental conductivity variations with w_o. A most remarkable feature was that both theoretical and experimental conductivity plots exhibited a maximum for value of w_o equal to 10 or so. The whole scheme can be transposed, at least qualitatively, because the surface active agent is not a single surfactant but an ionic surfactant/ alcohol combination, to the present case where the conductivity versus water content plots exhibit maxima corresponding to a value of w_o of 12 or so, as illustrated in Figure 6 and Table II. In this connection, it is important to mention here that it has been reported recently[64-66] that w/o Winsor IV media of the Type S ternary system water/AOT/dodecane exhibit a conductive behavior of the same kind as that of Type S quaternary w/o Winsor media containing either sodium dodecylsulfate or potassium oleate as the surfactant, the conductivity maxima occurring at a water/AOT ratio, w_o, equal to 11 or so. This suggests that the conductivity mechanism model proposed by Eicke and Denss[100] is quite general and can be applied to depict the conductive behavior of Type S w/o Winsor IV media, irrespective of the chemical nature of the ionic surfactant used.

The existence of a threshold value of w_o around 10 has been reported in the literature by several groups who considered it as marking the onset of the formation of particles with a central core of normal "free" water, irrespective of how they may be labeled, swollen inverted micelles, microdroplets, etc... For instance, Sjöblom and Friberg[60] found, from correlated density, light scattering and electron microscopy experiments carried out on

quaternary systems involving potassium oleate as the surfactant,
1-pentanol as the cosurfactant and either benzene, decane or phenyl-
dodecane as the hydrocarbon, a value of the water/surfactant molar
ratio w_0 of 8-10, below which there was no evidence for the
existence of swollen inverted micelles. It was concluded by these
authors that, at water/surfactant molar ratios smaller than 8-10,
the systems consisted of dispersions of surfactant aggregates with
highly bonded water molecules. A Rayleigh scattering study·
performed by Bellocq and Fourche[61] on the two quaternary systems
water/potassium oleate/1-hexanol/dodecane and water/sodium dodecyl-
sulfate/1-butanol/toluene yielded as well for w_0 a value of 8-10
at which a micellization process appeared to take place. It is
interesting to note that Bellocq and Fourche[61] determined from
their experiments, for both of the systems investigated, a "micellization line" quite comparable to the linear portion of the lines
Γ_1 shown during the present study. Threshold values of w_0 around
10 were also determined in the case of w/o Winsor IV media containing Aerosol OT as the unique surface active agent: 10 by Eicke and
co-workers[14-17], for systems involving isooctane; 8-10 by Rouviere
et al.[45] and Wong et al.[58] for systems involving heptane; 12 by
Wong et al.[59] for systems involving heptane or dodecane.

All these data converge to indicate that, for systems incorporating ionic surfactants used alone or in combination with an
alcohol, the low water concentration region does not correspond to
actual inverted swollen micelles or microdroplets. In light of
this conclusion, it appears that the line Γ_1 shown from conductivity
studies represents a structural transition locus separating, within
the w/o Winsor IV area, a region of pre-micellar entities from a
region corresponding to the existence of w/o microemulsions of the
Schulman's type (Figure 6). The existence of line Γ_1 is a very
general feature which has been shown for many Type S systems[64,65].
It may exist alone or associated with a second line Γ_2, detectable
at higher water/surfactant ratios from conductivity local minima.
Γ_2 could be considered as marking the onset of a micelle clustering process [62,63]. By assuming that along Γ_1, the system compositions correspond to monodisperse populations of swollen w/o spherical micelles, all the surfactant molecules being engaged in the
interfacial shells surrounding the micelles, it is possible to
evaluate the micelle water core radius r_w from the following
equation

$$r_w = 3 \; \frac{M}{N\rho\bar{\sigma}} \; w_0. \qquad\qquad (1)$$

where M and ρ are, respectively, the molar and volumic mass of
water, w_0 the water/surfactant molar ratio and $\bar{\sigma}$ the mean area per
surfactant polar head at the interface, N being Avogadro's number.
In the fifth column of Table II are reported the values of r_w,

computed from Equation (1) by taking for $\bar{\sigma}$ the minimum value $20\overset{\circ}{A}^2$ reported by Pethica and Fen[101] for sodium dodecylsulfate in planar layers, and the values of r_m, deduced from those of r_w by adding $21\overset{\circ}{A}$, the length of the extended sodium dodecylsulfate molecule. The values of r_m thus obtained are compatible with data reported by other authors for similar systems. It is to be noted that the values of r_w and r_m reported in Table II are maximum ones because of the choice made for $\bar{\sigma}$. If, instead of $20\overset{\circ}{A}^2$, the value $25\overset{\circ}{A}^2$ reported by Graciaa[102] is taken, the figures for r_w and r_m would be smaller by 20%, which would be a more realistic evaluation. It appears from Table II that, as p_h decreases, the constancy of r_w and r_m is no longer observed and that both radii increase drastically along the bent part of Γ_1. This can be interpreted either straightforwardly, as resulting from a mere increase of the micelle size, owing to the hydrocarbon content decrease in the system, or, more subtly, as reflecting the formation of swollen micelles out of non-spherical structures that exist in the surface active agent rich corner formed by the SW side of the phase diagram and the upper branch of the boundary Γ of the Winsor IV domain. This second interpretation would be consistent with both theoretical considerations[55] and experimental data[56] (already commented upon in the preceding section) that tend to indicate that the microscopic geometrical structure of surface active agent rich and both water and hydrocarbon poor media is probably not spherical rather is of the lamellar type.

Type U* Systems

Figure 7 shows the conductivity variations versus p_w, the water weight fraction, as recorded along an r-type experimental path, $(r = 1/9)$, that crosses the non-Winsor IV protuberance displayed in the phase diagram of the system water/sodium dodecylsulfate/3-methyl-1-butanol/benzene system, (Figure 3g). As already mentioned, contrary to what is observed in the case of Type S systems, the conductivity rapidly takes high values as p_w increases. After an initial non-linear increase, σ increases linearly with p_w. For values of p_w greater than p_w^b, (see Figure 7), this linear regime is followed by a non-linear one until the boundary Γ of the non-Winsor IV protuberance is reached. Beyond the protuberance, over the water-rich region of the Winsor IV domain, the conductivity decreases regularly and non-linearly from a maximum, until the boundary Γ is reached again. The composition point corresponding to the p_w^m value of p_w at which the conductivity maximum takes place is located in the close vicinity of the boundary Γ. Over the p_w range corresponding to the non-Winsor protuberance, no reliable conductivity measurements could be taken, owing to conductivity drifts with time. Identical observations were made for higher values of r. So it appears that on both sides of the non-Winsor IV protuberance the conductivity variation

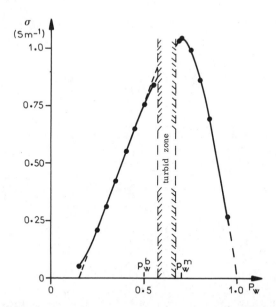

Figure 7. Variations of the electrical conductivity σ, (in Sm^{-1}),
versus p_w, the water weight fraction, as recorded at T = 25°C over
the Winsor IV domain of the system water/sodium dodecylsulfate/
3-methyl-1-butanol/benzene, along a composition path corresponding
to r = 1/9.

trends are quite dissimilar. Below the protuberance, i.e. over
the water-rich region of the Winsor IV domain, the conductivity
decrease with p_w can be ascribed to the progressive dilution of
the external aqueous phase of o/w microemulsion type media upon
addition of water. The interpretation of the conductivity variations above the non-Winsor IV protuberance, i.e. over the oil-rich
region of the Winsor IV domain, is less straightforward. As it
will be explained with greater details later, the conductivity
steep linear increase with p_w can be considered as reflecting a
percolative conduction phenomenon whose exact mechanism at the
molecular level is still to be understood clearly.

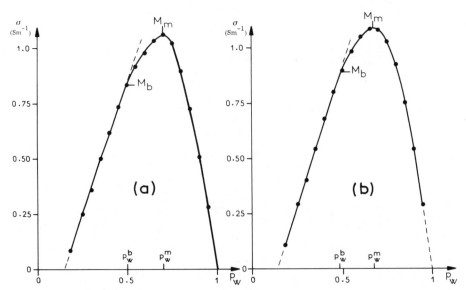

Figure 8. Variations of the electrical conductivity σ, (in Sm^{-1})
versus p_w, the water mass fraction, as recorded at T = 25°C over
the Winsor IV domain of the system water/sodium dodecylsulfate/
3-methyl-1-butanol/benzene, along the composition paths corresponding to low values of r.

        (a)   r = 0   (b)   r = 1/19

     At lower values of r (r < 0.08 or so) the composition paths
do not cross the non-Winsor IV protuberance and are stretched
entirely across the Winsor IV domain, as they go from the oil-rich
to the water-rich one through the channel that connects the two
regions (Figure 3g). Figure 8b shows a conductivity curve obtained for such a composition path, characterized by r = 1/19.
This curve is a continuous one that does not present any gap,
since, in this situation, the medium remains of the Winsor IV
type along the entire composition path. But it shares common
features with the curves obtained at higher values of r. In
particular, it exhibits a linear ascending branch, as p_w increases
up to p_w^b, and a non-linear descending branch as p_w increases
beyond p_w^m, that corresponds to the conductivity maximum, up to 1.
As explained in the preceding paragraph, the descending branch
depicts the medium conductivity decrease, (in this case down to

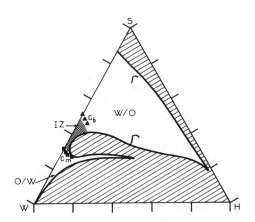

Figure 9.   Partition of the Winsor IV domain of the water/sodium dodecylsulfate/3-methyl-1-butanol/benzene system. C_b and C_m are composition lines corresponding, respectively, to the linear/non-linear transition and to the maximum appearing on conductivity plots such as those of Figure 8.
W/O: oil-external region; IZ: phase inversion zone; O/W: water-external region.

the value of the conductivity of the water used, that can be considered as equal to zero compared, for instance, with the conductivity maximum value) as it results from the dilution of the external phase of o/w microemulsion type media with added water. On the other hand, the ascending linear branch can be considered as characteristic of a percolative conduction phenomenon. Over the p_w range delimited by p_w^b and p_w^m, the two branches are connected by a non-linear portion that depicts a moderate non-linear increase of the conductivity σ with p_w. It is readily seen from Figure 8a that the same observations apply when $r = 0$, i.e. along the SW side of the phase diagram.

In the phase diagram, the composition points determined from the different values of p_w^b and p_w^m define two curves C_b and C_m that, inside the Winsor IV domain, appear as continuations of the side branches of the portion of the boundary Γ that delimits the non-Winsor IV protuberance, (Figure 9). The composition zone delimited by C_b and C_m,(dotted area in Figure 9), appears to be a continuation of the non-Winsor IV protuberance and corresponds exactly to the channel that connects the oil-rich and water-rich

Winsor IV regions. On the basis of the observations made on Type S
systems, the non-Winsor IV protuberance can be regarded as a
composition zone over which the w/o ↔ o/w phase inversion takes
place by involving the build-up of long range organized media.
Consequently, it can be argued that the dotted area, labeled IZ
in Figure 9, is the composition zone over which, within the Winsor
IV domain, the w/o ↔ o/w transformation occurs in a progressive
and diffuse way, as the medium undergoing the phase inversion process retains its Winsor IV characteristics, i.e. stability, fluidity, transparency and isotropy. This IZ zone separates the w/o
and o/w regions in the same way as the non-Winsor IV protuberance
does, the difference being in the nature of the phase inversion
process that affects differently the medium structure depending
on whether the value of the hydrocarbon to surface active agent
combination weight ratio r is low (diffuse process) or high (process involving long range ordered media).

All the above findings and conclusions appear to bear some
generality since phenomena quite similar to those described in
the case of the water/sodium dodecylsulfate/3-methyl-1-butanol/
benzene system have been shown for other Type U* systems, incorporating a different alcohol (water/sodium dodecylsulfate/2-methyl-
1-butanol/benzene[64]) or both different alcohol and surfactant
(water/potassium oleate/3-methyl-2-butanol/benzene[64,99]).

Type U Systems

Figure 10 shows the variations of the electrical conductivity
σ as the water content increases along an r-type composition path
(r = 1/9). Similar plots were obtained for other values of r,
especially r = 0, which corresponds to the SW side of the phase
diagram. As in the case of Type U* systems, the conductivity
rapidly takes fairly high values as Φ_w increases over the oil-rich
region of the Winsor IV domain. By comparing Figures 10 and 8, it
is readily seen that the plots obtained in the Type U case for any
value of r are quite similar to those recorded in the Type U* case
at low values of r. In both cases, the plots can be segmented into
four parts representing respectively, as the water content increases, an initial smooth non-linear increase, a steep linear
increase, a subsequent smooth non-linear increase up to a maximum,
and eventually a non-linear decrease of the conductivity σ.

Because of this conductive behavior similarity, the procedure
used for Type U* systems may be applied to the present case. By
plotting in the phase diagram the composition points corresponding,
on the conductivity plots, to the linear/non-linear transition
(point M_b on the plot of Figure 10b) and to the maximum (point M_m)
it is possible to define two curves, C_b and C_m, that can be considered as internal boundaries of the Winsor IV domain, as illus-

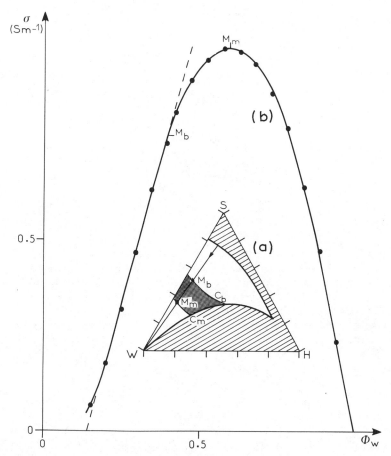

Figure 10. Variations of the electrical conductivity σ (in Sm^{-1}) versus Φ_w, the water volume fraction, as recorded at $T = 25°C$ over the Winsor IV domain of the system water/sodium dodecylsulfate/2-methyl-2-butanol/benzene, along an r-type composition path $(r = 1/9)$.

(a) Composition path followed in the phase diagram.
(b) Conductivity plot.

trated in Figure 11. By analogy with and an extension of the results obtained in the case of Type U* systems, it may be advanced tentatively that, within the Winsor IV domain, C_b and C_m separate the W/O and O/W type regions from an intermediary composition zone IZ over which the w/o to o/w transformation (phase inversion) occurs in a diffuse and progressive way that does not alter at the macroscopic level the Winsor IV character of the medium. Additional conductivity measurements carried out along P_s-type composition paths support this conclusion. Figure 12b

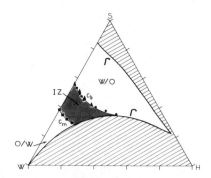

Figure 11. Partition of the Winsor IV domain of the water/sodium dodecylsulfate/2-methyl-2-butanol/benzene system. C_b and C_m are composition lines corresponding, respectively, to the linear/non-linear transition and to the maximum appearing in the conductivity plots such as that of Figure 10.
W/O: oil-external region; IZ: phase inversion zone; O/W: water-external region.

shows a conductivity plot recorded along the composition path defined by $p_S = 0.60$ and that, consequently, does not cross the curve C_b determined through r-type experiments (see Figure 12a). As the water content increases, the conductivity begins to increase non-linearly, and after a Φ_w value of 0.18 or so, it undergoes a steep linear increase until the SW side of the phase diagram is reached. Similar results were obtained for p_S-type composition paths with values of p_S higher than 0.57 or so. For values of p_S lower than 0.57, the composition paths cross the curve c_b determined through r-type experiments, as exemplified by Figure 13a. In that case, the p_S-type conductivity plots exhibit in their upper part (for $\Phi_w > \Phi_w^b$) an additional marked curvature that follows the linear part, (Figure 13b). This feature indicates a change in the conductivity variation trend with Φ_w, which may be interpreted as reflecting that modifications of the medium structure have occurred. As shown by the filled circles in Figure 11, the composition points defined from the values Φ_w^b of Φ_w at which the linear/non-linear transition occurs (point M_b' on the conductivity

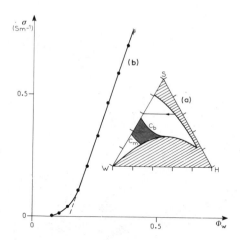

Figure 12:   Variations of the electrical conductivity σ (in Sm^{-1}) versus Φ_w, the water volume fraction, as recorded at T = 25°C over the Winsor IV domain of the system water/sodium dodecylsulfate/ 2-methyl-2-butanol/benzene, along a p_S-type composition path (p_S = 0.60).

    (a)   Composition path followed in the phase diagram.
    (b)   Conductivity plot.

plot of Figure 13b) fall exactly on the curve C_b previously determined through r-type experiments. This finding is quite gratifying and significant since it proves that r-type and p_S- type results are mutually consistent and converge to demonstrate that the curve C_b is a conductivity transition locus that can be considered as an internal boundary of the Winsor IV domain of Type U systems. Quite similar results were obtained for other Type U systems[64, 99] as well.

    So, it appears that the analysis of the electrical conductivity variations with composition allows to define two curves C_b and C_m that partition the Winsor IV domain of both Type U* and Type U systems into three adjacent regions (labeled W/O, IZ and O/W in

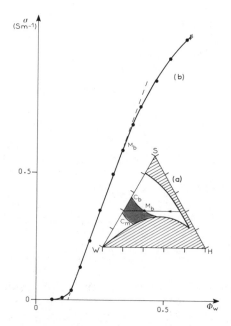

Figure 13. Variations of the electrical conductivity σ (in Sm^{-1}) versus Φ_w, the water volume fraction, as recorded at T = 25°C over the Winsor IV domain of the system water/sodium dodecylsulfate/2-methyl-2-butanol/benzene, along a p_S-type composition path (p_S = 0.40).

   (a) Composition path followed in the phase diagram.
   (b) Conductivity plot.

Figures 9 and 11). The fact that, in the case of Type U* systems, the IZ region appears to be a continuation of the non-Winsor IV protuberance (see Figure 9) suggests strongly that IZ represents the composition zone over which the w/o to o/w transformation (phase inversion) occurs in a progressive and diffuse way, which conclusion can be tentatively extended to the case of Type U systems (see Figure 11), owing to the conductive behavior similarities observed in both situations (see Figures 8 and 10). An additional support to this conclusion can be derived from a phenomenological analysis of the conductive behavior of Type U* and Type

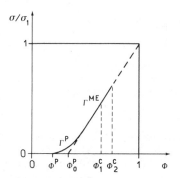

Figure 14.   Variations of the reduced electrical conductivity σ/σ_1 as Φ, the conductor volume fraction, increases in a randomly heterogeneous binary conductor-insulator composite sample (σ: sample conductivity; σ_1: conductor conductivity).

Φ^P: Percolation threshold; Γ^P: conductivity curve near Φ^P, Equation (2); Γ^{EMT}: conductivity curve beyond Φ^P, Equation (3); Φ_0^P: threshold for the EMT, Equation (3), Φ_1^C: critical volume fraction corresponding to the random close packing of identical disperse hard spheres, ($\Phi_1^C = 0.637$); Φ_2^C: critical volume fraction corresponding to the cubic close packing of identical disperse hard spheres, ($\Phi_2^C = 0.741$).

U Winsor IV media over the W/O region.

    Figure 14 shows the theoretical variations of the electrical conductivity of a randomly heterogeneous binary conductor-insulator composite sample, as the conducting component volume fraction increases. This conductive behavior can be described by means of equations derived from the Percolation[103,104] and Effective Medium[105-107] theories. The Percolation theory predicts, in a crude way, that the sample remains non-conducting as long as Φ remains lower than a threshold value Φ^P, called the percolation threshold. Then , in the upper vicinity of Φ^P, the conductivity increase with Φ is described by the following scaling law

$$\sigma \propto (\Phi - \Phi^P)^\gamma \qquad\qquad (2)$$

γ is a critical exponent whose value depends only upon the dimensionality of the system, contrary to ϕ^P whose value depends upon the dimensionality and upon the physical details of the system. For three dimensional systems, Kirkpatrick[104] has proposed the value of γ 1.5-1.6, as extracted from numerical simulations of the Percolation phenomenon. For ϕ^P, the values of direct relevance to the case of microemulsion type systems are those concerning populations of spherical particles. A value of 0.29 has been proposed by Kirkpatrick[104] from a numerical study of a continuum percolation model in which the allowed volumes surrounding percolation sites consist of identical spheres permitted to overlap and whose centers are distributed at random. Close values, ranging from 0.27 to 0.31, have been reported by Janzen[108] for populations of spherical particles with low polydispersity. A value of 0.25 has been suggested by Kirkpatrick[104] for systems in which the conducting and non-conducting regions have, on the average, similar shapes. As ϕ increases further beyond ϕ^P, it has been demonstrated by Kirkpatrick[103,104] that the conductivity variations can be fitted with the following equation

$$\sigma = \frac{3}{2} \sigma_1 \ (\ \phi - \phi_o^P) \qquad\qquad (3)$$

where $\phi_o^P = 1/3$. This equation is a simplified form, obtained by taking $\sigma_2 = 0$, of the Effective Medium Theory (EMT) formula

$$\frac{\sigma - \sigma_2}{3\sigma} = \frac{\sigma_1 - \sigma_2}{\sigma_1 + 2\sigma} \ \phi \qquad\qquad (4)$$

as it was derived by Bruggeman[105] and Bottcher[106], or

$$\frac{\sigma_1 - \sigma}{\sigma_1 + 2\sigma} \phi + \frac{\sigma_2 - \sigma}{\sigma_2 + 2\sigma}(1 - \phi) = 0 \qquad\qquad (5)$$

as it was formulated by Bruggeman[105] and Landauer[107]. It has been shown that several electrical properties of micro-heterogeneous materials can be described correctly by means of these equations[109,110]. The Percolation phenomenon can be pictured qualitatively as follows. When ϕ is lower than ϕ^P, the material consists of isolated conductor particles or clusters of particles and, consequently, behaves like and insulating medium. When ϕ reaches ϕ^P, an "infinite cluster" of conductor particles forms that acts as a conducting path stretched throughout the sample which then becomes conducting. As ϕ increases beyond ϕ^P, more and more infinite clusters form and the medium conductivity increases regularly upon its enrichment with added conductor

component. Two remarks ought to be made here, which are of particular relevance when liquid media such as microemulsions are under
consideration. The first one is that, owing to the existence of
fluctuations, the conductivity is not strictly equal to zero just
below the percolation threshold Φ^P but varies according to the
following scaling law

$$\sigma \propto (\Phi^P - \Phi)^{-s} \qquad (6)$$

This behavior can be predicted from the Phase Transition Theory[111],
with values of s in the range of 0.5 - 0.78[112,113]. The second
remark is that the conductive behavior cannot be depicted by means
of Equation (3) in the higher range of Φ, because of the existence
of limiting values of the disperse phase volume fraction that
correspond to close packing of the disperse particles. For instance, as indicated in Figure 14 in the case of populations of
identical hard spheres, the disperse phase volume fraction cannot
exceed the upper limit of $\Phi_2^C = 0.741$, in the case of a cubic close
packing[114]; and even the limit $\Phi_1^C = 0.637$, in the case of a random close packing[115]. Beyond such a limit, the medium cannot
accommodate any further addition of the disperse phase, unless
its organization be completely modified. In the case of liquid/
liquid dispersions, this organization change is the phase inversion
phenomenon by which both of the components exchange their respective
role as the continuous or disperse phase.

Without going into a detailed analysis which has been given
elsewhere[64,78], it is clear from the comparison of the theoretical
conductivity plot of Figure 14 with the experimental conductivity
plot of Figure 12 that the latter exhibits features characteristic
of a percolative conduction phenomenon, i.e., at lower water contents, a typical non-linear toe with a percolation threshold equal
to 0.09 in terms of Φ_w (water volume fraction), and, at higher
water contents, a typical linear section. The same considerations
apply to the left-hand part (up to the break-point marking the
linear/non-linear transition, i.e. over the W/O region), of the
conductivity plots of Figures 13 and 10 (Type U systems) and of
Figures 7 and 8 (Type U* systems). Similar observations were made
for other Type U* and Type U systems incorporating another surfactant[64,99] or another hydrocarbon[64,76]. So it appears that Type
U* and U systems can be defined, from the viewpoint of their electrical properties, as systems that exhibit a percolation type conduction phenomenon over their W/O Winsor IV region, contrary to
Type S systems that do not display any percolative conductive
behavior over their W/O region (see Figures 4 and 5). Other
authors have reported the existence of percolative conduction
phenomena in Winsor IV w/o type media[116-119] or in polyphasic
media[120]. It is worth mentioning here that Laguës[117,118] developed
a model of "stirred" Percolation that was found more suitable than
the classical Percolation model to account for the conductive

behavior just below the percolation threshold of the systems investigated (water/sodium dodecylsulfate/1-pentanol/cyclohexane)[119].
Although the exact molecular mechanism promoting it is still to be
understood, the percolative conduction phenomenon observed over the
W/O Winsor IV region of Type U* and Type U systems could be ascribed
to charge transfers occurring during "sticky" collisions between
disperse swollen w/o micelles. According to this tentative explica-
tive scheme, the "infinite clusters" of the Percolation Theory are
highly fluctuating dynamic strings of colliding and partially
interpenetrating swollen w/o micelles, which accounts for the existence of a non-zero conductivity whose increase with water content
proceeds from the collision and charge transfer growing probability,
until the micelle volume fraction reaches a close packing limit
beyond which, as explained in the preceding paragraph, the medium
undergoes phase inversion so as to be able to accommodate further
additions of water. So, in the light of this interpretation, it
can be considered that the line C_b, determined in the phase diagram from the break points marking in the conductivity plots the
end of the percolative type conduction regime, defines compositions
at which the volume fraction Φ of the population of w/o swollen
micelles reaches a close packing limit. Then, assuming that the
micelle population is monodisperse, it is possible, on the basis of
structural data available in the literature, to evaluate different
micellar parameters along the curve C_b. The aqueous core radius
r_w can be calculated by means of the following equation

$$r_w = \frac{e}{(\Phi^b/\Phi_w^b)^{1/3} - 1} \qquad (7)$$

where Φ^b, equal to a close packing limit, is the micelle population volume fraction corresponding to the water volume fraction
Φ_w^b and e is the difference between r_w and the actual micellar
radius. From the values of r_w, those of σ, the mean area per
surfactant molecule polar head at the interface, can be deduced
by using Equation (1) and also those of r_m, the theoretical micelle
radius. Two different alternatives ought to be considered separately prior to the calculations. The first one concerns the choice
of the value of Φ^b, which can be taken equal either to $\Phi_1^c = 0.637$
(random close packing), or to $\Phi_2^c = 0.741$ (cubic close packing).
The second one concerns the choice of e, which can be equal either
to ℓ_s, the length of the extended sodium dodecylsulfate molecule,
if the hard sphere volume is considered to be equal to the theoretical micelle volume, or to $r_c - r_w$, if the hard sphere volume is
defined by the "chemical" radius r_c. With this second assumption,
e can be assigned a value of 9 Å, in agreement with structural
results reported by different authors for systems incorporating
medium chain length alcohols[40-44]. The four sets of data obtained
by introducing in Equation (7) the four possible (Φ^b, e) couples
are listed in Table III, for different values of p_s, the weight

Table III. Evaluation of Micellar Parameters (Aqueous Core Radius r_w, Theoretical Micelle Radius r_m, and Mean Area at the Interface per Surfactant Molecule Polar Head $\bar{\sigma}$) along the Curve C_b Determined at 25°C for the Type U System Water/Sodium Dodecylsulfate/2-Methyl-2-Butanol/Benzene (Figure 11).

$\Phi^b = \Phi_1^C = 0.637$

		$e = 21$ Å			$e = 9$ Å		
P_s	Φ_w^b	r_w(Å)	r_m(Å)*	$\bar{\sigma}$(Å²)**	r_w(Å)	r_m(Å)	$\bar{\sigma}$(Å²)**
0.45	0.415	135	156	31	58	79	73
0.47	0.416	136	157	30	58	79	69
0.50	0.416	136	157	28	58	79	66
0.53	0.417	137	158	26	56	77	64

$\Phi^b = \Phi_2^C = 0.741$

		$e = 21$ Å			$e = 9$ Å		
P_s	Φ_w^b	r_w(Å)	r_m(Å)*	$\bar{\sigma}$(Å²)**	r_w(Å)	r_m(Å)*	$\bar{\sigma}$(Å²)**
0.45	0.415	99	120	43	42	63	101
0.47	0.416	99	120	41	42	63	96
0.50	0.416	99	120	38	42	63	91
0.53	0.417	100	121	36	43	64	83

* $r_m = r_w + 21$ Å

** $\bar{\sigma}$ was calculated by means of Equation (1).

fraction of the surfactant/alcohol combination of the water/sodium dodecylsulfate/2-methyl-2-butanol/benzene system. An analysis of these data shows that the most realistic set of data result from the double choice $\Phi^b = \Phi_1^C = 0.637$ and e = 9 Å which yields plausible values simultaneously for r_w and $\bar{\sigma}$. These values can be compared favorably with results obtained on close systems. The values of r_w around 58 Å are in good agreement with the values 53–56 Å obtained, through ultracentrifugation and small angle neutron scattering experiments by Dvolaitzky et al.[40], in the case of the water/sodium dodecylsulfate/1-pentanol/cyclohexane system, for a water/surfactant weight ratio of 2.5 that is close to the mean water/surfactant mass ratio of 2.7 characterizing the four compositions appearing in Table III. For $\bar{\sigma}$, the values deduced from those of

r_w by means of Equation (1) are in the range of 64–73 $\overset{\circ}{A}^2$, which is
comparable to the value of 68 $\overset{\circ}{A}^2$ found by the same group of authors
for the same system[40]. In addition, it can be remarked that $\bar{\sigma}$
increases slightly with the water/surfactant ratio, a feature which
has been mentioned in the literature by others[14,40]. The good
agreement between the r_w and $\bar{\sigma}$ values calculated with the $\phi^b =$
0.637 and e = 9 Å choice and literature data concerning systems of
the same kind is a fair indication that the curve C_b of Figure 11
defines compositions that correspond to the random close packing
of the hard-sphere cores of w/o swollen micelles whose peripheric
parts appear to be compressible, owing to the surfactant molecule
lipophilic tail flexibility. This conclusion, which is consistent
with the structural model proposed by different authors[40-44] as a
decisive improvement on Schulman school's original composite globule model, can be extended to other Type U and U* systems investigated and for which a curve C_b has been shown as well[99].

   All these finding are gratifying in that they tend to confirm that the curve C_b can be considered as a structural transi-
tion locus that forms an internal boundary separating, within the
Winsor IV domain of Type U and U* systems, the composition region
IZ, over which the medium internal geometry would not be spherical.
Similarly, on the basis of the results obtained in the case of
Type U* sustems, the curve C_m can be considered as a structural
transition locus separating IZ from the composition region O/W,
corresponding to the existence of o/w microemulsion type media.
The structure of the medium within the IZ region is a problem that
is still to be solved. As already pointed out, the coexistence
of inverse and direct swollen micelles suggested by Winsor[86] can be
ruled out as highly unrealistic and, according to the above scheme,
contradictory with the existence of the curves C_b and C_m considered
as close packing loci. Results obtained using various methods by
several groups of authors[56,68-70,97,98,121-123] seem to indicate
the absence of well defined structures over intermediate composition regions of the Winsor IV domain of several Type U systems.
This general observation led the authors to suggest that the medium
could be in a dynamic bicontinuous state which can be viewed as an
interspersion of aqueous and organic microdomains separated by highly flexible films of mixed surfactant and cosurfactant molecules.
The model of dynamic equilibrium bicontinuous structures introduced by Scriven[92,93] could prove to be quite instrumental concept
for the modelization of the structure and phase behavior of microemulsion type media[94-96,124]. A recent spin-label study performed
on water/sodium dodecylsulfate/1-pentanol/cyclohexane Winsor IV
media[71] has yielded results that were interpreted by the authors
as reflecting a progressive inversion of the curvature of the mixed
surfactant and cosurfactant interfacial film, though whose local
structure seemed not to be seriously affected[71,124]. A significant
feature of the phenomenon observed was that the interfacial film
appeared to be in a stationary state over a fairly broad water

content range, which would be compatible with the existence of bicontinuous type structures whose interfacial film separating the
aqueous and organic microdomains shows on the average, no marked
curvature towards either the aqueous or the organic phase. The
results of the present study of the electrical conductive behavior
of Type U and U* Winsor IV phases are consistent with the preceding
picture, since the transition curves C_b and C_m correspond, not to
discontinuities of the conductivity itself, but to changes in the
trend of the conductivity variations with composition, which indicates that the structural changes occurring when C_b and C_m are
crossed result in subtle modifications of the medium organization.
Although the exact nature of the so-called dynamic equilibrium bicontinuous structures is still to be clarified, the good agreement
established between the present conductivity results and literature
data obtained using other techniques on Type U systems similar to
those considered here is a strong justification for concluding that,
over the IZ composition region delimited in the Winsor IV domain
of Type U* and U systems by the curves C_b and C_m, the medium can
be depicted as being in a dynamic equilibrium bicontinuous state
related to a progressive and diffuse phase inversion process.

CONCLUSIONS

From the present study, which can be defined as an alcohol
chemical structure based scan of phase diagram properties and
electrical conductive behavior of quaternary Winsor IV media incorporating benzene as the hydrocarbon and involving as surface
active agents combinations of a given surfactant,(SDS), with various
alkanols, several general conclusions can be put forth that may
prove of importance for both future fundamental investigations
and practical applications concerning the so-called microemulsions.
These conclusions are in agreement with literature structural data
obtained by means of other methods and techniques.

Other things equal, the chemical structure of the alcohol
surfactant appears to have a strong influence upon both the configuration of the Winsor IV domain and the electrical conductive
properties of the Winsor IV media, which leads to distinguish between two main types of systems, Type S and Type U. Type S systems
are characterized by a phase diagram in which the W/O and O/W areas
are disjointed and separated by a composition zone over which viscous
and turbid media are formed. In contrast, for Type U systems, the
"oil-rich" and "water-rich" regions merge into each other so as to
form a unique Winsor IV domain. Hybrid configurations (Type U*)
are evidenced by using pentanol isomers as cosurfactants. Depending
on the value of r, the benzene to surface active agent combination
weight ratio, the "oil-rich" and "water-rich" regions are either
separated by a zone of viscous turbid media, (higher values of r),
or connected by a channel over which the medium remains fluid,

transparent and isotropic (lower values of r). The distinction between Type U and Type S systems, based on the features of the Winsor IV domain configuration, is paralleled by a distinction between percolating (Type U) and non-percolating (Type S) systems, based on the electrical conductive properties of both Type S and Type U "oil-rich" Winsor IV media.

The study of the electrical conductive behavior of the Winsor IV domain allows to delineate, for both types of systems, composition regions corresponding most probably to different medium structures. In the case of Type S systems, the W/O Winsor IV region is partitioned into two sub-regions by a well defined composition line (corresponding to conductivity local maxima) into two adjacent sub-regions. The sub-region characterized by low water/surfactant molar ratios represents most probably the realm of existence of suspensions of pre-micellar entities consisting of hydrated sur-factant aggregates; the other sub-region, characterized by water/surfactant molar ratios above 10 or so, being the realm of existence of proper W/O microemulsion type media depictable as populations of water-swollen micelles. In the case of Type U (and U*) systems, two internal boundaries partition the Winsor IV domain into three adjacent regions, W/O, IZ and O/W. The W/O and O/W regions are the realms of existence of inverted and direct microemulsion type media. The intermediary region IZ is the composition zone over which the w/o to o/w transformation (phase inversion) takes place in a progressive and diffuse way, without inducing the formation of long range organized media, contrary to what is observed in the case of Type S systems. The medium structure over the IZ region could be of the bicontinuous type.

The distinction established in the present paper between Type U and Type S systems, from the viewpoint of the Winsor domain configuration, or between percolating and non-percolating systems, from the viewpoint of the electrical conductive behavior, appears to bear some generality since it is applicable as well to different systems incorporating another hydrocarbon[76], or another surfactant[99], or both another hydrocarbon and another surfactant[63]. However, a general criterion for the formulation of either type of systems is still to be found because the dividing line between the two types depends upon subtle parameters such as the chemical structure of the alcohol, surfactant and hydrocarbon, or the alcohol/surfactant ratio[125], all these factors having intricate effects upon the form-ation and the composition of the interfacial film[73,74]. The re-sults reported here illustrate the influence of the cosurfactant chemical structure upon the interfacial film fluidity which De Gennes and Taupin[124] consider as "absolutely needed to maintain a microemulsion phase in the central region of the phase diagram". According to this statement, the obtention of Type U systems can be ascribed to the greater fluidity induced in the interfacial film by shorter alcohols. On the other hand, the incorporation of longer alcohols promotes the formation of more rigid interfacial films which results in the obtention of Type S systems whose both w/o and o/w Winsor IV areas shrink as the alcohol chain-length increases. Moreover, it has been shown by Cazabat and Langevin[42] that the attractive interactions between w/o swollen micelles, as estimated from the value of the second virial coefficient of the osmotic compressibility, become stronger when the alcohol chain-length decreases, other things equal. This finding, combined with the idea that an interfacial film incorporating short chain alcohol molecules is more fluid, provides a very sound structural basis to the distinction established between percolating (Type U) and non-percolating (Type S) systems.

ACKNOWLEDGEMENTS

The authors wish to convey their thanks to the D.G.R.S.T., France, for its financial support, more particularly for the postgraduated scholarships granted to J. Heil and L. Nicolas-Morgantini. The final version of this article was prepared by Dr. M. Clausse during his stay as NATO and CIES sponsored Visiting Professor in the Laboratory of Dr. Shah (University of Florida, Department of Chemical Engineering). Thanks are due to both NATO and CIES for their financial supports, to Dr. Shah for all the facilities provided, and to those who participated in the preparation of the typescript.

REFERENCES

1. P. A. Winsor, Trans. Faraday Soc., 44, 376 (1948); 44, 451 (1948).
2. P. A. Winsor, Trans. Faraday Soc., 46, 762 (1950).
3. T. P. Hoar and J. H. Schulman, Nature (London), 152, 102 (1943).
4. J. H. Schulman and T. S. McRoberts, Trans. Faraday Soc., 42B, 165 (1946).
5. J. H. Schulman and D. P. Riley, J. Colloid Sci., 3, 383 (1948).
6. J. H. Schulman and J. A. Friend, J. Colloid Sci., 4, 497 (1949).
7. H. E. Bowcott and J. H. Schulman, Z. Elektrochem., 59, 283 (1955).
8. J. H. Schulman, W. Stoeckenius and L.M. Prince, J. Phys. Chem. 63, 1677 (1959).
9. W. Stoeckenius, J. H. Schulman and L. M. Prince, Kolloid-Z., 169, 170 (1960).
10. C. E. Cooke and J. H. Schulman, in "Surface Chemistry", P. Ekwall, K. Groth and V. Runnström-Reio, Editors, pp. 231-251, Munksgaard, Copenhagen, 1965.

11. P. Edwall, L. Mandell and K. Fontell, J. Colloid Interface Sci., 33, 215 (1969).

12. H. Kunieda and K. Shinoda, J. Colloid Interface Sci., 70, 577 (1979).

13. H. Kunieda and K. Shinoda, J. Colloid Interface Sci., 75, 601 (1980).

14. H. F. Eicke and J. Rehak, Helv. Chim. Acta, 59, 2883 (1976).

15. M. Zulauf and H. F. Eicke, J. Phys. Chem., 83, 480 (1979).

16. P. E. Zinsli, J. Phys. Chem., 83, 3223 (1979).

17. H. F. Eicke, Pure Appl. Chem., 52, 1349 (1980).

18. K. Shinoda, in "Solvent Properties of Surfactant Solutions", K. Shinoda, Editor, pp 27-63, Surfactant Science Series, Vol. 2, Marcel Dekker, New York, 1967.

19. K. Shinoda, J. Colloid Interface Sci., 24, 4 (1967).

20. K. Shinoda and T. Ogawa, J. Colloid Interface Sci., 24, 56 (1967).

21. K. Shinoda and H. Kunieda, J. Colloid Interface Sci., 42, 381 (1973).

22. S. Friberg and I. Lapczynska, Progr. Colloid Polymer Sci., 56, 16 (1975).

23. S. Friberg, I. Buraczewska and J. C. Ravey, in "Micellization, Solubilization and Microemulsions", K. L. Mittal, Editor, Vol. 2, pp. 901-911, Plenum Press, New York, 1977.

24. L. M. Prince, in "Microemulsions. Theory and Practice", L. M. Prince, Editor, pp. 21-32, Academic Press, New York, 1977.

25. S. Friberg, Informations Chimie, 198, 235 (1975).

26. S. Friberg, Chem. Tech., 6, 124 (1976).

27. V. K. Bansal and D. O. Shah, in "Microemulsions. Theory and Practice", L. M. Prince, Editor, pp. 149-173, Academic Press, New York, 1977.

28. V. K. Bansal and D. O. Shah, in "Micellization, Solubilization and Microemulsions", K. L. Mittal, Editor, Vol. 1, pp. 87-113, Plenum Press, New York, 1977.

29. K. D. Dreher and S. C. Jones, in "Solution Chemistry of Surfactants", K. L. Mittal, Editor, Vol. 2, pp. 627-658, Plenum Press, New York, 1979.

30. G. De Lamballerie, La Recherche, 12 (119), 148 (1981).

31. C. Taupin, in "Physicochimie des Composés Amphiphiles", pp. 255-259, Editions du CNRS, Paris, 1979.

32. L. M. Prince, "Microemulsions. Theory and Practice", Academic Press, New York, 1977.

33. D. O. Shah and R. S. Schechter, "Improved Oil Recovery by Surfactant and Polymer Flooding", Academic Press, New York 1977.

34. K. L. Mittal, "Micellization, Solubilization and Microemulsions" Vol. 1 and 2, Plenum Press, New York, 1977.

35a. K. L. Mittal, "Solution Chemistry of Surfactants", Vol. 1 and 2, Plenum Press, New York 1979.

35b. K. L. Mittal and E. J. Fendler, Editors, "Solution Behavior of Surfactants: Theoretical and Applied Aspects", Vol. 1 and 2, Plenum Press, New York, 1982.

36. M. Rosoff, in "Progress in Surface and Membrane Science",
     J. F. Danielli, M. D. Rosenberg and D. A. Cadenhead, Editors,
     Vol. 12, pp. 405-477, Academic Press, New York,1978.
37. D. O. Shah, "Surface Phenomena in Enhanced Oil Recovery",
     Plenum Press, New York, 1981.
38. P. Becher, Editor, "Encyclopedia of Emulsion Technology",
     Vol 1, Marcel Dekker, New York, 1983.
39. W. G. M. Agterof, J. A. J. Van Zomeren and A. Vrij, Chem.
     Phys. Lett., 43, 363 (1976).
40. M. Dvolaitzky, M. Guyot, M. Lagües, J. P. Le Pesant, R. Ober,
     C. Sauterey and C. Taupin, J. Chem. Phys., 69, 3279 (1978).
41. R. Ober and C. Taupin, J. Phys. Chem., 84, 2418 (1980).
42. A. M. Cazabat and D. Langevin, J. Chem. Phys., 74, 3148 (1981).
43. D. J. Cebula, L. Harding, R. H. Ottewill and P. N. Pusey,
     Colloid Polymer Sci., 258, 973 (1980).
44. D. J. Cebula, R. H. Ottewill, J. Ralston and P. N. Pusey, J.
     Chem. Soc., Faraday Trans. I, 77, 2585 (1981).
45. J. Rouviere, J. M. Couret, M. Lindheimer, J. L. Dejardin and
     R. Marrony, J. Chim. Phys.-Phys. Chim. Biol., 76, 289 (1979).
46. R. A. Day, B. H. Robinson, J. H. R. Clarke and J. V. Doherty,
     J. Chem. Soc., Faraday Trans. I, 75, 132 (1979).
47. P. G. Nilsson and B. Lindman, J. Phys. Chem., 86, 271 (1982).
48. A. Graciaa, J. Lachaise, A. Martinez and A. Rousset, C. R.
     Acad. Sc. Paris, 285B, 295 (1977).
49. A. Graciaa, J. Lachaise, P. Chabrat, L. Letamendia, J. Rouch
     and C. Vaucamps, J. Phys. (Paris) Lett., 39, L-235 (1978).
50. E. Gulari and B. Chu, in "Surface Phenomena in Enhanced Oil
     Recovery", D. O. Shah, Editor, pp. 181-197, Plenum Press,
     New York, 1981.
51. M. Podzimek and S. Friberg, J. Dispersion Sci. Technol., 1,
     341 (1980).
52. H. L. Rosano, T. Lan, A. Weiss, J. H. Whittam and W. E. F.
     Gerbacia, J. Phys. Chem., 85, 468 (1981).
53. D. Oakenfull, J. Chem. Soc., Faraday Trans. I, 76, 1875
     (1979).
54. D. J. Mitchell and B. W. Ninham, J. Chem. Soc., Faraday Trans.
     II, 77, 601 (1981).
55. J. Biais, P. Bothorel, B. Clin and P. Lalanne, J. Colloid
     Interface Sci., 80, 136 (1981); J. Dispersion Sci. Technol.,
     2, 67 (1981).
56. A. M. Bellocq, J. Biais, B. Clin, P. Lalanne and B. Lemanceau,
     J. Colloid Interface Sci., 70, 524 (1979).
57. M. Wong, M. Grätzel and J. K. Thomas, Chem. Phys. Lett., 30,
     329 (1975).
58. M. Wong, J. K. Thomas and M. Grätzel, J. Am. Chem. Soc., 98,
     2391 (1976).
59. M. Wong, J. K. Thomas and T. Nowak, J. Am. Chem. Soc., 99,
     4730 (1977).
60. E. Sjöblom and S. Friberg, J. Colloid Interface Sci., 67, 16
     (1978).

61. A. M. Bellocq and G. Fourche, J. Colloid Interface Sci., 78, 275 (1980).

62. C. Boned, M. Clausse, B. Lagourette, J. Peyrelasse, V.E.R. McClean and R. J. Sheppard, J. Phys. Chem., 84, 1520 (1980).

63. M. Clausse, C. Boned, J. Peyrelasse, B. Lagourette, V. E. R. McClean and R. J. Sheppard, in "Surface Phenomena in Enhanced Oil Recovery", D. O. Shah, Editor, pp. 199-228, Plenum Press, New York, 1981.

64. J. Heil, Thesis, Université de Pau, France, 1981.

65. J. Heil, M. Clausse, J. Peyrelasse, and C. Boned, Colloid Polymer Sci., 260, 93 (1982).

66. C. Boned, J. Peyrelasse, J. Heil, A. Zradba and M. Clausse, J. Colloid Interface Sci., 88, 602 (1982).

67. G. Bakale, G. Beck and J. K. Thomas, J. Phys. Chem., 85, 1062 (1981).

68. B. Lindman, P. Stilbs and M. E. Moseley, J. Colloid Interface Sci., 83, 569 (1981).

69. C. Tondre and R. Zana, J. Dispersion Sci. Technol., 1, 179 (1980).

70. J. Lang, A. Djavanbakht and R. Zana, J. Phys. Chem., 84, 1541 (1980).

71. M. Dvolaitzky, R. Ober and C. Taupin, C. R. Acad. Sc. Paris, 293II, 27 (1981).

72. D. O. Shah, V. K. Bansal, K. Chan and W. C. Hsieh, in "Improved Oil Recovery by Surfactant and Polymer Flooding", D. O. Shah and R. S. Schechter, Editors, pp. 293-337, Academic Press, New York, 1977.

73. V. K. Bansal, K. Chinnaswamy, C. Ramachandran and D. O. Shah, J. Colloid Interface Sci., 72, 524 (1979).

74. V. K. Bansal, D. O. Shah and J. P. O'Connell, J. Colloid Interface Sci., 75, 462 (1980).

75. S. S. Atik and J. K. Thomas, J. Phys. Chem., 85, 3921 (1981).

76. M. Clausse, J. Heil, J. Peyrelasse and C. Boned, J. Colloid Interface Sci., 87, 584 (1982).

77. M. Clausse, in "Encyclopedia of Emulsion Technology", P. Becher, Editor, Vol. 1, Chapter 9, pp. 481-715, Marcel Dekker, New York, 1983.

78. B. Lagourette, J. Peyrelasse, C. Boned and M. Clausse, Nature, 281, 60 (1979).

79. M. Clausse, J. Peyrelasse, J. Heil, C. Boned and B. lagourette, Nature, 293, 636 (1981).

80. P. M. Ginnings and R. Baum, J. Am. Chem. Soc., 59, 1111 (1937).

81. S. Friberg, in "Microemulsions. Theory and Practice", L. M. Prince, Editor, pp. 133-146, Academic Press, New York, 1977.

82. S. Friberg and I. Buraczewska, in "Micellization, Solubilization and Microemulsions", K. L. Mittal, Editor, Vol. 1, pp. 791-799, Plenum Press, New York, 1977.

83. S. Friberg and I. Buraczewska, Progr. Colloid Polymer Sci., 63, 1 (1978).

84.   F. Candau, J. Boutillier, J. C. Wittmann and S. Candau, in
      "Physicochimie des Composés Amphiphiles", pp. 179-184, Editions
      du CNRS, Paris, 1979.
85.   A. M. Bellocq, G. Fourche, P. Chabrat, L. Letamendia, J. Rouch
      and C. Vaucamps, Optica Acta, 27, 1629 (1980).
86.   P. A. Winsor, J. Colloid Sci., 10, 88 (1955).
87.   D. O. Shah and R. M. Hamlin, Science, 171, 483 (1971).
88.   D. O. Shah, A. Tamjeedi, J. W. Falco and R. D. Walker, AIChE
      J., 18, 1116 (1972).
89.   J. W. Falco, R. D. Walker and D. O. Shah, AIChE J., 20, 510
      (1974).
90.   M. Clausse, R. J. Sheppard, C. Boned and C. G. Essex, in
      "Colloid Interface Science", Vol. 2, M. Kerker, Editor, pp.
      233-243, Academic Press, New York,1976.
91.   A. H. Roux, G. Roux-Desgranges, J. P. E. Grolier and A.
      Viallard, J. Colloid Interface Sci., 84, 250 (1981).
92.   L. E. Scriven, Nature, 263, 123 (1976).
93.   L. E. Scriven, in "Micellization, Solubilization and Micro-
      emulsions", K. L. Mittal, Editor, Vol. 2, pp. 877-893, Plenum
      Press, New York, 1977.
94.   Y. Talmon and S. Prager, Nature, 267, 333 (1977).
95.   Y. Talmon and S. Prager, J. Chem. Phys., 69, 2984 (1978).
96.   E. W. Kaler and S. Prager, J. Colloid Interface Sci., 86, 359
      (1982).
97.   B. Lindman, N. Kamenka, B. Brun and P. G. Nilsson, in "Micro-
      emulsions", I. D. Robb, Editor, pp. 115-129, Plenum Press,
      New York, 1982.
98.   J. Lang, A. Djavanbakht and R. Zana, in "Microemulsions", I.
      D. Robb, Editor, pp. 233-255, Plenum Press, New York, 1982.
99.   J. Peyrelasse, C. Boned, J. Heil and M. Clausse, J. Phys. C:
      Solid State Phys., 15, 7099 (1982).
100. H. F. Eicke and A. Denss, in "Solution Chemistry of Sur-
      factants", K. L. Mittal, Editor, Vol. 2, pp. 699-706, Plenum
      Press, New York, 1979.
101. B. A. Pethica and A. Fen, Disc. Faraday Soc., 258, 18 (1954).
102. A. Graciaa, Thesis, Université de Pau, France, 1978.
103. S. Kirkpatrick, Phys. Rev. Lett., 27, 1722 (1971).
104. S. Kirkpatrick, Rev. Mod. Phys., 45, 574 (1973).
105. D. A. G. Bruggeman, Ann. Phys., 24, 636 (1935).
106. C. J. F. Bottcher, Rec. Trav. Chim. Pays- Bas, 64, 47 (1945).
107. R. Landauer, J. Appl. Phys., 23, 779 (1952).
108. J. Janzen, J. Appl. Phys., 46, 966 (1975).
109. M. H. Cohen and J. Jortner, Phys. Rev. Lett., 30, 696 (1973).
110. I. Webman, J. Jortner and M. H. Cohen, Phys. Rev. B, 11, 2885
      (1975).
111. J. W. Essam, C. M. Place and E. M. Sondheimer, J. Phys. C:
      Solid State Phys., 7, L-258 (1974).
112. J. P. Straley, Phys. Rev. B, 15, 5733 (1977).
113. M. J. Stephen, Phys. Rev. B, 17, 4444 (1978).
114. K. Günther and D. Heinrich, Zeits. Phys., 185, 345 (1965).

115. J. L. Finney, Nature, <u>266</u>, 309 (1977).
116. M. Lagües, R. Ober and C. Taupin, J. Phys. (Paris) Lett., <u>39</u>,
      L-487 (1978).
117. M. Lagües, J. Phys. (Paris) Lett., <u>40</u>, L-331 (1979).
118. M. Lagües, C. R. Acad. Sc. Paris, <u>288B</u>, 339 (1979).
119. M. Lagües and C. Sauterey, J. Phys. Chem., <u>84</u>, 3503 (1980).
120. K. E. Bennett, J. C. Hatfield, H. T. Davis, C. W. Macosko
      and L. E. Scriven, in "Microemulsions", I. D. Robb, Editor,
      pp. 65-84, Plenum Press, New York, 1982.
121. B. Lindman, N. Kamenka, T. M. Kathopoulis, B. Brun and P. G.
      Nilsson, J. Phys. Chem., <u>84</u>, 2485 (1980).
122. F. Larche, J. Rouviere, P. Delord, B. Brun and J. L. Dussossoy,
      J. Phys. (Paris) Lett., <u>41</u>, L-437 (1980).
123. C. T. Meyer, Y. Poggi and G. Maret, J. Phys. (Paris), <u>43</u>, 827
      (1982).
124. P. G. De Gennes and C. Taupin, J. Phys. Chem., <u>86</u>, 2294 (1982).
125. A. Zradba, Thesis, Université de Pau, France, 1983.

FLUORESCENCE PROBE STUDY OF OIL IN WATER MICROEMULSIONS

Raoul Zana, Jacques Lang and Panagiotis Lianos

Centre de Recherches sur les Macromolécules - CNRS
6, rue Boussingault, 67083 Strasbourg-Cedex, France

Fluorescence probes have been used to investigate the changes of size and shape of the aggregates when generating oil in water (O/W) microemulsions by first increasing the surfactant concentration in a micellar solution, then adding an increasing amount of alcohol (thereby forming mixed alcohol+surfactant micelles) and finally adding oil to the mixed micelle solution. The surfactant aggregation number n of the aggregates, and the rate constant k_E for interaggregate excimer formation have been obtained from the analysis of the fluorescence decay of micelle-solubilized pyrene. The determination of the fluorescence spectra of micelle solubilized monomeric pyrene and dipyrenylpropane allowed the assessment of the changes of effective polarity and viscosity ($\bar{\eta}$), respectively, of the aggregates, at the site of solubilization of the probe, upon generation of the microemulsion. The effect of the chain length and/or nature of the surfactant (sodium dodecyl and tetradecyl sulfates, SDS and STS, respectively), alcohol (n-butanol, n-pentanol and its isomers, and n-hexanol), and oil (alkanes, from n-hexane to n-hexadecane and two isomers of n-octane, and arenes) on n, k_E, $\bar{\eta}$ and the effective polarity was investigated.

Additions of pentanol to concentrated (C>0.2 $\underline{M}$) SDS solutions result in an increase of n, and a decrease of k_E, $\bar{\eta}$ and effective polarity. These effects have been associated with the transformation of the initially spherical SDS micelles into possibly disk-shaped, large micelles upon dissolution of n-pentanol.

1627

Upon addition of alkane to mixed alcohol+surfactant micelles n may increase, remain constant or decrease depending on the chain lengths of the alcohol, surfactant and alkane, as well as on the alkane isomerism. For a given set of experimental conditions, the increase of the surfactant alkyl chain length (from SDS to STS) results in an increase of the alkane chain length (from n-heptane to n-nonane) for which the variation of n with the alkane volume fraction ϕ_0 is reversed. Contrary to the complex changes of n, $\bar{n}$ is always decreased upon addition of alcohol and oil.

For long chain alkanes, n goes through a minimum and then increases at high ϕ_0. This variation is attributed to a rapid change of aggregate shape, from disk-like to nearly spherical as ϕ_0 is increased, i.e., when an oil core is formed in the aggregates.

The various results show the importance of geometric factors in determining the changes of size and shape of the aggregates when generating an O/W microemulsion. A model based on geometric considerations is presented in an attempt to qualitatively explain the experimental findings.

INTRODUCTION

Microemulsions are complex, homogeneous systems containing in most instances four, if not five, components : water, oil, surfactant, cosurfactant (most often an alcohol) and salts dissolved in the water. Microemulsions are usually divided into oil in water (O/W) and water in oil (W/O) systems. The interest that microemulsions have aroused among workers both from the fundamental and applied research fields is well illustrated by the number of recent books and reviews dealing with these systems[1-5]. Nevertheless, a number of problems remain associated with microemulsions. With the O/W systems, which are the ones considered in this paper, the main problem arises from the large electrical charge of the surfactant+alcohol layer coating the oil microdroplets dispersed in the water-rich medium (in the following the words droplet or aggregate refer to the oil droplet plus the interfacial layer). The intense electrostatic repulsions between droplets prevent the determination of their characteristics (size, molecular weight) in the actual experimental conditions, that is, at finite droplet volume fraction, by means of the usual physicochemical methods. Measurements at sufficiently low volume fraction, where interactions become negligible, are then necessary. However,

in the process of dilution of the system one may drastically
modify the structure, mass and composition of the droplets.
Thus the study of O/W microemulsions requires methods which
can yield information on the droplets in the actual experimental
conditions, that is at given concentrations of oil, surfactant,
cosurfactant and salt. Recently, several fluorescence methods
have been proposed[6-10] which in principle allow the determination of the droplet concentration and, thus, of their molecular
weight, and of the number of surfactant per droplet, without
requiring a dilution of the microemulsion under investigation.
Among these methods we have selected the one based on the
analysis of the decay of fluorescence of a probe nearly insoluble
in water but solubilized by the droplets. This method encounters
minor problems in the case of polydisperse systems[11,12].
Nevertheless it has permitted us to measure the change of
the surfactant aggregation number n per droplet when an O/W
microemulsion is progressively generated by (i) concentrating
a dilute micellar solution up to a surfactant concentration
of 0.5 $\underline{M}$; (ii) adding alcohols (up to a concentration of
1-1.5 M) of varying chain length and isomerism, thus forming
a concentrated solution of mixed surfactant+alcohol micelles
and (iii) adding oils (up to an oil volume fraction of 0.13)[13].

In this paper we briefly outline the principle of the method
used to obtain n and summarize the main results concerning the
changes of n when forming an O/W microemulsion as described above.
It should be emphasized that such results are of great interest
for workers involved in studies of chemical reactions in micelles
and microemulsions. Indeed the distribution of the reactants
and thus the reaction kinetics strongly depend on the state of
association of the surfactant which must therefore be known.

FLUORESCENCE PROBING METHODS

Consider an O/W microemulsion where the droplet concentration is [M]. Let [P] be the stoichiometric concentration of a fluo
rescence probe P nearly insoluble in water and which has been
solubilized by the droplets ([P] is assumed to be much larger than
the solubility of the probe in water) and R=[P]/[M].

If it is assumed that the droplets are all identical (monodisperse) the distribution of the probes among the droplets is of
the Poisson's type[8,14-18], and the probability p_i of findings
droplets M_i containing i solubilized probes is :

$$p_i = [M_i]/[M] = e^{-R} R^i/i! \qquad (1)$$

The decay of fluorescence of the probe is then given by[9,14]

$$I(t) = I(0) \exp\left\{-k_o t + R\left[\exp(-k_E t) - 1\right]\right\} \tag{2}$$

where k_o and k_E are the rate constants for the reactions (3) and (4) of monomer probe decay and excimer formation, respectively.

$$P^* \xrightarrow{k_o} P \tag{3}$$

$$P^* + P \xrightarrow{k_E} (P^*P) \tag{4}$$

Note that reaction (3) includes all deactivation processes of the excited probe (emission, internal conversion and intersystem crossing). In micellar systems and microemulsions, several studies have shown that excimer formation can be treated as a pseudo first order reaction[19-23].

When $R \ll 1$, that is when the droplets contain no probe or only one solubilized probe, eq. 2 reduces to :

$$\ln I(t)/I(0) = - k_o t \tag{5}$$

which describes the decay of the monomeric probe in the droplet microenvironment.

When R is not too small, the experimental decay curve monitored at a wavelength of monomer emission usually shows a fast component associated with the consumption of excited probes through excimer formation in droplets containing two or more than two probes and a slow component associated with droplets containing only one probe. A three-parameter computer fitting of eq. 2 to the decay curve yields $I(0)$, k_E and R (k_o is obtained from an experiment where $R \ll 1$). The value of R yields the number n of surfactant per droplet according to :

$$n = (C - CMC)/[M] = R(C - CMC)/[P] \tag{6}$$

As will be seen below, information on the microemulsion can also be obtained from the value of k_E.

Two limiting cases must be considered[24]
(i) $k_o \ll k_E$. Eq. 2 reduces to

$$\ln I(t)/I(0) = -k_o t - R \tag{7}$$

thus R can be obtained by a simple graphical procedure, as the

intercept of the part of the decay curve at long t with the
ordinate axis, in a normalized plot[9]. The problem associated with
the value of I(0) has been discussed by Almgren and Löfroth[11] and
the same procedure was adopted in our studies.[24]
(ii) $k_o >> k_E$. Eq. 2 reduces to

$$\ln I(t)/I(0) = -(Rk_E + k_o)t \qquad (8)$$

Thus a linear plot is obtained as if R<<1 because an excited probe
in a droplet containing several probes has now enough time to decay
before reacting with ground state probes to form excimers. In this
case it is impossible to obtain R and k_E independently and so the
method cannot be used[24].

It is clear from the above that the range of n-values which
can be determined by this fluorescence decay method can be increased by selecting a probe having a small decay rate constant,
that is a long lifetime $\tau_o = 1/k_o$. Following Atik et al[9] we have
adopted pyrene for which $k_o \simeq 2.5 \times 10^6 s^{-1}$ in organic solvents and
which has been extensively used to study micellar systems and
microemulsions[6,8,9,13,15,24-28]. We have checked[28-30] a number
of micellar solutions and found that the n-values obtained with
pyrene agree with those determined by other methods, within
the experimental errors. These results indicate that the method
is valid and that the presence of a small amount of pyrene in
the micelles brings about a negligible change of the micelle
aggregation number.

The range of n-values accessible to the fluorescence decay
method also depends on the pseudo first order rate constant k_E
for excimer formation. k_E depends on the local concentration of
the probe, that is on n and R, and on the properties of the
medium (essentially its microviscosity $\bar{\eta}$) in which excited and
ground state probes diffuse to form excimers.

Decays typical of situations where $k_o << k_E$, $k_o \sim k_E$ and $k_o >> k_E$
have been observed with the micellar solutions and microemulsions investigated so far[13,24,28-30]. The case where $k_o << k_E$
is most often associated with micellar solutions with $\bar{n}$ values
below a limit of about 100-150, that is for ionic surfactants
in the absence of salt or at low salt or alcohol concentration[28-30]. For large anisotropic micelles (such as those formed
by ionic surfactants in the presence of salt and non ionic sur-
factants[2]) k_o is generally comparable to or larger than k_E as
n can increase very much. In solutions of mixed alcohol+surfac-
tant micelles as well as in microemulsions the n-values can
increase very rapidly upon addition of alcohol and oil. One
would have thus expected the fluorescence decay method not to
be very useful for the study of these systems. However it will
be seen below that the low microviscosity of the probe environ-

ment in its solubilization site results in an increase of k_E,
even though the local probe concentration is decreased. This
makes it possible to determine n-values up to 500-1000, as the
ln I(0)/I(t) vs t plots are not linear[13].

All the above refers to monodisperse systems. This appears
to be the case for micellar solutions of ionic surfactants in
the absence of salt[31], of mixed alcohol+surfactant micelles
at low alcohol concentration[32], and of microemulsions at large
oil content[33]. There are, however, a number of systems where
the aggregates may be very polydisperse, particularly the solutions of ionic surfactants in the presence of a large excess
of salt[34], of non-ionic surfactants[35] and of some zwitterionic
surfactants[36]. The difficulties introduced by the aggregate polydispersity in the analysis of the fluorescence decay data have
been recently thoroughly discussed by Almgren and Löfroth[12]. These
authors showed that in the case of very large polydispersity these
difficulties may result in large errors in the n-values. Thus,
because of polydispersity effects, our previously reported n-values for sodium dodecylsulfate micelles[29] at concentrations of
added NaCl above 0.5M are probably underestimated. The error,
however, is probably not very large because all of our measurements[13;21,28-30] were performed at R~1. For the very large micelles the ratio $\eta = R/n$, which is an essential quantity in Almgren
and Löfroth[12] calculations, was therefore always very small
($\eta = 3 \times 10^{-3}$ for n=300). In this case the calculations indicate
that the n-values obtained by the fluorescence decay method are
close to the true values, even for polydisperse systems[12]. Moreover, for most of the investigated systems it is likely that
the polydispersity was not as large as required by the simulations[12] to result in large errors in n. For instance for systems
similar to many of the mixed alcohol+surfactant micelles and
microemulsions used in the present study, independent quasi-elastic light scattering did not reveal a large polydispersity[37],
in agreement with the results of other studies using this last
technique[33]. Therefore, even if polydispersity resulted in some
errors in the values of n reported in the next section, these
errors are likely to be small, probably within the overall experimental error of the method. Thus the changes of n with the
system composition and nature presented below give the correct
trends and, at the very least, a semi-quantitative picture of
the changes undergone by the aggregates when going from a micellar solution to a mixed micellar solution and finally to a
microemulsion.

Before going into the experimental results it should be
added that fluorescence probing methods provide two additional
informations on the systems studied :

(i) The fluorescence emission spectrum of monomeric pyrene is

strongly dependent on the polarity of the solvent[6,25,38]. Thus
a polarity calibration curve can be prepared by plotting the value
of the ratio I_1/I_3 of the first and third vibronic peaks in this
spectrum against any of the quantities used to characterize the
polarity of the solvent. From the value of I_1/I_3 in the micellar
solution or microemulsion one can then estimate the polarity sen-
sed by the probe in its solubilization site, and thus conclude
about the location of the latter.

(ii) Furthermore, the ratio between the monomer and intramolecular excimer emissions of dipyrenylpropane (DPyP) strongly depends on the viscosity $\bar{\eta}$ of the medium[39,40]. As above a calibration curve can be prepared by plotting the ratio I_{IE}/I_M of the
intensity maxima for the intramolecular excimer and the monomer
against $\bar{\eta}$ for a series of paraffin oil-hexadecane mixtures or
glycerol-methanol mixtures[39-42]. An estimate of the microviscosity of the environment of DPyP in micelles and microemulsions
was obtained by measuring I_{IE}/I_M in these media. The chemical
similarity of pyrene and DPyP makes it likely that their solubilization sites in these systems are nearly identical. It must be
recalled that pyrene is generally believed to be solubilized in
the palisade layer of micelles[25]. As will be seen below the
solubilization behavior of pyrene in microemulsions is much more
complex.

It should be emphasized that the above procedure for estimating the polarity and viscosity of the probe microenvironment are
mostly useful for following changes of polarity and viscosity at
the solubilization site of the probe, induced by changes of
composition of the system. They should not be used for absolute
value determinations as the organized assemblies investigated
cannot be approximated by the homogeneous solvents or solvent
mixtures used for preparing the calibration curves. Also the
latter depends to some extent on the solvents used. This is
particularly true for the viscosity.

RESULTS AND DISCUSSION

Effect of the surfactant concentration on the micelle aggregation number

The results of Figure 1 in the absence of pentanol ($c_A=0$)
show that n remains nearly constant up to C$\approx$0.3$\underline{M}$, and equal to
65 ± 4, that is its value at the CMC[43]. Then, n increases with C.
Similar results have been found for several other ionic surfactants[28,30,44]. This behavior appears to be characteristic of
ionic surfactants with a single alkyl chain and a well defined
head group. Microviscosity as well as effective polarity remained
nearly constant in the whole range of surfactant concentration.

Effect of alcohol addition on the aggregation number of ionic
surfactant micelles

The effect of 1-pentanol on the aggregation number of SDS
micelles is shown in Figure 1. At low C, n decreases when the
alcohol concentration c_A is increased; whereas at higher C, n
first decreases, goes through a minimum and then increases with
c_A. At still higher C the minimum vanishes and at c_A above about
0.2$\underline{M}$ n increases very rapidly with c_A.

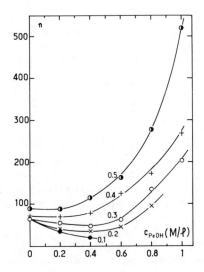

Figure 1. Effect of the 1-pentanol concentration on the aggregation number of SDS, in the mixed SDS-pentanol micelles, at various SDS concentrations, indicated on each plot, in mole/1
(Adapted from ref. 32. Copyright 1982 American Chemical Society).

Similar results have been found for the surfactant tetrade-
cyltrimethylammonium bromide (TTAB)[30]. For both SDS and TTAB the
effect of the alcohol chain length and isomerism on n have been

found to be extremely important. These results indicate that the most important parameter which determines the changes of n with c_A is the partition coefficient K_p of the alcohol between the micelles and the bulk and thus the amount of alcohol solubilized by the micelles at given values of C and c_A. K_p is known to increase by a factor of about 3 per additional methylene group. Thus 1-butanol which is fairly soluble in water (small K_p) brings about a decrease of n even at C=0.5$\underline{M}$ and c_A=1$\underline{M}$, in contrast to the large increase of n found under the same experimental conditions with 1-pentanol, with respect to the n-value for SDS micelles in the absence of alcohol. The isomers of pentanol which all have solubilities in water and thus K_p's intermediate between those of 1-butanol and 1-pentanol, give intermediate n-values, as can be seen in Table I. For 1-hexanol the increase of n is so large that the system forms an emulsion.

Table I. Effect of medium chain length alcohols on the aggregation number of SDS micelles and on the fluorescence emission properties of solubilized probes[1].

Alcohol	Solubility in $H_2O(\underline{M})$[2]	I_1/I_3	$10^{-6}k_o$ (s^{-1})	n	$10^{-7}k_E$ (s^{-1})	$\bar{\eta}$ (cP)
No alcohol	–	1.20	2.71	90	1.8	22
1-butanol	1.05	1.05	2.42	55	1.57	11.5
2-pentanol	0.55	1.04	2.50	126	0.68	11
2-methyl-1-1-butanol	0.35	1.02	2.54	223	0.38	11
4-methyl-1-butanol	0.32	1.03	2.55	253	0.44	11
1-pentanol	0.30	1.06	2.50	514	0.20	95
1-hexanol[3]	0.06	very large				

1. Alcohol concentration n=1$\underline{M}$; SDS concentration – 0.5$\underline{M}$; T = 25°C
2. From "Solubilities of Organic and Inorganic Compounds", H. Stephen and T. Stephen, Eds., Vol. I, Pergamon Press, MacMillan Co., New York, 1963
3. This mixture was in the form of an emulsion and remained emulsified even at c_A=0.5$\underline{M}$.

It should be noted that at low C the effect of alcohols, irrespective of the alcohol chain length, is to decrease the micelle aggregation number both for TTAB and SDS. A similar effect has been reported by others[46,47], using different methods to obtain n. This last result, in conjunction with the known increase of the micelle ionization degree upon addition of alcohol at low C[30],

provides an explanation for the effect of alcohol. At low C
where intermicellar electrostatic repulsions are weak, the addition of alcohol results in the formation of mixed alcohol+sur-
factant micelles, and in an increase of the micelle ionization
degree[30], which in turn brings about a breakdown of the micelles
into smaller ones, in order to reduce electrostatic repulsions
between head groups (micelles with small n have a larger surface
area per head group than micelles with large n[48]). This effect
is more pronounced the larger the value of K_p, as is indeed
observed. At higher surfactant concentration the intermicellar
repulsions become important and must be taken into account when
evaluating the free energy of the system. As alcohol is addeed
to the system the increased micelle ionization results in increased repulsions between head groups and, in turn, in a micelle
breakdown which increases the intermicellar repulsions. The minimum in the n vs c_A curves can then be interpreted as that alcohol concentration where intermicellar repulsions, which were
smaller than repulsions between head groups at low c_A, start
becoming predominant. As far as the free energy of the system is
concerned, it is then more favorable for the micelles to grow in
size and thus increase the intermicellar distance and reduce the
intermicellar repulsion as their ionization is increased by the
additions of pentanol. In this process the increase of repulsion
between head groups is partly compensated by the hydrophobic
interactions between alkyl chains.

Some remarks should be made concerning the shape of the
mixed SDS-pentanol micelles at high concentration. The large
values of n indicate that these micelles are not spherical (the
minimum spherical SDS micelles have an aggregation number of
about 62[45]). The dissymmetry of the light scattered by such
solutions, and the anisotropy of this scattered light were found
to be fairly small. Had the micelles been rod-like with aggregation number of 500, much larger values of these two quantities
should have been found. On the contrary, for the same large n-
values disk-like micelles would be characterized by much smaller
values of the anisotropy and dissymmetry of scattered light than
rod-like micelles. Another indication that the micelles may
indeed be disk-like is provided by the combination of the k_E, n
and $\bar{n}$ values. Indeed the excimers are formed through the
diffusion of a ground state probe and an excited state probe
towards each other in the micellar environment. The average displacement x of the probe during the time $t \sim 1/k_E$ required for
excimer formation can be written as

$$x^2 \propto Dt \propto D/k_E \qquad (1)$$

The probe diffusion coefficient is proportional to $\bar{n}^{-1}$. Also x
is related to the maximum dimension of the aggregate. Thus x
should be roughly proportional to n for rod-like micelles and to

$n^{\frac{1}{2}}$ for disk-like micelles. Equation 1 then predicts the near constancy of $n^2 k_E \bar{\eta}$ for rods and of $nk_E \bar{\eta}$ for disks. The results for micellar solutions containing 1 $\underline{M}$ pentanol and increasing amounts of SDS show a large increase of $n^2 k_E \bar{\eta}$, but a much smaller change of $nk_E \bar{\eta}$ with the SDS concentration again in agreement with a possible disk shape of the micelles.

The results in Table I indicate a decrease of the effective polarity and of the microviscosity of the probe microenvironment in the micelle palisade layer where the probes are preferentially solubilized[25] upon addition of pentanol. Two explanations can be given for these changes:

(i) the alcohol solubilized in the palisade layer replaces some of the water present in this layer, thereby reducing its polarity. At the same time the solubilized alcohol increases the molecular disorder in the palisade layer[49,50], and thus, its fluidity[51].

(ii) on a time average basis, the fluorescence probes would be solubilized deeper in the micelles in the presence than in the absence of alcohol. In this new site which is more hydrocarbon-like than the palisade layer the microviscosity and effective polarity would be lower than in the palisade layer.

Finally, the results in Table I show that the rate constant for pyrene excimer formation is always decreased upon addition of alcohol. It thus appears that the increase of n, and thus of micelle size, overcomes the decrease of $\bar{\eta}$ in determining the changes of k_E, upon addition of alcohol, for the systems investigated.

Effect of addition of oil to micellar solutions of alcohol-surfactant on the aggregation state of the surfactant

Starting from fairly concentrated alcohol+surfactant solutions the changes of n, $k_E \bar{\eta}$ and effective polarity have been studied upon addition of a series of oils differing in nature (arenes and alkanes), chain length (from hexane through hexadecane) and isomerism. In addition the effect of the nature of the alcohol, and chain length of the surfactant have also been investigated.

The effect of the nature of the oil is well illustrated by the results in Figure 2. For the 0.2$\underline{M}$ SDS+0.6 $\underline{M}$ pentanol system where the oil-free micelles have a fairly small aggregation number (47) and are probably nearly spherical the addition of oil gives rise to an increase of the surfactant aggregation number. The curves 1 and 2 relative to additions of dodecane and toluene are coincident but the maximum solubility of dodecane is about half that of toluene (maximum oil volume fraction 0.025 against

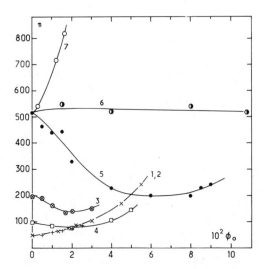

Figure 2 : Variation of the SDS aggregation number upon addition
of dodecane (+) and toluene (X) to the 0.2M SDS + 0.6M pentanol
system (curves 1 and 2) ; of dodecane to the 0.2M SDS + 0.6M
pentanol + 0.1 M NaCl system (⊗,curve 3) and to the 0.2M SDS +
0.8M pentanol system (□ , curve 4) ; and of dodecane (●), butylbenzene (◐) and toluene (O) to the 0.5M SDS + 1M pentanol system
(curves 5,6 and 7, respectively).

0.055). On the contrary, for the 0.5M SDS+1M pentanol system,
where the initial micelles are probably disk-like, the changes
of n with the oil volume fraction ϕ_0 (with respect to the solution
total volume) show striking differences depending on the nature
of the oil, increasing for toluene, remaining nearly constant for
butylbenzene, and decreasing, going through a minimum and then
increasing for dodecane. The solubilities of toluene ($\phi_0 \sim 0.015$),
dodecane ($\phi_0 \sim 0.09$) and butylbenzene ($\phi_0 \sim 0.13$) differ strongly.
Notice that in this fairly concentrated system the maximum amount of
solubilized toluene is smaller than that of dodecane, contrary to what
was found with the more dilute system. The decrease of n upon dodecane
addition is rather surprising. Indeed, an increase of n with ϕ_0 is ex-

pected to be associated with the swelling of the micelles upon
dissolution of oil, as for the more dilute 0.2M SDS + 0.6 M pentanol
system. The explanation of the difference of behavior between
the systems 0.2M SDS + 0.6M pentanol and the 0.5M SDS + 1M pentanol must therefore be sought in the difference of shape of the
micelles present in these two systems, in the absence of oil.
The three following results confirm this conclusion.

(i) The presence of 0.1M NaCl in the 0.2M SDS + 0.6M pentanol
system brings about an increase of the aggregation number of the
oil free micelles from 47 to 200. This last value corresponds to
non-spherical micelles. Figure 2, curve 3 shows that in this case
the addition of dodecane results in a n $\underline{vs}$ ϕ_0 curve qualitatively
similar to that for the 0.5M SDS + 1M pentanol system

(ii) The addition of dodecane to the 0.2M SDS + 0.8M pentanol
system where, in the absence of oil, n = 100 (slightly non spherical micelles) also yields a n $\underline{vs}$ ϕ_0 curve qualitatively similar
to that for concentrated systems, but with a much less marked
minimum.

(iii) The effect of dodecane additions to three 0.5M SDS + 1M
alcohol systems, differing in the nature of the alcohol (1-penta-
nol and two of its isomers) is shown in Figure 3. It can be seen

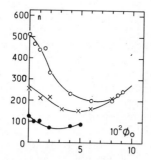

Figure 3 : Effect of addition of dodecane to 0.5M SDS + 1M alcohol
systems((O) : pentanol ; (x) :isopentanol ; and (●) : 2-pentanol),
on the SDS aggregation number . (Reprinted with permission from
ref. 13. Copyright 1982 American Chemical Society).

that the minimum in the n vs ϕ_0 curves becomes less and less
pronounced, as the value of n in the oil-free system decreases
that is when the oil-free micelles are closer and closer to
spheres.

From these results it is thus clear that the initial decrease
and the minimum in the n vs ϕ_0 curves correspond to some change
of shape of the non spherical oil-free micelles upon solubilization of dodecane. The fact that for oil-free spherical micelles
the addition of oil results in a monotonous increase of n with ϕ_0
leads to conclude that the oil-free non spherical micelles first
become spherical or nearly spherical upon addition of dodecane,
with a decrease of n. Once a nearly spherical shape is reached,
further dodecane additions bring about an increase of n, as for
the more dilute systems.

Before examining additional results which concern the complex
dependence of the shape of the n vs ϕ_0 curve on the nature of
the oil some indication will be given on the changes of I_1/I_3
(which reflect the changes of polarity in the micelle site of
solubilization of pyrene), microviscosity $\bar{\eta}$ and rate constant
for excimer formation k_E. The values of I_1/I_3 decreased upon dodecane
addition irrespective of the alcohol+surfactant concentration. At
high ϕ_0 the value of I_1/I_3 remained significantly larger than
in pure dodecane (0.9 to 1 against 0.7) suggesting that in the
dodecane containing microemulsions, pyrene is partitioned between
the dodecane core and the surfactant+alcohol interfacial layer.
On the contrary, at high toluene content pyrene appears to be
essentially solubilized in the toluene core as the I_1/I_3 value in
the microemulsion (1.21) is then very close to that in pure
toluene (1.25). The microviscosity $\bar{\eta}$ always decreased upon oil
additions to the mixed alcohol+surfactant micelles, irrespective of the
overall concentration of the system in agreement with the results
of others[27,52,53]. Finally, k_E was found to decrease (0.5M + 1M
pentanol system and additions of toluene) or increase (0.2M + 0.6M
pentanol system and additons of dodecane). In some instances the
initial increase of k_E was followed by a maximum and then a
decrease (0.5M SDS + 1M pentanol system and additions of dodecane;
0.5M SDS + 1M butanol system and additions of toluene). This
variety in the changes of k_E can be understood when it is
recalled that this quantity depends on the micelle size and on
the viscosity of the probe environment. The fact that pyrene can
be partitioned between various parts of the microemulsion aggregates adds another parameter to the changes of k_E. Nevertheless
k_E is usually found to decrease monotonously for the systems
where, upon oil additions, the effect of the micelle size increase
overcomes that of the viscosity decrease (0.5 M SDS + 1M pentanol
system, addition of toluene). For systems where n initially
decreased upon increasing ϕ_0, k_E always increased. For these
systems, k_E would be expected to always go through a maximum

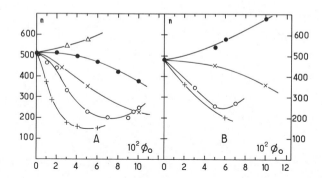

Figure 4. Variation of the surfactant aggregation number upon al-
kane addition to the 0.5$\underline{M}$ SDS + 1$\underline{M}$ pentanol system (Figure 4A)
and to the 0.5$\underline{M}$ STS + 1$\underline{M}$ pentanol system (Figure 4B). Addition
of n-hexane (Δ) ; n-octane ($\bullet$) ; n-decane (x) ; n-dodecane (o) ;
and n-hexadecane (+). (Reprinted with permission from ref. 13.
Copyright 1982 American Chemical Society).

at a ϕ_0 somewhat larger than that for which n would go through
a minimum, had the oil been surfficiently soluble. Indeed at the
ϕ_0 where n is minimum $\overline{n}$ would decrease only slightly upon further
addition of oil, which would increase n, as the microviscosity
is then already close to its minimum value of 4 or 5 cP. When
the effect of the increase of n becomes predominant, k_E decreases.
In many instances however the maximum of k_E and the minimum of n
could not be observed owing to the limited solubility of the oil.

    The next results to be examined concern the effect of the
alkane chain length (Figures 4A and 4B) and isomerism (Figure 5).
The first effect has been investigated for the two surfactants :
SDS, and its higher homolog, STS (sodium tetradecylsulfate) at
the concentration 0.5$\underline{M}$, in the presence of 1$\underline{M}$ pentanol. The
results of Figure 4 show that the n $\underline{vs}$ ϕ_0 curves present an increase for the short chain alkanes, and a decrease for the long
chain alkanes. The strong effect of the isomerism of octane on the
type of n $\underline{vs}$ ϕ_0 curve is also analogous to an effect of chain
length. Indeed, in terms of effective chain length the three
alkanes of Figure 5 range in the order : n-octane>isooctane>cyclo-
octane.

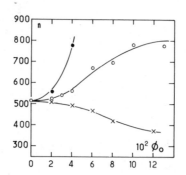

Figure 5. Variation of the SDS aggregation number upon addition
of n-octane (x) ; isooctane (o) , and cyclooctane (•) to the
0.5$\underline{M}$ SDS + 1$\underline{M}$ pentanol system. (Reprinted with permission from
ref. 13. Copyright 1982 American Chemical Society).

It should be noted that the change of n with ϕ_O is reversed
in going from n-hexane to n-octane for the SDS system (Figure 4A)
and from n-octane to n-decane for the STS system (Figure 4B).
Also, on a same plot, the n vs ϕ_O curves for the STS+n-octane system and for the SDS+n-hexane system would be nearly coincident.
The same remark holds for the results for the STS+n-decane and
SDS+n-octane systems. It thus appears as if an increase of surfactant chain length can be compensated by an equal increase of oil
chain length in order to retain the same n $\underline{vs}$ ϕ_O curve. This rule
however does not apply as well for the very long chain alkanes.
The results of Figures 4A and 4B suggest for SDS and STS, nearly
no change of n would be found upon n-heptane and n-nonane additions, respectively. In these two instances where n would be
almost independent of ϕ_O, the sum of the number of carbon atoms
of the alcohol N_A and of the oil N_O is equal to that of the surfactant, N_S. This rule (near constancy of n upon oil addition when
$N_A + N_O = N_S$) also applies to the case of additions of butylbenzene
(effective chain length N_O=7-7.5) to the 0.5$\underline{M}$ SDS+1M pentanol system (N_S=12, N_A=5).

The above rule must be extended to other systems before
its range of validity can be properly assessed. Nevertheless the

above findings emphasize the extreme importance of geometric
factors in determining the changes of n with ϕ_0, and thus, of the
changes of all the properties of the alcohol+surfactant system
related to the micelle size, such as the rheological properties,
as they come in contact with oil. Note that geometric factors
have repeatedly been assigned a very important role in determining
the structure of micelles, and of the aggregates present in microemulsions[49,54-56].

A qualitative interpretation of the above results, based on
purely geometric considerations,is as follows. For the sake of
clarity the oil-free alcohol+surfactant micelles are assumed to be
disk-like but the same reasoning would hold for rod-like micelles.
The half-thickness of the disks is smaller than the length of
the extended surfactant chain (see Figure 6A). Indeed, as the
alcohol is shorter than the surfactant and has its head group
anchored at the micelle surface, it creates voids in the micelle
interior which must be filled by the surfactant chains at the cost
of an increased chain folding and thus a decrease of the micelle
thickness (For rod-like micelles this would correspond to a
reduction of the rod diameter). The effect of short chain alcohols
on the thickness of organized assemblies is well established for
bilayers, through X-ray studies[57]. It is further assumed that the
chains have a preferential direction, perpendicular to the
micelle surface, as suggested by NMR studies [58,59] and that the
added alkane molecules retain this direction.

Consider first the case of short chain alkanes, such that
$N_S > N_A + N_0$. The first added alkane molecules will fill the voids in
the alcohol+surfactant packing, and be positioned, on a time-
average basis, as shown in Figure 6B. As more alkane is solubilized, the thickness of the disk, and to a lesser extent its
radius, increase slowly with ϕ_0, and n is also increased. Of
course, as more and more alkane is solubilized an alkane core
would form and the micelles would become nearly spherical or
spheroidal. It appears, however, that this situation may not
occur because the case where $N_S > N_A + N_0$ was always found to
correspond to a fairly low solubility of the alkane in the large,
disk-shaped micelles present in the concentrated alcohol+surfac-
tant solutions. The rate constant k_E shows a small increase or
decrease as the effect of increasing aggregate size is nearly completely compensated by that of decreasing $\bar{n}$.

Consider now the case where $N_S < N_0 + N_A$. The first added alkane
molecules bring about a considerable disorder in the original
micellar packing, creating bulges as shown on Figure 6C. This
disorder in a given part of the micelle may favor the solubilization of more alkane in the same micellar part, giving rise to a
nucleus for an alkane core. As ϕ_0 is increased nuclei grow in
different parts of the micelles. The evolution of the micelles

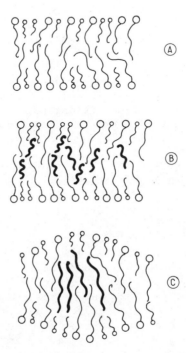

Figure 6. Model for the explanation of the changes of the surfactant aggregation number upon addition of oil to mixed surfactant-
alcohol micelles. (A) cross section of the micelle in the absence
of oil ; (B) micelle with a dissolved short chain alkane ; (C)
micelle with a dissolved long chain alkane. (The alcohol is represented with a small head group and a short chain and the surfactant with a large head group and a longer chain).

can be described by two possibly complementary mechanisms. In the
first one the nuclei merge into an alkane core which imposes a
spherical shape to the micelles owing to interfacial forces. This
process is·accompanied by a release of surfactant (and thus, a
decrease of n) and alcohol, as the sphere is the body having
the smallest surface area for a given volume. In the second mechanism the micelles containing several nuclei breakdown into
smaller ones (decrease of n) which take a spherical shape and grow
larger upon increasing ϕ_0. These two mechanisms probably occur
simultaneously, one or the other being predominant depending on
the alkane characteristics. Both mechanisms suggest an increase of
the width of the aggregate size distribution, at least at intermediate oil content. They also predict that the initial decrease
of n at low ϕ_0 should be the more pronounced the longer the
alkane chain as is indeed observed. Finally the decrease of n
through micelle breakdown, and of $\overline{n}$, through oil solubilization
both tend to increase k_E, as is indeed observed in all instances where n initially decreased upon oil addition.

When N_S is close to $N_A + N_0$ the change of n should be intermediate between the two preceding cases, that is n should change
only very little with ϕ_0 as the micelles become progressively
spherical.

No explanation was given, yet, for the striking differences
in the changes of surfactant aggregation number upon additions
of toluene, butylbenzene and dodecane to the 0.5$\underline{M}$ SDS+1$\underline{M}$ pentanol
system (Figure 2, curves 5 to 7). It is clear, however, that part
of the differences lie in the strongly differing effective chain
length of these three oils(4-4.5 ; 7-7.5 ; and 12, respectively).
Another difference is the fact that at low ϕ_0 (< 0.01), most of
the toluene may be dissolved in the micelle palisade layer[25]
where it would act as pentanol, and thus strongly increase n. It
is only when enough toluene has been added to the system that a
core starts forming.

Finally, it is clear that from the values of n, and the
weighing-in concentrations and partial molal volumes of the components of the systems, and on the assumption of spherical
aggregates, one can easily calculate the overall aggregate
radius, and the radius of the oil core. Such calculations have
been performed for several of the studied systems[13]. In all instances, the results indicate that the oil core extends over a
certain length of the surfactant, alkyl chain, nearly reaching the
terminal methyl group of the alcohol chain. It is noteworthy
that an identical result has been reached in structural studies
of water in oil microemulsions[60,61]. Thus the structure of the
interfacial film separating oil and water in microemulsions
appears to be independent of its curvature, and thus of the
microemulsion type.

CONCLUSIONS

This work constitutes the first attempt to study the complex changes of surfactant aggregation number and shape of the micelles when generating an O/W microemulsion by a stepwise procedure, first increasing the surfactant concentration ; second, adding alcohol ; and, third, adding oil.

In this process the aggregates undergo large changes of size and shape which are very sensitive to the chain length and concentrations of the alcohol, surfactant and oil, as well as to the nature of the oil, and to the isomerism of the oil and alcohol. These parameters are the same as those involved in the optimization procedure of the formulations used in terriary oil recovery.

A simple model purely based on geometric considerations has permitted us to explain the main results obtained in the present study. The range of validity of this model is presently checked on other systems, including non-ionic surfactants.

REFERENCES

1. K. L. Mittal, Editor, "Micellization, Solubilization and Microemulsions", Vols. 1 and 2, Plenum Press, New York, 1977.
2. L. M. Prince, Editor, "Microemulsions. Theory and Practice", Academic Press, New York, 1977.
3. K. L. Mittal, Editor, "Solution Chemistry of Surfactants", Plenum Press, Vols. 1 and 2, New York, 1979.
4. D. O. Shah, Editor, "Surface Phenomena in Enhanced Oil Recovery", Plenum Press, New York, 1981.
5. I. D. Robb, Editor, "Microemulsions", Plenum Press, New York, 1982.
6. R.C. Dorrance and T.F. Hunter, J.Chem.Soc. Faraday Trans. I, $\underline{68}$, 1312 (1972) and $\underline{70}$, 1572 (1974).
7. N.J. Turro and A. Yekta, J.Am.Chem.Soc., $\underline{100}$, 5951 (1978) ; A. Yekta, M. Aikawa and N.J. Turro, Chem.Phys.Lett., $\underline{63}$, 543 (1979).
8. P. Infelta, Chem.Phys.Lett., $\underline{61}$, 88 (1979) ; P. Infelta and M. Grätzel, J.Chem.Phys., $\underline{70}$, 179 (1979).
9. S. Atik, M. Nam and L. Singer, Chem.Phys.Lett., $\underline{67}$, 75 (1979).
10. P.K. Koglin, D.J. Miller, J. Steinwandel and M. Hauser, J.Phys. Chem., $\underline{85}$, 2363 (1981).
11. M. Almgren and J.E. Löfroth, J.Colloid Interface Sci., $\underline{81}$, 486 (1981).
12. M. Almgren and J.E. Löfroth, J.Chem.Phys., $\underline{76}$, 2734 (1982).
13. P. Lianos, J. Lang, C. Strazielle and R. Zana, J.Phys.Chem., 86, 1019 (1982) and manuscript submitted for publication.
14. M. Tachiya, Chem.Phys.Lett., $\underline{33}$, 289 (1975) ; J.Chem.Phys., $\underline{76}$, 340 (1982).

15. D.J. Miller, U.K. Klein and M. Hauser, Ber.Bunsenges.Phys. Chem., 84, 1135 (1980).
16. M.A. Rodgers, Chem.Phys.Lett., 78, 509 (1981).
17. A. Watkins and B. Selinger, Chem.Phys.Lett., 64, 250 (1979).
18. F. Grieser and R. Tausch-Treml, J.Am.Chem.Soc., 102, 7258 (1980).
19. M. Van der Auweraer, J.C. Dederen, E. Geladé and F.C. De Schryver, J.Chem.Phys., 74, 1140 (1981).
20. M. Tachiya, Chem.Phys.Lett., 69, 605 (1980).
21. U. Gosele, U.K. Klein and M. Hauser, Chem.Phys.Lett., 68, 291 (1979).
22. M.D. Hatlee and J.J. Kozak, J.Chem.Phys., 72, 4358 (1980) and 74, 1098 (1981).
23. M.D. Hatlee, J.J. Kozak, G. Rothenberger, P.P. Infelta and M. Grätzel, J.Phys.Chem., 84, 1508 (1980).
24. P. Lianos, M. Dinh-Cao, J. Lang and R. Zana, J.Chim.Phys., 78, 497 (1981).
25. J.K. Thomas, Chem.Revs., 80, 283 (1980) and references therein.
26. S.S. Atik and J.K . Thomas, J.Am.Chem.Soc., 103, 3550 (1981) and 103, 4367 (1981).
27. S.J. Grigoritch and J.K. Thomas, J.Phys.Chem., 84, 1491 (1980).
28. P. Lianos and R. Zana, J.Colloid Interface Sci., 84, 100 (1981).
29. P. Lianos and R. Zana, J.Phys.Chem., 84, 3339 (1980).
30. R. Zana, S. Yiv, C. Strazielle and P. Lianos, J.Colloid Interface Sci., 80, 208 (1981).
31. E.A.G. Aniansson, S.N. Wall, M. Almgren, H. Hoffmann, I. Kielman, W. Ulbricht, R. Zana, J. Lang and C. Tondre, J.Phys. Chem., 80, 905 (1976).
32. S. Yiv, R. Zana, W. Ulbricht and H. Hoffmann, J.Colloid Interface Sci., 80, 224 (1981).
33. A number of quasi-elastic light scattering studies of microemulsions have shown single or nearly single exponential behavior of the autocorrelation function of the scattered intensity, which indicate low polydispersity : A.M. Bellocq and G. Fourche, J.Colloid Interface Sci., 78, 275 (1980) ; A.M. Bellocq, G. Fourche, P. Chabrat, L. Latemendia, J. Rouch, and C. Vaucamps, 27, 1629 (1980) ; G. Fourche, A.M. Bellocq and S. Brunetti, J.Colloid Interface Sci., inpress; A.M.Bellocq D. Bourdon, B. Lemanceau and G. Fourche, J.Colloid Interface Sci., in press ; A. Graciaa, J. Lachaise, P. Chabbrat, L. Letamendia, J. Rouch, C. Vaucamps, M. Bourrel and C. Chambu, J.Physique Lett., 38, L 253 (1977) ; A.M. Cazabat and D. Langevin, J.Chem.Phys., 74, 3148 (1981) ; A.M. Cazabat, D. Langevin and A. Pouchelon, J.Colloid Interface Sci., 73, 1 (1980) ; A.M. Cazabat, D. Chatenay, D. Langevin and A. Pouchelon, J.Physique Lett., 41, L441 (1980). There has been one report of non single exponential behavior of the autocorrelation function of the scattered light : E. Gulari, B. Bedwell, S. Alkhafaji, J.Colloid Interface Sci., 77, 202 (1980). This however may be due to the presence of dust in

the microemulsion, or to the existence of very strong interaction between particles which may also result in non exponential behavior of the autocorrelation function (see for
instance : T. Tsang and H.T. Tang, J.Chem.Phys., $\underline{76}$, 3873
(1982)).

34. P.J. Missel, N. Mazer, G.B. Benedek, C.Y. Young and M.C.
    Carey, J.Phys.Chem., $\underline{84}$, 1044 (1980).
35. C. Tanford, Y. Nozaki and M. Rohde, J.Phys.Chem., $\underline{81}$, 1555,
    (1971) and references therein.
36. J.M. Corkill, J.F. Goodman, T. Walker and J. Wyer, Proc.Roy.
    Soc., A $\underline{312}$, 243 (1969) ; J.M. Corkill, K.W. Gemmell, J.F.
    Goodmann and T. Walker, Trans.Faraday Soc., $\underline{66}$, 1817 (1970).
37. E. Hirsch, S. Candau and R. Zana, unpublished results.
38. P. Lianos and S. Georghiou, Photochem.Photobiol., $\underline{30}$, 355
    (1979).
39. K. Zachariasse, Chem.Phys.Lett., $\underline{57}$, 429 (1978) and referen-
    ces therein.
40. M. Viriot, M. Bouchy, M. Donner and J.C. André, paper
    submitted for publication. These authors are thanked very
    much for supplying us with highly purified dipyrenylpropane
    and the I_{IE}/I_M vs $\bar{n}$ calibration curve obtained with paraffin
    oil-hexadecane mixtures.
41. J. Emert, C. Behrens and M. Goldenberg, J.Am.Chem.Soc., $\underline{101}$,
    771 (1979).
42. N. Turro, M. Aikawa and A. Yekta, J.Am.Chem.Soc., $\underline{101}$, 774,
    (1979).
43. J. Kratohvil, J.Colloid Interface Sci., $\underline{75}$, 271 (1980).
44. P. Lianos and R. Zana, J.Colloid Interface Sci., in press.
45. K. Hayase and S. Hayano, Bull.Chem.Soc.Jpn., $\underline{50}$, 83 (1977).
46. K. Birdi, S. Backlund, K. Sorensen, T. Krag and S. Dalsager,
    J.Colloid Interface Sci., $\underline{66}$, 118 (1978).
47. F. Grieser, J.Phys.Chem., $\underline{85}$, 928 (1981).
48. C. Tanford, J.Phys.Chem., $\underline{78}$, 2469 (1974).
49. V. Bansal, K. Chinnaswamy, C. Ramachandran and D. Shah, J.
    Colloid Interface Sci., $\underline{72}$, 524 (1979).
50. B. Lemaire and P. Bothorel, Macromolecules, $\underline{13}$, 311 (1980).
51. M. Dvolaitzky and C. Taupin, Nouveau J. Chimie, $\underline{1}$, 355 (1977).
52. M. Almgren, F. Grieser and J.K. Thomas, J.Am.Chem.Soc., $\underline{102}$,
    3188 (1980).
53. Y. Tricot, J. Kiwi, W. Niederberger and M. Grätzel, J.Phys.
    Chem., $\underline{85}$, 862 (1981).
54. V. Bansal, D. Shah and J.O'Connell, J.Colloid Interface Sci.,
    $\underline{75}$, 462 (1980).
55. D. Oakenfull, J.Chem.Soc. Faraday Trans.I , $\underline{76}$, 1875 (1980).
56. J. Mitchell and B. Ninham, J.Chem.Soc.Faraday Trans.II, $\underline{77}$,
    601 (1981).
57. J. François, B. Gilg, P. Spegt and A. Skoulios, J.Colloid
    Interface Sci., $\underline{21}$, 293 (1966).
58. B. Cabane, J.Physique, $\underline{42}$, 847 (1981).
59. K. Dill and P. Flory, Proc.Natl.Acad.Sci. USA, $\underline{78}$, 676 (1981);

K. Dill, J.Phys.Chem., <u>86</u>, 1498 (1982).

60. M. Dvolaitzky, M. Guyot, M. Lagües, J-P. Le Pesant, R. Ober,
    C. Sauterey and C. Taupin, J.Chem.Phys., <u>69</u>, 3279 (1978).

61. D. Cebula, L. Harding, R. Ottewill and R. Pusey, Colloid
    Polym.Sci.,<u>258</u>, 973 (1980).

CHARACTERIZATION OF MICROEMULSION STRUCTURE USING MULTI-COMPONENT SELF-DIFFUSION DATA

B. Lindman[+] and P. Stilbs[o]

[+]Physical Chemistry 1, Chemical Center
University of Lund, S-220 07 Lund, Sweden
[o]Institute of Physical Chemistry, University of Uppsala
S- 751 21 Uppsala, Sweden

Multi-component self-diffusion studies can provide a
direct insight into the presence of water- or hydrocarbon-
continuous regions and can, therefore, be used to test
models of the structure of various phases formed in complex amphiphilic systems. For isotropic solution phases
in surfactant systems, which contain at the same time large
amounts of water and hydrocarbon, the self-diffusion coefficients were determined for several components to obtain
information on solution structure. A novel Fourier transform pulsed gradient spin-echo ^{1}H and ^{13}C NMR method is
particularly useful in simultaneously providing multi-
component self-diffusion data. Different microemulsions,
defined here as thermodynamically stable isotropic solutions of oil, water and surfactant(s), are structurally
very different. In systems of a typical micelle-forming
surfactant, oil and water, pronounced O/W microemulsions
are formed at high water contents. In other cases, like
systems of Aerosol OT, hydrocarbon and water, the solutions are distinctly of the W/O character. For microemulsions with a medium-chain alcohol (butanol, pentanol)
as co-surfactant, one cannot reconcile the observations
with either water or oil confined in closed regions.
Thus over wide composition ranges, both hydrocarbon and
water diffusion are rapid, demonstrating that the solutions are hydrocarbon- and water-continuous. The results
do not show areas with oil-in-water and water-in-oil microemulsions with a sharp transition between the two types
of aggregates. Thus it is only at quite high water and
hydrocarbon concentrations that typical normal and reversed

1651

micellar solutions, respectively, are encountered. Microemulsions are argued to contain very flexible structures
with low-order internal interfaces. Inter alia because
of the miscibility of cosurfactant and water, one is
probably not far from the rather structureless situation
applicable to simple solutions.

INTRODUCTION

The structure of microemulsion systems is at present intensely
discussed since it is fundamental to a general understanding of
microemulsions and to numerous phenomena in the field. (For recent
reviews, see References 1-3.) Knowing the structure is also a prerequisite to meaningful theoretical investigations. It is, therefore, not surprising that considerable efforts are made using a
large number of experimental approaches to investigate microemulsion structure. However, the problem is made difficult since many
experimental methods are not directly applicable and, furthermore,
confusion arises since different workers often consider different
microemulsion systems and because the structural variability seems
appreciable.

Self-diffusion studies appeared as one possible approach
which could add to our knowledge of microemulsion structure. The
situation improved greatly with the development of the Fourier
Transform NMR method, which provides ready access to multi-component
self-diffusion data[4-6]. This article discusses microemulsion structure on the basis of self-diffusion data for a range of different
systems. The experimental results have been published, in part,
previously[7], so only some general features will be considered here. In
the discussion we will, in addition to the FT NMR data, also refer
to tracer self-diffusion data obtained in collaboration with N. Kamenka and B. Brun[8].

Often in discussions of microemulsions, one considers as structural alternatives only schematized O/W and W/O structures (using
"oil" in a broad sense). An O/W microemulsion has water as the
continuous medium and therein are dispersed small oil droplets with
surfactant (and perhaps also cosurfactant) at the surface. A W/O
microemulsion has an "oil"-continuous medium with small water droplets with surfactant (cosurfactant) at the surface. It is very
difficult to reconcile a large number of experimental observations
with the occurrence of only these two structures and as a result,
a large number of alternative structural models have been suggested.

In this article, we will restrict the discussion to two points,
i.e.:

1) To what extent are the two idealized structures applicable at
   all for surfactant systems?

2) Is the structure of microemulsions of the cosurfactant-type, i.e. those composed of surfactant, cosurfactant, oil and water, compatible with the model of W/O or O/W structures?

A PROPOSED DEFINITION OF MICROEMULSION

In the classic work of Schulman and co-workers, it was observed that on adding a so-called cosurfactant to a dispersion of oil, water and surfactant, one may, under certain conditions, get a clear sample. The cosurfactant is often a C_4-C_6 alcohol. It was suggested that one obtains a very stable emulsion with very small drops. There has been a lot of debate on several aspects of microemulsions, in particular, on their stability and structure. The term 'microemulsion' has been introduced to describe many phenomena in surfactant and surfactant-less systems, creating considerable confusion in the literature. This problem has been discussed at length   among Scandinavian scientists on various occasions. As a result of this, it was agreed to propose that a microemulsion is to be defined as follows[9]: A microemulsion is a single phase composed of at least water, oil and surfactant, and this phase is an optically isotropic and thermodynamically stable solution. This definition excludes real emulsions, i.e., all multi-phase systems, liquid crystalline phases and surfactantless systems. The definition is certainly not without problems. and Friberg[10] has criticized the above definition to the extent that it requires thermodynamically stable solutions and proposes instead a requirement of spontaneous formation.

In the present article, we follow the above definition and consider mainly thermodynamically stable isotropic solutions composed of a typical surfactant, a medium-chain alcohol, hydrocarbon and water.

MULTI-COMPONENT SELF-DIFFUSION AND MICROEMULSION STRUCTURE

A characteristic feature of the microemulsion phase in such systems is that it has a very large existence region often from essentially pure water to pure oil. The structurally most interesting part is that where one has simultaneously large concentrations of both oil and water. As elaborated on in the original papers[7,8], studies of the molecular self-diffusion may provide a good insight into whether a particular phase is oil- or water-continuous or both or neither.

In the use of self-diffusion for structural studies on surfactant systems, one makes two general assumptions, i.e.
- microscopically, all molecular motions are very rapid, "liquid-like"
- there is a distinct division into hydrophobic and hydrophilic domains separated by (hard) well-defined interfaces.

For an O/W type structure, i.e. with water continuity and oil discontinuity one predicts then water self-diffusion to be rapid and oil diffusion to be slow and in particular $D_{water} \gg D_{oil}$. For a W/O structure one expects analogously $D_{oil} \gg D_{water}$.

CASES OF O/W STRUCTURE

A situation with pronounced O/W character is well established for normal micellar solutions[11],[12]. Performing self-diffusion studies on solutions of ionic surfactants with solubilized oil gives as expected a water self-diffusion which is more rapid than that of oil by one to two orders of magnitude. This is illustrated in Fig. 1

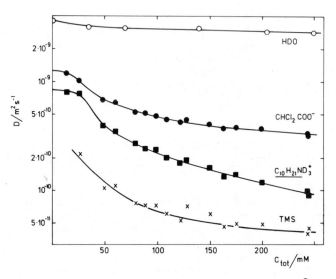

Figure 1. Observed self-diffusion coefficients at $40^{\circ}C$ in the $C_{10}H_{21}NH_3^+CHCl_2COO^-$ – D_2O system with trace amounts of solubilized $(CH_3)_4Si$ (TMS) present. The water proton diffusion has been corrected as described in Reference 7 for averaging with the ammonium proton sites. (Data mainly from Reference 13.) The CMC is in the 20 mM range.

for the case of tetramethylsilane (TMS) in micelles of decylammonium dichloroacetate (except for water self-diffusion, data are from Ref. 13). For tetramethylsilane in aqueous solutions of sodium dodecyl sulfate the TMS and water self-diffusion coefficients differ by a factor of ca. 50 or more[14]. Tracer self-diffusion data have established clear differences (factor of ca. 20) between the self-diffu-

sion coefficient of water and that of solubilized decanol for solutions of sodium dodecyl sulfate[11], sodium p-octylbenzene sulfonate[15]
and sodium octanoate[16]. One may note, furthermore, from these examples that counterion diffusion is relatively rapid and that surfactant
diffusion approaches oil diffusion at high concentrations.

The examples cited demonstrate the occurrence of microemulsions
of the O/W type with a distinct separation into hydrophilic and hydrophobic domains.

COSURFACTANT SOLUBILIZATION AND DISRUPTION OF MICELLES

Closely connected with the problem of the cosurfactant-type
microemulsions is that of the effect of cosurfactant on surfactant
micelles in water-rich solution. In particular we need information
on the distribution of cosurfactant between the micelles and the
intermicellar solution and on the effect of cosurfactant on micelle
stability. These aspects can conveniently be monitored by the FT
NMR self-diffusion technique. For, inter alia, solutions of sodium
dodecyl sulfate the solubilization equilibria have been quantified
for a large number of different alcohols[6]. The equilibrium constants
are such that for moderate to concentrated surfactant solutions,
butanol and pentanol show a roughly even distribution between the
micelles and the intermicellar solution. Thus typical cosurfactants
have the maximal compatibility with both the aqueous and nonaqueous
compartments of surfactant systems.

Addition of higher concentrations of cosurfactant to surfactant
micellar solutions has, as illustrated in Fig. 2, marked influences
on the self-diffusion coefficients[14]. In the absence of cosurfactant
or at low cosurfactant concentrations, a nonpolar solubilizate like
tetramethylsilane has a self-diffusion coefficient which is very
low and is lower than that of the surfactant. This corresponds to
a rather complete confinement of solubilizate to the micelles which
have a low translational mobility. At higher cosurfactant concentrations, the TMS self-diffusion coefficients may increase very strongly
and become much larger than the surfactant self-diffusion coefficient.
The latter also shows a marked increase. The facile motion of TMS
demonstrates that the TMS molecules are no longer confined to distinct well-defined micelles. Instead there appears to be a breakdown of the micelles. From the phase diagrams of surfactant systems[17]
one can infer that this break-down of organized surfactant structures
is by no means restricted to micelles but occurs also for, for example, lamellar liquid crystalline phases. These observations have
clear implications as regards the effect of cosurfactant on "interface" organization in microemulsions and thus for discussions of
the structure of cosurfactant-type microemulsions.

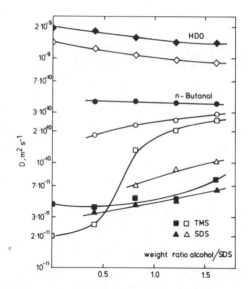

Figure 2. Component self-diffusion coefficients in the n-butanol-SDS-D$_2$O-TMS system. (Data from Reference 14.) Filled symbols = 0.073 and empty symbols = 0.291 weight ratio SDS/D$_2$O, respectively. The water proton self-diffusion coefficient is corrected for site exchange with the butanol hydroxyl group as described in Reference 7.

CASES OF W/O STRUCTURE

A situation with pronounced W/O character occurs for the isotropic hydrocarbon-rich solution phases of three-component systems surfactant-hydrocarbon-water. In Fig. 3 are given self-diffusion data for the system Aerosol OT/p-xylene/water[7]. As can be inferred $D_{oil} \gg D_{water}$ over a very wide range of composition. The rapid oil self-diffusion and the slow water self-diffusion demonstrate that these microemulsions are distinctly of the W/O type with a very marked separation into hydrophilic and hydrophobic domains.

The W/O structure applies also with good approximation to three-component systems surfactant-long-chain alcohol-water but not always so pronouncedly as in the previous system. For the system sodium octanoate-decanol-water, decanol self-diffusion in the water-poor solution phase is more rapid than water self-diffusion by a factor of ca. 6 over a wide concentration range[18]. Dividing the self-diffusion coefficients with those of the pure solvents they differ by a factor of ca. 40. In this case, the solutions consist of water droplets in a continuum of decanol but with a non-negligible amount of water in the continuous medium. The separation into hydrophilic and hydrophobic domains is thus less distinct. It should be remembered that sodium octanoate is a rather weakly self-associating surfactant and that the situation could be rather different with a long-chain surfactant.

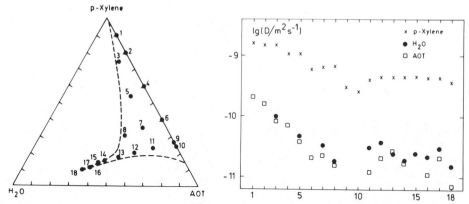

Figure 3. In the extensive isotropic solution phase of the system Aerosol OT-p-xylene-water, hydrocarbon self-diffusion is very much more rapid than water self-diffusion.

COSURFACTANT-TYPE MICROEMULSIONS

Multi-component self-diffusion data were obtained as a function of composition for several microemulsion systems. In Fig. 4 are given FT ^{1}H NMR results for the system sodium octylbenzene sulfonate-pentanol-decane-sodium chloride-water. (Tracer self-diffusion data and phase extension for this system are given in Ref. 8). In Fig. 5 are presented data for the system sodium dodecylsulfate-butanol-toluene-water.

As illustrated here in studies of "classic" microemulsions composed of surfactant, co-surfactant (butanol or pentanol), hydrocarbon and water one thus observes for several systems a quite rapid self-diffusion of all components. In particular one observes that $D_{oil} \simeq D_{water}$, both being large. For the highest water and hydro-

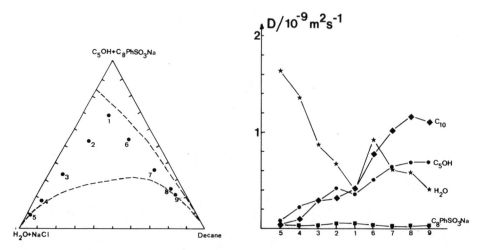

Figure 4. Self-diffusion coefficients for the system sodium octyl-
benzenesulfonate-pentanol-decane-sodium chloride-water.
(From Ref. 7).

carbon concentrations, the difference in magnitude between the
values of D_{water} and D_{oil} becomes appreciable and thus one approaches
a situation of type O/W and W/O, respectively.

It is immediately clear, however, that over extensive regions
of stability of the isotropic solutions one can not reconcile these
observations with either water or oil confined in closed regions
like the traditional views of microemulsions as having either an
O/W or W/O structure. What emanates instead is that
- either there is some type of bicontinuous structure. One periodic
  regular structure has been suggested by Scriven[19] and structures
  of this type are well-known for cubic mesophases[20].
- or there is no distinct separation into hydrophobic and hydrophilic domains. There may be no well-defined interfaces, they
  may instead be labile and rapidly deforming. Perhaps there is a
  large fraction of small aggregates? An alternative way of ex-
  pressing the notion that there is no distinct separation into
  hydrophobic and hydrophilic domains is that, for example, there
  may be a structure with droplets containing essentially water (+
  counterions and some cosurfactant) in a medium which is essentially

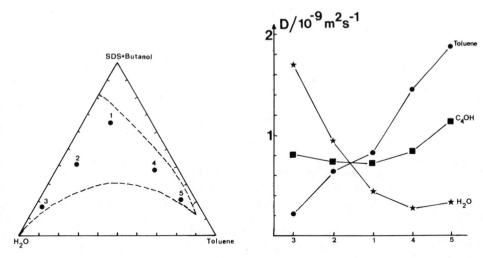

Figure 5. Self-diffusion coefficients for the system sodium dodecylsulfate-butanol-toluene-water. (From Ref. 7).

composed of oil and cosurfactant. However, there is a nonnegligible
amount of water in the continuous medium. From an estimate of the
value for the droplet self-diffusion, this amount of water can be
deduced.

COMMENTS ON MICROEMULSION STRUCTURE

In attempts to develop a structural model of microemulsions
there are a number of observations in addition to the self-diffusion
results which merit consideration. If we restrict ourselves to the
systems composed of ionic surfactant-cosurfactant-hydrocarbon-water
one notes that
- viscosity is quite low
- the NMR spin relaxation of, for example, the protons or the ^{13}C
  nuclei in the chains is quite slow
- from theoretical studies[21,22] it appears that alcohol cosurfactants
  have the tendency of lowering the interfacial energy between the

hydrophobic and the hydrophilic domains
- addition of alcohol cosurfactants to micellar solutions of an
  ionic surfactant causes a breakdown of micelles as manifested
  by a very much facilitated translational motion of oil. (See
  above and Ref. 14). This observation suggests that the separation
  into hydrophobic and hydrophilic domains is considerably reduced.
- in kinetic studies one observes that alcohol cosurfactant accele-
  rates considerably the exchange of surfactant between different
  domains[23]. This suggests labilization of the internal interfaces.
- studies of cosurfactant partitioning by different techniques[6,24,25]
  demonstrate a rather even distribution of cosurfactant between different domains.

    The low viscosity and the slow NMR relaxation demonstrate that
there are no extended aggregates with ordered interfaces. The low
viscosity rules out extended ordered bicontinuous structures. It
should be recalled that cubic mesophases having such structures are
extremely viscous.

    The low interfacial energy, which is also suggested by the even
distribution of cosurfactant between different domains, implies a
small energy difference between different aggregate geometries and
also that small aggregates (favoured by the entropic term) become
energetically favoured with respect to large aggregates. In the
limit where this tendency is very pronounced one may encounter
microemulsions approaching simple molecularly disperse solutions.
The observed self-diffusion coefficients are indeed consistent
with a picture of only small kinetic entities.

    When discussing microemulsion structure it is important to
consider structural features of similar systems which are more
easily investigated. Three-component systems surfactant-oil-water
generally display a rich phase behaviour with micellar solutions
and different liquid crystalline phases. The structures of these
are easy to study and have been well characterized[17]. There is here
a quite distinct separation into water-rich and water-poor domains
and we could somewhat loosely talk about a high degree of structure.
On the other extreme, as regards structure, we may take simple co-
surfactant-water isotropic solutions like butanol-water etc. It is
clear that we here have a quite low degree of structure over distances of the order of 10^{-9} m. The different types of microemulsions
apparently fall between these examples of high and low "degree of
structure". For systems like Aerosol OT-hydrocarbon-water one is
quite close to the highly structured cases. With a surfactant and
a long-chain alcohol, the degree of structure is also quite high.
However, with cosurfactants like butanol and pentanol, the situation is clearly quite different, with a quite low degree of structure.

Of course, it is not unexpected that in the presence of large amounts of butanol or pentanol, having a rather high miscibility with water, one may come quite close to the rather structureless limit.

At present this discussion has several deficiencies. For instance, we have no quantitative measure of the degree of structure, our measurements are restricted to rather few systems and we are not able to suggest a structural model which can be tested by other methods. However, it appears to be clear that future work on microemulsion structure should,inter alia,proceed along the following lines:

1) There is a need for much theoretical work which can provide models of microemulsion systems which are amenable to test by a wide range of physico-chemical methods.

2) It is important to study the same well-defined microemulsion system with a range of experimental approaches. The initiative taken by M. Kahlweit at this meeting leading to collaborative work of a number of groups within the "Lund Project" is highly welcomed. This work which concentrates on two systems, i.e. sodium octylbenzenesulfonate-butanol-water-toluene or dodecane, will bring together several types of experimental studies.

3) The present self-diffusion studies have to be extended to several other systems. In particular we want to systematically vary the chemical structure of the different components such as the alkyl chain length of surfactant, cosurfactant and oil,and compare aliphatic and aromatic hydrocarbons etc. It would be expected that even minor changes, such as changing the length of the surfactant or especially the cosurfactant by one carbon atom, could have quite marked effects on microemulsion structure.

REFERENCES

1. I. Robb, Editor, "Microemulsions", Plenum Press, New York, 1982.
2. Th. F. Tadros, these proceedings.
3. P.G. De Gennes and C. Taupin, J. Phys. Chem., 86, 2294 (1982).
4. P. Stilbs and M.E. Moseley, Chem. Scr., 15, 176 (1980).
5. P. Stilbs and M.E. Moseley, Chem. Scr., 15, 215 (1980).
6. P.Stilbs, J. Colloid Interface Sci., 87, 385 (1982).
7. B. Lindman, P. Stilbs, and M.E. Moseley, J. Colloid Interface Sci., 83, 569 (1981).
8. B. Lindman, N. Kamenka, T.-M. Kathopoulis, B. Brun, and P.-G. Nilsson, J. Phys. Chem., 84, 2485 (1980).
9. I. Danielsson and B. Lindman, Colloids and Surfaces, 3, 391 (1981).
10. S. Friberg, Colloids and Surfaces, 4, 201 (1982).
11. H. Wennerström and B. Lindman, Phys. Reports, 52, 1 (1979).
12. B. Lindman and H. Wennerström, Topics in Current Chemistry, 87, 1 (1980).

13. P. Stilbs and B. Lindman, J. Phys. Chem., 85, 2587 (1981).

14. P. Stilbs, J. Colloid Interface Sci., in press.

15. B. Lindman, M.-C. Puyal, N. Kamenka, B. Brun, and G. Gunnarsson, J. Phys. Chem., 86, 1702 (1982).

16. B. Lindman and N. Kamenka, to be published.

17. P. Ekwall, Adv. Liquid Cryst., 1, 1 (1975).

18. H. Fabre, N. Kamenka, and B. Lindman, J. Phys. Chem., 85, 3493 (1981).

19. L.E. Scriven, "Micellization, Solubilization and Microemulsions," K.L. Mittal, Editor, Vol. 2, p. 877, Plenum Press, New York, 1977.

20. K. Fontell, Mol. Cryst. Liquid Cryst., 63, 59 (1981).

21. B. Jönsson and H. Wennerström, J. Colloid Interface Sci., 80, 482 (1981).

22. B. Jönsson, "The Thermodynamics of Ionic Amphiphile-Water Systems. A Theoretical Analysis," Thesis, Lund, 1981.

23. R. Zana and J. Lang, in "Solution Behavior of Surfactants: Theoretical and Applied Aspects", K. L. Mittal and E. J. Fendler, Editors, Plenum Press, New York, 1982.

24. J. Biais, P. Bothorel, B. Clin, and P. Lalanne, J. Colloid Interface Sci., 80, 136 (1981).

25. J. Biais, P. Bothorel, B. Clin, and P. Lalanne, J. Dispersion Sci. Techn., 2, 67 (1981).

PHOTON CORRELATION TECHNIQUES IN THE INVESTIGATION OF WATER-IN-OIL

MICROEMULSIONS

J.D. Nicholson and J.H.R. Clarke

Chemistry Department
University of Manchester Institute of
Science and Technology
Manchester M60 1QD, U.K.

Water-in-oil microemulsions formed with surfactant
Aerosol OT have been extensively investigated by the
technique of dynamic light scattering. For various mole
ratios of water to surfactant, the volume fraction
dependence of the translational diffusion coefficient
and the so-called index of polydispersity have been
determined to obtain information concerning particle
interactions present in these systems. The range of
volume fractions covered in this study (0.05 to 0.4) is
much larger than previously attempted. The precision
and limitations of photon correlation techniques in
investigations of this nature are discussed.

1. INTRODUCTION

Several studies have been reported in which dynamic light
scattering (DLS) has been used to investigate water-in-oil micellar
and microemulsion systems[1-7]. These experiments yield translational
diffusion coefficients which have been mainly interpreted in terms
of particle hydrodynamic radii, micellar molecular masses and
(with the aid of additional ultracentrifuge data) micelle aggregation numbers. The dependence of diffusion coefficients on
volume fraction and its relation to particle interactions has also
attracted attention[6,7]. Recently the interpretation of non-
exponential correlation functions has been discussed in terms of
polydispersity fluctuations[7-9].

In addition to their practical applications these systems are
of considerable interest from a more fundamental point of view.

In contrast to the related oil-in-water systems, the particle interactions are predominantly van der Waals' in nature and the differential scattering cross-sections are rather small so that concentrated systems (up to volume fractions, ϕ, of about 0.5) are often quite transparent and can be examined by DLS without complications due to multiple scattering.

In this article we present and discuss some DLS results on micelle diffusion in the system water/Aerosol OT/n-heptane. This system is advantageous in that no co-surfactant is required to stabilize the microemulsions. Virtually all the water is contained in the microemulsion droplets and essentially all the AOT in the surfactant 'coat' (the c.m.c. for this system is very low so that the concentration of free AOT monomers should be very small). In Figure 1 is shown that part of the phase diagram which has been covered in this study.

After an outline of the relevant current theory for interpreting DLS we discuss, in Section 3, the determination of experimental correlation functions and the precautions that must be taken in the analysis of data. We then describe the volume fraction dependence of mean diffusion coefficients, in Section 4a, pointing out that reliable hydrodynamic radii can only be obtained in the limit of infinite dilution. After discussing the mechanism of micelle growth in Section 4b, we give some attention to deviations from the 'ideal' exponential form of the correlation functions and suggest possible interpretations.

2. THEORY

In dynamic light scattering laser light is scattered through an angle θ by the system under observation without absorption but with a Doppler broadening of the incident frequency due to the motions of the scattering particles. In practice it is easier to measure the normalised correlation function of the scattered light intensity fluctuations, $g^{(2)}(t)$[10].

$$g^{(2)}(t) = <I(0)I(t)><I>^{-2} \tag{1}$$

where the angle brackets denote a time average. For a large number of scattering particles Gaussian field statistics apply and $g^{(2)}(t)$ can be related to the electric field correlation function $g^{(1)}(t)$[10].

$$g^{(2)}(t) = 1 + C|g^{(1)}(t)|^2 \tag{2}$$

where the experimental constant C is close to unity for correctly chosen experimental conditions.

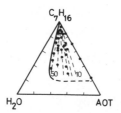

Figure 1. The phase diagram of the ternary system water/AOT/n-heptane at 20 °C (J. Rouviere, J. Couret, M. Lindheimer, J. Dejardin and R. Marrony, J. Chim. Physique 76, 289 (1979)), showing the points included in this study.

The field correlation function describes the interference fluctuations at a remote detector due to scattering through an angle θ from the moving particles in the sample and[9]

$$g^{(1)}(t) = F(\vec{q},t)/S(\vec{q}) \qquad (3)$$

where $S(q)$ is defined as the static structure factor at wave vector $|q| = (4n\pi/\lambda)\sin\theta/2$. $F(\vec{q},t)$ is the so-called intermediate scattering function.

$$F(\vec{q},t)=(1/N\langle A^2(\vec{q})\rangle) \sum_{i,j=1}^{N} \langle A_i(\vec{q},0)A_j(\vec{q},t)\exp i\vec{q}\cdot[\vec{r}_j(t)-\vec{r}_i(0)]\rangle \qquad (4)$$

and

$$S(\vec{q}) = F(\vec{q},0)$$

$A_j(\vec{q})$ defines the amplitude of the scattered field from particle j at position $\vec{r}_j$ at time t. N is the total number of particles. For the systems discussed here the particle sizes are much smaller than 1/q so that the q-dependence in $A(\vec{q})$ can be suppressed.

Various assumptions can be made to simplify the expression (4). In a dilute monodisperse system $A_j = A_j$ and in addition all cross correlations between the motions of different particles can be ignored. Upon the assumption of Brownian translational diffusion

$$F(\vec{q},t) = \exp(-D_s q^2 t) \tag{5}$$

where D_s is the particle self diffusion coefficient. In this case the Stokes-Einstein relation is valid, relating D_s to the particle hydrodynamic radius r_H;

$$D_s = k_B T (6\pi r_H \eta)^{-1} \tag{6}$$

where η is the solvent viscosity.

For systems studies here, equation (5) is only strictly valid for small volume fractions where the average distance between particles is $\gg 1/q$. At higher volume fractions cross correlations in equation (4) become significant and one observes mutual diffusion of the particles which occurs in response to concentration gradients. For a monodisperse system, D_s in equation (5) is replaced by D_m, the mutual diffusion coefficient.

The above theoretical summary is the basis of particle size determination using DLS. The conditions, however, rarely hold exactly and significant complications occur[9,10]. If the system is polydisperse but still dilute then F(q,t) is a weighted sum of exponentials from different particle types. Several methods of analysis have been discussed[10-12]. We have employed the cumulants expansion[11], for which

$$\ln g^{(1)}(t) = -\bar{\Gamma}t + \tfrac{1}{2}(\mu_2/\bar{\Gamma}^2)(\bar{\Gamma}t)^2 - \tfrac{1}{3!}(\mu_3/\bar{\Gamma}^3)(\bar{\Gamma}t)^3 + \dots \tag{7}$$

where $\bar{\Gamma}$ is the mean inverse relaxation time and the μ's are moments of the normalised distribution of Γ. With the available precision of DLS data statistical significance can usually be given only to the first two terms. $\mu_2/\bar{\Gamma}^2$ is a variance which gives a measure of 'non-ideality' or departure from single exponential behaviour.

In a concentrated polydisperse system there is another possible mode of relaxation due to polydispersity fluctuations[8,9], characterised by a mean self diffusion coefficient. This has been identified in some microemulsion systems at high volume fraction

as a separate slow exponential contribution to $g^{(1)}(t)$[6,7]. For
the systems and volume fractions discussed here it was not
possible satisfactorily to resolve the correlation functions in
the above way.   Non-exponential behaviour has been interpreted in
terms of equation (7).

3.   EXPERIMENTAL

Microemulsions were prepared using Fluka AG Purum grade AOT,
Analar grade organic solvents and triply distilled water.
Correlation functions determined from such samples were compared,
for a water-to-AOT mole ratio of 10 and for volume fractions up to
0.30, with samples made up from specially purified AOT obtained
from the University of Basel[13].   Results from the two sets of
measurements were identical and we conclude that our results are
not significantly affected by small amounts of salt impurities in
Fluka AOT.   Particulate impurities were removed by filtration
through a sintered glass disc and finally by centrifuging of
samples in the sealed 1 cm square section light scattering cells.
Particular precautions were taken to minimise changes in sample
composition (due to solvent evaporation, etc) during the
preparation procedure.

All the results reported here were obtained using 488 nm argon
ion laser illumination on samples maintained at $25+0.5$ ^{o}C for at
least two hours prior to measurements in a thermostatted light
scattering cell holder.   All data reported here are for $\theta = 90+0.3^{o}$.
Scattering from less than a single coherence area within the sample
was transferred using a lens and aperture stops onto an ITT FW 130
photomultiplier.   Photon count fluctuations were analysed with a
Malvern Instruments K7025 128-channel multibit correlator.

The configuration of the correlator channels was modified as
follows in order to give a measurement and check on the experimental
DC background level.   88 correlation channels (time steps on the
correlation function) were used to characterise the decay curves.
A 64-channel delay was then inserted followed by 8 correlation
channels to obtain an intermediate background.   A further 64-
channel delay was followed by a final 16 correlation channels used
to fix the experimental value of the background.   In this way we
were able to check for the existence of any long time decay
(spurious or otherwise) on $g^{(2)}(t)$.

The importance of precise determination of the DC background
in the analysis of DLS data cannot be too strongly emphasised. This
background has to be subtracted from the experimental correlation
function and any error produces a large uncertainty in the form of
ln $g^{(1)}(t)$.   The procedure adopted in this work was as follows.
The experimental background was compared with the theoretical

value, calculated from the total number of photon counts and the total number of time increments sampled during the run. Provided the two values agreed to within the standard deviation of the background then the data was accepted with confidence, yielding values of D to within ±5%. If the difference between the two quantities was greater than three times the standard deviation than the data was discarded. An additional check on the appropriateness of the background was provided by a comparison of

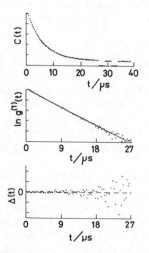

Figure 2. Examples of data obtained for microemulsions in the water/AOT/n-heptane system at 25 °C. The volume fraction is 0.06 and R=10. C(t) is the intensity correlation function. The two 64 channel delays are shown as breaks of 8 channels. A quadratic best fit for ln $g^{(1)}(t)$ is shown with $\mu_2/\bar{\Gamma}^2 = 0.06$. $\Delta(t)$ are the deviations of the experimental points from this line, showing no systematic effects.

the intermediate 8 and final 16 channels, the requirement being
that they should agree to within the combined standard deviation.

The apparatus was checked and calibrated using monodisperse
suspensions of Dow latex spheres, which gave single exponential
correlation functions. Although not discussed here, the q^2
dependence of inverse mean relaxation times (expected for
diffusional motion) was verified as previously described[1,6].
Examples of an experimentally measured correlation function for a
water-in-oil microemulsion and also a weighted least mean square
quadratic fit to ln $g^{(1)}(t)$ are given in Figure 2.

4. RESULTS AND DISCUSSION

4a. Volume Fraction Dependence of $\bar{D}_m$

The dependence of the mean mutual diffusion coefficient $\bar{D}_m$
($=\bar{\Gamma}q^{-2}$) on volume fraction for water/AOT/n-heptane microemulsions
and for various values of R (the mole ratio of water to AOT) is
shown in Figure 3. Volume fractions are calculated here from the
volume of n-heptane required to make up a microemulsion of given
total volume.

For all the samples shown (and for similar microemulsions in
different organic solvents) the diffusion coefficient decreases
with increasing volume fraction. This decrease is linear up to
$\phi=0.52$, within the experimental error, and fits the expression

$$\bar{D}_m = \bar{D}_m(\phi \rightarrow 0)[1+\alpha\phi] \tag{8}$$

For n-heptane microemulsions $\alpha=-2.1\pm0.4$ and there is no systematic
variation with R.

It has been predicted that D_s should show similar behaviour
to that of D_m in equation (8) in dilute systems as a result of
particle interactions[14,15]. For hard spheres with purely
repulsive interactions α is positive whilst it is predicted to
be negative in the case of attractive interactions. Since D_s is
usually less than D_m at a given volume fraction, the observed
result may be indicative of net attractive forces between
microemulsion particles. There is an implicit assumption here
that the particle size does not vary significantly with volume
fraction but there is some support for this from neutron scattering
data[16].

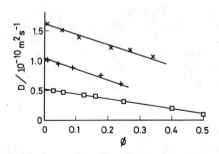

Figure 3. The volume fraction dependence of experimental diffusion coefficients in the water/AOT/n-heptane system at 25 $^{\circ}$C. x – R = 11.2, + – R = 30, □ – R = 50.

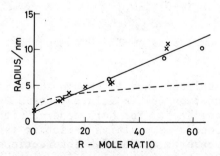

Figure 4. The dependence of mean hydrodynamic radii on the mole ratio of water to AOT at infinite dilution. Best fit lines for the model of droplet growth by aggregation (solid line) and by particle swelling (broken line) are also shown. Circles refer to data from Ref. 4.

4b. Hydrodynamic Radii and the Dependence on R

One advantage of data such as are presented in Fig. 3 is that one can extrapolate to $\phi = 0$ where $D_m = D_s$ and the Stokes-Einstein relation (equation (7)) can be used to calculate $\bar{r}_H$, the mean particle hydrodynamic radius. This is of course a 'z-averaged mean' not a true size mean.

The $\bar{r}_H$ values obtained for the water/AOT/n-heptane system are plotted against mole ratio of water to surfactant, R, in Figure 4 for the range 0<R<30. The following relationship is obeyed.

$$r_H/nm = 0.175 R + 1.5 \qquad\qquad (9)$$

where the gradient of 0.175 is a least mean squares fit.

The predictions of two simple theoretical models are included in Figure 4. In both of these the simplifying assumption is that all of the water and AOT present in the system is contained in the microemulsion droplets. In the first model it is assumed that the total number of microemulsion droplets remains constant. In this case $\bar{r}_H$ would vary as $R^{1/3}$ and the droplets would grow by swelling, analogous to an expanding balloon. The area covered by each molecule of the surfactant sheath would have to increase. In the second model it is assumed that the interfacial area per surfactant molecule remains constant (i.e. the surfactant sheath probably maintains the same thickness for differing size droplets) while the number of microemulsion droplets decreases as R increases. This model predicts a linear dependence of $\bar{r}_H$ on R and corresponds to growth by droplet aggregation.

The experimental results indicate that $\bar{r}_H$ is linearly dependent on R for 0<R<30, with a gradient of 0.175. A recent theoretical study suggests that the gradient should be 0.128 for the aggregation model of these microemulsions[17].

Perhaps it might have been expected that the smaller micelles would take up water by swelling initially until the limit is reached where the surface area per surfactant molecule is optimal and that above this limit the particles would grow by aggregation when taking up additional water. Such behaviour has not been observed in this or in previous[1,6] studies even at low mole ratios. For R>30, however, the values for the hydrodynamic radii are greater than as predicted by the droplet aggregation model. One possible explanation is that as the maximum microemulsion drop size is approached some monomeric AOT transfers to the bulk medium. This decrease in AOT at the interface will necessarily lead to an increased mean droplet size. Small inverse micelles could be formed by self aggregation of such AOT in the bulk medium.

4c. 'Non-Ideality' of the Correlation Functions

All of the correlation functions measured in this study were
found to depart from the 'ideal' exponential form to an extent
which varied with R and volume fraction. It is important to
emphasise here that observation of these deviations was only made
possible by accurate determination of the long time limiting
background levels of the correlation functions using the procedure
described in Section 3. Without this procedure most of the data
could easily be fitted to single exponentials using a 'freely
fitted' background[1,2]. In this work ln $g^{(1)}(t)$ was fitted to a

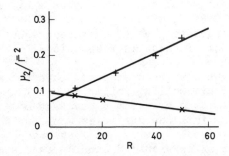

Figure 5. The dependence of the 'polydispersity factor', $\mu_2/\bar{\Gamma}^2$,
on the mole ratio R of water to AOT at constant volume fraction,
and at 25 °C. + - volume fraction = 0.24, x - volume fraction
= 0.06.

quadratic function using a correctly weighted least mean squares procedure[6]. The deviations of the experimental points were then plotted to search for systematic deviations. For 10<R<52 all of the quadratic fits were satisfactory for volume fractions up to about 0.3, the observed deviations being randomly scattered. For some systems with volume fractions greater than 0.3 systematic deviations were observed and a quadratic fit was obviously inappropriate. These cases will, however, not be discussed in the present article. In the present discussion we relate the quadratic fits to the first two terms in the cumulant expansion for $\ln g^{(1)}(t)$, equation (7) of Section 2 and interpret the quadratic coefficient in terms of polydispersity effects.

A plot of $\mu_2/\bar{\Gamma}^2$ against R is shown in Figure 5 for both low volume fraction (0.06) and high volume fraction (0.24) micro-emulsions in the water/AOT/n-heptane system. The mean separation of spherical particles at a volume fraction of 0.06 is about two particle radii so that the direct short range van der Waals' interactions are likely to be fairly weak. It is seen from Figure 5 that the values of $\mu_2/\bar{\Gamma}^2$, which are a measure of the polydispersity, decrease as the value of R increases. At first sight this may seem a strange result since one intuitively expects that the larger swollen particles might show the larger size fluctuations. For small particles, however, the fractional difference made by the addition or subtraction of a single surfactant molecule is far greater than for larger particles. The same would apply to variations in the amount of solubilised water in the droplet cores. On this basis a wider distribution in sizes relative to the mean, for smaller particles, is understandable.

The behaviour observed at the higher volume fraction is quite different. In this case the values of $\mu_2/\bar{\Gamma}^2$ show a dramatic increase as R increases. At this volume fraction the mean interparticle spacing is much less than one particle diameter and we must expect significant direct and hydrodynamic interactions between the particles. Nevertheless the fact that the value of $\mu_2/\bar{\Gamma}^2$ for low R is very close to that observed at much lower volume fractions gives a basis for concluding that size polydispersity is again the predominant cause of the non-ideality. The error bars at high volume fractions are quite large and this reflects an increasing difficulty in fitting a quadratic function to $\ln g^{(1)}(t)$. The alternative two exponential fit[6] was certainly not possible, however, for any of the data on Figure 5.

If the high values of $\mu_2/\bar{\Gamma}^2$ for large particles at high volume fractions are due to polydispersity then this is understandable on the following basis. Large swollen particles are likely to be much more responsive to the strong direct particle interactions at large packing densities and this may lead to an interaction-induced polydispersity. It is well known that three dimensional space can be filled much more efficiently by particles of variable size. Such an explanation would require that the size fluctuations occurred on a time scale long compared to the observation time (1-10 microseconds). If this were not the case then a single relaxation would be observed in DLS corresponding to the mean particle size in a fast fluctuating structure. The mean time between collisions at this volume fraction is much shorter than the observation time. If the polydispersity interpretation is correct than it implies that the microemulsion droplets are still rather stable with respect to droplet collisions.

ACKNOWLEDGEMENTS

The authors wish to thank Esso Chemical Research for financial support.

REFERENCES

1. R.A.Day, B.H.Robinson, J.H.R.Clarke and J.V.Doherty, J. Chem. Soc. Faraday I, 75, 132 (1979)
2. M.Zulauf and H.F. Eicke, J. Phys. Chem., 83, 480 (1979)
3. A.M.Cazabat, D.Langevin and A.Pouchelon, J. Colloid Interface Sci., 73, 1 (1980)
4. E.Gulari, B.Bedwell and S.Alkhafaji, J. Colloid Interface Sci., 77, 202 (1980)
5. R.Finsy, A.Devriese and H.Lekkerkerker, J. Chem. Soc. Faraday II, 76, 767 (1980)
6. J.D.Nicholson, J.V.Doherty and J.H.R.Clarke, in 'Microemulsions' I.D.Robb, editor, pp.33-47 Plenum, New York, 1982
7. D.J.Cebula, R.H.Ottewill, J.Ralston and P.N.Pusey, J. Chem. Soc. Faraday I, 77, 2585 (1981)
8. M.B.Weissman, J. Chem. Phys., 72, 231 (1980)
9. P.N.Pusey and R.J.A.Tough, Adv. Colloid Interface Sci., 16, 143 (1982)
10. P.N.Pusey and J.M.Vaughan, in "Dielectric and Related Phenomena", Chemical Society Specialist Periodical Report, 2, 48 (1975)
11. D.E.Koppel, J. Chem. Phys., 57, 4814 (1973)
12. F.C.Chen, A.Yeh and B.Chu, J. Chem. Phys., 66, 1290 (1977)
13. The authors wish to express thanks to Professor H.F.Eicke for the gift of this sample.
14. G.K.Batchelor, J. Fluid Mech., 74, 1 (1976)
15. B.U.Felderhof, J. Phys., A11, 929 (1978)
16 P.D.I.Fletcher, B.H.Robinson, F.Bermejo-Barrera, D.Oakenfull, J.C.Dore and D.C.Steytler, in 'Microemulsions', I.D.Robb, editor, pp.221-232 Plenum, New York, 1982
17. D.Oakenfull, J. Chem. Soc., Faraday I, 76, 1875 (1980)

WATER/OIL MICROEMULSION SYSTEMS STUDIED BY

POSITRON ANNIHILATION TECHNIQUES[†]

S. Millán, R. Reynoso, J. Serrano,
R. López and Luz Alicia Fucugauchi*

Química del Positronio
Instituto Nacional de Investigaciones Nucleares
06140 México, D. F., México

The positron annihilation technique was applied to
the study of water/oil microemulsion formation process
in sodium stearate (or oleate)-alcohol-solvent-water systems. At certain water/oil ratios (R_M-points) drastic
changes in the number of thermal o-Ps atoms formed can
be observed which can be associated with the onset of
microemulsion formation. The effect of the nature of the
solvents and cosurfactants as well as surfactant concentration were further investigated. Dynamic laser light
scattering has been utilized for substantiating the
phase transitions determined in the different microemulsion systems by positron annihilation. Difference in
the behavior between saturated and unsaturated surfactants was observed. This behavior has been rationalized
by considering packing and kink presence in microemulsion formation.

[†]This research is in furtherance of the U.S.A.-
Mexico Cooperative Science Program through the National
Science Foundation and the Consejo Nacional de Ciencia
y Tecnologia (Project BCCBCEU-010569).

INTRODUCTION

Positron annihilation has been found to be an extremely sensitive technique for investigating aggregation mechanisms in both aqueous[1,2] and reversed micelles[3-6] as well as in microemulsions[7]. The application of this technique is based on the very high dependence of the positronium atom (Ps) (the bound state of a positron and an electron) formation and its subsequent reactions on phase changes occurring in the aggregates.

The mechanism of Ps formation in condensed systems has been described in terms of various models[8,9] in which translationally excited positron abstracts an electron from the surroundings or in which the positron recombines with an ejected electron at the end of the positron track.

Positronium formation and positronium reactions can both be observed by measuring the positron lifetimes.[1-8,10-13]   The positron annihilation parameters can be correlated with changes occurring in the solutions as a function of their composition. The drastic decrease in positronium formation probability in both aqueous[1,2] and reversed systems[3-6] as well as in microemulsions[7] has been explained by an effective trapping of (energetic) positrons or Ps atoms, thereby, reducing the number of (thermal) Ps atoms formed.

Effect of increasing the amount of water solubilized in reversed micelles on positronium formation probabilities has been studied. A more quantitative investigation is reported in the present work. Sodium stearate-alcohol-oil-water microemulsion systems were investigated by varying the solvent (oil), the cosurfactant (alcohol) and the surfactant concentrations. Isooctane, hexadecane and cyclohexane were used as solvents. N-butanol, n-pentanol, isopentanol and n-hexanol were the cosurfactants. In order to assess the effect of unsaturation on microemulsion formation mechanism, sodium stearate was replaced by sodium oleate in all the solutions under study. The observed phase changes, detected by positron annihilation, have been substantiated by dynamic laser light scattering.

EXPERIMENTAL SECTION[†]

a) Materials. Sodium oleate and sodium stearate (ICN Inc.) were of pharmaceutical grade and were used without further purification. Solvents such as isooctane, hexadecane, cyclohexane and alcohols (n-butanol, n-pentanol, isopentanol and n-hexanol) were spectro-

---

[†]Experimental assistance provided by C. Rodríguez F.

scopıc grade (Merck). The hydrocarbons were further dried by distillation over metallic sodium.

b) Preparation of solutions for positron lifetime measurements.
The various solutions studied in this investigation were prepared
by mixing the components in the indicated proportions to which different amounts of triply distilled water were added. Composition
of the solutions was surfactant (2.5 or 5.0 g), solvent (25 ml),
cosurfactant (10 ml) and water.

c) Positron lifetime measurements and preparation of samples.
Positron lifetime measurements were carried out by a fast coincidence technique[7] (figure 1). The 1.27-MeV photons resulting from
the deexcitation of the excited state of ^{22}Ne, formed in the positron decay of ^{22}Na, are detected by a plastic scintillation detector (Naton 136, 2x1 in.) mounted on a photomultiplier connected to
a base. The output signal is processed in a differential constant
fraction discriminator, which permits only the passage of signals
which correspond to a photon energy in the 0.8-1.3-MeV range. These

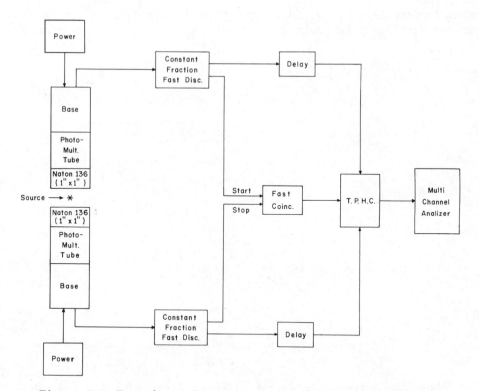

Figure 1.   Experimental arrangement using the principle of
"fast coincidence".

signals provide the starting signal in a time-pulse height converter (TPHC). The 0.51-MeV photons resulting from the positron annihilation are similarly processed, except that the window in the differential constant fraction discriminator is set to allow the passage of pulses corresponding to a 0.3-0.52-MeV range. These signals provide the stop signal in the TPHC. Since the energy discrimination is done on the fast signals, only valid pulses arrive at the TPHC, and the dead time of the unit is greatly reduced, leading to a much higher coincidence counting rate and thus, to significantly shorter counting times?

Measurements under these experimental conditions required typically 15-20 minutes. Random coincidences were further suppressed by requiring coincidence between the output pulses of the discriminators, utilizing a fast coincidence unit, whose output pulse was used as gating signal at the TPHC. The output of the TPHC was stored in the usual way in a multichannel analyzer. The resolution of the system, as measured by the prompt time distribution of a ^{60}Co source and without changing the 1.27-0.511-MeV bias, was found to be 0.392 ns fwhm. Corrections for the source component, which had an intensity of less than 3%, were made in the usual way by using conventional computational methods.[1-7,14] Specially designed cylindrical sample vials (Pyrex glass 100 mm long and 10 mm i. d.) were filled with about 2 ml of the sample solution. The posi-

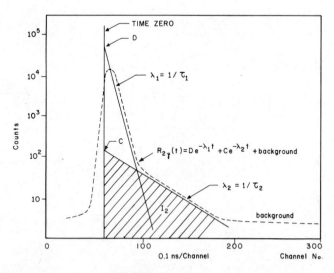

Figure 2. Typical positron lifetime distribution curve as acquired by "fast coincidence techniques".

tron sources were placed inside the vials and completely immersed
in the liquid sample. The vials were degassed and subsequently
sealed off and counted at room temperature.[1-7]

d) General method of data analysis. Positron lifetimes and
distributions were obtained by standard computational techniques[1,4]
The lifetime spectra were resolved, as previously described[1,4] into
two components: a short-lived component, which is the result of
p-Ps annihilation, free positron annihilation and epithermal Ps interactions, and the long-lived component, with a decay constant λ_2
and its associated intensity I_2 which originates from the reactions
and subsequent annihilation of thermalized or nearly thermalized
o-Ps (figure 2).

e) Dynamic laser light scattering experiments. The dynamic
laser light scattering measurements were carried out by using a
Malvern 2000 spectrometer in conjunction with a 80 mW Spectra-
physics argon ion laser.

RESULTS AND DISCUSSION

Complexity of microemulsions became painfully apparent during the course of the present study. Many parameters affect the
onset of microemulsion formation and phase diagrams which describe
them. Effects of altering the solvent (isooctane, hexadecane and
cyclohexane), the type and concentration of the surfactant (sodium
stearate and oleate), and the cosurfactants (n-butanol, n-pentanol,
isopentanol and n-hexanol) had to be examined systematically. Relatively small alterations of each of these parameters caused pro-
found changes in the composition of R_M-points*.

Differences in the behavior between saturated and unsaturated
surfactants is the most remarkable result of the present investigation. Thus, replacing sodium stearate by sodium oleate in the sur-
factant-n-hexanol-isooctane-water system obviated microemulsion
formation.

Effects of the various components on the R_M-points in these
two systems will be discussed sequentially.

Effect of the solvent. Microemulsion formation has been studied as a function of the solvent composition in sodium stearate
and sodium oleate systems which contained n-butanol, n-pentanol or
n-hexanol in the presence of isooctane, hexadecane or cyclohexane.

---

* In order to avoid the cumbersome "water/oil ratio at which microemulsion formation occurs" the abbreviation R_M-point or R_M-value
is introduced and is used throutout this paper.

All the solutions investigated in the present work contained
2.5 g sodium stearate or 5.0 g sodium oleate, 10 ml alcohol, 2.5
ml solvent and various amounts of water. The data are shown in
figure 3 and Table I. I_2, the long-lived component in the positron annihilation spectra is plotted as a function of the water
content. In sodium stearate-n-pentanol systems, the plots reveal
an increase of I_2 up to R_M-points of 0.20 and 0.35 in the presence
of isooctane and hexadecane, respectively; whereas in sodium stear-
ate-n-hexanol dispersions R_M-points reach values of 0.15 and 0.30
when isooctane or hexadecane is present (Table I).

If the water-oil ratios (W/O) increase, I_2 shows a significant
drop which can be associated with the onset of microemulsion formation[7]. At still higher values of W/O ratios, I_2 increases again
signaling additional structural changes.

In sodium stearate systems, R_M-values are seen to increase

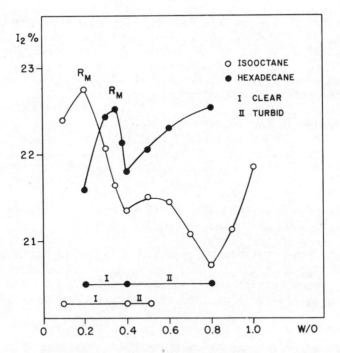

Figure 3 . I_2 vs.water content in sodium stearate-n-pentanol-
oil-water systems at room temperature.

Table I.- Water/Oil Ratios at Which Microemulsions are Formed.

System	Surfactant	Water/Oil (v/v)	Water/Oil molar ratio	Water/Oil clear range upper limit
Surfactant				
Alcohol 10 ml	n-Butanol	Sodium Stearate 2.5 g	0.25	2.31	0.40
Isooctane 25 ml	n-Pentanol	Sodium Stearate 2.5 g	0.20	1.85	0.45
		Sodium Oleate 2.5 g	0.20	1.83	0.80
Water	Isopentanol	Sodium Stearate 2.5 g	0.20	1.85	0.45
	n-Hexanol	Sodium Stearate 2.5 g	0.15	1.38	0.50
		Sodium Oleate 2.5 g	0.10	0.92	0.30
Surfactant	n-Butanol	Sodium Oleate 5.0 g	0.50	4.59	0.80
Alcohol 10 ml	n-Pentanol	Sodium Stearate 5.0 g	0.40	3.70	0.60
Isooctane 25 ml		Sodium Oleate 5.0 g	0.50	4.59	0.60
Water	n-Hexanol	Sodium Stearate 5.0 g	0.30	3.70	0.60
		Sodium Oleate 5.0 g	no formation		
Surfactant	n-Butanol	Sodium Stearate 2.5 g	no formation		
Alcohol 10 ml		Sodium Oleate 5.0 g	no formation		
Hexadecane 25 ml	n-Pentanol	Sodium Stearate 2.5 g	0.35	5.67	0.40
		Sodium Oleate 5.0 g	0.40	6.50	0.90
Water	n-Hexanol	Sodium Stearate 2.5 g	0.30	4.86	0.50
		Sodium Oleate 5.0 g	0.30	4.86	1.00
Surfactant	n-Butanol	Sodium Stearate 2.5 g	0.20	1.20	0.70
Alcohol 10 ml		Sodium Oleate 5.0 g	0.40	2.40	0.70
Cyclohexane 25 ml	n-Pentanol	Sodium Stearate 2.5 g	0.20	1.20	0.50
		Sodium Oleate 5.0 g	0.20	1.20	0.30
Water	n-Hexanol	Sodium Stearate 2.5 g	0.20	1.20	0.50
		Sodium Oleate 5.0 g	0.40	2.40	1.00

with increasing chain length of the hydrocarbon (figure 3; Table I). Apparently larger average intermolecular distances are created between the surfactant molecules[15,16] which require more water to fill the available space. This, in turn, leads to increased R_M-values.

Conversely, in cyclohexane-containing systems, microemulsion formation occurs at an R_M-value of 0.2 in the presence of any co-surfactant (i.e., n-butanol, n-pentanol or hexanol; Table I).

Microemulsion is not seen to form in sodium oleate-n-butanol-hexadecane dispersions (Table I). In solutions containing cyclohexane, microemulsion formation occurs at an R_M-point of 0.4 in the presence of n-butanol and n-hexanol. In the n-pentanol-containing system, microemulsions are formed at an R_M-point of 0.2. It seems that, except in the case of the n-pentanol solution, when cyclohexane is present in the microemulsion mixtures the hydrocarbon chain length of the alcohol has no influence on the R_M-value in sodium oleate systems. Data for sodium stearate dispersions have been rationalized analogously. For sodium oleate-n-butanol or n-hexanol-cyclohexane systems the R_M-value is higher (0.4) than that in sodium stearate-n-butanol or n-hexanol-cyclohexane systems (0.2). Conversely, if pentanol is the cosurfactant, microemulsion formation is the same in the saturated and unsaturated systems (0.2; Table I).

In sodium oleate-n-pentanol-isooctane or hexadecane-water systems the R_M-values decrease with increasing hydrocarbon chain length of the solvent. In hexanol-containing solutions microemulsion does not form in isooctane but it is formed in hexadecane at 5.0 g surfactant concentration at an R_M-value of 0.3. This trend is opposite to that observed in sodium stearate systems.

Influence of the cosurfactant. Phase changes for sodium stearate-alcohol-oil-water microemulsion systems have been investigated as a function of the hydrocarbon chain length of the cosurfactant. N-butanol, n-pentanol, isopentanol and n-hexanol have been used as cosurfactants at different water concentrations.

R_M-points reached values of 0.25, 0.20 and 0.15 in the presence of n-butanol, n-pentanol and n-hexanol. Maxima are reached at the same value for n-pentanol and isopentanol (0.2; Table I).

A consistent trend is seen to emerge which depends on the chain length of the alcohol. In each system, the value of the R_M-point decreases with increasing chain length of the alcohol. This behavior can be rationalized by the lesser alcohol partitioning into the oil phase with decreasing chain length of the cosurfactant. Longer chain length alcohols partition more favorably

in the oil phase than shorter ones.[15,16] The average intermolecular distance is larger for shorter than longer chain alcohols
(compare maxima for butanol (R_M = 0.25), pentanol (R_M = 0.2),
hexanol (R_M = 0.15) in Table I. Similar results have been obtained in hexadecane (Table I). Conversely, R_M-values have been found
to be independent of the chain length of the alcohol cosurfactant
in the cyclohexane solvent (Table I).

In sodium oleate-alcohol-isooctane-water systems microemulsion formation occurs at an R_M-point of 0.5 in n-butanol or n-
pentanol-containing systems. Conversely, in the n-hexanol containing dispersion, microemulsion formation does not occur since I_2 is
seen to reach the first maximum outside the clear range ratio of
these systems[7] (figures 3-5; Table I).

The R_M-value decreases with increasing chain length of the
alcohol in sodium oleate-alcohol-hexadecane-water systems for n-
pentanol and n-hexanol. Similar behavior has been observed in
sodium stearate dispersions (*vide supra*) However, for isooctane-

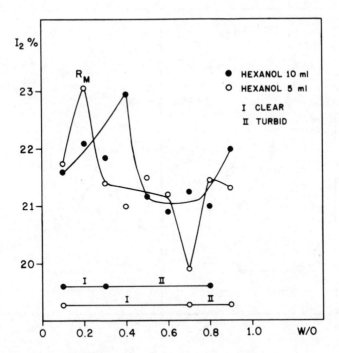

Figure 4. I_2 *vs.* water content in sodium oleate-n-hexanol-
isooctane-water systems at room temperature.

containing mixtures with n-butanol or n-pentanol, microemulsion
formation occurs at the same R_M-value (0.5); in contrast, there is
no microemulsion formation in the n-hexanol-containing system.

Similarly, no microemulsion formation is seen in sodium-oleate
-n-butanol-hexadecane solutions. Lack of microemulsion formation
cannot be attributed to unsaturation since sodium oleate and sodium
stearate systems behave similarly in the presence of butanol. Combination of short chain alcohol with long chain solvents (or vice
versa: combination of long chain alcohols with short chain solvents) is generally expected to preclude microemulsion formation[15]
In sodium oleate-n-hexanol-isooctane-water systems microemulsions
do not form. If the concentration of sodium oleate is changed from
5.0 to 2.5 g in this mixture, microemulsion formation occurs at an
R_M-point of 0.1 Same phenomenon is observed if the amount of alcohol is changed from 10 to 5 ml in the same dispersion at 5.0 g surfactant concentration at an R_M-point of 0.2 (figure 4).

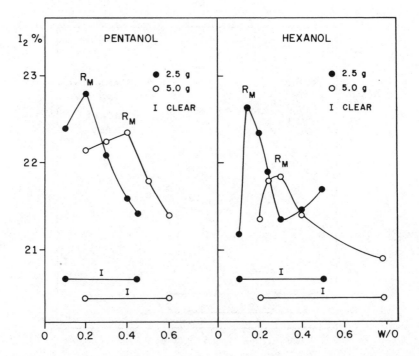

Figure 5. I_2 vs. water content in sodium stearate-alcohol-
isooctane-water systems at room temperature.

In sodium stearate-alcohol-cyclohexane system there is no influence of the hydrocarbon chain length of the alcohol on microemulsion formation; whereas in sodium oleate-cyclohexane dispersions, microemulsions are formed at the same R_M-point for butanol and hexanol-containing solutions (0.4) and at a lower R_M-point for the pentanol-containing system (0.2).

In sodium stearate systems, the effect of the hydrocarbon chain length of the solvent in microemulsion formation is the opposite to that caused by the chain length of the alcohol. Conversely, in sodium oleate systems the influence of the chain length of the solvent is the same as that observed for the alcohol chain length. This fact might be attributed to unsaturation.

Effect of altering the surfactant concentration.
Effects of changing the surfactant concentration from 2.5 to 5.0 g in sodium stearate-hexanol or pentanol-isooctane-water systems are shown in figure 5 and Table I. In n-pentanol-containing systems, R_M-points reach values of 0.2 and 0.4 at 2.5 and 5.0 g surfactant concentrations, respectively. In n-hexanol dispersions, R_M-points reach values of 0.15 and 0.30 at 2.5 and 5.0 surfactant concentrations, respectively (Table I, figure 5). These results are explicable in terms of the water solubilizing capacity of microemulsions. At higher surfactant concentrations there are more aggregates which require more water for the microemulsion formation. The observed values of I_2, which is related to the number of Ps atoms formed, might explain this fact since I_2 reaches smaller values at higher surfactant concentrations (i.e., 5.0 g) than at lower ones (i.e., 2.5 g) which suggests that at higher surfactant concentration there may be more aggregates which trap the energetic positrons or Ps atoms, thus reducing the number of Ps atoms formed[1-7] (compare I_2 values in plots of figure 5).

Effect of altering the surfactant concentration in sodium oleate systems from 2.5 to 5.0 g has been assessed. In sodium oleate-n-pentanol-isooctane-water systems R_M-points occur at 0.2 and 5.0 g surfactant concentration, respectively. Surfactant concentration influences microemulsion formation more drastically. In hexanol-containing solutions, in the presence of 2.5 g surfactant, microemulsion formation occurs at an R_M-point of 0.1 whereas in the presence of 5.0 g surfactant the R_M-point does not occur. Similar behavior has been observed when the amount of hexanol was decreased from 10 to 5 ml (R_M = 0.2; figure 4).

PARTITIONING AND FREE ENERGY TRANSFERS

Partitioning of the cosurfactant between the continuous phase (solvent) and the microemulsion has been determined by titration.[18]

Table II.- Partitioning and Free Energy Transfers

SOLVENT		(b) $(na/ns)_I$	(m) $(na/ns)_S$	ΔG(Kcal/mol)
		SODIUM STEARATE 2.5 g		
ISOOCTANE	n-Butanol	5.60	0.050	-2.748
	n-Pentanol	6.51	0.235	-1.935
	Isopentanol	5.71	0.185	-1.998
	n-Hexanol	5.31	0.109	-2.263
CYCLOHEXANE	n-Butanol	6.08	0.025	-3.196
	n-Pentanol	4.02	0.043	-2.646
	n-Hexanol	6.94	0.073	-2.655
		SODIUM OLEATE 2.5 g		
ISOOCTANE	n-Butanol	3.28	0.026	-2.185
	n-Pentanol	2.78	0.046	-2.391
	n-Hexanol	47.49	0.379	-2.813
HEXADECANE	n-Pentanol	2.43	0.155	-1.604
	n-Hexanol	2.92	0.080	-2.094
CYCLOHEXANE	n-Butanol	2.71	0.026	-2.694
	n-Pentanol	1.87	0.027	-2.455
	n-Hexanol	46.85	0.238	-2.738
		SODIUM OLEATE 5.0 g		
ISOOCTANE	n-Pentanol	1.69	0.078	-1.792
	n-Hexanol	1.33	0.060	-1.802

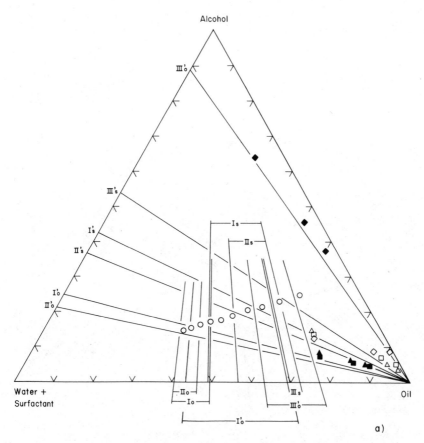

a)

Figure 6a. Partitioning determined by titration[18] and the
microemulsion range studied by the positron annihilation
technique are shown for sodium stearate or sodium oleate-
isooctane systems in the presence of different alcohols.
Initial compositions of the solutions were: 25 ml solvent,
2.5 g surfactant, 10 ml (o), 7 ml (·) and 5 ml (x) alcohol
(expressed in w/w). Microemulsion range is indicated for
n-butanol (I'), n-pentanol (II'), n-hexanol (III') and isopentanol (IV'). I, II, III and IV correspond to I', II',
III', and IV', except that these contain 5.0 g surfactants.
Indexes o and s indicate sodium oleate and sodium stearate,
respectively. The partitioning $(na/ns)_I$ is indicated on
the alcohol-water + surfactant axis by the same symbols as
used for microemulsion ranges. Compositions of the solutions w/w titrated with n-butanol (Δ), n-pentanol (□), n-
hexanol (◇) and isopentanol (*) are indicated by the appropriate symbols. The closed symbols correspond to sodium
oleate and the open ones to sodium stearate.

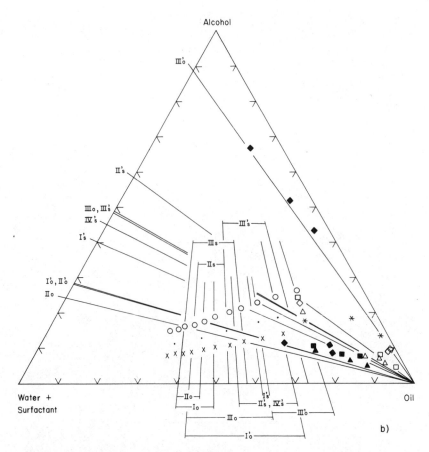

Figure 6b. Partitioning determined by titration[18] and the
microemulsion range studied by positron annihilation technique are shown for sodium stearate or sodium oleate-cyclohexane dispersions in the presence of different alcohols.
Initial compositions of the solutions were: 25 ml solvent,
2.5 g surfactant, 10 ml (o) alcohol (expressed in w/w).
Microemulsion range is indicated for n-butanol (I'), n-
pentanol (II') and n-hexanol (III'). I, II, and III correspond to I', II', and III', except that these contain
5.0 g surfactants. Indexes o and s indicate sodium oleate
and sodium stearate, respectively. The partitioning
$(na/ns)_I$ is indicated on the alcohol - water + surfactant
axis by the same symbols as used for microemulsion ranges.
Compositions of the solutions w/w titrated with n-butanol
(△), n-pentanol (□) and n-hexanol (◇) are indicated by the
appropiate symbols. The closed symbols correspond to
sodium oleate and the open ones to sodium stearate.

A typical solution contains 2.5 g or 5.0 g surfactant, 25.0 ml solvent and 2.5 ml water. The initially turbid solutions were titrated
to clarity with the alcohol. A minimum amount of solvent was
then added to cause turbidity. Clarity was reestablished by re-
titrating with alcohol. This procedure was repeated several times
until sufficient number of points were obtained in the plots of
moles of alcohol/mole of surfactant $vs.$ moles of oil/mole of sur-
factant. Intercepts of these plots |(b) values in Table II| were
taken to be the number of moles of alcohol at the interphase/mole
of surfactant. Slopes of the plots |(m) values in Table II| rep-
resented the solubility of the alcohol in the continuous phase.
(Values are given in Table II and in the corresponding diagrams
shown in Figures 6a and 6b). From the values obtained, the free
energy per mole of alcohol absorption into the interphase from the
continuous phase was calculated by using the formula.
$\Delta G = -RT\ln|(Xi)_I/(Xs)_S|$ where $Xi = (na/ni)_I$ and $Xs = (na/ns)_S$ are
the mole fraction of the alcohol in the interphase and the continuous phase, respectively.

Composition of the solutions (in weight %), studied by the
positron annihilation technique is indicated by open circles in
the phase diagrams in Figures 6a and 6b.[19,20] These points have
the same coordinates in the alcohol-oil axis due to the similarity
of molecular weight of the surfactants (sodium stearate and oleate)
and those of the densities of alcohols. The sodium stearate-hexa-
nol-isooctane system was reinvestigated by the positron annihilation technique by changing the amount of the alcohol in the initial
composition of the solution (Figure 6a). The lower limit of the
microemulsion range in the oil-water + surfactant axis in Figure
6a, corresponds to the first maximum in the curve of I_2 $vs.$ w/o
ratios in the positron annihilation experiments (Figure 3 Table I).
The upper limit of the microemulsion range corresponds to the upper
limit of the clear range ratio of the solutions determined by tur-
bidimetry.

DYNAMIC LASER LIGHT SCATTERING

Dynamic laser light scattering has been utilized for substantiating the phase transitions determined in the different microemulsion systems by positron annihilation.

Diameters of microemulsion droplets range from 500 to 2000 Å
(isopentanol-isooctane), from 1347 to 1656 Å (n-pentanol-hexa-
decane) for sodium stearate systems (Figure 7), and from 780 to
550 Å for the sodium oleate-n-pentanol-hexadecane solution (Figure
8). These diameters correspond to values generally attributed to
microemulsions. Significantly, no diameter smaller than 10,000 Å
were obtained in the sodium oleate-n-hexanol-isooctane system (Fig-

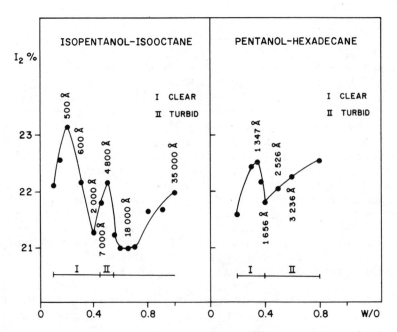

Figure 7. Microemulsion droplet diameters measured by dynamic laser light scattering in sodium stearate systems.

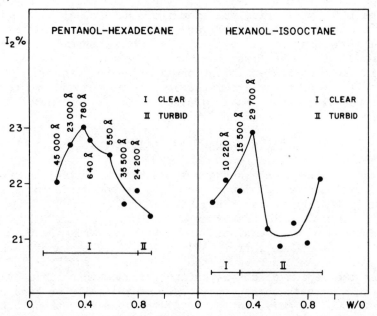

Figure 8. Microemulsion droplet diameters measured by dynamic laser light scattering in sodium oleate systems.

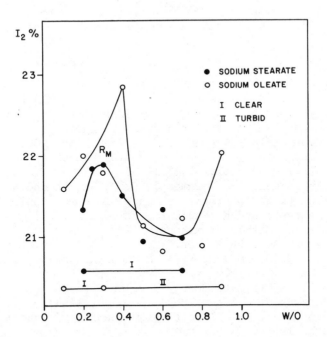

Figure 9. I_2 vs. water content in surfactant-n-hexanol-iso-octane solutions at room temperature.

ure 8). Positron annihilation indicated, of course, no microemul-sion formation here. This once again substantiates the power of the positron annihilation technique for determining subtle phase transitions.

EFFECT OF UNSATURATION ON MICROEMULSION FORMATION

Differences in the behavior of saturated and unsaturated surfactants can be rationalized by considering packing and kink presence[21,22] in microemulsion formation (compare Figures 3, 4 and 9; Table I). Assuming one kink formation for a surfactant with hydrocarbon chain length of 18 carbon atoms and one more for the presence of the double bond leads to volumes of $2(25 \rightarrow 50)$ $Å^3$ and $3(25 \rightarrow 50)$ $Å^3$ occupied by sodium stearate and sodium oleate, re-spectively[21] Sodium oleate occupies, therefore, larger volume than its saturated analog. This larger volume may well prevent packing the long chain cosurfactant n-hexanol into the microemulsion as-sembly. Inability of packing the cosurfactants precludes micro-emulsion formation. Conversely, shorter chain cosurfactants n-butanol and n-pentanol pack easier and can therefore, form micro-emulsions (Figure 9).

ACKNOWLEDGMENT

The authors wish to express their gratitude to D. Sc. Janos H. Fendler for his valuable advice and the discussion of this paper.

REFERENCES

1.   Y. C. Jean and H. J. Ache, J. Am. Chem. Soc. 99, 7504 (1977).
2.   E. D. Handel and H. J. Ache, J. Phys. Chem. 71, 2083 (1979).
3.   Y. C. Jean and H. J. Ache, J. Am. Chem. Soc. 100, 984 (1978).
4.   L. A. Fucugauchi, B. Djermouni, E. D. Handel and H. J. Ache, J. Am. Chem. Soc. 101, 2841 (1979).
5.   L. A. Fucugauchi, B. Djermouni, E. D. Handle and H. J. Ache, in "Proc. Fifth International Conf. Positron Annihilation", R. R. Hasiguti and K. Fujiwara, Editors, p. 857, The Japan Institute of Metals, Yamanaka, 1979.
6.   H. J. Ache, Adv. Chem. Ser. 175, 1 (1979).
7.   Boussaha, B. Djermouni, L. A. Fucugauchi and H. J. Ache, J. Am. Chem. Soc. 102, 4654 (1980).
8.   V. I. Goldanskii, At. Energy Rev. 6, 3 (1968).
9.   O. E. Mogensen, J. Chem. Phys. 60, 998 (1964).
10. J. Green and J. Lee, "Positronium Chemistry", Academic Press, New York, 1964.
11. J. D. McGervey, in "Positron Annihilation", A. T. Stewart and L. O. Roelling, Editors, p. 143, Academic Press, New York, 1964.
12. J. A. Merrigan, S. J. Tao and H. J. Green, in" Physical Methids of Chemistry", D. A. Weissberger and B. W. Rossiter, Ediors, Vol. I, Part III, Wiley, New York 1972.
13. H. J. Ache, Angew. Chem. Int. Ed. Engl. 11, 179 (1972).
14   W. J. Madia, A. L. Nichols and H. J. Ache, J. Am. Chem. Soc. 97, 5441 (1975).
15. V. K. Bansal, C. Ramachandran and D. O. Shah, J. Colloid Interface Sci. 72, 524 (1979).
16. V. K. Bansal, D. O. Shah and J. P. O'Connel, J. Colloid Interface Sci. 75, 462 (1980).
17. H. J. Schulman, W. Stoeckenius and L. M. Prince, J. Phys. Chem. 11, 169 (1960).
18. W. Gerbacia, H. L. Rosano, J. Colloid Interface Sci. 44, 242 (1973).
19. L. M. Prince, Editor, "Microemulsions", p. 147, Academic Press, New York , 1977.
20. S. Friberg in "Microemulsions". L.M. Prince, Editor, p. 131, Academic Press, New York, 1977.
21. G. Lagaly, Angew. Chem. Int. Ed. Engl. 15, 575 (1976).
22. J. H. Fendler, "Membrane Mimetic Chemistry", Wiley, New York, 1982.

ZETA POTENTIAL AND CHARGE DENSITY OF MICROEMULSION DROPS FROM

ELECTROPHORETIC LASER LIGHT SCATTERING - SOME PRELIMINARY RESULTS

S. Qutubuddin[a], C.A. Miller[b], G.C. Berry[c], T. Fort, Jr.[d],
and A. Hussam[e].

[a]Chemical Engineering Department, Case Western Reserve
University, Cleveland, Ohio 44106
[b]Chemical Engineering Department, Rice University,
Houston, TX
[c]Chemistry Department, Carnegie-Mellon University
Pittsburgh, PA
[d]Academic Affairs, Cal. Poly. State University, San
Louis Obispo, CA
[e]Chemistry Department, University of Pittsburgh,
Pittsburgh, PA

Electrophoretic laser light scattering has been
developed as a technique for determining the zeta potential and interfacial charge density of microemulsion drops.
Some preliminary results are reported for dilute oil-in-
water microemulsions in systems containing an anionic
surfactant, a short chain alcohol, a straight chain hydrocarbon, and sodium chloride brine. Zeta potentials
calculated from the measured electrophoretic mobilities
using Henry's equation were in the range of 90 mV. Drop
sizes were obtained from dynamic light scattering experiments and used with the zeta potentials to calculate
interfacial charge densities. The technique is promising for developing an improved understanding of microemulsions and other charged colloidal systems.

INTRODUCTION

Microemulsions have received much attention in recent years because of their interesting thermodynamic properties[1] and practical applications in areas such as enhanced oil recovery[2], drug delivery[3], and reactions[4]. The term microemulsion as used in this paper denotes a translucent stable dispersion of two immiscible fluids, generally water and oil. The dispersion is stabilized with surfactant molecules and, often, also a cosurfactant such as a short chain alcohol. The microemulsion may also contain an electrolyte as in this work.

The continued interest in microemulsions is evident from the number and diversity of papers dealing with microemulsions presented in this Symposium. Various techniques employed in investigating microemulsions include laser light scattering[5-9], small angle neutron scattering[5], NMR[10], and fluorescence quenching[11].

Phase behavior of microemulsions has been studied extensively, primarily with reference to surfactant flooding a process which is based on ultralow interfacial tensions for efficient oil displacement. Model ionic microemulsion systems, both pH-independent[12-14] and pH-dependent[15], have been developed. These exhibit ultralow interfacial tensions with brine and oil. The phase continuity and structure of microemulsions depend on various parameters, e.g. salinity, temperature and in some cases, pH. Ionic microemulsion systems are generally water-continuous at low salinities and oil-continuous at high salinities. The structure may be bicontinuous at intermediate salinities where significant amounts of both oil and water are solubilized. This work involves dilute water-continuous microemulsions with dispersed oil droplets. Of course, electrophoretic laser light scattering (ELLS) is not limited to such systems[28,29].

Phase transitions and ultralow interfacial tensions are related to the drop size as well as attractive (van der Waals) and repulsive (electrostatic) interactions between drops[15]. The drop size has been measured by several techniques such as ultracentrifugation[2,17-18], light scattering[9,20-25] and membrane diffusion[26]. Direct measurements of the charge density or surface potential in model microemulsion systems have not been reported in the literature to our knowledge. Of course, it is possible to calculate values for the charge density or surface potential indirectly on the basis of other measurements. Such calculations have been done for microemulsions using solubilization and ultracentrifugal data[5,17] and for micellar systems using quasi-elastic light scattering results[7,27]. However, various assumptions are involved in these treatments. Direct measurement of the electrophoretic mobility, and hence charge or surface potential, in microemulsions is expected to be useful in evaluating existing thermodynamic models for phase behavior of microemulsions and in improving the understanding of electrostatic interactions both in the charged palisade layer and between drops.

Further relevance of the present work is in surfactant flooding. A novel concept for reducing sensitivity of microemulsion phase behavior to salinity has been proposed[15]. The idea is based on making a microemulsion system more hydrophilic by increasing pH, and hence, the degree of ionization of the surfactant if it is a fatty acid or similar compound. This effect can counteract that of increasing salinity which is to make the system more hydrophobic. Direct measurements of charge or surface potential will be useful in predicting the effect of pH on such microemulsion systems.

ELLS is a powerful technique. However its applications to date have been limited primarily to proteins[28], polyelectrolytes[29,30] and biological systems[31]. There has been little done in terms of colloidal dispersions[32,33]. The preliminary results reported here suggest that further application of ELLS to various colloidal systems, particularly microemulsions, should be fruitful.

ELECTROPHORETIC LASER LIGHT SCATTERING (ELLS)

ELLS is a recent development which combines electrophoresis with laser light scattering or photon correlation spectroscopy. In principle it enables the simultaneous determination of particle size and electrophoretic mobility, and hence, the charge or zeta potential. This technique has been reviewed by Ware[28] and more recently by Uzgiris[31]. The determination of the electrophoretic mobility is based on the Doppler effect.

The homodyne dynamic light scattering experiment, which involves no reference beam or electric field, yields values of the particle diffusion coefficient. Particle size can be calculated using the Stokes-Einstein equation under appropriate conditions. The heterodyne or optical beating experiment is required for the electrophoretic mobility measurements. There are several ways of introducing the reference beam needed for this experiment. Stationary scattering elements in the observed region of the cell can provide the reference beam automatically. Alternatively, as in this work, a portion of the incident light may be directed around the moving scatterers and then redirected down the axis of optical detection with the closest possible match of scattered and reference beam wavefronts. A third method is to use an optical fiber. Perhaps the most reliable technique is to use two crossed laser beams and produce a differential Doppler signal[33].

Assuming that the scatterers are statistically independent, identical and spherical, the diffusion coefficient, D, can be obtained from the light intensity autocorrelation function, $C(\tau)$, for a homodyne experiment[26]:

$$C(\tau) = A \ e^{-i\omega_o \tau} \ e^{-2DK^2\tau} \tag{1}$$

where A is a constant related to the number of scatterers and the intensity of scattering, ω_o is the incident frequency and K is the scattering vector. K is equal to $(4\pi n/\lambda_0) \sin \theta/2$ where n, λ_0 and θ denote the refractive index, incident wavelength and scattering angle respectively. Thus, for a monodisperse system, the intensity autocorrelation function $C(\tau)$ decays exponentially with a decay time of $1/(2DK^2)$. The spectrum of the scattered light is obtained by Fourier transformation.

The decay time for the heterodyne experiment if $1/DK^2$. When an electric field E is applied in the heterodyne mode, the auto-correlation function is given by[26]:

$$C(\tau) = A \ e^{-i\omega_o \tau} \ e^{-DK^2\tau} \ e^{iuEK \cos(\theta/2) \ \tau} \tag{2}$$

where u is the electrophoretic mobility. It can be obtained directly from the time period $\Delta\tau$ required for one complete oscillation of the correlation function:

$$u = 2\pi/(\Delta\tau EK \cos \theta/2) \tag{3}$$

The resolution of ELLS, defined as the ratio of the Doppler shift to the half-width of the spectrum, increases with increasing electric field or particle size and with decreasing scattering angle. The applied electric fields were limited to low values in order to prevent gas evolution due to electrolysis of brine. Very low scattering angles could not be used due to instrumentation limitations.

EXPERIMENTAL

Light Scattering Set-up

The set-up used for ELLS is shown schematically in Figure 1. A description of the basic apparatus has been reported by Lee[34] et al. Major components of the set-up include a Birnboim Digital Correlator used as a data acquisition and processing system (Science Research System of Tory, Model DAS3), an argon-ion laser with etalon for single frequency operation (Lexel Model 85), a photomultiplier with a S-2 photocathode (ITT, FW-120-RF), an electric field genera-tor, and optical accesories. Optical components allow for beam alignment, rotation of the polarization plane of the incident beam (one-half wave plate), definition of the scattering volume through

horizontal and vertical slits (S_H and S_v), and collimation of the
accepted rays (pinhole). An analyzer is used to select the polarization of the scattered beam. The set-up is mounted on an air
supported optical table.

The data acquisition and processing system acquires pulses (up
to 2^{15}) in 4,736 intervals of length Δt each. The autocorrelation
function is computed over the primary data base of $T = 2^{12}$ intervals to give a correlation (up to 512 points) spaced at intervals
of Δt. Details of the computation of the correlation function are
given in Reference 34.

Electrophoresis Cell

A schematic of the electrophoresis cell used is shown in Figure 2. The scattering region in this simple design is a rectangular
glass capillary 8 mm high. Two different capillary thicknesses
were used, 1 mm and 10 mm. The electrodes were metal rods placed in
cylindrical tubes on the sides of the cell. Silver electrodes were
preferred over platinum since in the systems investigated the tendency for gas evolution was comparatively less with silver. The
distance between the electrodes was either 15 or 30 mm depending on
the cell type. Constant voltage electric fields were applied using
a regulated DC supply source (Kepco). The polarity of the electrodes was reversed after every application of field in order to
minimize concentration polarization. Also there was a substantial
time interval between the pulses.

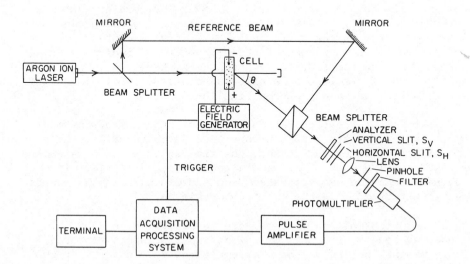

Figure 1. Experimental set-up for electrophoretic laser light
scattering.

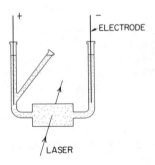

Figure 2.   A schematic of the electrophoresis cell.

Microemulsion Systems

Two different microemulsion systems were investigated, one
showing pH-dependent phase behavior, and the other pH-independent.
The pH-dependent system contained 0.6% (v/v) oleic acid (> 99% pure,
Sigma) as the surfactant, 6.5% (v/v) 2-pentanol as the cosurfactant,
and n-octane as the hydrocarbon. Sodium hydroxide (.025M) was used
to partially neutralize the acid. The pH-independent system was
formulated with 1.0% (v/v) TRS 10-410 (Witco), a commercial petroleum sulfonate, as the surfactant, 3.0% (v/v) isobutanol as the
cosurfactant, and n-decane as the hydrocarbon. Sodium chloride
was used as the electrolyte in both the systems. Alcohols, hydrocarbons and electrolyte were all reagent grade. Deionized distilled
water (Ricca) was used in preparing the solutions.

Details of the preparation and phase behavior of the microemulsion systems are given in Reference 15. Basically, the aqueous
solution containing surfactant, cosurfactant, and salt was mixed
with an equal volume of oil and allowed to equilibrate. The resulting microemulsion was separated from the excess oil phase before
being introduced into the electrophoresis cell.

The microemulsion systems were investigated at 23°C using a
laser wavelength of 0.5145 μm and an apparent scattering angle of
25°. The salinities of the microemulsions were 0.38 M for the

oleic acid system and 0.10 M for the Witco TRS 10-410 system. Both salinities were chosen to be significantly lower than the corresponding critical salinities for phase separation, i.e., formation of a "middle phase" microemulsion. This procedure was necessary to avoid effects of critical fluctuations[35,36]. Within this limitation, however, compositions were chosen to maximize drop size and hence improve the resolution of the ELLS experiment.

RESULTS AND DISCUSSION

Oleic Acid System

Figure 3 shows the decay of the homodyne autocorrelation function for the oleic acid system. The data can be fitted quite well using a single exponential. The apparent diffusion coefficient was calculated to be 0.61×10^{-7} cm^2/sec.

The volume fraction of the dispersed oil phase being appreciable (about 10%), the diffusion coefficient is probably affected by interparticle interactions[6,9,20,37]. A simple hard sphere correction could be used, but it is not sufficient owing to strong attractive interactions which have been found in similar systems[35]. Due to lack of a comprehensive treatment of the diffusion coefficient in concentrated systems with interacting particles, the hydrodynamic radius r was calculated from the apparent diffusion coefficient using the simple Stokes-Einstein relationship:

$$D = kT/6\pi\eta r \tag{4}$$

where k is the Boltzmann constant, η the viscosity of the solvent, and T the absolute temperature. As the viscosity of the continuous phase was not directly measurable, the value for brine was used, i.e., effects of dissolved alcohol were neglected. The value of r obtained was 37.2 nm.

The autocorrelation function in the heterodyne mode with an applied electric field is shown in Figure 4 for the oleic acid system. The observed Doppler shift was 60 Hz for an electric field of 11.4 V/cm. The electrophoretic mobility was calculated to be 6.5×10^{-4} cm^2/V-sec using Equation (3).

The zeta potential ζ for the particle may be obtained from the mobility u by means of Henry's equation[38]:

$$\zeta = \frac{6\pi\eta u}{\varepsilon X_1(\kappa r)} \tag{5}$$

Here ε is the dielectric constant of the continuous phase, κ^{-1} is the Debye length, a measure of electrical double layer thickness, and $X_1(\kappa r)$ is a function which can be obtained from Henry's theory. In the oleic acid experiment $\kappa r = 77$ and $X_1(\kappa r) = 1.44$, which is very near its limiting value of 1.5 for very large values of κr. With ε taken as the value for water, i.e., again neglecting the effect of dissolved alcohol, ζ is found to be 93 mV.

Owing to the relatively large values of κr and ζ, the surface charge density σ is best calculated from the Gouy–Chapman theory for a plane surface[39]:

$$\sigma = \frac{\varepsilon \kappa kT}{2\pi z e_o} \quad \sinh \left(\frac{z e_o \zeta}{2kT} \right) \tag{6}$$

where z is the valence of the ions in solution and e_o is the electronic charge. From this equation, the area (e_o/σ) per ion on the surface is found to be 0.71 nm²/ion (71 Å²/ion). Both this value and that of the zeta potential are within reasonable ranges for the given experimental conditions.

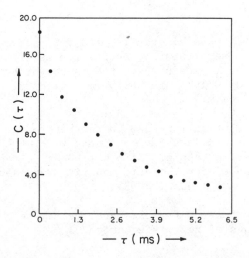

Figure 3.   Decay of homodyne autocorrelation function for oleic acid microemulsion system.

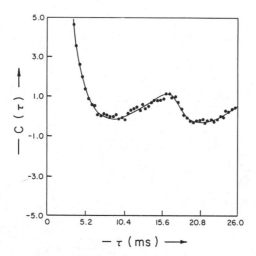

Figure 4. Heterodyne autocorrelation function for oleic acid micro-emulsion system with an applied electric field of 11.4 V/cm.

Petroleum Sulfonate System

The above results are for a pure microemulsion system. Similar experiments were performed using the Witco petroleum sulfonate TRS 10-410 which is only 60% active. The volume fraction of the dispersed phase in the microemulsion was about 0.05.

Behavior different from the oleic acid system was observed in both the homodyne and electrophoresis experiments. Figure 5 shows the decay of the homodyne autocorrelation function. A plot of log $C(\tau)$ versus τ is not linear. Factors which can contribute to the non-linearity include polydispersity and particle-particle inter-actions. Polydispersity and strong interactions in some microemulsion systems have been reported by several investigators[6,25,35,40]. The decay shown in Figure 5 can be fitted with a bimodal particle size distribution. However, with the present preliminary data a choice between polydispersity and interactions would be arbitrary.

From the initial decay rate, however, it is possible to calculate an average value of the diffusion coefficient, and, hence, an average hydrodynamic radius. The resulting diffusion coefficient and hydrodynamic radius were 0.5×10^{-7} cm^2/sec and 44.0 nm, respectively.

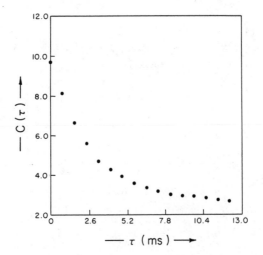

Figure 5. Decay of homodyne autocorrelation function for Witco
(TRS) 10-410) microemulsion system.

The heterodyne autocorrelation function with an electric field
of 25.4 V/cm is shown in Figure 6. The observed behavior is different from that of the oleic acid system in that $C(\tau)$ starts to
show sinusoidal behavior before reaching the base value. Such
behavior could result if the particles are polydisperse and carry
different charges. The homodyne experiment in this system, discussed earlier, also indicated polydispersity. In future work spectrum
analysis will be performed to distinguish the different electrophoretic species and perhaps measure their relative concentrations.

Examination of Figure 6 shows that the period of oscillation
is approximately constant during the time of observation. This
behavior suggests that the curve of Figure 6 might be considered as
the sum of two terms having the form of Equation (2). One term
would have a period of oscillation equal to that observed, the other
a much longer period so that it simplifies to a simple exponential
decay for the relatively short duration of the experiment. In this
case the observed period is that of the more mobile type of particle. By the same method as above, we obtain a zeta potential of
90 mV and a surface concentration of 1.46 nm^2 per ion (146 $\overset{\circ}{A}^2$ per
ion) for this system.

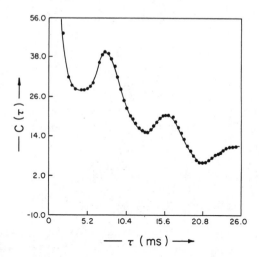

Figure 6. Heterodyne autocorrelation function for Witco (TRS 10-410) microemulsion system with an applied electric field of 25.4 V/cm.

Electroosmosis

As discussed by Ware[28] and Uzgiris[31], ELLS experiments should be designed to avoid introduction of error by effects such as electroosmosis, Joule heating, gas evolution, and vibration of the experimental apparatus. We believe that such errors are small in our experiment because the same value of the zeta potential was obtained in each system within experimental error for both the 1 mm and the 10 mm thick cells described above.

Independent calculations are also possible for some effects. We consider, in particualr, electroosmosis. The negatively charged drops in the micoremulsion move toward the positive electrode in Figure 2. On the other hand, if the cell walls are negatively charged, the usual situation, the positively charged electrical double layers near the walls move toward the negative electrode. The flow of fluid induced by double layer motion causes the liquid level in the tube containing the negative electrode to increase. The result is a gravitational driving force for fluid flow toward the positive electrode.

At steady state these effects balance so that there is no net flow of fluid toward either electrode. As Uzgiris[31] has shown, the velocity profile in the rectangular portion of the cell in Figure 2 has the form

$$v(x) = \frac{3}{2} \frac{v_o}{a^2} (x^2 - a^2) + v_o \tag{7}$$

where 2a is the thickness of the cell, x is the distance from the centerline of the cell, and v_o is the velocity induced by electro-osmotic motion at the cell walls $x = \pm a$. The latter is given by

$$v_o = - \frac{\varepsilon E \zeta_w}{4\pi\eta} \tag{8}$$

with ζ_w the zeta potential of the wall.

According to Equation (7) the velocity v(o) at the tube center-line is $(-v_o/2)$ at steady state. Since ζ_w is likely to be comparable to the drop zeta potential ζ, the correction for electroosmotic effects should be significant if steady state is reached. In this event drops near the centerline move toward the positive electrode more rapidly than in the absence of electroosmotic effects while drops near the walls move less rapidly.

It remains to be seen if steady state is actually approached during the short duration of the present experiments. Starting with the liquid at rest, the electric field is applied, and all useful data are obtained in a few milliseconds.

To answer this question, we first calculate the amount of liquid which must flow to the negative electrode to produce the gravitational driving force for back flow. The pressure gradient producing the flow described by the first term of equation (7) is given by

$$\frac{\Delta P}{L} = \frac{3\eta v_o}{a^2} = \frac{\rho g \Delta h}{L} \tag{9}$$

where ρ is the density of the liquid, L the length of the rectangular cell, and Δh the difference in height between liquid columns surrounding the two electrodes. The volume of liquid flow required to produce this Δh is

$$V = \pi R^2 \frac{\Delta h}{2} \tag{10}$$

with R the radius of the tube containing the electrode. Combining (9) and (10), we obtain

$$V = \frac{3\pi R^2 \eta v_o L}{2\rho g a^2} = -\frac{3\varepsilon E R^2 L \zeta_w}{8\rho g a^2} \qquad (11)$$

Taking a plausible ζ_w of -50 mV and dimensions of the smaller cell of R = 0.4 cm, L = 3.0 cm, and a = 0.05 cm, we find V = 3.7 x 10^{-5} cm^3 when E = 11.4 V/cm as in the oleic acid system.

The flow produced by electroosmosis within the rectangular cell may be modeled as flow near a wall where lateral motion with velocity v_o suddenly starts. The velocity profile in this case is known[41]

$$v = v_o \left(1 - \text{erf} \frac{a-x}{(4\nu t)^{\frac{1}{2}}} \right) \qquad (12)$$

where (a-x) is the distance from the wall, ν is the kinematic viscosity (η/ρ), and the error function erf u is defined by

$$\text{erf } u = \frac{2}{\pi^{\frac{1}{2}}} \int_o^u e^{-y^2} dy \qquad (13)$$

If we assume, as a rough estimate, that the velocity has an average value ($v_o/2$) over a region of thickness $(4\nu t)^{\frac{1}{2}}$ at time t, the volumetric flow rate Q(t) toward the positive electrode is given by

$$Q(t) = 2v_o W (\nu t)^{\frac{1}{2}} \qquad (14)$$

where W is the width of the rectangular cell and the factor of two accounts for the fact that electroosmotic flow is induced at both walls of the cell. Integration of (14) yields

$$V' = \int_o^{t_o} Q(t) dt = \frac{4}{3} v_o W (\nu t_o^3)^{\frac{1}{2}} \qquad (15)$$

With W = 0.8 cm and t_o = 10 ms we obtain V' = 4.3 x 10^{-7} cm^3.

Since the actual volume V' of liquid moved by electroosmotic flow during the experiment is almost a factor of 100 less than the volume V needed to produce the steady state backflow, we conclude that the backflow is negligible. Electroosmosis does induce flow in a region having a thickness of some 0.1 mm near each wall, but otherwise liquid in the tube may be considered static. Hence, our results should not be significantly influenced by electroosmosis.

CONCLUSION

Preliminary results reported here indicate the electrophoretic laser light scattering is a promising technique for determining the zeta potential and interfacial charge density of microemulsion drops and other colloidal particles. Zeta potentials of about 90mV were found for oil-in-water microemulsions in two different systems containing anionic surfactants.

ACKNOWLEDGEMENTS

Thanks are due to W.J. Benton, H.M. Cheung, and S. Mukherjee for stimulating discussion. The assistance of Dr. F. Lin, Ms. R. Furukawa and Mr. R. Frye in experimental set up is very much appreciated. This work was support in part by the U.S. Department of Energy and in part by grants from Amoco Production Company, Gulf Research and Development Company, Exxon Production Research Company, and Shell Development Company. The experiments were performed at Carnegie-Mellon University.

REFERENCES

1. I.D. Robb, Editor, "Microemulsions," Plenum Press, New York (1982).
2. C.A. Miller, S. Mukherjee, W.J. Benton, S. Qutubuddin and T. Fort, Jr., AIChE Symposium Series, $\underline{78}$ (212), 28 (1982).
3. J. Ziegenmeyer and C. Fuhre, Paper presented at the International Symposium on Surfactants in Solution, Lund, Sweden, June 27 - July 2, 1982.
4. R.A. Mackay, Adv. Colloid Interface Sci., $\underline{15}$, 131 (1981).
5. L.J. Magid, R. Trioli, E. Gulari, and B. Bedwell, These Proceeding, Vol. 1.
6. J.D. Nicholson and J.H.R. Clarke, These Proceedings, Vol.3.
7. D.F. Nicoli, R. Dorshow, and C.A. Bunton- These Proceedings, Vol. 1.
8. N.A. Mazer, Paper presented at the International Symposium on Surfactants in Solution, Lund, Sweden, June 27 - July 2,1982.
9. D. Roux, A.M. Bellocq, and P. Bothorel, These Proceedings,Vol.3.
10. B. Lindman and P. Stilbs, These Proceedings, Vol. 3.
11. R. Zana, J. Lang, and P. Lianos, These Proceedings, Vol. 3.
12. W.J. Benton, J. Natoli, S. Qutuubddin, S. Mukherjee, C.A. Miller, and T. Fort, Jr., Soc. Pet. Eng. J., $\underline{22}$, 53 (1982).
13. K.E. Bennett, C.H.K. Phelps, H.T. Davis, and L.E. Scriven, Soc. Pet. Eng. J., $\underline{21}$, 747 (1981).
14. A.M. Bellocq, J. B s, B. Clin, A. Gelot, P. Lalanne, and B. Lemanceau, J. Colloid Interface Sci., $\underline{74}$, 311 (1980).

15. S. Outbuddin, Ph.D. Thesis, Carnegie-Mellon University (1983).
16. C.A. Miller, R. Hwan, W.J. Benton, and T. Fort, Jr., J. Colloid Interface Sci., 61, 554 (1977).
17. R.N. Hwan, C.A. Miller, and T. Fort, Jr., J. Colloid Interface Sci., 68 221 (1979).
18. M. Dvolaitzky, M. Guyot, M. Lagues, J.P. Le Pesant, R. Ober, C. Snaterey, and C. Taupin, J. Chem. Phys., 69, 3279 (1978).
19. S. Mukherjee, Ph.D. Thesis, Carnegie Mellon University (1981).
20. A.A. Calje, W.G.M. Agterof, and A. Vrij in "Micellization, Solubilization, and Microemulsions," K.L. Mittal, Editor, Vol. 2, p.779, Plenum Press, New York, 1977.
21. A.H. Cazabat and D. Langevin, J.Chem. Phys., 74, 3148 (1981).
22. A.M. Cazabat, D. Chatenay, D. Langevin, and A. Pouchelon, J. Phys. Lett., 41 L-441 (1980).
23. M. Corti and V. Degiorgio in "Light Scattering in Liquids and Macromolecular Solutions," V. Degiorgio, M. Corti and M. Giglio, Editors, Plenum Press (1980).
24. E. Gulari, B. Bedwell, and S. Al-Khafaji, J. Colloid Interface Sci., 77 202 (1980).
25. J. van Neiuwkoop and G. Snoei, "Phase Behavior and Structure of a Pure Component Microemulsion System," to be presented at the Second European Symposium on Enhanced Oil Recovery, Paris, November 1982.
26. S.I. Chou and D.O. Shah, J. Colloid Interface Sci., 78, 249 (1980).
27. M. Corti and V. Degiorgio, J. Phys. Chem., 85, 711 (1981).
28. B.R. Ware, Adv. Colloid Interface Sci., 4, 1 (1974).
29. J.P. Meullenet, A. Schmitt, and M. Drifford, J. Phys. Chem., 83, 1924 (1979).
30. M. Drifford, P. Tivant, F. Benkreich-Larbi, C. Rochas, and M. Rinando, "Light Scattering in Aqueous Solutions of Kappa Carrageenan," paper submitted to Biopolymers.
31. E.E. Uzgiris, Adv. Colloid Interface Sci., 14, 75 (1981).
32. V. Novotny and M.L. Hair, in "Polymer Colloids II," R.M. Fitch, Editor, Plenum Press, New York, (1980).
33. H.M. Cheung, R. Edwards, and J.A. Mann, unpublished results.
34. C. Lee, G.C. Berry, and C.-G. Chu, "Studies of Dilute Solutions of Rodlike Macroiions I," J. Polymer Sci., Polymer Physics Edition (in press).
35. A.M. Cazabat, D. Langevin, J. Meunier, and A. Pouchelon, J. Physique Lett., 43, L-89 (1982).
36. J.S. Huang and M.W. Kim, SPE Paper #10787 presented at the 1982 SPE/DOE Third Joint Symposium on Enhanced Oil Recovery, Tulsa, April 1982.
37. D.R. Bauer in "Polymer Colloids II," R.M. Fitch, Editor, Plenum Press, New York (1980).
38. D.C. Henry, Proc. Roy. Soc. (London), A133, 106 (1931).

39. P.C. Heimenz, "Principles of Colloid and Surface Chemistry,"
    p. 378 Marcel Dekker, New York (1977).
40. D.J. Cebula, R.H. Ottewill, J. Ralson, and P.N. Pusey, J. Chem.
    Soc., Faraday Trans.I, 77, 2585 (1981).
41. R.B. Bird, W.E. Stewart, and E.N. Lightfoot, "Transport Phen-
    omena," p. 124 Wiley, New York, (1960).

LOW TEMPERATURE DIELECTRIC PROPERTIES OF W/O MICROEMULSIONS

AND OF THEIR HIGHLY VISCOUS MESOPHASE

D. Senatra and C.M.C. Gambi

Liquid State Physics Laboratory,GNSM Group of the CNR
University of Florence
Largo E.Fermi,2 (Arcetri),50125 Florence, Italy

The dielectric properties of a w/o microemulsion
were investigated by means of two different experimental approaches. We measured,in the concentration intervals (C,mass fraction),0.14-0.30 and 0.68-0.80,the temperature dependence (70°K-293°K) of both the real and
the imaginary part of the relative complex permittivity; in the interval 0.68-0.80,the thermally stimulated
dielectric polarization release (TSD),at different polarizing temperatures in the range 188°K-293°K. In the
concentration interval 0.13-0.30,both ε' and ε'' increase with decreasing temperature,going through a
sharp maximum at T=234.8°K (C=0.248,ν =160 KHz). In
the range 0.68-0.80,a TSD current peak at T=198°K was
observed which follows a first order relaxation kinetics ; its maximum current peak intensity is an inverse
function of the polarizing temperature.

INTRODUCTION

In previous papers we investigated the structural transitions
that develop in a w/o microemulsion upon increasing water content
by studying at room temperature its dielectrical[1-5], optical[6],
electro optical [7-10] and viscosity [6] properties.

In the present work we report,for the first time,the results
obtained by studying the low temperature dielectric behavior of the
microemulsion in a temperature interval which extends from 293°K
down to 70°K.

The microemulsion is composed of: dodecane,hexanol,potassium
oleate, and water. The proportions,by weight,of the different constituents are: 58.6% dodecane,25.6% hexanol and 15.8% K-oleate;
with K-oleate/dodecane=0.4(g/ml) and hexanol/dodecane=0.2(ml/ml).
The system's water content,C,is expressed by the weight ratio:
water/(water + oil).

The phase diagram of the actual microemulsion against water
addition,within the temperature interval (253°K-363°K),is plotted
in Figure 1. By observing the samples between two crossed polaroids
with a white light source, two main concentration regions were distinguished: an optically isotropic "microemulsion" region and a
birefringent liquid crystalline "LC region". These are divided by
an interval,"1st Gap",where the system separates into two different phases,an isotropic w/o microemulsion and a birefringent LC
mesophase.[5]   In the LC region two mesomorphic structures of lyotropic type develop: an inverted middle phase (water cylinders)
and a neat phase (lamellae). These are confined in the concentration intervals 0.480< C< 0.582 and 0.582< C< 0.650,within the temperature domains 288< T°K< 315 and 272< °K< 315,respectively.[9-10]
The two structures coexist in the narrow interval 0.582< C< 0.590,
"2nd-Gap",where the system exhibits also a thermotropic mesomorphism .

The low temperature analysis was carried out in the concentration intervals 0.143< C< 0.30 and 0.65< C< 0.80. In the former,the
system is an optically isotropic,perfectly fluid w/o microemulsion,
in the latter it exhibits a highly viscous,not birefringent mesophase that occurs at the end of the LC region (See Figure 1).

Two experimental approaches were adopted: 1) dielectric measurements in the frequency range 1 KHz-50 MHz; and 2) the thermally
stimulated depolarization method (TSD).

The general dielectric features of the system,at room temper-

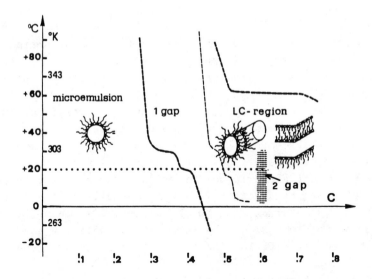

Figure 1. The phase diagram of the actual w/o microemulsion vs. water content, C, mass fraction. See text for explanations.

ature ,as a function of increasing water content over the whole range
analyzed so far (0.024-0.80) will be summarized.

A synthesis of the main results obtained by applying the TSD
analysis at room temperature to microemulsion samples with concentration in the intervals 0.143-0.30 and 0.65-0.80 respectively,
will be reported for the sake of comparison with the low temperature
study. Since the latter method was applied to a w/o microemulsion,
for the first time,by the authors of the present paper[3,5] ,the
fundamental principles that characterize this technique will be
briefly outlined.

In order to help understanding the difference between the TSD
and the usual dielectric relaxation measurements,the former method
will be described first.

EXPERIMENTAL

The TSD Method. Historically the TSD analysis is linked with
two main mechanisms:the permanent electrification of a substance
"via electret effect" and the thermally activated release of dielec-

tric polarization. [11-14,15,16]

An electret is a permanently electrified solid substance exhibiting electrical charges of opposite sign at its extremities. Therefore,an electret is the exact electric counterpart of a permanent
magnet.

Typical substances capable of permanent electrification are:
carnauba wax and the mixtures with carnauba,resin and beeswax,as
well as those with compounds containing polar hydrocarbons. Non polar substances,like paraffin wax,do not form electrets.

The permanent electric polarization of an electret is produced
by means of a strong electric field (E_p),usually of the order of 10
kV/cm.The sample alignment in the E_p direction is then "fixed" by
allowing the molten mass of the substance to solidify under the action of the electric field. The orientation induced in the sample is
therefore "frozen-in".

The electrification of a true electret is a volume effect extending through its whole mass. Therefore,by cutting an electret
body into sections,each section still exhibits electret characteristics.

Upon heating an electret,a discharge current is measured since
the stored charge is released on reheating. This "thermally activated" release of polarization is characterized by the intensity of the
electric field E_p applied for an interval of time t_p at the temperature T_p as well as by the heating rate employed. The subscript "p"
indicates the polarizing conditions under which the given material
is oriented.

The systematic measure of discharge currents from electrets,
using a linear heating rate,for various polarizing fields,tempera -
tures and times,was originally devised by Gross [12-14] for studying
the nonisothermal behavior of carnauba wax electrets.

The method was,later on,extended by Bucci and Fieschi [15,16] to
investigate the mechanism of polarization in solid dielectrics exhibiting a uniform polarization due to dipole orientation or migration of charges (ions) over microscopic distances with trapping.

Actually the method,called Ionic Thermal Current (ITC) or Thermally Stimulated Depolarization (TSD),has become a powerful tool
for the dielectric analysis of the mechanism of charge storage and
polarization in a great variety of inorganic materials such as ionic crystals [15-17] , molecular solids [18] ,polymers [19] ,glasses [20] ,
ice [21] ,liquid crystals [10,22], as well as in many organic substances such as collagen and tendon [23] ,and bone tissue [24] ,to mention just a few.

One of the most striking applications of the TSD method is in the study of dielectric relaxation of solid polar dielectrics. The latter study is usually carried out by means of a series of isothermal measurements of the system's complex permittivity vs.frequency in the range (Hz-GHz),over a very broad temperature interval. If a dipolar relaxation is involved,the activation energy (ξ) and the relaxation time constant (τ_o) that characterize the process,can be evaluated by measuring the relaxation time $\tau(T)$ at the different temperatures,using the relation $\tau(T) = \tau_o \exp(\xi/KT)$,(K, Boltzmann's constant;T,absolute temperature). However this can be done only when the resonance frequency,inversely proportional to the relaxation time of the dipoles at the measuring temperature,is known for several temperatures within the chosen interval.

With the TSD method,by means of a single experimental run,the measure of the depolarization current gives a complete picture of the temperature dependent relaxation processes occurring in the system and allows also the direct determination of their parameters, activation energy (ξ) and relaxation time constant (τ_o).

If the dielectric can be described in terms of a single relaxation process,the discharge current $J(T)$ vs.increasing temperature is an asymmetric peak characterized by its maximum peak temperature. If more relaxation processes are involved,the resultant curve is a "spectrum" with a series of current peaks. Each relaxation process is distinguished by the temperature at which the maximum of the current occurs.

When these peaks are due to "uniform polarization" caused by dipole or pseudodipole orientation (ions migrating over microscopic distances),the theory developed by Bucci and coworkers may be used to account for them. In both cases the temperature at which the maximum of the current peak occurs is indpendent of both the polarizing temperature and the polarizing field strength. If the discharge current is caused by "space charge" build up or charge transfer at the electrode sample surface,the orientation of the sample at different polarizing temperatures offers an experimental means of isolating the contribution of the latter from the TSD spectrum. Moreover,the stored charge due to space charge does not increase linearly with the applied field,E_p,on the contrary,the dipole or pseudodipole stored polarization is a linear function of E_p .

Thereof,by means of the above technique,it is possible,"a posteriori" to check whether the polarizaton of the sample was uniform as well as to distinguish one mechanism of polarization from the other one.

Analytical   expression of the J(T) current.   The discharge current J(T),interpreted as the rate of change of polarization,may be
expressed in terms of the orientation parameters P (polarization)
and τ(relaxation time) by:

$$J = dP/dt = -P/ \tau \tag{1}$$

where: $\tau = \tau_0 \exp (\xi/KT)$ $\tag{2}$

The second and the third term in Equation(1) represent the rate
at which P decays with time if the mechanism of polarization follows
a first order relaxation kinetics.

Upon heating the "frozen" sample with a linear heating rate
b= dT/dt,the relaxation time becomes a function of temperature. The
latter,in turn,is a function of time. Therefore,by solving Equation
(1) we have:

$$J(T) = - P_o/ \tau . \exp(- \int dt/ \tau)=-P_o/\tau . \exp(-\int dT/\tau b) \tag{3}$$

In Equation(3) P_o is the initial polarization induced at saturation by applying the E_p field for an interval of time $t \gg \tau(T)$.
In a dipolar dielectric for which Langevin function applies,[25] ,
$P_o=Np^2E_p a/KT$,where N is the dipole concentration and'a'a geometric
factor depending on the possible dipole orientation. For instance,
for freely rotating dipoles,a=1/3.

Since the area delimited by the current peak is proportional to
the initial dipolar moment per unit volume[12], the polarization
induced at saturation is: $P_o= \int_0^\infty \sigma(t)dt$,where σ is the stored charge
density. The total charge Q released during the discharge process is
simply Q= $\int_c^\infty J(T)dt$.

Substitution of Equation (2) into (3),leads to the final expression of the depolarization current J(T):

$$J(T)=-P_o/\tau_o . \exp (- \xi/KT- (b \tau_o)^{-1}\int_{T_o}^T \exp (- \xi/KT)dT) \tag{4}$$

The values of the activation energy and of the relaxation time
can be evaluated by:

$$\tau(T)= P(T)/J(T)= (\int_{t(T)}^\infty J(t')dt') / J(T) \tag{5}$$

$\tau(T)$ being determined from the experimentally measured current peak.
In addition the following relation will hold:

$$\ln \tau(T)= \ln \tau_0 + \xi/KT = \ln \left(\int_{t(T)}^{\infty} J(t')dt'\right) - \ln J(T) \qquad (6)$$

The expression to the right in Equation (6) is determined from the integration of the area delimited by the given depolarization current peak. The plot of $\ln\tau$ vs. $1/T$ is a straight line, whose slope and intercept are the activation energy and the relaxation time constant.

From the behavior of the latter plot or the fit of the $J(T)$ function (4) to the experimental points it is possible to verify whether any particular experimental current peak agrees with the theory and thus represents a uniform mechanism of polarization.

Microemulsion TSD study. By applying the TSD method to the microemulsion it was possible to identify both the concentration intervals and the temperature domains within which each phase may exist, since typical TSD spectra were found to characterize the different structure regions shown in Figure 1. In addition, the former study has offered an accurate and so far unique description of the temperature dependent relaxation processes occurring in w/o microemulsion samples in the concentration interval from 0.024 up to 0.65.

We should note here that the method did not work for samples exhibiting thermotropic mesomorphism (Figure 1,2-Gap). In this case the development of the phase transitions hinders the depolarization process.[8-10,22]

The experimental conditions adopted for the TSD analysis of the microemulsion at room temperature were: $T_p=293°K, E_p=33$ V/cm, $b=0.1°K/s, t_p=5$ minutes. The freezing process was realized by immersing the samples in liquid nitrogen. Since at room temperature the system is in the liquid state, neither the melting of the sam - ples nor the imposition of a strong electric field were necessary. Further details may be found in references 5,9 .

In the present paper we report, for the first time, the results obtained by orienting the microemulsion samples at low polarizing temperatures in the interval from 293°K down to 188°K.

The low temperature T_p was obtained by means of a cryogenic set up. The freezing of the field induced orientation and the heating processes were the same as those described in reference 5 .

Depending on T_p, the intensity of the static electric field was increased from 33 V/cm up to 10 kV/cm, in order to achieve the complete orientation of the samples.

The sample temperature values were monitored with thermocouples calibrated by the Italian Calibration Service, Metrological Commission (CNR),with a class of accuracy GRI-UNI 7938.

The low temperature TSD analysis was applied to samples belonging to the concentration interval 0.65-0.80 as reported in Section I.

Dielectric study. The dielectric properties of the w/o microemulsion against water addition were investigated at room temperature by studying the concentration dependence of ϵ' and ϵ'' at
different frequencies in the range 1 KHz-50 MHz.

The values of both ϵ' and ϵ'' were determined by measuring the
impedance magnitude and the phase angle of a two terminal cell filled with the sample. The method,whose technical details were reported
earlier [1,2,10],was specifically devised to follow the continuous evolution of the system under test,by recording the variation of the
complex impedance of the dielectric filled cell,at any given concentration,as a function of frequency.

The temperature dependence of both ϵ' and ϵ'' at different
concentrations (0.134<C<0.30) and frequencies (1KHz-50MHz) is reported here for the first time. The temperature interval extends from
293°K down to 70°K. The study was carried out by slightly modifying
the aforementioned procedure.

The temperature dependence of the sample dielectric properties
is followed by recording the variation of $\underline{Z}$ at any given concentration and frequency,as a function of linearly decreasing and/or increasing temperature.

The cooling and heating rates were performed with a Neslab Cry-
ocool CC-100,cold finger,controlled by an Exatrol Unit and an ETP-3
temperature programmer. Different cooling and heating rates were tested. Experimentally the main difference between the dielectric measure at room temperature and the study of the temperature dependence
of the sample dielectric properties resides in the sample holder device.

Two sample holder devices were used,a constant volume cell and
a volume independent cell. In the former,the plane parallel,circular
gold electrodes enclosed in a very thin Teflon jacket were kept at a
constant distance of 2mm and the sample occupied the whole volume
between them. In addition,the little Teflon box was placed into a
0.3mm thick stainless steel cylindrical container that could directly be dipped into a thermostatic bath or in liquid nitrogen.

The sample temperature was measured with a thermocouple placed
on the upper electrode of the cell,outside the sample,but into the
stainless steel container. With the above sample holder arrangement
very fast temperature gradients were possible,up to 4.5°K/s,but it

was not possible to distinguish,in the impedance values,the amount
of variation eventually due to a change of the sample volume from
that due to the temperature dependence of the sample dielectric characteristics. Moreover, had the sample expanded upon the temperature
changes,it would have found itself in a "clamped" condition.

In the second sample holder device,the electrode geometry,mate-
rial and distance were the same as in the former cell,but the electrodes were surrounded by a thin concentric golden ring with a
height of 6mm. The ring and the bottom electrode were Teflon insulat-
ed. The upper electrode and the ring were,instead,at the same potential;therefore the latter behaved as a guard ring. The upper electrode was a gold disk attached to a 3mm long shaft which projected
from the cover of the cell. In this way upward the latter electrode,
between the golden disk and the cover of the cell,there was some
room for the sample to expand. Such a variation would not affect the
impedance values since the upper electrode and the guard ring were
kept at the same potential. With this arrangement,the electrode distance variation technique could also be applied to avoid electrode
polarization impedance. Also the second type of cell was enclosed
into a stainless steel container that could be immersed in a thermostatic bath.

The sample temperature was measured with a calibrated thermocouple placed directly inside the sample,held vertically and insulat-
ed by a very thin glass capillary tube that was inserted in the narrow gap between the electrodes and the guard ring. The average uncertainty, over the whole temperature range ($77.14^{\circ}K$-$372.98^{\circ}K$), was
of the order of $0.08^{\circ}K$.

The electrodes of both sample holder devices were carefully
rubbed in one direction with calibrated Carborundum 20 μm powders,
as explained in references 5-10 . The cell constant of the sample
holder devices were evaluated as reported in references 1,2

RESULTS

General dielectric features at room temperature. In the interval $0.024 < C < 0.80$ at $T=293^{\circ}K$,the dielectric behavior of the w/o
microemulsion as a function of increasing concentration,at the different frequencies,is characterized by two main features (Figure 2):
an abrupt increase of the ε' and ε'' values at C=0.31 that strongly
depends on the measuring frequency and a frequency independent di-
vergent behavior as the concentration approaches the value C=0.582.

The former,hardly detectable at frequencies higher than 10 MHz,is
quite well observable in the low frequency range (16 Hz-5 MHz) on
samples with water content in the interval 0.30 <C <0.42.

As reported earlier [9,10] the divergent behavior at C=0.582 that
falls inside the LC region (Figure 1) was ascribed to the structural
transition from cylinders to lamellae.

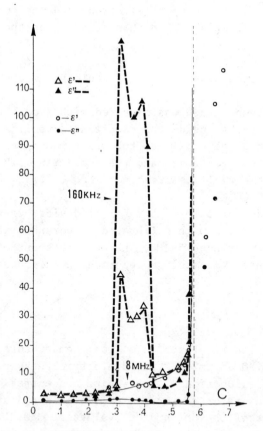

Figure 2. Behavior of ε' and ε'' vs.C at two frequencies. Room
temperature measurements.

The phenomenon was interpreted as due to the change in the organization of the system's inner surface from the closed to the
open configuration. In the lamellar mesophase, in fact, there is no
surface enclosing the volume of the dispersed water phase.

With the TSD analysis at room temperature, T_p =293°K, we found
that optically isotropic microemulsion samples are characterized by
a single peak TSD spectrum with a maximum current peak temperature
at T=253°K (Figure 3;C=0.134). The 253°K peak does not depend on the
system's water content nor on the values of the polarizing parameters T_p, t_p, and E_p [3-5,9,10].

The first appearance of depolarization processes linked with
the system's water content was observed in the neighborhood of C=
=0.31 as shown in Figure 3 by the spectra with C=0.315,0.334 and
0.353, respectively.

In the interval 0.36 C 0.42, apparently homogeneous, transparent
and isotropic microemulsion samples are distinguished by the complex TSD spectrum shown in Figure 4, curve 1. A part from the first
relaxation process that occurs always at T=253°K, the latter is characterized by the depolarization peaks at T=273°K and 288°K, whose
presence strongly depends on the concentration. Moreover these are
confined not only within a concentration range of existence but also
in a temperature domain (Figure 4, curve 2).

Considering the above results, the 253°K peak was interpreted as
due to the orientation of the water-oil interface of the dispersed
system that represents the only common feature shared by all the
samples and called "interface peak".

The process defined by both a concentration range of existence
and a temperature domain was instead interpreted as due to the orientation of some kind of structure, of lyotropic nature, which is destroyed if the samples are oriented at a T_p that falls outside the
temperature interval within which the given structure may exist. [3,4]
This type of depolarization process was called "structure peak" [5].

Both the interface and the structure peaks follow a first order relaxation kinetics and fit quite well the theoretical function
given in Equation (4). (Figure 5).

For the actual system, the concentration C=0.31 represents, in
our opinion, the end of the truly isotropic w/o dispersion and the
beginning of the prestructural region where an early structural ordering of the microemulsion liquid crystalline phase occurs [10].

Low temperature dielectric results. The temperature depend -

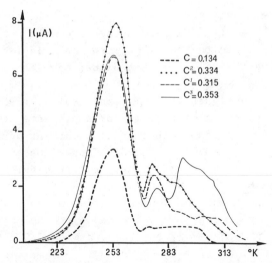

Figure 3. C=0.134,single peak TSD spectrum of an isotropic w/o microemulsion sample; ξ =0.483+0.004,ln τ_o=-17.25+1.90."Interface peak"
at T=253°K. First appearance of orientation processes linked with the
system's water content for C>0.3.Interface peak parameters are in the
order:$\xi^1$0.705+0.016,ln τ_o=-27.70+0.78; ξ^2=0.715+0.026,ln τ_o=-27.97+
1.24; ξ^3=0.683+0.013,ln τ_o=-26.76+0.64 . (ξ values are in eV).

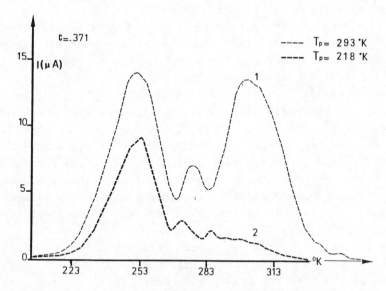

Figure 4. Differentiation between the "interface" and the "structure" peak. The latter is no longer detectable if the sample is
oriented at a T_p that falls outside the temperature domain of the
given structure.

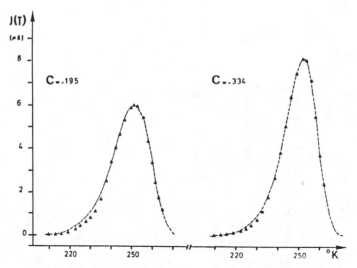

Figure 5.   Theoretical fit (continuous line) of the J(T) function
given in Equation (4) to the experimental points.

ence of both ε' and ε" in isotropic microemulsion samples (0.13<C<
0.3) does not follow the usual pattern exhibited by the normal dipolar liquids [25]. In the actual system the real and the imaginary
part of the relative complex permittivity increase with decreasing
temperature,going through a sharp maximum at a temperature that depends upon the system's water content. For example,with C=0.248,at
160 KHz,the maximum occurs at T=234.8°K. In the frequency range
1 KHz-50 MHz,the above temperature shifts from 235.4°K to 233°K.
    The direct recordings of the sample impedance magnitude and
phase angle vs. time are shown in Figure 6. The calculated values
of ε' and ε" during the cooling process are summarized in Table I .
We should note here that the /Z/ and φ values are within an error
of 2% and +2° respectively.
    From the experimental results we could ascertain that neither
the low temperature sharp dielectric increase nor its frequency behavior depend on the particular sample holder used. However the
second dielectric cell,because of the guard ring and the variable
electrode spacing,allows the control of the stray field effects as
well as the elimination of the electrode polarization impedance.
Therefore,if slow cooling (or heating) rates are used,of the order
of 0.1°K/s,this cell is the most suitable for an accurate determination of the temperature dependence of the values of ε' and ε ";
while it fails if faster rates are employed.

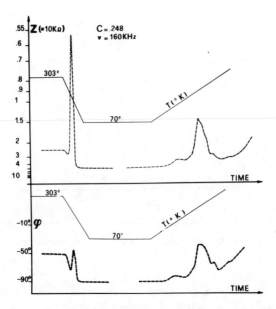

Figure 6. Typical low temperature dielectric behavior of an isotropic w/o microemulsion with C=0.245. Direct recordings of the total impedance /Z/ and of the phase angle ϕ of the dielectric filled
cell. Cooling rate 4.5°K/s;Heating rate 0.1°K/s as that used in the
TSD method. Both ϵ' and ϵ'' increase with decreasing temperature,go-
ing through a sharp maximum at T=234.8°K.

Table I.   Low Temperature Dielectric Values. (Figure 6)

T°K	293	234.8	70
ϵ'	3.64	23.81	1.64
ϵ''	3.82	281.14	0.04

From a general point of view, the low temperature dielectric
behavior of any concentration in the interval 0.15-0.3 is frequency
and temperature-rate dependent. At any given T in the range 70°K-
293°K, besides the point where the sharp dielectric peak occurs, the
frequency dependence of the system is what is expected for a dispersed system. For a given temperature gradient, the order of magnitude
of the impedance value at T=234.8°K does not change with frequency;
whereas it decreases if slower cooling rates are employed. The same
applies to the absolute values of the phase angle (i.e. the resis-
tive component of the system increases upon lowering the tempera-
ture-gradient). Identical cooling and heating rates lead to a different dielectric behavior of the system. During the heating process,
a shift to higher temperatures of the anomalous dielectric peak occurs. The peak itself is not as well defined as in the cooling run
(Figure 6) . The optimum cooling rate for a given sample does not
correspond to its optimum heating rate. Research is in progress to
obtain a more quantitative information on the latter topic.

The dielectric study on samples with high water content (C>0.68)
will be reported in the next Section for the sake of comparison
with the TSD results.

Low temperature TSD results. The water rich phase of the system, C>0.68, appears optically isotropic, strongly viscoelastic and
"metastable"; it collapses into a transparent, perfectly fluid dispersion after 30-40 days which is not an o/w type of microemulsion
because the latter exhibits a completely different TSD spectrum
(See reference 5, Figure 3).

The TSD spectrum obtained by orienting a viscoelastic sample
with C=0.752, at T_p =293°K, is plotted in Figure 7, curve 1, where the
corresponding spectrum of the collapsed sample, 40 days later, is also shown (curve 2). The spectra were obtained by applying a +10 V,
140 KHz square wave that was found to be the most suitable frequency for detecting the "viscoelastic peak" at T=273°K.

Since highly viscous samples do not follow a linear temperature
dependence, despite the linear freezing and heating rates applied,
the TSD spectra plotted in Figure 7 must not be compared with those
obtained on samples belonging to completely different phases of the
system [9,10].

Due to the high frequency of the E_p field as well as the non
linear sample temperature dependence, the above spectra could not be
analyzed with the procedure described in Section II.

However, the dielectric measurements on viscoelastic samples

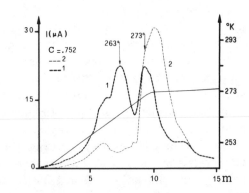

Figure 7. Typical TSD spectra of the highly viscous microemulsion isotropic phase polarized at room temperature. Curve 1:the spectrum of a sample in the strongly viscous state (C=0.752). Curve 2:the corresponding spectrum performed 40 days later on the collapsed, fluid sample. The temperature dependence of the sample against time, (minutes),is also shown (continuous line).

proved that the nonlinear temperature behavior is confined in a narrow interval in the neighborhood of T=273°K where a phase transition
occurs. Outside the above interval,the temperature dependence of the
dielectric properties of these samples follows the behavior that
characterize the normal polar liquids [25] . No sharp increase was
detected in the ε' and ε'' trend vs. temperature,in the interval
70°K-293°K.

The frequency dependence of highly viscous samples dielectric
properties was indeed rather surprising. As a matter of fact,such
a water rich phase,with plenty of ions and charges,displays,at extremely low frequencies (5 Hz-1 KHz),a very low dissipative component and a liquid crystalline type of frequency dependence,with a
relaxation frequency in the KHz range. The latter topic will be
reported elsewhere.

On the basis of the above results,the TSD method was applied
to viscoelastic samples by using low polarizing temperatures(<273°K).

The most remarkable feature of the TSD study is represented by the current peak at T=198°K shown in Figure 8. The peak follows a first order relaxation kinetics,but its maximum current peak intensity is an inverse function of the polarizing temperature. The activation energy ξ and the relaxation time constant $\ln \tau_0$ of the 198°K peak,calculated by means of Equation (6) are: ξ =0.481 eV$\pm$ 0.007 eV and $\ln \tau_0$=-26.34$\pm$ 0.044.   The high temperature contribution is a depolarization current band centered around 240°K. At low temperature,no macroscopic difference was observed between the TSD spectra of strongly viscous samples and the corresponding collapsed ones.   It was verified that the polarization induced in the samples is proportional to the intensity of the applied field.

The decrease of the 198°K current peak intensity upon increasing T_p ,was found to correspond to the increase of the current band (See Figure 8,curves 1,2,3).   In other words,a rising of the temperature favours the structural ordering of the viscoelastic component of the system while it inhibits that of the solid phase and vice versa.

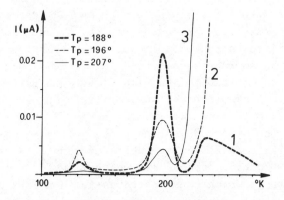

Figure 8.   Low temperature TSD study on a highly viscous sample with C=0.752. The first current peak does not follow a first order kinetics,while the 198°K peak does.   Its maximum current peak intensity is an inverse function of the polarizing temperature. The ξ and $\ln\tau_0$ values are reported in the text

DISCUSSION

By means of the isothermal study of the relative complex permittivity at room temperature,we obtained a general description of
the microemulsion dielectric behavior as a function of increasing
concentration (Figure 2).

With the TSD method we investigated at T_p=293°K the temperature dependent relaxation processes that characterize the system in
the different concentration intervals. It was found that the mechanism of polarization of the microemulsion is of dipolar nature and
follows a first order relaxation kinetics (Equation 1). The agreement between the depolarization current peaks recorded experimentally and the analytical expression of the discharge current giv-
en in Equation (4) proved also that a uniform polarization occurs
in microemulsion samples oriented at room temperature. Thus,for each
relaxation process ,the activation energy and the relaxation time
constant were calculated according to Equation (6).

Furthermore,by orienting the samples at low polarizing temperatures,the nonisothermal behavior of "solid" microemulsion samples
(0.68-0.80) was analyzed in order to understand,possibly,the role
played by the system's counterions and also to distinguish their
contribution to the total stored charge.

The latter study demonstrated that upon lowering T_p,two new
current peaks develop in the interval (100°K-200°K);these can be
detected only by orienting "solid" microemulsion samples.The lower the
T_p,the higher becomes their current intensity. As soon as T_p reaches
the temperature at which the highly viscous and isotropic phase may
exist,both depolarization processes vanish. By extending the concentration interval investigated,we could check that the above peaks
occur unchanged also in the isotropic microemulsion region.

On the basis of the aforementioned characteristics,we interpreted the 198°K peak which follows a first order kinetics,as due
to the migration over microscopic distances of the system's counter
ions. These may be oriented and therefore evidenced when they stop
acting as interface stabilizers. Such a circumstance is verified
only when the samples,independently on their water content,are kept
at a temperature at which none of the system's phases may exist as
a liquid or as a liquid crystal.

The study of the behavior of ε' and ε'' as a function of decreasing temperature was undertaken with the aim to investigate the
modification of the system's dielectric properties during the "freezing" process .

We reported earlier [5,9,10] that upon heating frozen micro-emulsion samples in the absence of any electric field,pyroelectric currents were detected in the temperature interval 173°K–230°K.

Since pyroelectricity is due to a spontaneous polarization which is solely a function of the temperature, it follows that the presence of pyroelectric currents is linked with an interior field.

We believe the effect to be an indirect pyroelectric effect caused by the mechanical strains induced in the system by the freezing process which is enhanced by the electric field applied at room temperature. When isotropic,liquid microemulsions are frozen in the electric field,they become a pyroelectric body with a polar axis in the direction of the latter; thus a transition from a pyroelectric to a non pyroelectric phase occurs. The sharp but finite maximum exhibited by the dielectric constant in the polar direction on transition is consistent with the above interpretation [26]. The latter,in turn, is consistent with the TSD results mostly,if one takes into account the assumptions implicit in the derivation of the $J(T)$ function given in Equation (4).

In order to investigate whether the observed low temperature dielectric properties are restricted to the particular microemulsion studied or they represent some intrinsic characteristic of microemulsion systems,in general,research is in progress in our laboratory on a four component microemulsion with a nonionic surfactant.

ACKNOWLEDGEMENTS

The authors express their gratitude to Mrs.M.Magini,V.Bertini and S.Bigi of the "Officina Meccanica S.Salvadori" of Florence for their very kind and competent technical assistance during the preparation of the sample holder devices. Financial support of this work by the "Ministero della Pubblica Istruzione" and the "Gruppo Nazionale di Struttura della Materia" (G.N.S.M.) of the CNR is gratefully acknowledged.Our warmest thanks are also due to the "Thermofas", "Sedas" and "Hewlett and Packard Italiana" companies.

REFERENCES

1. D. Senatra and G. Giubilaro, J. Colloid Interface Sci., 67, 448 (1978).

2. D. Senatra and G. Giubilaro, J. Colloid Interface Sci., 67, 457 (1978).

3. D. Senatra, C. M. C. Gambi and A. P. Neri, Lett.Nuovo Cimento
    28, 433 (1980).
4. D. Senatra, C. M. C. Gambi and A. P. Neri, Lett.Nuovo Cimento
    28, 603 (1980).
5. D. Senatra, C. M. C. Gambi and A. P. Neri, J.Colloid Interface
    Sci., 79, 443 (1981).
6. S. Ballaro',F. Mallamace, F. Wanderlingh, D. Senatra and G. Giu-
    bilaro, J. Phys.,C, Solid State Physics, 12, 4729 (1979).
7. D. Senatra, M. Vannini and A. P. Neri, Lett.Nuovo Cimento, 28,
    453 (1980).
8. D. Senatra, M. Vannini and A. P. Neri, Lett.Nuovo Cimento, 28,
    608 (1980).
9. D. Senatra, Il Nuovo Cimento B, 64, 151 (1981).
10. D. Senatra, J. Electrostatics, 12, 383 (1982).
11. M. Eguchi, Phil.Mag., 49, 178 (1925).
12. B. Gross and S. F. Denard, Phys.Rev., 67, 253 (1945).
13. B. Gross, Phys. Rev., 94, 1545 (1954).
14. B. Gross and R. J. De Moraes, J. Chem. Phys., 37, 710 (1962).
15. C. Bucci, and R. Fieschi, Phys. Rev. Lett., 12, 16 (1964).
16. C. Bucci, R. Fieschi and G. Guidi, Phys. Rev., 148, 816 (1966).
17. R. Cappelletti and R. Fieschi, in " Electrets ",M. Perlman, Edi-
    tor, p.1, Electrochemical Soc.Inc., Princeton, New Jersey, 1973
18. M. Campos, S. Mascarenhas and G. Leal Ferreira, Phys. Rev. Lett.,
    27, 1432 (1971).
19. M. Perlman and C. Reedyck, J. Electrochem. Soc., 115, 45 (1968).
20. B. Gross, Phys. Rev., 107, 368 (1957).
21. S. Mascarenhas, in "Physics of Ice", Riehl,Bullemer and Engel-
    hardt, Editors,p. 483, Plenum Publishing Corporation,New York,
    1969.
22. S. Bini and R. Cappelletti, in "Electrets", M. Perlman,Editor,
    p. 66, Electrochemical Soc. Inc.,Princeton, New Jersey, 1973
23. S. B. Lang, Nature (London), 212, 704 (1966).
24. E. Fukada and I. Yasuda, J. Phys., Soc. Japan, 12, 1158 (1957).
25. C. P. S. Smyth, in "Dielectric Behavior and Structure", McGraw
    Hill Book Company, Inc., New York, 1955.
26. H. V. R. Jaffe, Phys. Rev., 53, 917 (1938).

MUTUAL AND SELF DIFFUSION COEFFICIENTS OF MICROEMULSIONS FROM SPONTANEOUS AND FORCED LIGHT SCATTERING TECHNIQUES

A. M. Cazabat, D. Chatenay, D. Langevin, J. Meunier
Ecole Normale Supérieure, 24, rue Lhomond
75231 Paris Cedex 05, France

L. Léger
Collège de France, 11, place Marcelin Berthelot
75231 Paris Cedex 05, France

Several microemulsion systems have been investigated along a dilution line using both spontaneous and forced scattering techniques. In the first case, the intensity results were interpreted following the model developed by A. Vrij for spherical particles. The dynamical behaviour of the system was accounted for using theories by Felderhof for mutual diffusion coefficients.

The agreement is quite satisfactory in the domains where the exchanges between droplets are not the main feature. If the exchanges become important: concentrated systems, vicinity of critical consolute points, a different description of the system must be used (see following paper).

For polydisperse systems, Pusey's treatment has been used to deduce both mutual and self diffusion coefficients from spontaneous light scattering data. These self diffusion coefficients are in good agreement with those measured with forced light scattering experiments.

The interpretation of experimental studies in microemulsions requires in most cases a model for the structure of the medium. In this paper, we present the information provided by such models for the analysis of light scattering results and we discuss its limitations.

. Vrij was the first one to propose that a microemulsion could be represented as a dispersion of <u>permanent identical Brownian spheres</u> in a continuous medium. The osmotic pressure Π of the droplets is assumed to be the sum of a hard sphere contribution Π_{HS} and a perturbative terme Π_A

$$\Pi = \Pi_{HS} + \Pi_A \qquad\qquad \Pi_A = \frac{k_B T}{v} \frac{A}{2} \phi^2$$

where v is the droplet volume, ϕ the droplet volume fraction, A a constant related to the strength of the perturbative potential. More recent models [2] allow to take into account a possible <u>polydispersity</u> of the spheres.

. A theoretical treatment of the Brownian motion of <u>interacting identical permanent spheres</u> was given recently by Batchelor [3] and Felderhof [4], among others ,at low volume fraction ϕ. In this concentration range, the collective diffusion coefficient of the spheres can be written as :

$$D = D_0 (1 + \alpha\phi) \qquad\qquad D_0 = kT/6\pi\eta R_H$$

D_0 is the diffusion coefficient at zero volume fraction given by the Stokes-Einstein formula ; η is the viscosity of the continuous medium, R_H the hydrodynamic radius of the spheres. Knowing the interaction between spheres, Felderhof's theory allows calculation of α [4]. In particular, it is possible to deduce α from A using the Vrij model for osmotic pressure [1].

. The dynamical properties of a system of slightly <u>polydisperse permanent</u> spheres has been investigated by Weissman [5] and Pusey [6]. They showed that two different diffusion processes had to be taken into account in this case : a collective mean diffusion coefficient D and a self diffusion coefficient D_s ; the latter usually can only be observed with tracer experiments. The contribution of this second process, usually very small, becomes visible at fairly high $\phi (> 0.1, 0.2)$ particularly in repulsive systems.

. Let's now explore the possibility to use such a model to describe microemulsions and, more precisely, to interpret the results obtained by light scattering techniques.

. First, the microemulsion droplets are not <u>permanent</u>
ones : exchanges of constituents take place between droplets and continuous phase and between droplets themselves during the collisions.
Usually, the exchange times are in the range 10^{-8} - 10^{-6}s, which is
shorter than the time scale used in light scattering experiments
($\tau > 10^{-5}$ s). Thus light scattering techniques give an average picture of the medium, averaged over times longer than 10^{-5} s. Although
an instantaneous picture of the medium would probably show complicated amoebic structures, this mean picture should show quite well defined droplets, whose dynamical behaviour is well accounted for by
equations of <u>Brownian</u> motion [7] ★.

However, the dynamical properties of these mean droplets
do not necessarily correspond to their static properties. In other
words, knowing A is not enough to calculate α except if there are
very few exchanges. As previously noted [9,10], the α values deduced
from measured A values using Felderhof's theory are never in perfect
agreement with experimental data in microemulsions. At the best,
α_{calc} - α_{exp} ∿ 2 or 3, for not too attractive droplets (A ⪎ −4 typically). Let us mention that the agreement between calculated and
measured α is quite good for silica particles [11] which are truly
permanent spheres.

If the interaction between droplets is strongly attractive
(A < −10), sticky collisions with formation of transient dimers, trimers, ... are more probable [12], in contrast with the preceding case,
where almost all collisions were elastic ones (therefore very short).

If sticky collisions become frequent, there is no reason
to expect structures of sizes comparable to isolated droplet size by
averaging. Large scale structures may appear. Moreover, if the lifetimes of these structures become comparable to the time scale studied,
a large polydispersity will be observed.

Let us now turn to the experimental results.

We will first present original results pertaining to an
oil in water microemulsion, studied along a dilution line by both
spontaneous and forced Rayleigh scattering. The composition of the
droplets is 0.4 g SDS, 0.5 cm³ toluene, 0.17 cm³ butanol. The continuous phase composition is 100 cm³ salted water (1g/ℓ NaCℓ), 4.86 cm³
butanol. Spontaneous light scattering techniques are well known
now [1,9,10]. They allow measurements of scattered intensity I, which
is related to the osmotic pressure Π :

★ The reactive diffusion effects discussed recently by Phillies [8]
for micelles are certainly negligible here since microemulsion
droplets are very large.

$$I \propto \phi \left(\frac{\partial \, \Pi}{\partial \, \phi} \right)^{-1}$$

and its autocorrelation function $\mathscr{C}(t)$, t being the time. As predicted for monodisperse permanent spheres, in many microemulsion systems, $\mathscr{C}(t)$ is a single exponential with characteristic decay time :

$$\tau = 1/Dq^2$$

q is the sacattering wave vector, D the collective diffusion coefficient.

The droplets in this system behave as repulsive spheres
(A = +8) of hydrodynamic radius 46 Å. The calculated value of α is
6.2 and the experimental one +4 , a reasonable agreement. Exchanges
are not too important in this system, which is easily understood in
view of the repulsive interaction between droplets. So we are allowed
to use models of permanent droplets to check other features of this
system :

Up to $\phi = 0.1$, the correlation function $\mathscr{C}(t)$ is single-
exponential, which means no visible polydispersity. For $\phi > 0.1$, a
second exponential with longer decay time is observed. This time is
found to be proportional to the shear viscosity of the sample. Similar results obtained by other authors [13,14], have been interpreted
in terms of self diffusion process, following Pusey's theory.

So we tried to check this theory by measuring directly
the self diffusion coefficient of droplets using forced Rayleigh
scattering [15], which is a tracer technique.

A schematic picture of the experimental set up is given
in Figure 1. A short pulse of UV laser beam is divided into two beams
which interfere in the sample. A small amount of hydrophobic photochromic molecules (spiropyran), previously added to the microemulsion
(2×10^{-5} wt), is embedded in the oil core of the droplets. During
the light pulse, the photochromic molecules located in bright interference fringes change their conformation. The polarizability of the
excited molecules being different from that of the nonexcited ones,
a diffraction grating is built into the sample. As the lifetime of
the excited state is large (about one minute), the grating disappears
by diffusion of the droplets which restores an uniform distribution
of excited and nonexcited molecules. The time evolution of the grating is detected by measuring the intensity of a diffracted He-Ne
beam. The fringe spacings were chosen between 1 and 3 μm ; thus the
measured times were in the range 50 ms - 2s, much shorter indeed than
the lifetime of excited configuration.

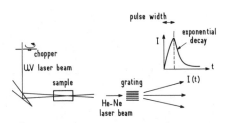

Figure 1.   Principle of forced Rayleigh scattering experiment.

        The diffracted intensity decreases exponentially with
characteristic decay time :

$$\tau = 1/D_s q^2 \qquad\qquad q = 2\pi/i$$

where i is the fringe spacing and D_s is the self diffusion coefficient of the droplet.

As can be seen in Figure 2, the measured values of D_{self} are in
fairly good agreement with the values deduced from   the long
decay time observed in spontaneous measurements for $\phi > 0.1$. The
accuracy on D_s measurements is usually 2 or 3% in forced Rayleigh
experiments.   The data spread as indicated in Figure 2 is larger
and we are presently trying to improve the D_s determination. Despite
their still preliminary character, all these measurements are in
good agreement with Pusey's theory.

        Let us turn now to w/o microemulsion and recall briefly
the results obtained on two typical systems for comparison [9]. They
were (as in the o/w case) studied along dilution lines. In the first
one, referenced as ATP, the composition of the droplets is 0.4 g SDS,
0.5 cm^3 water, 0.33 cm^3 1-pentanol. The continuous phase composition
is 100 cm^3 toluene, 17 cm^3 pentanol, 0.4 cm^3 water. As evidenced by
spontaneous light scattering, the droplets behave as slightly attractive spheres (A = -4) of hydrodynamic radius 45 Å. The calculated
value of α is 0, the experimental one -2 , again in reasonable agreement : exchanges are not important in this system. Figure 3 shows the
measured intensity, collective diffusion coefficient and electrical

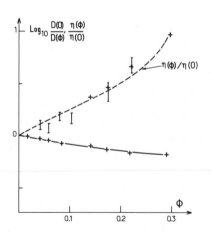

Figure 2.. Logarithmic plot of $D_{collective}$, D_{self} and the viscosity
η for o/w microemulsion ;
dotted line : reduced viscosity $\eta(\phi)/\eta(0)$;
crosses : inverse of the reduced collective diffusion
coefficients obtained by spontaneous scattering $D(0)/D(\phi)$
errors bars : inverse of reduced self diffusion coefficient obtained by forced scattering $D(0)/D_{self}(\phi)$

$D(0) = 4 \times 10^{-7}$ cm^2 s^{-1} $\eta(0) = 1.17$ cp

conductivity versus volume fraction in these microemulsions. In the
second system, referenced as ATB, the composition of the droplets is
0.4 g SDS, 0.5 cm^3 water, 0.17 cm^3 butanol. The continuous phase composition is 100 cm^3 toluene, 13.5 cm^3 butanol, 0.4 cm^3 water. The
droplet radius is 42 Å. The attractive term is so strong that it
would lead to a value of A smaller than −25 , which in fact has little meaning because of the perturbative treatment used in Vrij's
theory. Figure 4 shows the main features observed in this system :
very large variation in intensity, diffusion coefficient and electrical conductivity with volume fraction ϕ. A steep increase in conductivity takes place close to $\phi \sim 0.08$ where are observed the extrema in I and D.
Except for very low $\phi(\phi < 0.04)$, it is clear that a model of permanent droplets cannot be used to interpret light scattering data in
the ATB system. Surprisingly, the correlation functions $\mathcal{E}(\tau)$ are
exponential whatever ϕ is, in the two systems. This is probably due
to the fact that interactions are not repulsive as in the o/w system, which makes polydispersity fluctuations difficult to observe [6].
We have not been able to extract the self diffusion coefficient from
either spontaneous or forced light scattering experiments. In the
last case, the spyropyran is partitioned between droplets and continuous phase and the interpretation of the measurements is not straight
forward.

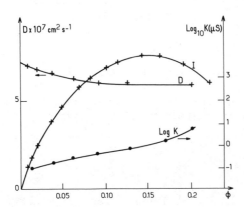

Figure 3.   ATP system. Scattered intensity I : arbitrary units.
Collective diffusion coefficient D : left scale. Electrical conductivity κ : right scale (logarithmic plot).

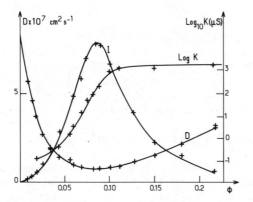

Figure 4.   ATB system. Scattered intensity I : arbitrary units.
Collective diffusion coefficient D : left scale. Electrical conductivity κ : right scale (logarithmic plot).

Further experiments with different techniques on theses systems were then needed, they will be presented in the following paper : in this paper the ATB and ATP microemulsions are used as model systems for discussing the effects of exchanges between droplets on various physical properties.

ACKNOWLEDGEMENTS

It is a real pleasure to acknowledge M. Veyssié and C. Guglielmetti for their gift of spyropyran.

REFERENCES

1. W.G.M. Agterof, J.A.J. Van Zomeren and A. Vrij, Chem. Phys. Lett. 43, 369 (1976).
2. A. Vrij, J. Chem. Phys. 71, 3267 (1979).
    A. Vrij, J. Colloid Interface Sci., 90, 110 (1982).
3. G.K. Batchelor, J. Fluid Mech. 52, 245 (1972), 74, 1 (1976).
4. B.U. Felderhof, J. Phys. A 11, 929 (1978).
5. M. Weissman, J. Chem. Phys. 72, 231 (1980).
6. P.N. Pusey, in "Light Scattering in Liquid and Macromolecular Solutions", V. DeGiorgio, M. Corti, M. Giglio, Editors, Plenum Press, New York (1980).
    M.M. Kops-Werkhoven, H.J. Mos, P.N. Pusey and H.M. Fijnaut, Chem. Phys. Letters, 81, 365 (1981).
    P.N. Pusey, H.M. Fijnaut and A. Vrij, J. Chem. Phys. 77, 4270 (1982).
7. P.N. Pusey, J. Phys. A 8, 1433 (1975).
8. G.D. Phillies, J. Colloid Interface Sci. 86, 226 (1982).
9. A.M. Cazabat and D. Langevin, J. Chem. Phys. 74, 3148 (1981).
10. W.N. Brouwer, E.A. Nieuwenhuis and M.M. Kops-Werkhoven, to be published in J. Colloid Interface Sci.
11. M.M. Kops-Werkhoven and H.M. Fijnaut, J. Chem. Phys. 74, 1618 (1981).
12. B. H. Robinson, in "Microemulsions", I. D. Robb, Editor, Plenum Press (1982).
    H. F. Eicke, J. C. W. Shepherd and A. Steineman, J. Colloid Interface Sci., 56, 168 (1976).
13. D.J. Cebula, R.H. Ottewill, J. Raltson and P.N. Pusey, J. Chem. Soc. Faraday Trans. I. 77, 2585 (1981).
14. M.M. Kops-Werkhoven, private communication.
15. D.W. Pohl, S. Schwartz and V. Irniger, Phys. Rev. Lett. 31, 32 (1973).
    H. Eichler, G. Salje and H. Stahl, J. Appl. Phys. 44, 5383 (1973).

PERCOLATION AND CRITICAL POINTS IN MICROEMULSIONS

A.M. Cazabat, D. Chatenay, P. Guering, D. Langevin,
J. Meunier and O. Sorba

Ecole Normale Supérieure, 24 rue Lhomond,
75231 Paris Cedex 05 , France

J. Lang and R. Zana

C.R.M., 6 rue Boussingault, 67083 Strasbourg, France

M. Paillette

Groupe de Physique des Solides de l'ENS
2, place Jussieu, 75251 Paris Cedex 05 , France

Electrical percolation in w/o microemulsions is
interpreted as a transition from individual droplets to
large scale open structures. This model accounts for various features previously explained as due to the vicinity
of critical consolute points. The relation between critical
points and percolation is discussed.

In the preceding paper, we investigated the possibility of using a model of individual permanent droplets to describe micro-emulsion phases. As long as direct exchanges between droplets through sticky collisions are not important this model was found suitable. In that case, light scattering results are well accounted for by assuming hard sphere-like behaviour of droplets. (Conversely, this feature proves the suitability of the model).

In this paper, we turn to the opposite situation where direct exchanges between droplets through collisions become important. This is the case in all concentrated systems (volume fraction of droplets ϕ above 30%) and also in some systems even at low volume fraction $\phi < 10\%$. Light scattering experiments performed at low ϕ in these systems show large attractive interaction between droplets.

In these two cases, the droplet model becomes inadequate, leading us to use a more collective description of the medium. In the following, we will be mainly concerned with the peculiar behaviour of microemulsions with strongly attractive droplets. Such a microemulsion is studied and compared with a hard sphere-like microemulsion. Light scattering results [1] on these systems have been discussed in the preceding paper.

Let us summarize them briefly here.

Both systems are studied along a dilution line. For the first one, referred to as ATP, the composition of the droplets is 0.4g SDS, 0.5 cm^3 water, 0.33 cm^3 1-pentanol. The continuous phase composition is 100 cm^3 toluene 17 cm^3 1-pentanol, 0.4 cm^3 water. As evidenced by spontaneous light scattering, the droplets behave as hard spheres of radius 45 Å. Scattered intensity I and diffusion coefficient D show only smooth variations versus volume fraction ϕ in this system. In the second one, referred to as ATB, the composition of the droplets is 0.4g SDS, 0.5 cm^3 water, 0.17 cm^3 1-butanol. The continuous phase composition is 100 cm^3 toluene, 13.5 cm^3 1-butanol, 0.4 cm^3 water. A strong attraction exists between droplets whose radius is 42 Å. Opposite behaviour is observed in this system, where the sharp maximum in I versus ϕ coincides with a deep minimum in D. At the same volume fraction ϕ_c, a large and steep increase of electrical conductivity is observed [3].

This steep increase of the electrical conductivity occurs if there is an attractive interaction between droplets, even if the interaction is not very strong : A $\lesssim$ –5 (for notations, see preceding paper). It was first observed by M. Lagües [2] and interpreted as geometrical percolation of droplets. However, no percolation arises in the hard sphere-like ATP system at similar volume fractions [3].

The characteristic features of light scattering results become more and more marked as the attractive interaction increases.

For very low values of A ($A \lesssim -20$), both intensity I and diffusion coefficient D show angular dependence at $\phi \sim \phi_c$ [3-5]. This angular dependence is very well accounted for in terms of theories of critical phenomena : the dilution line passes close to a critical demixion point in the phase diagram. Many such critical points in micellar and microemulsion systems have been studied recently [6-8].

However, critical behavior in these systems is not yet fully understood. The meaning of measured critical exponents is not clear [6-9] and basic questions about the true structure of the medium remain unanswered. This last point is specially difficult because polydispersity of the microemulsion droplets can give rise to angular variations of both D and I in quite good agreement with critical point theories [10-11]. So even close to a critical consolute point, the interpretation of experimental results is not straightforward. As critical-like behaviour is usually observed even very far from the critical consolute point [3-9] in multicomponent systems, the situation is really intricate.

Let us note finally that for oil-external critical-like systems, electrical percolation takes place at the same volume fraction as critical-like behaviour (see the last Figure in the preceding paper). This particular point will be discussed later. However, it makes the interpretation of the results even more difficult : indeed the occurrence of droplet clusters near the percolation threshold can explain some features of light scattering experiments [11], as previously noted.

Clearly additional studies were needed for a better understanding of percolation and critical phenomena in microemulsions. We have therefore studied an oil-external critical-like system (ATB) using different techniques : electrical conductivity [3], viscosity [12], ultrasonic absorption, electrical transient Kerr effect. We also studied a hard sphere-like microemulsion as reference system (ATP).

The electrical conductivity measurements are shown in Figure 1, together with viscosity measurements for both ATB and ATP systems. Viscosity measurements were performed using an Ubbelhode flow viscosimeter with shear rates between 100 and 200 s^{-1}, small compared to reciprocal characteristic times in the system (see later). As can be seen from Figure 1, an anomalous increase in viscosity takes place in the ATB system at about the same volume fraction as electrical percolation. On the contrary the ATP system shows no peculiar behavior of viscosity or conductivity.

Figure 2 shows an example of ultrasonic absorption measurements performed on these systems. Excess ultrasonic absorption is obvious in the ATB system at about $\phi \sim 0.1$ compared to the ATP system. This effect is observed mainly at lower frequencies (< 30 MHz). More detailed results on these systems will be published later.

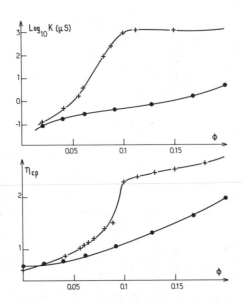

Figure 1. top : logarithmic plot of electrical conductivity κ versus volume fraction φ. Upper curve : ATB system ; lower curve ATP system – bottom : linear plot of viscosity η versus φ. Upper curve : ATB ; lower curve : ATP.

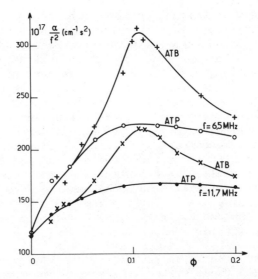

Figure 2. ultrasonic absorption measurements versus volume fraction φ at two different frequencies. For each frequency, the upper curve corresponds to the ATB system, the lower one to the ATP system.

Finally, transient electrical Kerr measurements have been performed very recently. Preliminary results will be briefly discussed here.

Measurements were made in both systems up to ϕ =10% . The electrical pulses have short rise and fall times (< 35ns) allowing us to measure decay times larger than 100ns. The height and width of the pulses have been varied between 0-2kV and 1-300 µs respectively. Each experimental curve is an average of 32 or 64 recordings. The reproductibility of the results was satisfactory.

A large difference in birefringence was found between the two systems. For instance, if the birefringence of ATP system at ϕ =10% is taken as unity, the values 43 , 190 and 1.7×10^4 are obtained in ATB system for the volume fractions 1.74% , 2.9% and 4.2%   respectively. At higher volume fractions, the birefringence increases very rapidly and great care must be taken to obtain fully saturated effect without disturbing the system.

A similar behaviour is observed for the characteristic decay times. In the ATP system, the measured times are shorter than 1µs. In the ATB system, a short time in the same range is observed, but longer times with a broad distribution are the dominant feature. The width of this distribution increases from 10 µs at ϕ=2% up to more than 300µs at ϕ =9%. Further experiments are currently under way to check more precisely the behaviour of the percolating system in the long time range.

Let us now turn to the viscosity measurements. First of all, all, it is obvious that electrical percolation in microemulsions is not a purely geometric effect. Indeed, geometrical percolation always occurs in hard sphere systems at $\phi \sim$ 14 %. Transient infinite clusters appear at this point, but they have no influence on the physical properties of the medium if no specific exchange is possible between the interconnected droplets during the lifetime of the cluster. As electrical conductivity means transport of charges, sticky collisions [13] are required to allow ion exchanges. This is why no electrical percolation is observed in the hard sphere ATP system at $\phi \sim$14%.

Finally, a crude picture of the phenomena is given in Figure 3. On the left, we see transient clusters in a hard sphere-like system, and on the right, exchanges take place between connected droplets. Transient open structures with large spatial scale appear in the system. This picture is quite similar to the onset of bicontinuous structures as first proposed by Scriven [14] . From the theoretical point of view, it is clear that electrical percolation in microemulsions can only be accounted for in terms of site-bond percolation models [15] , the bond probability increasing with attractive interaction between droplets. This last point is in perfect agreement with present pictures of collisions between droplets [16]

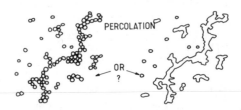

Figure 3. left : geometrical percolation
right : the same system with exchanges beween connected
droplets

Let us now turn to the viscosity measurements. First of all,
it must be noted that in both systems the measured values are fairly low : all structures change very rapidly in microemulsions and
there are no giant effects as in some micellar systems [17].

It is not clear if the excess viscosity in the ATB system has
to be accounted for in terms of critical or percolation effect.
There is a good qualitative agreement between experiment and critical point theories [12] but the quantitative comparison is less
satisfactory. On the other hand, a percolation model gives a good
picture of the phenomenon, the large scale structures of the
medium being "torn" due to the shear [18] . This last explanation
is in agreement with recent work of the Minnesota group [19].

Ultrasonic absorption data also lead to a similar situation : a
good qualitative agreement with critical point theories is observed,
but not a quantitative one. Further experiments on various systems
have been performed and will be published shortly. The final answer
is that, except perhaps very close to the critical points, the excess
ultrasonic absorption reflects the onset of open structures inducing
strong surfactant rearrangement. It follows that this model is probably good also for interpreting viscosity anomalies.

It also provides straightforward interpretation of electrical
birefringence measurements : the short time corresponding to isolated droplets or small droplets aggregates, the long times to a
collective response of the structure. This picture is in good agree-

ment with results obtained in micellar system [17]. However, in a percolating system a broad distribution of relaxation times is observed, due to the broad distribution of clusters sizes.

As previously noted, some particular distributions of cluster sizes lead to angular variations in scattered intensity I and diffusion coefficient D which are not distinguishable from their variations close to a critical point [20].

The analogy between percolation and critical points can be carried very far. However, a peculiar point in microemulsion systems is that the percolation is dynamic, or stirred (Note also that for spherical particles typical values of the critical concentration ϕ are between 10% and 20%, in the same range as percolation threshold values). This will perhaps allow one to explain why in these systems percolation threshold and critical points are always very close, which is not the case in other systems [21].

In summary, it appears that percolation in microemulsions is not a purely geometrical effect. It also corresponds to the onset of open structures. It is always present in critical like water in oil microemulsions and accounts for various anomalies previously explained as due to the vicinity of critical consolute points.

ACKNOWLEDGMENTS

We would like to acknowledge helpful conversation with Professor H. Hoffmann.

REFERENCES

1. A.M. Cazabat, D. Langevin, J. Chem. Phys. 74 , 3148 (1981)
2. M. Lagües, J. de Phys. Lettres , 40 , L-331 (1979)
3. A.M. Cazabat, D. Chatenay , D. Langevin and A. Pouchelon,
   J. de Phys. Lettres, 41 , L-441 (1980)
4. A. . Cazabat, D. Langevin, J. Meunier, and A. Pouchelon , J. de
   Phys. Lettres, 43 , L-89 (1982)
5. G. Fourche, A.M. Bellocq and S. Brunetti, J. Colloid Interface Sci.,
   88 , 302 (1982).
6. M. Corti and V. Degiorgio, Phys. Rev. Lett. 45 , 1045 (1980)
7. J.S. Huang and M.W. Kim , Phys. Rev. Lett. 47 , 1962 (1981)
8. R. Dorshow, F. de Buzzaccarini, C.A. Bunton and D.F. Nicoli,
   Phys. Rev. Lett. 47 , 1336 (1981)
9. J. Lang , these proceedings
10. M.E. Fisher, Physics, 3, 255 (1967)
11. B.J. Ackerson, C.M. Sorensen, R.C. Mockler, and W.J. Sullivan,
    Phys. Rev. Lett. 34, 1371 (1975)
12. A.M. Cazabat, D. Langevin and O. Sorba, J. de Phys. Lettres, 43,
    L-505 (1982)

13. H.F. Eicke, J.C.W. Shepherd and A. Steineman, J. Colloid
     Interface Sci., 56, 168 (1976).
14. L.E. Scriven, Nature (London), 263, 123 (1976)
15. A. Coniglio, H.E. Stanley and W. Klein, Phys. Rev. Lett. 42 ,
     518 (1979)
16. H. Fletcher, A.M Howe, N.M. Perrins, B.H. Robinson and
     C. Toprakcioglu these proceedings
17. H. Hoffmann, H. Rehage, W. Schorr, and H. Thurn these
     proceedings
18. P.G. De Gennes, private communication
19. K.E. Bennett, J.C. Hatfield, H.T. Davis, C.W. Macosko and
     L.E. Scriven, in "Microemulsions", I.D. Robb,Editor, Plenum
     Press, New York 1982.
     Y. Talmon and S. Prager, J. Chem. Phys. 69, 2984 (1978)
20. A. Coniglio and W. Klein, J. Phys. A 13 , 2775 (1980)
21. T. Tanaka, G. Swislow and L. Ohmine, Phys. Rev. Lett. 42 ,
     1556 (1979).

STRUCTURAL AND DYNAMIC ASPECTS OF MICROEMULSIONS

P.D.I. Fletcher, A.M. Howe, N.M. Perrins,
B.H. Robinson, C. Toprakcioglu and J.C. Dore[†]

Chemical Laboratory and Physics Laboratory[†]
University of Kent at Canterbury
Canterbury, U.K.

Water-in-oil microemulsions stabilized by Aerosol-OT (AOT) have been investigated using both structural and kinetic techniques. The size and polydispersity of the water droplets have been investigated by means of the small angle neutron scattering (SANS) technique. Preliminary SANS data on the solubilization of the enzyme α-chymotrypsin in w/o microemulsions is also discussed.

Results are reported for the kinetics of hydrolysis of AOT in hydroxide-containing water droplets, studied using a pH-indicator spectrophotometric method.

The mechanism of exchange (transfer) of ions between the water cores of droplets, which occurs following collision of the droplets, has been further investigated by fast flow methods with $Fe(CN)_6^{4-}$ as the transferring ion. Additional information is obtained from the temperature dependence of the exchange process.

The kinetics of metal-ligand complex formation involving $Ni^{2+}(aq)$, which takes place at the surfactant-water interface, has been investigated in aqueous SDS micellar solutions and AOT reversed micelle/microemulsion systems. The kinetic treatment proposed by Berezin has been used to interpret the results.

INTRODUCTION

The main theme of this contribution is to consider how simple
kinetic studies can be used to probe structural aspects of water-
in-oil microemulsion systems. This paper, however, will only be
concerned with Aerosol-OT (AOT)-stabilised microemulsions.
Structural information obtained using the small angle neutron
scattering (SANS) technique will also be reported in support of
the general picture which emerges from the kinetic studies. Away
from the phase transition region, these systems are already quite
well understood from a structural viewpoint[1-3] and so we are now
at the stage where detailed information is being sought.

The following topics are considered in this paper:

a) The size, size range and size profile of water droplets
   dispersed in various hydrocarbons as a function of concentration
   (variation of mole ratio of water to surfactant) and solvent
   composition (variation of the hydrocarbon chain length in a
   series of n-alkanes).

b) The distribution and state of organization of the surfactant
   in the system especially as the phase boundary for maximum
   water solubilization is approached.

c) The nature of the interaction between droplets. In particular,
   what happens when two droplets collide with each other?

d) What structural changes occur in an AOT-stabilized system on
   solubilization of the water-soluble enzyme α-chymotrypsin?
   Preliminary results obtained using the SANS technique will be
   reported.

e) Charged micelles in aqueous solution can dramatically affect
   the rates and equilibrium positions of reactions, especially
   when charged reactants are involved.[4-8] Does this also occur
   in microemulsions? The kinetics of a metal-ligand complexation
   reaction occurring exclusively at the AOT-water interface are
   compared with the same reaction at the SDS-water interface.

DETERMINATION OF WATER DROPLET SIZES
USING SANS

The results reported are for the system $D_2O/H(AOT)/H(alkane)$.
Substituting D_2O for H_2O and the good 'match' between the hydrocarbon chain of the surfactant and the alkane enables the size of
the water pool to be determined. For a system of non-interacting
monodisperse spherical particles of radius r, the intensity of
scattered neutrons I(Q), of wavelength λ, as a function of
scattering angle (θ) is given by Equation (1) which incorporates

only the form factor $F(Q)$:

$$I(Q) = 9(\Delta\rho)^2 V^2 n \left\{ \frac{\sin Qr - Qr \cos Qr}{(Qr)^3} \right\}^2 \qquad (1)$$

$\Delta\rho$ is the scattering contrast, V is the volume of the particle and n is the number of particles per unit volume.

$$Q = (4\pi/\lambda) \sin \theta/2 \qquad (2)$$

Since $I(Q) = F(Q).S(Q)$, the structure factor $S(Q)$ has been taken to be unity. This would seem to be a reasonable assumption, as coulombic interactions are much weaker than in aqueous charged micellar solutions[9] and we are working with a relatively dilute dispersion.

From an analysis of the neutron data, the radius of the water pool (r_w) can be obtained as a function of R_T, where R_T is defined as the mole ratio of water to AOT in the system (i.e. $R_T = [H_2O]_T/ [AOT]_T$ where the subscript 'T' refers to total or weighed-in concentrations). For heptane as continuous phase and a temperature of 20°C, the neutron data fit reasonably well to Equation (1) in the low R_T range (5-30). However, the fit is by means perfect in this region; for example, the predicted minimum in the plot of $I(Q)$ vs (Q) is never observed, and so it would appear that one (or more) of the assumptions on which Equation (1) is based must be invalid. The most likely candidate is that the system has been taken to be 'monodisperse'; we believe this is only true as a first approximation. It is also clear that for R_T values > 30, the fit becomes progressively worse and a plot of r_w versus R_T has an increasing positive slope as R_T is increased (Figure 1). One explanation for the curvature in the plot is that the surface area/head group is decreasing with increasing R_T, but another explanation, which we prefer, is that not all the AOT is located at the interface of the droplets.

A comparison of the scattering profiles at R=20 in heptane, decane and dodecane is shown in Figure 2. It is clear that a progressively less good fit to Equation (1) is obtained as the chain length of the alkane is increased. Results for hexane and octane are also included for comparison. The explanation for the particularly poor 'fit' in the case of the dodecane system is that 20°C is quite close to the upper phase transition temperature. There is evidence from the profile for the existence of large structures, manifested by the behaviour at low Q values.

Extrapolation of the r_w data in Figure 1 to $R_T=0$ indicates
the likely existence of a core radius which is > 10Å. This
would probably correspond to a domain containing the sodium ions,
the sulphonate head groups and the carbonyl groups of the
surfactant. A value for the hydrodynamic radius (r_h) at $R_T=0$ of
15.0 ± 3Å has previously been obtained by photon correlation
spectroscopy.[1] Neutron scattering experiments using the system
$H_2O/H(AOT)/D$ (heptane) gave values for the overall radius r_{ov} of
the droplet of 16.1Å ($R_T=0$), 19.2Å ($R_T=3$), 25.8Å ($R_T=5$) and
32.6Å ($R_T=10$). The differences between r_{ov} and r_w are 4.6Å at
$R_T=10$ and 7.3Å at $R_T=20$. It should be pointed out that, as with
the 'dry' micelle discussed above, there is some uncertainty
concerning the extent of the water-core region defined by r_w since
the sulphonate group will certainly be in the water pool and water
is probably also associated with the oxygen atoms in the two
branched alkyl-chains.

Neutron results also indicate that r_w does not change
significantly with [AOT] at a fixed value of R_T over the range
$R_T=15$-30 and [AOT] from 1×10^{-2} mol dm^{-3} to 5×10^{-2} mol dm^{-3}. Similar
results were obtained by Zulauf and Eicke using photon correla-
tion spectroscopy[1], although their data suggested that there was
a tendency for r_h to decrease somewhat for larger values of R_T
(20,38) and low concentrations of AOT (< 3×10^{-2} mol dm^{-3}).

There are various possible explanations for the approximate
fit of the experimental data to Equation (1). The effect of a
'diffuse' interface adding an extra contrast region is not likely
to eliminate the predicted minimum in I(Q) at 4.41/r (Equation (1)).
The introduction of elliptical distortion (shape fluctuations) of
the microemulsion changes this situation slightly, but does not
have a major effect on the results unless large eccentricities
are used. The absence of a zero-scattering region suggests that
polydispersity of droplet size is the most likely cause of the
discrepancy between the experimental data and the behaviour
predicted by Equation (1).

It must be stressed that the analysis based solely on polydispersity would only be appropriate when the microemulsion system
is located well away from the upper phase boundary for stability,
as already discussed for the dodecane system on the previous page.
Critical phenomena become important and affect the neutron
scattering profile at a temperature corresponding approximately to
$T_c - T < 20°C$, where T_c corresponds to the upper phase transition
temperature.

The neutron profiles also provide clear evidence that the low
temperature phase transition is much sharper and of a different
nature to the upper temperature one.

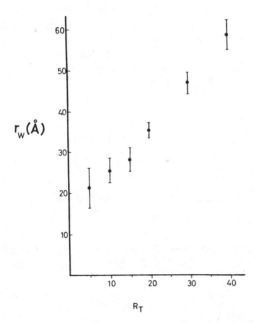

Figure 1.   Plot of r_W versus R_T.   [AOT] = 0.05 - 0.1mol dm^{-3} in
heptane.   Temperature = 20°C.   Values of r_W are obtained from the
"best-fit" to Equation (1).

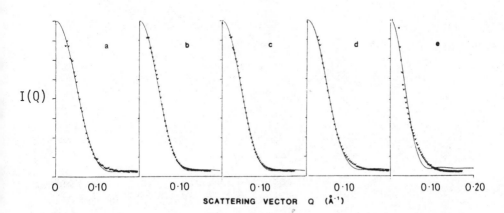

Figure 2.   I(Q) vs. Q for heptane(b), decane(d) and dodecane(e).
"Best-fit" values of r_W are 36Å (C_7H_{10}), 34Å ($C_{12}H_{26}$) and >50Å
($C_{12}H_{26}$).   Profiles are also shown for hexane(a) and octane(c).

Various models for polydispersity have been tested for the decane
system. The fit is quite sensitive to the model chosen and in
decane the 'best-fit' was obtained for a concave parabolic
distribution as shown schematically below. Such a profile may result from inelastic collisions between the droplets.

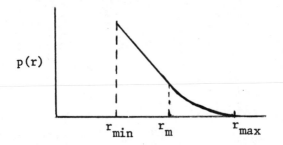

The value of r_m obtained was 32Å, and the index of polydispersity λ
$(=\{r_m-r_{min}\}/r_m)$ = 0.6. The 'fit' to the above function is shown
in Figure 3.

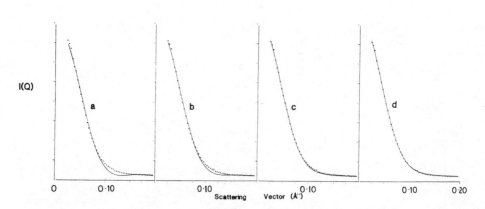

Figure 3. The experimental data and optimized fits corresponding
to different polydispersity models for water/AOT/decane (R=20);
[AOT]=0.1mol dm^{-3}.
a) Monodisperse; fit to Equation (1). b) Fit based on optimum
symmetric model (inverted parabola). c) Fit based on optimum
triangular model. d) Fit based on concave parabola (as discussed
in text).

HYDROLYSIS OF AOT IN MICROEMULSIONS/EVIDENCE
FOR PARTITIONING OF AOT.

It is difficult to measure directly the proportion of AOT present in the microemulsion which is actually located at the interface for a given value of R_T. Generally the AOT is assumed to be totally bound. An attempt has been made, from a study of the rate of hydrolysis of AOT (an ester) in microemulsions, to obtain information on this point.

The OH^--promoted hydrolysis of AOT leads to production of a carboxylate ion and an alcohol. Therefore we have:

$$R'-O-\overset{\overset{O}{\|}}{C}-R'' + OH^- \longrightarrow R'OH + R''CO_2^-$$

The rate of hydrolysis of AOT in water has been measured under conditions of excess OH^-. AOT has a half-life of about 3 days at pH~12, T=25°C. Since similar rates are found for AOT hydrolysis in microemulsions, the acid and alcohol products are always possible impurities in AOT microemulsions, particularly in H^+ and OH^--containing systems.

In OH^--containing water droplets the rate of hydrolysis is easily monitored spectrophotometrically by measuring the rate of decrease in pH by means of water-soluble pH indicators (pKa's 10-12; $[Ind]_{H_2O} \ll [OH^-]_{H_2O}$). A plausible kinetic scheme, which allows for partitioning of OH^- between the interface (S) and the water core (W), is:

$$AOT_s + OH_s^- \xrightarrow[k_s]{SLOW} Products.$$

$$FAST \downarrow K_{OH}$$

$$OH_w^- \qquad\qquad K_{OH}=[OH^-]_s/[OH^-]_w$$

Then, for $[AOT] \gg [OH^-]$,

$$\frac{-d[AOT]}{dt} = \frac{-d(OH^-)_T}{dt} = k_s[AOT]_s[OH^-]_s \qquad (3)$$

$$= k_{obs}[OH^-]_T \qquad (4)$$

$$\text{where} \quad k_{obs} = \left(\frac{k_s K_{OH}}{1+K_{OH}}\right)[AOT]_s \qquad (5)$$

A suitable measure of $[AOT]_s$ is given by $[AOT]_s'/[H_2O]_T$ where $[AOT]_s'$ is the concentration of AOT at the droplet interface expressed as per dm^3 of total solution. This will be equal to R_T^{-1} if all the AOT is present at the interface. A plot of k_{obs} vs R_T^{-1} should then pass through the origin.

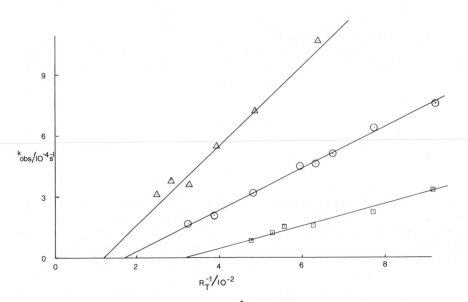

Figure 4. Plots of k_{obs} vs. R_T^{-1}. Indicator = sulpho-orange.
⊡ = 15°C; ⊙ = 25°C; △ = 35°C.

The plots are apparently linear but with an intercept on the
'x'-axis. The intercept value R_o^{-1} on the 'x'-axis corresponds to
the minimum value of R_T when no reaction takes place. In Equation
(5), it is reasonable that K_{OH} is a function of R (i.e. droplet
size) and k_s could conceivably be as well (due to steric effects
which might depend on droplet curvature) but these factors are
unlikely to explain the non-zero intercept. A tentative explanation is on the basis of partitioning of AOT away from the interface, which increases as R_T is increased. Then at the limit of R_o,
no AOT is bound at the surface of the droplets. For a variety of
solvents and temperatures a correlation has been found between R_o
and the value of R_T corresponding to the phase transition.

If the AOT is not at the interface, where is it? AOT is very
soluble in the organic phase in the absence of added water. It is
dissolved predominantly in the form of dry (or almost dry) reversed micelles. Therefore in the microemulsion system we have
the possibility of a dynamic equilibrium between reversed micelles
and microemulsion droplets.

INTERACTIONS BETWEEN WATER DROPLETS

It has previously been shown[10] that exchange of solubilized ions between water droplets occurs on the ms to μs time scale. The process can be represented in its simplest form as follows:

$$A = H^+, \; Zn^{2+}(aq), \; Fe(CN)_6^{4-}$$

Since the ions are very soluble in water and insoluble in the oil phase it seems highly likely that exchange of A can only occur when the droplets collide.

For several positive ions (eg. H^+, Zn^{2+}) at room temperature in heptane, k_{ex} has been determined in the range 10^6–10^7 $dm^3 mol^{-1} s^{-1}$, based on droplet concentrations.[11] If every encounter between droplets resulted in transfer between them, a value of ~10^{10} $dm^3 mol^{-1} s^{-1}$ would be obtained, so only ~1 in 1000 encounters is effective. To further our understanding of the nature of the 'collision' process, we have recently used, as an indicator for transfer, the fast electron-transfer reaction between $IrCl_6^{2-}$ and $Fe(CN)_6^{4-}$. This reaction has advantages over the proton-transfer and metal-ligand complexation reactions previously studied since it is irreversible and concentrations of the reactants can be chosen such that multiple occupancy of the pools is avoided. This greatly simplifies the kinetic analysis and enables k_{ex} to be determined independently of the need for information on precise droplet concentrations. Both the stopped-flow and continuous-flow methods were used, equipped with visible detection. The results are shown in Tables I and II.

Table I. Exchange of $Fe(CN)_6^{4-}$ in Microemulsions ([AOT] = 0.1 mol dm^{-3} in heptane)

Temp°C \ R_T	11	21	31
	($k_{ex}/10^6 dm^3 mol^{-1} s^{-1}$)		
5	2.4	1.3	0.68
10	4.0	2.0	1.3
15	6.5	3.5	2.3
19	–	–	5.0
25*	14	14	14

*Data obtained using the continuous-flow method.

$$R_T = [H_2O]_T/[AOT]_T$$

Table II. Activation Parameters for Solubilizate Exchange

R_T	$\Delta H^{\ddagger}/kJ\ mol^{-1}$	$\Delta S^{\ddagger}/JK^{-1}mol^{-1}$
11	59(60)	90(90)
21**	75(90)	143(180)
31**	105	244

**The Arrhenius Plots were slightly curved, and $\Delta H^{\ddagger}$ seems to decrease at lower temperature. Data in parenthesis were obtained previously for H^+ transfer.

It is significant for the detailed mechanism that $Fe(CN)_6^{4-}$ and H^+ are transferred at similar rates, independent of ion charge. To explain these observations we propose that an intermediate transient "water-droplet dimer" is formed. Exchange is effected as a result of formation of this dimer.

'Transient-dimer'

To prevent further association the transient dimer must be short-lived and its lifetime is probably of the order of 0.1-10 μs, although of course we have no direct information on this from our kinetic studies. The process of exchange is associated with a relatively large value of $\Delta H^{\ddagger}_{ex}$ compensated by a large positive $\Delta S^{\ddagger}_{ex}$. Both these activation parameters increase with R_T and hence droplet size. $\Delta H^{\ddagger}_{ex}$ will (in part) be associated with the energy required for the compression of the surfactant head groups on formation of the dimer. The most likely explanation for the $\Delta S^{\ddagger}$ values is that surfactant (probably associated with Na^+) is released from the interface as the dimer is formed. Collisions between droplets therefore cause surfactant to be released into the oil phase contributing to the partitioning of AOT as discussed in the previous section. Furthermore, the collision process would be expected to introduce polydispersity into the system as a result of the uneven splitting of the dimer.

ENZYMES IN MICROEMULSIONS

There is much current interest in enzyme kinetics in water-in-oil microemulsions[12-14] especially from the viewpoint of novel synthetic applications.[15] Water-soluble enzymes like α-chymotrypsin (dimensions 50x40x40 Å) are readily solubilized in AOT-containing microemulsion systems; the enzyme dissolves even when very small amounts of water are present. For structural considerations and the interpretation of kinetic data it is useful to ascertain if the location of the enzyme is inside a water droplet; if so then a further question arises concerning the extent of the water shell surrounding the enzyme. In addition, when the water core size of the droplet system before solubilization is less than the enzyme dimensions there may be a tendency for droplets to aggregate around the enzyme. A further general point of interest is the nature of enzyme-water and enzyme-surfactant interactions. So far very few structural studies have been undertaken although circular dichroism has been used to probe changes in secondary structure of the enzyme on solubilization.[12] A recent paper by Martinek et al. reports sedimentation velocity measurements using an ultra-centrifuge.[16] They conclude that when the water core of a micro-emulsion droplet is less than the effective size of the protein or enzyme to be solubilized, the macromolecule induces the formation of a reversed micelle around itself of the required size.

We have carried out some preliminary measurements using SANS as a probe for changes in the nature of the water droplets on addition of α-chymotrypsin. The system was D_2O/α-CT/H(AOT)/H (Heptane). An enzyme concentration was chosen such that the enzyme concentration overall was equal to the initial droplet concentration. Thus we have a system which may be represented most simply as:

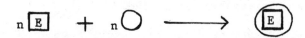

Surprisingly, there is hardly any change in the scattering profile (for R_T = 15 and 20) as compared with the reference system $D_2O/H(AOT)/H$ (Heptane). No quantitative interpretation concerning the enzyme location is therefore possible but it would appear that the number and shape of the water droplets are largely unperturbed on addition of α-chymotrypsin. The enzyme is extremely soluble in water so it would seem reasonable to conclude that it is not at all in contact with the organic phase. The most likely enzyme location would appear to be one in which it is surrounded by a thin skin (perhaps a monolayer) of water molecules so that the enzyme is in close association with the surfactant head groups, essentially in a 'damp' reversed micelle situation. However, the intra-droplet interactions in an enzyme-free and an enzyme-containing reversed micelle would be expected to be very different.

The location of the substrate and the enzyme are of importance in any detailed interpretation of the kinetics and this is currently being investigated using circular dichroism and nmr techniques for a series of hydrophilic oligo-peptides.[17]

METAL-LIGAND COMPLEXATION IN
AOT-STABILISED WATER-IN-OIL MICROEMULSIONS

The reaction between Ni^{2+}(aq) and the azo dye complexing
agent PAP has been studied in bulk water, in aqueous micellar
solutions formed by sodium dodecylsulphate (SDS) and in water-in-
oil microemulsions stabilised by AOT. (R_T=5-20: [AOT]=0.1mol dm^{-3}).

At the appropriate pH, the reaction is a simple 1:1 complexation involving the uncharged form of the ligand. That is:

$$Ni^{2+}(aq) + PAP \xrightarrow{k_f^M}$$

(PAP = pyridine-z-azo-p-phenol)

PAP was specially selected for detailed study since it is
almost insoluble in water and totally insoluble in n-heptane. It
is, however, easily dissolved in both SDS micellar and AOT w/o
microemulsion systems. Hence we conclude that the ligand must be
located at the surfactant-water interface in both media. Then
the kinetic analysis is considerably simplified.

The Berezin kinetic treatment[18] can be readily applied to
the reaction. Under pseudo-first-order conditions (i.e. $[Ni^{2+}]_T$
$\gg [PAP]_T$) and given that exchange or transport processes are
not rate-limiting (i.e. $k(s^{-1}) < 10^{+3}s^{-1}$) as discussed in a
previous section, we have:

$$-d[PAP]/dt = k[PAP]_T \qquad (6)$$

and $\quad k(s^{-1}) = (k_f^M [Ni^{2+}]_T)/(C.V) \qquad (7)$

where $\quad C = [SDS]_T - cmc;$ or $[AOT]_s (\sim [AOT]_T) \qquad (8)$

and $\quad V = $ contribution to the reaction volume
per mole of surfactant at the
interface.

The kinetic constant k_f^M is a second-order rate constant
(units-$dm^3 mol^{-1}s^{-1}$) for reaction in the micelle surface (M) or
interface region. Hence k_f^M is directly comparable with k_f^W - the
second order rate constant for reaction in bulk water (obtained
from $k = k_f^W [Ni^{2+}]_T$).

The stopped-flow method was used to follow the kinetics and in all cases single exponential transients were observed from which k was obtained directly.

It should be noted from Equation (7) that it is impossible to obtain independent values of k_f^M and V. However, the enthalpy of activation for the reaction, $\Delta H_f^{\ddagger M}$, can be obtained from the temperature dependence of k_f^M (assuming V to be independent of temperature) and derived values are shown in Table III.

Table III. Values of $\Delta H_f^{\ddagger}$ and V for the Reaction between Ni^{2+}(aq) and PAP in Different Media

	$\Delta H_f^{\ddagger}$ (kJmol^{-1})	V(if, $k_f^M = k_f^W$)* dm^{3}mol^{-1}
Aqueous solution	51±3	–
SDS/water interface	48±3	0.5±0.1
AOT/water interface	50±2	0.3±0.05

*See text for details.

The constancy of $\Delta H_f^{\ddagger}$ ($\Delta H_f^{\ddagger M} \sim \Delta H_f^{\ddagger W}$) indicates (i) that there is no significant "catalysis" of the reaction and (ii) that there is no change in mechanism for the reaction proceeding in bulk water and at the two types of interface. The data suggest that the nickel aquo-ion is fully solvated at the interface prior to ligand substitution and that the rate-determining step in complex formation is loss of a first water molecule from the inner co-ordination sphere of the metal ion.

Since there is no change observed in $\Delta H_f^{\ddagger}$, it is reasonable to make the assumption that $k_f^M = k_f^W$. Then a value for V can be calculated. The similarity of the two values in Table III indicates a 'thickness' of around 10Å for the 'shell' in which reaction takes place. The reaction, involving a neutral species as one reactant, is apparently insensitive to the curvature of the surface (possible steric effects) and the local electric field strength. It can also be inferred that the Ni^{2+}(aq) ion is strongly partitioned to the interface region in both systems. The kinetics involving the AOT system were studied at low R_T values (5-20) where the effects of surfactant partitioning do not seem to exert much influence on the reaction. (The kinetic results are rather insensitive to changes of up to 50% in any case.)

Much greater effects on the rates and equilibrium positions of the reaction are observed in the SDS micellar solutions. This is because of the very strong binding of Ni^{2+}(aq) to the SDS micelle surface.

ACKNOWLEDGEMENTS

The SANS work was carried out at AERE, Harwell, U.K., and Institut Laue-Langevin, Grenoble. We thank the Neutron Beam Research Committee of the SERC for financial support. We also thank Dr. Joseph Holtzwarth, Fritz-Haber Institut, W. Berlin, for use of his continuous-flow instrument, the DAAD for a grant (to A.M.H.) and Esso Chemicals and Shell Research for financial support (to N.M.P. and A.M.H., respectively).

REFERENCES

1.   M. Zulauf and H-F. Eicke, J. Phys.Chem., 83, 480 (1979).
2.   C. Cabos and P. DeLord, J.Appl.Cryst., 12, 502 (1979).
3.   R.A. Day, B.H. Robinson, J.H.R. Clarke and J.V. Doherty, J.C.S. Faraday 1, 75, 132 (1979).
4.   J.H. Fendler and E.J. Fendler, "Catalysis in Micellar and Macromolecular Systems", Academic Press, New York 1975.
5.   A.D. James and B.H. Robinson, J.C.S. Faraday 1, 74, 10 (1978).
6.   V.C. Reinsborough and B.H. Robinson, J.C.S. Faraday 1, 75, 2395 (1978).
7.   J. Holzwarth, W. Knoche and B.H. Robinson, Ber.Bunsenges. Phys.Chem., 82, 1001 (1978).
8.   P.D.I. Fletcher and V.C. Reinsborough, Can.J.Chem., 59, 1361 (1981).
9.   J.B. Hayter, Ber.Bunsenges.Phys.Chem., 85, 887 (1981).
10. S.S. Atik and J.K. Thomas, Chem.Phys.Lett., 79, 351 (1981).
11. P.D.I. Fletcher and B.H. Robinson, Ber.Bunsenges.Phys.Chem., 85, 863 (1981).
12. S. Barbaric and P.L. Luisi, J.Am.Chem.Soc., 103, 4239 (1981).
13. F.M. Menger and K. Yamada, J.Am.Chem.Soc., 101, 6731 (1979).
14. P. Douzou, E. Keh and C. Balny, Proc.Natl.Acad.Sci.USA, 76, 681 (1979).
15. K. Martinek, Y.L. Khmelnitsky, A.V. Levashov, N.L. Klyachko, A.N. Semenov and I.V. Berezin, Dokl.Akad.Nauk SSSR, 256, 1423 (1981) [Russ].
16. A.V. Levashov, Y.L. Khmelnitsky, N.L. Kyachko, V.Y. Chernyak and K. Martinek, J.Colloid.Interface.Sci., 88, 444 (1982).
17. L.M. Gierasch, J.E. Lacy, K.F. Thompson, A.L. Rockwell and P.I. Watnick, Biophys.J., 37, 275 (1982).
18. I.V. Berezin, K. Martinek and A.K. Yatsimirski, Russ.Chem.Rev., 42, 787 (1973).

STRUCTURE OF NONIONIC MICROEMULSIONS BY SMALL ANGLE NEUTRON SCATTERING

J. C. Ravey and M. Buzier

Laboratoire de Biophysique Moléculaire
Université de Nancy I, E.R.A. N° 828
B.P. N° 239, 54506 Vandoeuvre les Nancy Cedex, France

Structural studies of micellar aggregates of the nonionic surfactant tetra-ethylene glycol dodecyl ether, decane and water were performed by using small angle neutron scattering. For that purpose, a precise method has been developed which allows us to gain some insight into the structure of the micellar aggregates. At 20°C for $C_{12}(EO)_4$ system, results show that all the microemulsions are only of the "reverse" type: they consist of definite, large nonspherical water droplets inside a surfactant shell, even for fraction of disperse phase as high as 0.50. The structure of this interface is well defined, since we are able to describe the area per polar group, the conformation of the surfactant molecule and the extent of interpenetration of oil and water inside the surfactant film.

INTRODUCTION

At the present time, the question of the structure of nonionic microemulsions is largely open to discussion, owing to the high complexity of the problem which is largely due to the non dilute state of these three component mixtures. As far as nonionic surfactants are concerned, it can even be said that almost nothing is known about their true structural properties, although the problem has raised up much speculation. Let us remind of the debate about the structure of simple aqueous micelles when temperature is changed.[1,5,15,16]

The methods used for investigation allow one to form only a vague idea about these structures. For example, it is the case for the now well known NMR self-diffusion studies, from which clear evidence for typical reversed micellar solutions with close water cores was not obtained. Their data[14] prove the existence of highly labile aggregates, and they have contributed to suggest the existence of the so-called bicontinuous systems, for which theoretical models have already been proposed. It may be quoted here that the random internal geometry resulting from a Voronoi tessellation[2] can account for a progressive regular change from mainly oil/water to mainly water/oil systems. For example, as far as scattering data are concerned, it has been shown that the similarity of Voronoi calculation result with that from a sphere model does not allow a distinction to be made.[3] On the other hand, the so-called critical opalescence is also more and more frequently invoked to explain light scattering data.[4-5]

We want to show here how the small angle neutron scattering technique can be used to obtain a more precise structural description of some of the nonionic systems. Of course the conclusions of the present study may not be generalizable to every such system, since the following results concern only a dozen of samples and one type of surfactant. But most of these results, which concern only a small part of the general investigation we undertook a few years ago, can be directly compared to those from the NMR self diffusion measurements, which have been performed on exactly the same microemulsions.[14]

In this paper, we would like to consider the types of microemulsion which exist at temperatures just above the phase inversion temperature (PIT) (or HLB temperature).[7-8] For that purpose we chose the system $C_{12}(EO)_4$-water-decane, the PIT of which is about 15°C. On crossing this temperature, the surfactant molecules which were previously more soluble in water become preferentially soluble in the oil. More specifically, on the ternary diagrams determined by Friberg and coworkers,[6] the corresponding overall compositions are represented by the points in Figure 1 and are presented in Table I. It is obvious these samples are mixtures of comparable amounts of oil and water.

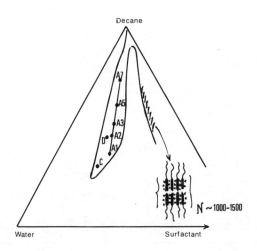

Figure 1. Phase diagram of water–decane – $C_{12}(EO)_4$ system at 20°C. The structural investigations concern the systems whose compositions are represented by the points A1–A7, C, D.

Table I. Composition in Water (W), Surfactant (S) and Decane (O) concentration of the systems noted A1–A7, C, D. [(a) expressed in g/100g; (b) volume ratios].

	% W (a)	$\frac{W+S}{W+S+O}$ (b)	$\frac{S}{S+W}$ (b)
A_1	42	.58	.36
A_2	38	.51	–
A_3	29	.36	–
A_5	22	.24	–
A_7	15	.17	–
C	48	.62	.30
D	32	.42	.35

At this time, we suppose there is no interparticle interactions.

We suppose from now on that certain microemulsions may possess such a layered structure (Table II). What are then the morphological parameters ?

Table II. List of the Morphological Parameters of an Aggregate.

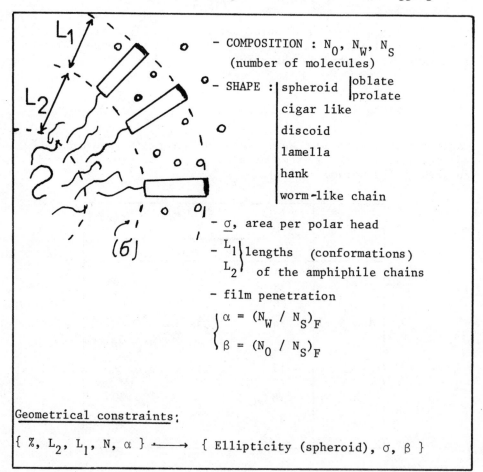

- COMPOSITION : N_0, N_W, N_S
  (number of molecules)
- SHAPE : spheroid oblate
                  prolate
  cigar like
  discoid
  lamella
  hank
  worm-like chain
- σ, area per polar head
- L_1 lengths (conformations)
  L_2 of the amphiphile chains
- film penetration
  $\alpha = (N_W / N_S)_F$
  $\beta = (N_0 / N_S)_F$

Geometrical constraints:

$\{ \%, L_2, L_1, N, \alpha \} \longrightarrow \{ \text{Ellipticity (spheroid)}, \sigma, \beta \}$

EXPERIMENTAL AND NEUTRON SCATTERING METHOD

(a) Materials and experimental

The nonionic surfactant tetra ethylene glycol dodecylether $(C_{12}(EO)_4)$ was purchased from Nikko Chemicals (Japan) and was used without further purification. Decane was Merck reagent grade. Small angle neutron scattering measurements were all carried out at the Institut Laue-Langevin (I.L.L.) Grenoble, using the neutron diffractometers D11 and D17.[9]

(b) the method

Neutron scattering is very similar to small angle X ray scattering, except that
- (i) the wavelength may be varied (between about 5 to 12 Å), and
- (ii) the contrast between different elementary volumes of the scattering particles can be changed by isotopic substitution of a part of the atoms. This constitutes the basis of the variation constrast method.
Here, the scattered intensity is the sum of an incoherent plus a coherent component. T is the transmission of the sample. The quantity of interest, I_{coh}, is given by the following relations :
(q is the transfer momentum)

$$I = T (I_{inc} + I_{coh})$$

$$I_{coh} \sim < \sum_i \sum_j b_i b_j e^{-\vec{q}.(\vec{r}_i - \vec{r}_j)} >$$

The double sum is extended to all atoms i or j whose position is defined by vector $\vec{r}_i$, and whose scattering length is b_i. If one component is such that it may be considered as the solvent, the sum is extended to the remaining components, and the scattering lengths can be replaced by excess quantities: (v represents the corresponding specific volumes)

$$b_i \rightarrow b_i - b_s \frac{v_s}{v_i} \quad , \quad s = solvent$$

If the scatterer has a stratified centrosymmetric structure, the coherent intensity may be expressed in terms of A_k, the Fourier transform of each k layer, and of Q, their scattering length per unit volume.

$$\overline{F^2 (q)} = < (\sum_i (Q_i - Q_{i+1}) \sum_{k=0}^{i} A_k (q))^2 >$$

$$Q_i = b_i / v_i$$

These are :
- the composition, which is assumed to be identical to the global
one, the isolated molecules having no influence,
- the general form, which can be chosen among several of the aggregate model shapes (prolate and oblate spheroids or better, ci-
gar-like or discoid, lamella, hank, worm-like chain),
- the area per polar head σ,
- the length and the conformation of both the hydrophilic and hydrophobic moieties of the surfactant molecule (L_1, L_2),
- the extent of interpenetration of oil and water inside the surfactant film on the molecular scale

$$\alpha = (N_W / N_S)_F$$

$$\beta = (N_O / N_S)_F$$

- N_S the number of surfactant molecules per aggregate,
- the specific volume of each component.

In fact we assume the existence of three homogeneous concentric zones, which then makes a four density system. Of course, all
these parameters are not independent, due to geometrical constraints : for example, for spheroids, if we fix the composition,
the lengths L_2 and L_1, the aggregation number N, and the rate of
penetration α, the values of β, area per polar head and the shape
of the aggregate are automatically determined. Nevertheless, the
number of independent variables remains important, which makes any
satisfactory interpretation of the results uneasy.

Moreover, since all these systems are generally concentrated
solutions, great additional difficulties arise due to the interparticle interactions : for the smaller q values of the spectra,
the data are drastically dependent on concentration. To avoid this
difficulty, very dilute systems should be used ; but at least for
nonionics, the possibility of dilution at constant size and shape
should not be anticipated as a rule. Besides, approximate theories
of these interparticle effects are available only in a very few
special cases.[10-12] Then, small q data are generally unusable
as far as morphological determinations are concerned. However,
they may be interesting in other respects, as will be explained
shortly.

Let us summarize the steps involved in our analysis, when
well-defined particles are thought to exist :
- For the calculation of theoretical spectra, we first have to
determine, if possible, the type of the aggregates (reverse or direct). Then, after a proper choice of values for all the parameters, a numerical integration is performed by computer, by using
a Gauss quadrature method. (The integration takes into account all
the orientations of the scatterer).

- For experimental curves, the intensities must be determined in absolute units and for q values as large as possible, taking into account the corrections due to extinction and noncoherent scattering effects.

The best fits between the theoretical and experimental curves, and for as many contrasts as possible for both oil and water, allow the determination of the most probable set of the parameter values. It must be emphasized that this best fit is to be investigated only for sufficiently high q values, at which the interparticle effects are negligible.

RESULTS AND DISCUSSION

We present here the results on the oil rich systems with $C_{12}(EO)_4$ at 20°C, whose compositions are given in Table I.

The first question is : do definite particles exist ? The second is : if they do, are they of the same type everywhere (oil / water or water /oil) ?

To answer these questions, we can use the contrast variation method (partial deuteration of the oil and water components). There may be an important objection concerning the effect of D/H substitution. Of course some alteration of the structures cannot be discarded a priori. But all the large effects cited in the literature exclusively concern quaternary ionic systems, and are ascribed to a change in the surface charge density (due to the change of free energy of hydration of ionic species) ; they are very dependent on the alcohol content.[17] On the other hand the lower consolute curves of aqueous nonionic surfactants are a pure demonstration of the hydration forces. However they are shifted by only about 1°C when D_2O is replaced by H_2O. That constitutes a very small effect, given the high sensitivity of cloud points towards the presence of any contaminant. Besides the molecular weights of the micelles have not been proved to be different according to the type of water.[18] Similarly, we noted that the exact delineation of our present microemulsion domain is almost insensitive to any D/H substitution, although it is very sensitive to any temperature change (mostly due to the change in hydration forces). We must also emphasize that the H_2O/D_2O ratio was kept constant throughout the present investigation : we only made isotopic substitutions on the decane component. And since the phase diagram is not noticeably modified we can believe that the small change in Van der Waals forces must influence only slightly the structure of that microemulsion. Nevertheless we have not studied samples along the demixing lines, in order to avoid such an eventual complication .

Whatever the extent of the interactions, the square root of the intensity scattered at zero angle must be proportional to Z, which is the resulting excess scattering length of the aggregate, normalized to one surfactant molecule. Depending on whether the system is inverse or direct, Z is given by one or the other of the relations of Table III. b_s, b_w and b_o stand for the molecular scattering lengths of the surfactant, water and oil components, which are, in fact, mixtures of deuterated and hydrogenated molecules.

If the conditions are favorable, Z should be zero for two sufficiently different couples of scattering lengths (oil and water), depending on whether the structure is inverse or direct, i.e. (b_w, b_o^i) and (b_w, b_o^d). If the system is really bicontinuous, there is no reason for such a matching point to be achieved for one of these particular couples.

In the following measurements, water was always a mixture containing 37 per cent of heavy water. Decane was a mixture containing between 0 and 50 % of deuterated decane. Four of the results are shown in Figure 2. Clearly, the scattered intensity always tends to be zero for exactly that oil scattering length (i.e. b_o^i) which corresponds to the inverse system, whatever the content of water : it only depends on the water/surfactant ratio. In other words, for that contrast the scattered intensity is practically zero at $q = 0.002$ Å^{-1} : therefore, inside the sample, all the elementary volumes of size $D \sim 2\pi/q\lambda > 300$ Å which produce the coherent scattering have compositions with identical water/surfactant ratio, whatever the oil concentration of the sample. Such a result cannot be fortuitous. It excludes :
- a variation of the particle sizes with the H/D substitution,
- the "classical" bicontinuous model,
- a large polydispersity if the water/surfactant ratio is not constant for all the globules. The least we can say is that everything happens as if all these microemulsions were inverse systems, containing relatively well definite aggregates.

Now a maximum of detailed structural information can be attained, thanks to the study of the angular repartition of the scattered intensities over a wide domain of q values. But, owing to the interactions, results for the smaller q values cannot be used. Then for what q values are the scattering intensities practically independent on the interparticle effects?

Table III. Expression for the excess scattering length of the aggregate, normalized to one surfactant molecule, for both types oil /
water (direct) and water / oil (inverse).

$$Z \sim \overline{I(q = 0)}$$

INVERSE	DIRECT
$Z = b_s + \dfrac{N_w}{N_s} b_w - b_o \dfrac{v_s + v_w\, N_w/N_s}{v_o}$	$Z = b_s + \dfrac{N_w}{N_o} b_o - b_w \dfrac{v_s + v_o\, N_o/N_s}{v_w}$
$Z = 0$ for $b_\alpha = b_o^i$	$Z = 0$ for $b_o = b_o^d$

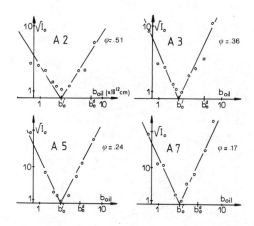

Figure 2. Square root of the scattered intensity for the systems
A2, A3, A5, A7 for several decane scattering lengths.

If we suppose that correct morphological properties are obtained only from the intensities at large q values as explained below, then it is possible to check a posteriori that the curves are effectively insensitive to the interactions in that q domain. For this purpose, the effects of the interactions have been evaluated to a first approximation by using the theory of fluids of hard spherical particles, which is the only one available at the present time.[10-12] For the volume of the equivalent hard sphere, we can take the volume of the water plus hydrophilic part of the surfactant film. Two very dissimilar examples are shown in Figure 3 which correspond to two scattering lengths of decane, and for the composition previously named D, the fraction of disperse phase being as high as 40 %. Curves 1 and 2 correspond to the intensity scattered by particles whose morphology has been determined from data at q greater than 0.04 $\mathring{A}^{-1}$, with or without the concentration effects, curve 3 represent the experimental results.

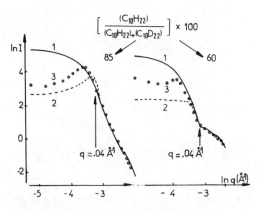

Figure 3. Two examples of interparticle effects 1 and 2 : theoretical curves for isolated and concentrated particles ; 3 : experimental curve (Sample D).

The results are self consistent, since all the three curves coincide for all q greater than 0.04 $Å^{-1}$. The "hard volume" fraction appears as if it were noticeably less than the total disperse phase fraction ϕ , and then the interactions are reduced to a lower level. (Interaction peaks do not exist for concentrations less than 0.2). This can be due to a small extra attractive term, to a possible interpenetration of the "micelles" or/and to their nonsphericity and deformability ; but it does not matter here : the essential point is the fact that the interaction effects on the scattering spectra are scarcely appreciable for q values greater than q_{min}, which obviously depends on ϕ and the size of the particles : 0.04 $Å^{-1}$ for sample A1, 0.03 $Å^{-1}$ for sample A7, 0.02 $Å^{-1}$ if ϕ = 0.1. From now on, only the data corresponding to q greater than these q_{min} values will be considered.

In Figure 4 are reported the results corresponding to the system which contains the largest amount of water. The small circles represent the experimental data. The full lines are the theoretical spectra which correspond to the best set of parameter values if we may suppose that aggregates have a well-definite morphology and are nearly monodisperse. The aggregation number is about 2000 or 1500. The particle seems to be a discoid, whose ellipticity is about 0.5 at the hydrophobe-hydrophile interface, where the area per polar head is 57 $Å^2$. Both moieties of the surfactant molecule are rather extended, and inside the surfactant film, there are three water molecules per oxyethylene group. Each curve corresponds to a given scattering length of the decane : the figures are the percentage of hydrogenated decane in the oil mixture. For convenience, these curves have been regularly shifted, both in ordinate and in abscissa. For comparison, the lines q^{-4} have also been drawn, which are generally not parallel to our experimental curves.

This value of the ellipticity may seem to be a little bit curious. At this point, it should be instructive to recall that along the first demixing line (see Figure 1), the micelles are bilayered lamellas of aggregation number about 1000, still with three water molecules for each oxyethylene group entrapped along the hydrophilic chains.[19] A further water solubilization makes the aggregates much more spherical, but with some memory of the initial disc-like shape. In fact, we think that the limiting factor is the area per polar head, which cannot accommodate too large a value (its initial value is about 47 $Å^2$).

And why not spherical particles ? Obviously from the curves of Figure 5, monodisperse spheres cannot account for the experimental data. The fits would be better for smaller spheres (N = 1000), for which the corresponding σ value could seem excessive (σ = 65 $Å^2$).

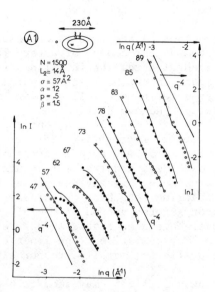

Figure 4. Experimental results (o , •) and theoretical curves (—)
corresponding to the best fit (sample : A1). Each curve corresponds
to a certain isotopic composition of the decane component.

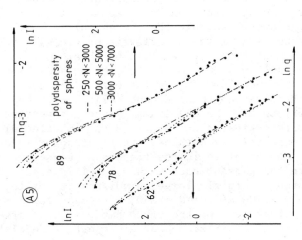

Figure 6. Comparison of the experimental results with the theoretical curves for polydisperse spheres (sample A5).

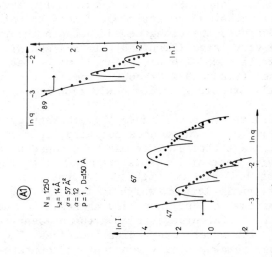

Figure 5. Comparison of the experimental results with the theoretical curves of monodisperse spheres (sample A1).

Polydisperse systems of spheres must be considered. But according
to the contrast variation result, all the aggregates should contain
the same number of water molecules per surfactant. Model systems
with rectangular size distribution have been examined : they introduce effects large enough to make the curves quite smooth, without
the need of very small or large unrealistic particles. Thus for
sample A5, an apparent good fit can also be obtained if the range
of particle size is [250-3000] (Mw = 2000, Mn = 1500). But for the
ranges [500-5000] , and [3000-7000] the quality of the fits depends on the scattering length of decane (Figure 6). The curves
do not necessarily tend to be q^{-4} lines, since the scattering
lengths change more or less sharply across the interfacial film.
These good fits mean that we obtained the right specific area of
the systems since the absolute intensities have been correctly
evaluated. But in the present case such a specific area can be
obtained only through a fairly large polydispersity of the area
per polar head [σ = 100 $\mathring{A}^2$ for N = 250, σ = 40 $\mathring{A}^2$ for N = 3000]
which may be questionable. On the other hand, light scattering
measurements[20] reveal the existence of an important depolarization,
even for less concentrated systems. This can be accounted for only
by a non spherical shape, given the small size of the aggregates
relative to the wavelength.

    It also seems unlikely that the particles could be more ani-
sometric, even for more dilute system, as shown in Figure 7: σ maintains a value about 55 to 57 $\mathring{A}^2$, and the aggregation number
cannot be 5000. In fact a value of about 2500 could be the best
choice, as actually found. Particles with smaller aggregation
number but still rather anisometric would lead to rather unrealistic σ values. Besides, the disagreement between theory and experiment would be very large (Figure 8). If instead of the discoid
model, the cigar like model would be used, no better agreement
would be obtained (Figure 9). Of course some type of polydispersity
of these slightly prolate globules could also be suggested.

    For more dilute systems (Figure 10), the aggregation number
is still about 3000, the shape does not vary, neither does α, but
σ seems to decrease a little. That may be a consequence of the more
hydrophobic environment of the surfactant. On the other hand, for
the most concentrated systems with the largest water/surfactant ratio,
σ may be significantly larger, which in turn may correspond to the
hydrophilic environment. But shape and aggregation number seem to
change very little. Of course, since in the present case more water
is incorporated, the water core radius increases. The fact that the
surfactant number remains nearly constant is counterbalanced by the
small variation of σ (Figure 11).

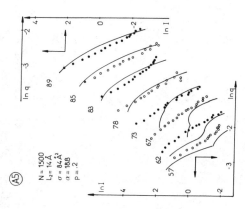

Figure 8. Comparison of the experimental results with the theoretical curves of very anisometric smaller spheroids (sample A5).

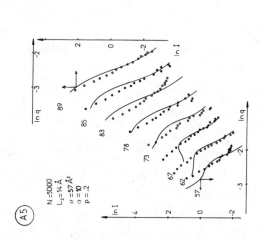

Figure 7. Comparison of the experimental results with the theoretical curves of very anisometric large spheroids (sample A5).

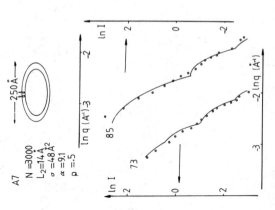

Figure 10. Comparison of the experimental
results with the theoretical curves of the
best oblate spheroid (sample A7).

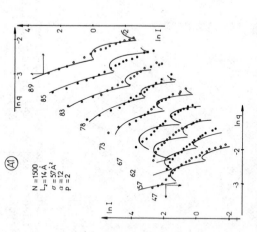

Figure 9. Comparison of the experimental results
with the theoretical curves of prolate spheroids
(sample A1).

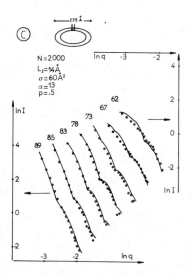

Figure 11. Comparison of the experimental results with the theoretical curves for the best oblate spheroid (sample C)

The last parameter which has to be checked is the conformation of the hydrophilic chain of the surfactant. From these curves it appears that the meander conformation, with $L_2 = 7$ Å leads to an agreement of poorer quality (Figure 12).

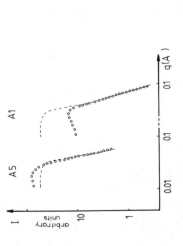

Figure 13. Small angle X ray scattering data for samples A1 and A5 (fully hydrogenated samples). Dotted lines are theoretical curve for oblate spheroid (p = 0.5, N = 2000) for the volume fraction φ = 0.32

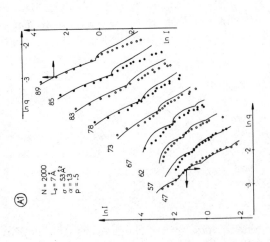

Figure 12. Effect of the conformation of the hydro-philic chain (meander) on the theoretical curves (sample A1) .

CONCLUSION

According to the type of information required, two ranges of q values may be used :
- for sufficiently small q, the intensities can be extrapolated to zero scattering angles ; from these values, the nonionic micro-emulsions studied here are shown be "water in oil" systems.
- the higher values of q, where there are no more concentration effects,which allow the morphology of the micellar aggregates to be determined in some detail.

Now let us emphasize once again the method we used throughout the analysis of our data. We want to find the best theoretical spectra corresponding to water in oil globules which could fit the experimental spectra for the whole range $q_{min} < q < 0.15$ Å^{-1}. But these fits necessitate the calculation of the intensities on the absolute scale, both theoretically and experimentally, and not only to consider some particular relative q^{-n} dependence: these fits are not at all dependent on the presence of any small "bumps" on the curves, which may be due to the interparticle effects or to the exact distribution of the atoms in the surfactant film. (As a matter of fact, the distribution of density is taken into account by the three homogeneous layers as a first approximation. As a second approximation, the true atomic distribution would affect the scattering curves only for $q > 0.2$ Å^{-1}). The goal is to obtain the good absolute level of the intensities over the whole q range described above and for each oil scattering length. This kind of fitting is by no means trivial, even if some incertitude remains. Moreover all the result are physically quite relevant and the models that we have found must at least be considered as possible models.

As a last (but not least) proof of that assertion, we can see on figure 13 that SAXS data (with H_2O, i.e; without any deuterium atom) are quite consistent with our interpretation of the neutron scattering data. So the "electronic contrast" can be considered as one particular oil-water contrast, and we believe that SAXS cannot by itself solve the problem of the microemulsion structure.

As a summary, all over the phase domain examined here, the aggregates look like discoids, the important parameter being the area per molecule of surfactant : all the small angle scattering spectra are consistent with a well-defined discoid-like aggrega-te. However, we have seen that a certain polydispersity of sphe-res or slightly prolate globules cannot be discarded on the ba-sis of these curves. So we could consider that we have in fact some mixture of nearly spherical globules (but mainly oblate), the pre-cise discoid model being just an "average shape" of the molecular aggregates.

Besides, the surfactant interface may be labile, otherwise definite constraints had to be introduced. On the other hand, such a
lability was shown by self diffusion NMR measurements,[14] and by
electric conductivity.[13] Recently such a flexibility of oil/wa-
ter interfaces has been theoretically recognized.[21]

A more accurate determination of the exact size of the globules
seems hopeless. The precision of our results is about 20-30%, as
far as the volume of the water core is concerned. Nevertheless,
the number of surfactant per globule seems to increase a
little with water incorporation (σ increases), and also when
decane is added (σ decreases). However, let us emphasize that
this low level of accuracy does not prevent the oil/water
interface to be well described (use of large q values), and is
independent of the determination of the type of these systems
(use of zero q values).

REFERENCES

1. M. Corti, V. Degiorgio and M. Zulauf, Phys. Rev. Letters (to
    be published).
2. K.Z. Bennet, J.C. Hatfield, H.T. Davis, C.W. Macosko and
    L.E. Scriven,in "Microemulsions", I.D. Robb Editor, Plenum
    Press, New York, (1982).
3. E.W. Kaler and S. Prager, J. Colloid Interface Sci., 86, 359
    (1982).
4. M. Corti and V. Degiorgio, Optics Communication, 14, 358
    (1975) ; also Phys. Rev. Lett., 45, 1045 (1980).
5. R. Triolo, L.J. Magid, J.S. Johnson,Jr.,and H.R. Child,
    J. Phys. Chem., 86, 3689 (1982).
6. S. Friberg and I. Lapczynska, Progr. Colloid Polymer Sci.,
    56, 16 (1975).
7. S. Friberg, I. Buraczewska and J.C. Ravey, in "Micellization,
    Solubilization and Microemulsions",K.L. Mittal Editor, Plenum
    Press, New York, (1977).
8. K. Shinoda and H. Saito , J. Colloid Interface Sci., 26, 70
    (1968)
9. Neutron beam facilities at the H.F.R. available for users
    (Institut Laue Langevin, Grenoble 1977).
10. M.S. Wertheim, Phys. Rev. Letters, 10, 321 (1963).
11. E.J. Thiele, J. Chem. Phys., 38, 1959 (1963).
12. L. Verlet and J.J. Weis, Phys. Rev. A, 5, 939 (1972).
13. M.H. Boyle, M.P. Mc Donald, P. Rossi and R.M. Wood, in
    "Microemulsions", I.D. Robb Editor, Plenum Press,New York
    (1982).
14. P.G. Nilsson and B. Lindman, J. Phys. Chem., in press.
15. M. Corti and V. Degiorgio, Phys. Rev. Lett. 45, 1045 (1980).
16. M. Zulauf and J.P. Rosenbusch, J. Phys. Chem. (to be publis-
    hed).

17. S. I. Chou and D. O. Shah, J. Colloid Interface Sci., 80, 49 (1982).
18. R. H. Ottewill, C. C. Storer and T. Walker, Trans. Far. Soc., 63, 2796 (1967).
19. J. C. Ravey, M. Buzier and C. Picot, J. Colloid Interface Sci., (to be published).
20. J. C. Ravey and S. Friberg, unpublished results.
21. P. G. De Gennes and C. Taupin, J. Phys. Chem., 86, 2294 (1982).

ACKNOWLEDGEMENTS

The authors wish to thank Dr. Cl. Picot, who introduced them to SANS. Acknowledgement is made to Dr. Oberthür (ILL, Grenoble) for enlightening discussions.

FLUCTUATIONS AND STABILITY OF MICROEMULSIONS

S. A. Safran

Corporate Research Science Laboratories
Exxon Research and Engineering Company
P. O. Box 45
Linden, New Jersey 07036

The magnitude of both size (polydispersity) and shape fluctuations for a single phase of spherical microemulsion droplets is calculated. The stability of the spherical globular phase to these fluctuations is determined as a function of the concentrations of the water, oil, and surfactant. The results are related to the phase diagrams and to recent experiments.

INTRODUCTION

Recent scattering experiments[1-5] have contributed to the understanding of the microscopic structure of and interactions between
microemulsions (e.g. micelles swollen with water in an oil continuous phase). Neutron scattering experiments, which measure the
structure of individual globules, have shown[1] departures from
spherical symmetry for the drops, as well as indications of a
distribution of drop sizes[5,6]. On the other hand, light scattering
experiments, which are insensitive to the details of the individual
globules, probe the collective interactions between droplets. Elastic light scattering experiments yield the correlation length which
describes the length scale for density fluctuations of the ensemble
of droplets. Near the single to two-phase transition, recent light
scattering measurements[2-4] have shown a diverging correlation length
at a particular point on the coexistence curve, indicating a critical point for phase separation.

This paper presents a unified model for studying the stability
of a single phase of spherical microemulsion droplets to size
and shape fluctuations. The model considers an idealized three component system of water, oil, and surfactant, but the presence of
salt and/or alcohol can be treated by a suitable renormalization of
the parameters. It is shown that the only important perturbations
of the droplets are polydispersity (size fluctuations) and one mode
($\ell=2$ spherical harmonic) of the possible shape fluctuations which
represents a crimping of the droplets. Each of these types of fluctuations can become large in a different region of the phase diagram.
The range of stability of the spherical globular phase is limited to
the region where these fluctuations are small and this region is
calculated as a function of the concentrations of oil, water, and
surfactant.

In addition to size and shape instabilities, the single phase
of spherical globules can become unstable to phase separation into
either (i) two phases of globules with different sizes and densities,
or (ii) a globular phase coexisting with excess water or oil. The
coexistence curve for the latter case has been previously discussed
in the literature[11-14]. Here, it is shown that the phase separation
into globules coexisting with an excess oil or water phase (type
(ii)) limits the magnitude of the polydispersity fluctuations of the
droplets.

Previous theoretical treatments of microemulsions have focused
either on the bicontinuous phase[8-10] where well-defined globules do
not exist, or on the spherical phase, but without[12-14] consideration
of the deviations from monodispersity (size fluctuations) or spher-

ical symmetry (shape fluctuations). The present work is most applicable to the dilute rather than bicontinuous (or percolation domin-
ated) regime, and focuses on the sizes and shapes of the globules
which are well-defined in this regime. Furthermore, previous theoretical studies have examined the drop sizes for microemulsions in
equilibrium with an excess phase and the free energies for various
phase equilibria[11,12]. In this paper, these coexistence curves are
related to the size and shape fluctuations.

SINGLE PHASE: SIZE AND SHAPE FLUCTUATIONS

In this Section, a summary is given of the calculation of the
magnitude of the size (polydispersity) and shape fluctuations of
spherical microemulsions in the single phase regime. The limits
of the spherical phase are discussed in terms of the concentrations
of water, oil, and surfactant where these fluctuations become large.
The model used in these calculations assumes that the surfactant
molecules form an interface between the pure oil and pure water
regimes. It is the statistical mechanics of these interfaces that
are of interest. Furthermore, it is assumed that the oil, water
and surfactant are nearly incompressible, so that a constraint of
constant volume per molecule is imposed. This volume might be
weakly temperature dependent. With these assumptions, which are
similar to those of References 9, 15, 16, the free energy consists
of the bending energy of the interface, and the entropy of dispersion of the globules.

In order to calculate the magnitude of the thermal fluctuations of both size and shape for spherical microemulsions, it is
necessary to consider an expression for the free energy, F, which
is applicable to globules of quite general shape. The attractive
interaction between the droplets that gives rise to the phase
separation into two globular phases (type (i)) with a critical
point, is discussed in Reference (18). The bending energy of a
single globule is

$$H = 4K \sum_{i=1}^{N} \int dS_i \left[\frac{1}{\rho_i(\vec{r})} - \frac{1}{\rho_1} \right]^2 \tag{1}$$

where S_i is the area of the ith interface. The normal ($\hat{n}$) to the
interface is equivalent to the local orientation of the surfactant

molecule and the local radius of curvature is $\rho(\vec{r})=2(\nabla \cdot \hat{n}(r))^{-1}$.
The tendency for the interface to locally bend to either the oil or
water region is accounted for by the parameter ρ_1 which is the
"natural" bending radius of the interface. The sign of ρ_1 determines the predisposition to forming oil-in-water or water-in-oil
globules. The convention taken here is for positive radii of curvature to represent water-in-oil globules.

Here it is noted that for $\rho_1^{-1}=0$, Equation (1) is identical to
the splay energy for a single layer of a smectic liquid crystal
where K is typically .1 eV, although the addition of co-surfactants
or the variation of the salinity of the water can reduce this value
for K. Previous thermodynamic treatments have also considered the
bending energy described by Equation (1)[11-14]. However, those
studies have assumed a particular structure, in contrast with the
statistical approach presented here which uses Equation (1) to test
the stability of an arbitrary structure to fluctuations (see Reference 15 for a treatment of the lamellar phase).

The bending energy, Equation (1), must be supplemented by constraints on the total number or volume fractions of two of the three
microemulsion components ($v_o + v_w + v_s = 1$). In particular, the
total surface area of and volume enclosed by the droplets is fixed.
In addition, the free energy contains an entropy term, S, (F=H-TS),
where

$$S = -N(\log \ (x) \ -1) \qquad (2)$$

and x is the volume fraction of globules (e.g. for water-in-oil
droplets $x = v_w + 1/2 \ v_s$ [17]).

For nearly spherical droplets, the equation which defines the
surface of the ith droplet with its center at the origin, is

$$R_i(\vec{r}) = r - \bar{\rho} \ [1 + g_i(\theta,\phi)] = 0 \qquad (3)$$

where $\bar{\rho}$ is the average droplet radius which is independent of angle
and the index i.

The dimensionless fluctuation amplitudes $\{g_i\}$ are expanded in
spherical harmonics as

$$g_i(\theta,\phi)= \sum_{\substack{\ell m \\ \ell \neq 1}} a_{\ell m}^i \ Y_{\ell m}(\theta,\phi) \qquad (4)$$

The details of the calculations have been presented in Reference (16). For an ensemble of approximately spherical globules with $\tau = T/16\pi K$, the free energy $F = H - TS$ is given by

$$F = F_o + \sum_{\ell m i} |a_{\ell m}^i|^2 \, \Delta F_\ell \tag{5a}$$

with

$$\Delta F_\ell = -12\tau(\log(x)-1) \, (1-b_\ell) + 4(3-4b_\ell+b_\ell^2) + 8\left(\frac{\rho_o}{\rho_1}\right)(b_\ell-1) \tag{5b}$$

$$b_\ell = \frac{1}{2}\,\ell(\ell+1) \tag{5c}$$

F_o is the free energy of a monodispersed set of spheres with radius $\rho_o = \dfrac{3\delta x}{v_s}$, where δ is the characteristic length of the surfactant molecule and it is assumed that $\rho_o > \delta$. Thus,

$$F_o = N_o \left[16\pi K(1-\frac{\rho_o}{\rho_1})^2 - T\,(\log x - 1)\right] \tag{6}$$

In Equation (6), the number of droplets (assuming monodispersed spheres) is $N_o = Vn_o$, where V is the volume of the system and $n_o = \dfrac{x}{36\pi}\left(\dfrac{v_s}{\delta x}\right)^3$. Equipartition and Equation (6) imply that the average value of the fluctuation, $\langle |a_{\ell m}|^2 \rangle$ is given by

$$\langle |a_{\ell m}|^2 \rangle = 8\pi \, \tau/\Delta F_\ell \tag{7}$$

Examination of Equation (7) reveals that for $\tau < 1$. the fluctuation amplitudes are large for $\ell = 0,2$. The $\ell = 0$ fluctuations represent polydispersity of the sphere sizes, while the $\ell = 2$ fluctuations result in a crimping of the spheres to an ellipsoidal shape[16]. The predicted form for the magnitude of the polydispersity can be tested in neutron scattering experiments. For fixed x (volume fraction of globules, e.g., $x = v_w + 1/2\, v_s$ for water spheres in oil), Equation (7) predicts for small polydispersity, the dependence of $\langle |a_{oo}|^2 \rangle$ upon ρ_o, which is proportional to the ratio x/v_s (see above). In particular,

$$\langle |a_{oo}|^2 \rangle = \frac{\pi\,\tau}{\left(-\dfrac{\rho_o}{\rho_1}+\dfrac{3}{2}\right)-\dfrac{3}{2}\,\tau\,(\log x - 1)} \tag{8}$$

Thus, for fixed x, the magnitude of the polydispersity is a linear function of $\rho_0 \sim x/v_s$, diverging near $\rho_0/\rho_1 \simeq 3/2$, i.e., when the ratio of the average sphere radius to the natural radius is equal to 3/2.

The divergence of the polydispersity fluctuation amplitude delimits the range of stability of the monodispersed, spherical phase. However, the single phase of droplets (e.g., water-in-oil) is unstable to phase separation into a phase of globules which coexist with excess water. This can be seen from Equation (7) by setting $\partial F/\partial x = 0$ at fixed droplet radius. It is assumed that $\tau \ll 1$ and the chemical potential of the excess water has been set to zero. Thus, the maximum value that the radius can assume is ρ_1. This limits the magnitude of the polydispersity fluctuations to a value of $< |a_{oo}|^2 > \simeq T/8K$. For typical value of $T/K \simeq 1/3$, the rms value of the maximum polydispersity is about 20%. This is consistent with recent neutron scattering experiments [6].

While polydispersity fluctuations are dominant for $\rho_0 \approx \rho_1$, shape fluctuations are large for small values of ρ_0. In particular,

$$< |a_{2m}|^2 > \simeq \frac{T/32K}{\rho_0/\rho_1} \qquad (9)$$

The concentrations at which amplitude of these fluctuations diverges represent values of ρ_0/ρ_1, which are small and negative. It is to be noted that negative values of ρ_1 refer to a natural radius which favors the creation of oil drops in water (by the conventional described above). It is thus plausible that for such values of ρ_1, the spherical drops with the "wrong" curvature (e.g., oil in water) would be unstable. If roles of the water and oil volume fractions are interchanged, the sign of ρ_1 is changed.

The present calculation tests only the stability of the spherical phase, but does not indicate the equilibrium shape of the phase which results from the aforementioned instability. A calculation of the non-spherical shapes for various values of the natural radius will be published later[18]. In addition, this work will discuss the "saddle-splay"[19] curvature energy of microemulsions.

DISCUSSION

In the previous Sections, the stability of a monodispersed phase of spherical microemulsion droplets to size and shape fluctuations as well as to phase separation has been examined. The functional dependence of the polydispersity on the concentration has been predicted. In addition, the limits of stability of the mono-dispersed, spherical phase to polydispersity fluctuations has been shown to occur when $\rho_0/\rho_1 \simeq 3/2$. This instability is preceded, however, by the type (ii) phase separation into globules coexist-ing with an excess phase. This phase separation limits the maximum value of the polydispersity fluctuations in the limit $T/16\pi K \ll 1$. The instability to $\ell = 2$ (crimping) shape fluctuations occurs at small, negative values of $\rho_0 < \rho_1$, corresponding to oil in water droplets for $\rho_1 > 0$ (which favors water in oil) and water in oil droplets for $\rho_1 < 0$ (which favors oil in water).

ACKNOWLEDGMENTS

The author acknowledges useful discussions with R. W. Cohen, J. S. Huang, M. W. Kim, P. Pincus, L. E. Scriven, L. A. Turkevich, and T. A. Witten.

REFERENCES

1. R. Ober and C. Taupin, J. Phys. Chem., 84, 2418 (1980).
2. J. S. Huang and M. W. Kim, Phys. Rev. Lett., 47, 1462 (1981).
3. A. M. Cazabat, D. Langevin, J. Meunier, and A. Pouchelon, J. de Phys. Lett., 43, 89 (1982).
4. G. Fourche, A. M. Bellocq and S. Brunetti, J. Colloid Inter-face Sci., 88, 302 (1982) and these proceedings.
5. B. H. Robinson, these proceedings.
6. M. Kotlarchyk, S. Chen and J. S. Huang, J. Phys. Chem., 86, 3273 (1982).
7. L. E. Scriven in "Micellization, Solubilization and Micro-emulsions", K. L. Mittal, Editor, Vol. 2, p. 877, Plenum Press, New York, 1977.
8. Y. Talmon and S. Prager, J. Chem. Phys., 69, 2984 (1978).
9. Y. Talmon and S. Prager, J. Chem. Phys., 76, 1535 (1982).
10. J. Jouffroy, P. Levinson, P. G. de Gennes, J. de Physique, 43, 1241 (1982).
11. C. A. Miller and P. Neogi, AIChE J., 26, 212 (1980).
12. E. Ruckenstein in Reference 7, p. 755, E. Ruckenstein and J. C. Chi, JCS Faraday Trans. II, 71, 1690 (1975).
13. C. Huh, J. Colloid Interface Sci., 71, 408 (1979) and unpublished.
14. M. Robbins, in Reference 7, p. 713.
15. P. G. de Gennes and C. Taupin, J. Phys. Chem., 86, 2294 (1982).
16. S. A. Safran, J. Chem. Phys., 78, 2073 (1983).
17. The volume of surfactant is split between that of the oil and the water.
18. S. A. Safran and L. A. Turkevich, to be published.
19. W. Helfrich, Z. Naturforsch, 28, 693 (1973).

INFLUENCE OF SALINITY ON THE COMPOSITION OF MICROEMULSION

PSEUDOPHASES: CORRELATION BETWEEN SALINITY AND STABILITY

J. Biais[*], B. Clin, P. Lalanne and M. Barthe

Centre de Recherche Paul Pascal
Domaine Universitaire
33405 Talence, France

In this paper we have studied the influence of salt on the stability of oil rich microemulsions made of water, oil, anionic surfactant and alcohol. By means of our model[1] (pseudophase assumption) we show that salinity doesn't affect the composition of oil and membrane pseudophases. An evaluation of the free enthalpy of micellization is given. The phase diagram modifications are related to the influence of salinity on the interfacial tension of the membrane of micelles.

[*] To whom correspondence should be addressed

INTRODUCTION

It is well known that microemulsions give rise to very low
interfacial tensions, particularly in the case of Winsor III
systems[2], and that this property could be of great interest in oil
recovery[3]. Such systems are generally observed for brine, oil, surfactant and cosurfactant mixtures. Phase behavior and microemulsion
stability have been the object of intensive studies[4-10]. It would
be very interesting to be able to predict the influence of composition, salinity and molecular properties of compounds on phase
diagram. The difficulties encountered in studying stability are
numerous and essentially related to

    i) the choice of relevant functions for free enthalpy
        calculations,

    ii) the theoretical (or graphical) determination of the
        stability condition in case of quaternary microemulsion
        system.

The goal of this work was to show that our model[1] (i.e. the
pseudophase assumption) makes such a study easier.

In the first section we briefly depict the model and thermodynamics which supports it. In the second section experimental
results for quaternary systems at different salinities are given.
In the third section these results are studied within the framework of the model. Finally free enthalpy calculations are performed for the pseudophase diagram and the stability of water in oil
microemulsions is discussed. We show that the phase diagram modifications observed when the salinity varies may be related to a
lowering of the interfacial tension due to the membrane when the
salinity increases.

THE MODEL

We will recall here only the main assumptions of the model
in case of water in oil microemulsions. These are :

    i) the continuous phase, the core and the membrane of the
        droplets can be considered as macroscopic phases (pseu-
        dophase assumption) : two volumic phases (continuous
        phase and core of droplets) one surface phase (the mem-
        brane) ;

    ii) these three pseudophases obey laws of thermodynamics. At
        equilibrium the compositions of the pseudophases are
        governed by :

          - the autoassociation constant K for the alcohol in oil
            pseudophase,

          - three partition constants which are depicted in
            Table I.

Table I. Partition Constants Between Pseudophases.

constant	for	between
k_W	alcohol	water and oil pseudophase
k_M	alcohol	membrane " "
ε	water	oil and water "

iii) the only approximation made concerning the composition
of microemulsion is to neglect the solubility of surfactant in water and oil pseudophases.

Experimental determination of the constants
In order to determine the values of the constants the following studies are performed :
i)   autoassociation constant K : N.M.R. chemical shift
       study of hydroxylic proton of alcohol in an organic
       phase (alcohol-oil).
ii)   partition constants k_W and ε : tie lines studies in two-
       phases domain of the ternary system(water oil alcohol).
iii)   partition constants k_M : determination of oil pseudo-
       phase composition by dilution technique[11,12] for limit
       microemulsions (i.e. : saturated microemulsions).

Important remarks
Let us consider a given macroscopic system the composition
of which is : VA, VO, VS, VW(the volume fractions of alcohol, oil,
surfactant and water respectively). The knowledge of the four
thermodynamical constants allowed us to calculate the composition
(and volume fractions) of the three pseudophases we will denote
O', M', W' (the oil, membrane and water pseudophases respectively).

In the space of the system states (tetrahedron VA, VO, VS,
VW in which any point represents the composition of a given macroscopic system) O', W', M' define a triangle we will call pseudophase triangle (see Figure 1). Such a triangle is the locus of
macroscopic systems that have the same pseudophase composition
O' M' W'. If it is possible to dilute water in oil limit microemulsions the microemulsion (one phase) domain is bound by a
straight line (toward O', see Figure 1b). Furthermore, if our
assumptions are valid, the compositions of the true phases of a
polyphasic system are represented by points that are located in
the same pseudophase triangle O', M', W'.

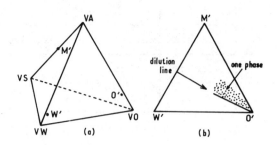

Figure 1.(a) representation tetrahedron for the states of the
             system.
         (b) pseudophase triangle and dilution line.

EXPERIMENTAL

Chemical
Chemicals used were all pro analysi qualities and were used
without further purification. Water was twice distilled.

Experimental techniques
Phase titrations, chemical shift studies were performed by
High Resolution N.M.R.. Spectra have been recorded at 294 K with a
270 MHz Bruker spectrometer acting in F.T. mode.

Theoretical calculations
Analysis of experimental data and calculations within the
framework of the model were performed with a DEC, VAX 11/780
computer.

Systems studied
We have studied the following systems
i) water (W), toluene (T), sodium dodecylsulfate (S),
    n-butanol (B) (WTSB system)
ii) water, octane (O), sodium dodecylsulfate, n-butanol
    (WOSB system)
at different salinities (sodium chloride).

EXPERIMENTAL RESULTS

Phase diagrams
1. Microemulsions
Boundaries of one phase regions were determined by titration.

Probes were kept for a long time in thermostat at 294 K. Results
for WTSB and WOSB are given in Figure 2 (for different salinities).
In these diagrams the alcohol to surfactant volume ratio is constant (VA/VS = 2.8).

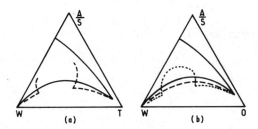

Figure 2. Phase diagram for :
        (a) WTSB system —— no NaCl
                            --- NaCl 20g/1
        (b) WOSB system —— no NaCl
                            --- NaCl 20g/1
                            ... NaCl 40g/1

      We can see in these diagrams that the single phase boundary
of the oil rich domain is lowered when salinity increases.

    2. Ternary systems
      In order to determine the value of the constants k_W and ε
we have studied the phase diagrams of ternary systems WTB and WOB.
However, we present here, in Figure 3, the results obtained for
the WOB system only at two different salinities.

Chemical shift
      In Figure 4 is given the result we obtained for the organic
phase of the WTB ternary system.

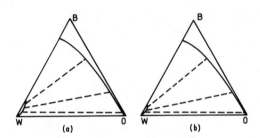

Figure 3. : --- tie lines for WOB system
(a) no NaCl ; (b) 40g/1 NaCl
The solubility of butanol in water varies from
5.7g/1 (no NaCl) to 4.6 g/1 (40 g/1 NaCl).

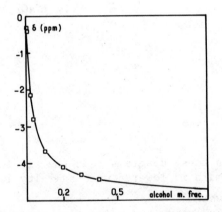

Figure 4. Chemical shift variation of hydroxylic proton of butanol
in WTB system.
☐ experimental points   —— theoretical curve.

Oil pseudophase composition

In order to determine the composition of oil pseudophase (and further the value of k_M) of limit(saturated)microemulsions we have used the so-called dilution technique[11],[12].

We have studied three "limit" microemulsions (at each salinity) for the WOSB system and twelve limit microemulsions for the WTSB system at 20g/1 NaCl. For this last system, free of salt, we have considered the Graciaa's results[12] (thirty microemulsions).

EXPERIMENTAL RESULTS AND MODEL

The analysis of the experimental results allowed us to determine the values of thermodynamical constants that are given in Table II.

Table II. Thermodynamical Constants for Studied Systems.

System	Salinity gr/1	K	k_M	k_W	ε
WTSB	0	85	60	4	0,06
	20	85	60	2,5	0,06
WOSB	0	140	120	9	0,03
	20	140	120	7	0,03
	40	140	120	4	0,03

For each system :
i) The autoassociation constant K is of course independent of the salinity. We have given in Figure 4 an example of calculated curve with the model used[1].
ii) The partition constant k_W decreases when the salinity increases. Figure 5 gives a comparison between calculated and experimental tie lines.

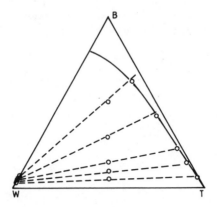

Figure 5. Experimental (o) and calculated (---) tie lines for WTB
system (no NaCl).

iii) The experimental accuracy didn't allow us to find a
variation of ε versus salinity.

iiii) The partition constant k_M appears to be independent of
the salinity. If we call ψ_0 the composition of the oil
pseudophase, defined by :

$$\psi_0 = \frac{V_A^o}{V_o}$$

where V_A^o = volume of alcohol in oil pseudophase
V_o = volume of oil in oil pseudophase.

the model, with the values of the constants of Table II,
allows the calculation of ψ_0.

In Figure 6 and 7 we give the comparison between experimental
and calculated values of ψ_0 for the WTSB system ; in Figure 8 the
same comparison for WOSB system.

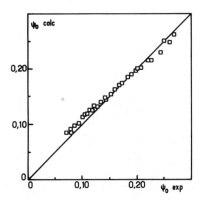

Figure 6. Calculated and experimental ψ_0 for WTSB system (no NaCl).

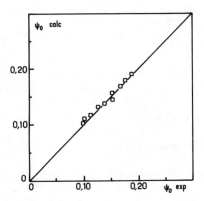

Figure 7. Calculated and experimental ψ_0 for WTSB system (20 g/1 NaCl).

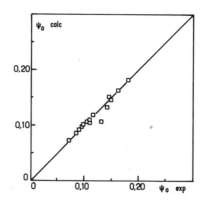

Figure 8. Calculated and experimental ψ_o for WOSB system (any salinity).

So, the conclusion we reach for oil rich system is that : when the salinity of brine used varies, there is no variation of the composition of oil and membrane pseudophases.

In order to understand the phase diagram modifications observed in Figure 2 we have to take into account some other properties of these systems.

STABILITY AND SALINITY

We will consider a pseudophase triangle W'O'M' (see Figure 1b). We already pointed out that in this triangle every point represent systems that have the same pseudophases composition W'O'M'. To discuss the stability we have to calculate the free enthalpy G for each point. The stability of a given system, in the oil rich domain, for example, is related to the value and variation of G in this domain but also to the value (and variation) of G in others domains (i.e. water rich domain). Nevertheless, in this first approach, we will focus our attention only on the oil rich region.

Free enthalpy of micellization

According to Ruckenstein's work[7] on microemulsion stability, the free enthalpy of micellization G_M is given by

$$G_M = \Delta G_1 + \Delta G_2 - T\Delta S_M \tag{1}$$

where :

ΔG_1 = surface enthalpy of membrane ;
ΔG_2 = Van der Waals interactions ;
ΔS_M = entropy of micellization.

We neglect here electrostatic interactions.

1. $\underline{\Delta G_1}$
This term may be written

$$\Delta G_1 = \gamma \cdot \Omega$$

where γ is the specific interfacial tension and Ω the surface area of the membrane pseudophase.

Miller and Neogi[13] have shown that γ is dependent upon the droplet radius. We will use the following relation

$$\Delta G_1 = \gamma^o \cdot \Omega \ (1 + (R^*/R)^2) \tag{2}$$

where R^* is a constant, R the radius of the droplet at the level of polar head. γ^o is the interfacial tension for a planar inter- face. Of course γ^o is dependent on the pseudophases compositions O', M', W' that we will not consider here. R and Ω can be calcula- ted by means of our model with the knowledge of the polar head area σ. In calculations we will consider a constant and mean value of σ for both alcohol and S.D.S. .
γ^o and R^* are arbitrarily chosen.

2. $\underline{\Delta G_2 \text{ and } T\Delta S_M}$
In order to calculate these terms we use the Overbeek rela- tion[10] :

$$\Delta G_2 - T\Delta S_M = k \ \frac{\phi \cdot V}{\frac{4}{3} \Pi (RH)^3} \ T \left\{ \log \phi - 1 + \frac{\phi(4-3\phi)}{(1-\phi)^2} + \log \frac{v_1}{\frac{4}{3} \Pi (RH)^3} \right\} \tag{3}$$

where ϕ is the volume fraction of micelles, RH is the hydrodynamic radius of droplet, v_1 is the molar volume of continuous phase, k is the Boltzman's constant and V is the whole volume of the micro- emulsion.

Calculation and results
We have performed calculations in W', O', M' triangle along straight lines A defined by a constant VW'/VM' ratios (see Fi- gure 9).

In Figure 10 we give the variation of G_M along A lines for different values of VM'/VW' and in Figure 11 the variation of G_M along A line for two different values of γ^o. The values of cons- tants are given in Table III.

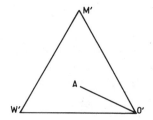

Figure 9. Straight line A in W', O', M' triangle.

Table III. Values of Constants.

Constants	γ^o	σ	R^*	v_1	T
Values	0,01 dyn/cm	20 $\overset{o}{A}{}^2$	70 $\overset{o}{A}$	166 $\overset{o}{A}{}^3$	300 K

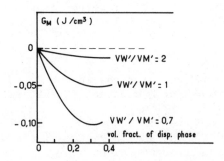

Figure 10. Variation of G_M versus ϕ (volume fraction of the dispersed phase) for different values of VW'/VM' along A lines (see Figure 9).

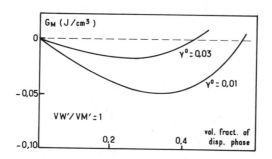

Figure 11. Variation of G_M versus ϕ (volume fraction of the dispersed phase along A line (see Figure 9) for different values of the interfacial tension.

Free enthalpy, phase diagram and discussion

The results obtained, according to previous results in literature (i.e. Ruckenstein[7]), show that microemulsions may be thermodynamically stable systems ; the stability increases when the interfacial tension decreases (see Figure 11). To really discuss the stability we have to consider the whole pseudophase diagram W'O'M'. For the moment we will consider only the following remark:

Along a dilution line A (Figures 1.b and 9.) the free enthalpy appears to be quite linearly dependent of the volume fraction of the dispersed phase ϕ ($\phi = VM' + VW'$) for $0 < \phi < .3$:

$$G_M \simeq a\,\phi \qquad (4)$$

where a is the slope of $G_M(\phi)$. So, the stability break condition is reached for a given value of a we call a^*

$$\frac{G_M}{\phi} = a^* \qquad (5)$$

By means of our model we have searched for the locus of points where Equation (5) is satisfied. Calculations have been performed for case of WTSB system in a diagram defined by a constant alcohol volume to surfactant volume ratio (VA/VS = 2.8). The values of parameters are those of Tables II and III.

Results are given in Figure 12.

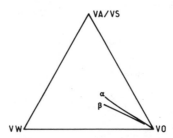

Figure 12. Demixing lines calculated for $a^* = -0.5$
    $\alpha : \gamma^0 = 3.10^{-2}$ dyne/cm $\beta : \gamma^0 = 10^{-2}$ dyne/cm.

One can see from Figure 12 that the calculated loci look like experimental demixing boundaries. The position of these calculated boundaries varies with the assumed value of the interfacial tension: their variation is consistent with experimental ones as far as interfacial tension lowering can be related to an increase of salinity (see Figure 2).

CONCLUSIONS

In this paper, we have essentially shown that for oil rich systems :

i)   In the framework of our model (pseudophase assumption) the salinity does not affect the partition constant of alcohol between the oil and membrane pseudophase and consequently the composition of the membrane.

ii)   The pseudophase (and pseudotriangle W'O'M') concept makes easier the study of stability in the case of quaternary system.

iii) The relation between stability and salinity may be interpreted by a decrease of the interfacial tension (at membrane level) when the salinity increases as it is well known in case of true polyphasic systems (i.e. Winsor III).

REFERENCES

1. J. Biais, P. Bothorel, B. Clin and P. Lalanne, J. Dispersion Sci. Technol. $\underline{2}$, 67 (1981).
2. P.A. Winsor, "Solvent Properties of Amphiphilic Compounds", Butterworths, London, 1954.
3. D.O. Shah and R.S. Schechter Editors "Improved Oil Recovery by Surfactant and Polymer Flooding" Academic Press, New York, 1977.
4. M.L. Robbins in "Micellization, Solubilization and Microemulsions", K.L. Mittal, Editor, Vol. 2, 713 , Plenum Press , New York, 1977.
5. A.M. Bellocq, J. Biais, B. Clin, A. Gelot, P. Lalanne and B. Lemanceau, J. Colloid Interface Sci., $\underline{74}$, 311, (1980).
6. J.L. Salager, Ph.D. Dissertation, University of Texas at Austin, Texas, 1977.
7. E. Ruckenstein and J.C. Chi, J. Chem. Soc., $\underline{71}$, 1690, (1975).
8. E. Ruckenstein, Chem. Phys. Lett., $\underline{57}$, 517, $\overline{(1978)}$.
9. E. Ruckenstein and R. Krishnan, J. Colloid Interface Sci., $\underline{75}$, 476, (1980).
10. J.Th.G. Overbeek, The First Rideal Lecture, Faraday Discussion of the Chem. Soc. N° 65, p. 7, (1978).
11. H.L. Rosano and R.C. Peiser, Rev. Fr. Corps Gras, $\underline{16}$, 249, (1969).
12. A. Graciaa, Thèse Doct. ès Sci., Université de Pau, 1978.
13. C.A. Miller and P. Neogi, AIChE J., $\underline{26}$, 212, (1980).

PHASE STUDIES AND CONDUCTIVITY MEASUREMENTS IN MICROEMULSION-FORMING SYSTEMS CONTAINING A NONIONIC SURFACTANT

T.A. Bostock and M.P. McDonald

Department of Chemistry
Sheffield City Polytechnic
Pond Street, Sheffield S1 1WB, U.K.

G.J.T. Tiddy

Unilever Research Port Sunlight Laboratory
Quarry Road East, Bebington, Wirral
Merseyside, U.K.

Binary and ternary phase diagrams have been con-structed for mixtures of tetraoxyethyleneglycol dodecyl ether ($C_{12}E_4$) + water, and $C_{12}E_4$ + water + heptane. Electrical conductivity measurements were made on all the isotropic single phase regions of the diagrams.

For the water continuous systems, comparison of the conductivities with values predicted by semi-empirical equations for dispersed systems show dis-crepancies which have been interpreted in terms of changes in ethylene oxide-water interactions.

The conductivity measurements on some of the hydrocarbon-rich systems indicate that enormous changes in phase structure are occurring over very small ranges of temperature and composition. Some systems which form water-continuous microemulsions at low temperatures invert to give a hydrocarbon-continuous phase of extremely low conductivity at higher temperatures. Other systems containing higher concentrations of surfactant only change from water-continuous to bicontinuous in structure since

they still retain a significant conductivity at
high temperatures.

It is found that the water-continuous microemulsions stable at highest temperatures contain approximately equal weights of surfactant and water as does
the highest melting liquid crystal structure in the
binary phase diagram.

INTRODUCTION

The phase diagrams of a number of ternary systems of water +
oil + nonionic surfactant have been investigated in recent years.
In many cases isotropic liquid phases have been found to occur
over considerable areas of the ternary diagrams at certain temperatures[1,2]. The fact that many of the phases consist of large
proportions of both water and oil, with relatively small amounts
of surfactant, means that they must be regarded as dispersions of
oil in water (O/W) or water in oil (W/O). The term microemulsion
has been much used in reference to such systems. Unfortunately
the physical properties of the systems are often such as to make
it extremely difficult to decide whether they are O/W or W/O. It
has even been necessary to invoke a bicontinuous structure in
order to explain some of their properties[3]. This bicontinuous
structure can be visualized as a rapidly changing three dimensional network of interconnecting volumes of each phase which are
completely intertwined. The surfactant molecules make up the
shape determining interfacial layers. An attempt to represent
this theoretically using statistical mechanics has been made by
Talmon and Prager[4]. Microemulsions also occur in systems
containing water + oil + ionic surfactant + cosurfactant[2]   and
in order to explain the unexpectedly high electrical conductance
of some high oil content mixtures a percolation mechanism has
been proposed[5,6,7]. According to this mechanism, conducting paths
occur as a result of collisions between inverted micelles in a
predominantly non-conducting organic medium. At a certain concentration of micelles there is a finite probability of continuous
paths through the organic medium occurring instantaneously thus
allowing the passage of current through the medium.

In this paper are reported phase studies on the water-
heptane-tetraoxy ethyleneglycol dodecyl ether ($C_{12}E_4$) system over
a wide range of temperatures. Conductance measurements have been
made in the isotropic liquid regions of this system and also in
similar regions of the $C_{12}E_4$-water system. The use of 0.01 M
sodium chloride solution instead of water has no significant
effect on the phase transition temperatures but enhances the
sensitivity of the conductance measurements. Conductivity values
at different compositions and temperatures have been correlated

with values predicted by semi-empirical equations for the conductivities of emulsions. Moreover, the possible existence of bicontinuous and other structures in certain cases is discussed.

EXPERIMENTAL

$C_{12}E_4$ was obtained from Nikkol Chemicals, Japan and used as
received. Heptane was obtained from BDH Chemicals Limited
(Laboratory Reagent grade - 99.5%). Distilled deionised water was
used and the added sodium chloride was Analar grade.

The phase studies were carried out using weighed mixtures in
test tubes with ground glass stoppers. The samples were observed
while being stirred with a thermometer as they were heated and
cooled to determine the temperatures between which the single
phase liquid regions existed. Any birefringence due to the presence of a liquid crystal phase in the bulk sample in the test tube
was observed when the tube was placed between crossed polaroid
sheets;whereas streaming birefringence, when present, could be
observed when the thermometer was moved briskly up and down in
samples showing no birefringence in the undisturbed state. For
the ternary systems thirty six samples at 10% composition intervals

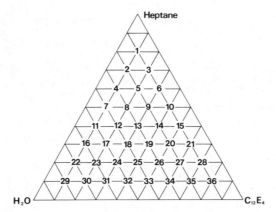

Figure 1. The numbering system used for samples in the ternary
diagram.

were examined, so covering the whole diagram and numbered as shown in Figure 1. Extra samples of intermediate composition were made up where necessary to delineate the isotropic liquid region more precisely.

Conductance measurements were made with a Wayne Kerr B224 Universal Bridge operating at 1592 Hz and a specially-constructed cell. Electrodes of bright platinum approximately 2 cm x 1 cm and 2 mm apart were mounted on a support with a ground glass joint which enabled measurements to be made on the small test tube samples of binary and ternary mixtures. Temperature variation was achieved by immersion of the sample tubes in thermostatically controlled water baths.

For all conductance measurements the samples were made up with 0.01 M sodium chloride solution instead of water. The conductivity of this electrolyte was found to vary with temperature according to

$$K/mSm^{-1} = 63.47 + 2.14t \text{ from } 0 \text{ °C to } 20 \text{ °C}$$

$$K/mSm^{-1} = 57.46 + 2.48t \text{ from } 20 \text{ °C to } 83 \text{ °C}$$

where t is the temperature in degrees Celsius.

RESULTS AND DISCUSSION

1. Binary system

The phase diagram of the $C_{12}E_4$-water system (Figure 2) is well known[8]. There are three isotropic liquid regions (designated L_1, L_2 and L_3) in all of which the surfactant is thought to be aggregated[8].

(a) L_1 region. Since it is known that Na^+ ions do not complex with ethylene oxide groups in aqueous solution[9], then in a system containing NaCl solution of constant concentration and differing concentrations of nonionic surfactant aggregates it seems permissible to discuss conductivity changes using equations developed to explain the conductivities of emulsions.

A number of emulsion systems have been found to give conductivities which agree with those predicted by the Bruggeman equation

$$\frac{K}{Km} = (1 - \phi)^{1.5} \tag{1}$$

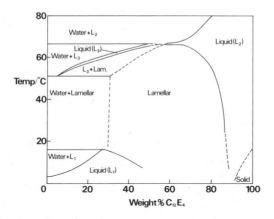

Figure 2.   Phase diagram of $C_{12}E_4$-water.

where K and Km are the conductivities of the emulsion and the continuous phase respectively and ϕ is the volume fraction of the disperse phase[10]. Mackay and Agarwal[11] found that their molar conductivity values for a four component system of water + oil + nonionic surfactant + n-pentanol varied more quickly with $(1 - \phi)$ and actually fitted an equation

$$\Lambda = \Lambda^\circ (1 - a\phi)^{1.5} \qquad (2)$$

where a (>1) is introduced to allow for the fact that hydration of the ethylene oxide chains of the surfactant may effectively increase the volume of the disperse phase.

Because of the way in which Mackay and Agarwal doped their system with sodium chloride to increase conductance, the above equation is equivalent to $K = Km (1 - \phi)^{2.5}$ for their conductivity values.

It can be seen in Figure 3 that whereas the conductivities of the $C_{12}E_4$-water samples do not follow the Bruggeman equation they do fit the equation

$$K = Km (1 - a\phi)^{1.5} \qquad (3)$$

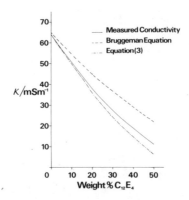

Figure 3. Conductivity in the L_1 region of the $C_{12}E_4$-water system at 0 °C. Samples made up with 0.01 M sodium chloride solution.

reasonably well if the parameter a = 1.5.

In this work ϕ has been taken as the volume fraction of the dissolved surfactant calculated from its weight fraction and measured density.

If one interprets 'a' in terms of ethylene oxide group hydration then the value of 1.5 implies that this group binds approximately its own volume of water. This is equivalent to ca. 2.5 molecules of water per ethylene oxide group which is within the range of values reported by other workers[12].

(b) L_2 and L_3 regions. The conductivities in the L_3 region are compared with values predicted by Equation (1) in Figure 4. These measurements cover a range of temperature values from 53.6 °C to 71.2 °C because the narrowness of the L_3 region at a given temperature and composition prevented more than a single measurement on each sample. The calculated values refer to the same temperature as each of the corresponding measured values. The agreement between measured and predicted values is not good.

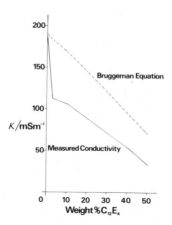

Figure 4. Conductivity in the L_3 region of the $C_{12}E_4$-water system. Samples made up with 0.01 M sodium chloride solution.

It has also been found that the self diffusion coefficients of water in the L_3 phase are much lower than those in pure water[8] and the dielectric permittivities are lower than those predicted by the Hanai equation for a spherically shaped oil phase dispersed in a continuous aqueous phase[13]. It was suggested[8] that all these results might be explained qualitatively by the formation of secondary aggregates of ellipsoidal micelles in the L_3 phase. The formation of these micelles presumably involves an increase in the surfactant head group area over that in the lamellar liquid crystal phase. This could arise from the exclusion of water adjacent to the alkyl chain/ethylene oxide group interface.

Returning to the L_2 region of the binary phase diagrams, Figure 5 shows how conductivity varies with temperature for the 75.2, 80.3 and 89.9% $C_{12}E_4$ samples. The binary phase diagram indicates that mixtures of $C_{12}E_4$ with up to 10% water seem to remain isotropic up to very high temperatures. This has been shown to be true for the $C_{10}E_4$-water system up to 300 °C[14]. It is not surprising therefore that the conductivity of the 89.9% $C_{12}E_4$ mixture increases steadily with temperature. The fact that $dK/dT \sim 2(dKm/dT)$ is due to the decrease in hydrogen bonding between the small amount of water present (2 moles per mole of

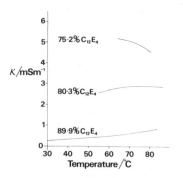

Figure 5. Variation of conductivity with temperature in the L_2 region of the $C_{12}E_4$-water system. Samples made up with 0.01 M sodium chloride solution.

$C_{12}E_4$) and the ethylene oxide groups as the temperature rises. At lower concentrations of surfactant it appears that some sort of structural change must be proceeding as the temperature increases since dK/dT becomes negative in 80.3% $C_{12}E_4$ and is negative throughout in 75.2% $C_{12}E_4$. In these mixtures lamellar phase is formed at low temperatures and water separates at higher temperatures. The unexpected conductance behaviour is possibly due to the gradual isolation of some of the water and ions in small regions, still within the one phase system, but separated from each other by increasingly large regions of lower ion-water concentration.

2. Ternary Systems

The isotropic liquid phases occurring in the water-heptane-$C_{12}E_4$ system at various temperatures are shown on the ternary diagrams in Figure 6. Values of conductivity (μSm^{-1}) at particular compositions and temperatures have been included in the diagrams. The different clear regions have been identified with the principal letters L_1, L_2, L_3 for easy reference to the binary

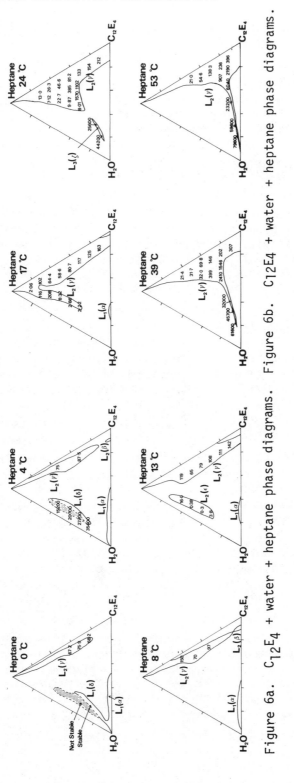

Figure 6a. $C_{12}E_4$ + water + heptane phase diagrams.   Figure 6b. $C_{12}E_4$ + water + heptane phase diagrams.

system and subsidiary Greek characters in order to distinguish
them from each other.

   The $L_1(\alpha)$ and $L_1(\delta)$ regions. $L_1(\alpha)$ is an extension of the L_1
region in the binary system $C_{12}E_4$-water. Only small amounts of
heptane are solubilized at concentrations of $C_{12}E_4$ above 20% before
lamellar phase separates. Most of the solutions exhibit streaming
birefringence at lower temperatures. Between 10% and 20% $C_{12}E_4$
however, there is a narrow region of very much enhanced heptane
solubility at 0 °C which eventually detaches itself from the $L_1(\alpha)$
region at 1 °C forming the $L_1(\delta)$ region which is stable up to ca
7 °C.

   This is the region of the phase diagram denoted as 'surfactant
phase', S, by a number of workers[17,18]. At the lowest temperatures
studied the samples are highly viscous and although isotropic at
rest they exhibit streaming birefringence indicating the presence
of easily deformable particles or anisotropic micelles which on
close approach to each other form a lamellar phase-like layer
under flow at the plane of contact. The presence of large aggregates is suggested by the blue translucence of the mixtures before
phase separation on the low surfactant boundary of the region.

   The conductivities on the phase diagram confirm that the
$L_1(\delta)$-region is water continuous, while the decrease with increasing heptane concentration is as expected for large swollen micelles
due to the increasing excluded volume. Application of the Bruggeman
equation to the series of samples containing 10% $C_{12}E_4$ at 4 °C shows
reasonable agreement (Figure 7).

   Above 20% (W/W) heptane the results lie between the Bruggeman
equation and that of Wagner[15](Equation 4).

$$K = Km \frac{2(1 - \phi)}{2 + \phi} \tag{4}$$

   The phase volume, ϕ, has been calculated on the basis of the
densities of the components so is open to some error in that slight
changes may occur when the components are mixed e.g. heptane in
micelles may not have exactly the same effective density as the
pure component.

   It appears that the conductivity in $L_1(\delta)$-region shows a much
greater resemblance to the conductivity in a normal emulsion than
it does to the conductivity in the L_1-region of the binary system.
This is probably due to to the fact that increasing the volume fraction of $C_{12}E_4$ in the binary system affects the volume fraction of
water directly by exclusion and indirectly by using some of it to

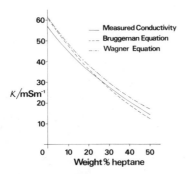

Figure 7. Conductivity in the $L_1(\delta)$ region of the $C_{12}E_4$ + water + heptane system at 4 °C. Samples made up with 0.01 M sodium chloride solution.

hydrate the ethylene oxide groups. In the ternary systems increase of heptane content, keeping the $C_{12}E_4$/water ratio $\leqslant 1$, is equivalent to changing the volume fraction of a 'normal' emulsion. Furthermore, the micelles containing heptane are larger and thus approximate more closely to emulsion particles than do surfactant micelles.

The $L_2(\beta)$ and $L_2(\gamma)$ regions. At very low temperatures $L_2(\gamma)$ is a narrow region, parallel to the heptane-$C_{12}E_4$ axis, all mixtures containing approximately 10% water. At 4 °C the $L_2(\beta)$ region appears as a small extension of the L_2 region in the $C_{12}E_4$-water system which by 8 °C is continuous with the $L_2(\gamma)$ region. As the temperature increases this combined region stretches back to the heptane-$C_{12}E_4$ axis and outward into the centre of the diagram, joining up with other regions. Thus there is likely to be considerable structural diversity within this region which is confirmed by the conductance measurements.

Figure 8 shows the variation of conductivity with temperature for the series of samples containing 10% doped water. At their lowest temperatures all the mixtures except number 1 have conductivities of similar magnitude to that of 10% doped water.

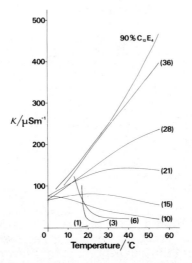

Figure 8. Variation of conductivity with temperature in the $L_2(\gamma)$ region of the $C_{12}E_4$ + water + heptane system. Samples made up with 0.01 M sodium chloride solution and sample numbers (see Figure 1) in brackets.

It may be concluded that mixtures, 10, 15, 21, 28 and 36 are composed of micellar solutions containing solubilized heptane in a hydrogen bonded ethylene oxide-water continuum at their lowest temperatures. This agrees with the conclusions of Boyle et al. from measurements of the dielectric permittivities of these mixtures[16]. As the temperature increases there is an increasing tendency to depart from the expected increase in conductivity as the water/$C_{12}E_4$ weight ratio of the samples increases from 1/8 in mixture 36 to 1/4 in mixture 10. This must be due to the dehydration effect discussed in connection with the behaviour of the L_2 samples. Normal micellar solutions are not likely in samples 6 and 3 because of the small volume fraction of water and surfactant. It has been proposed by Boyle et al.[16] that large lamellar aggregates breakdown to give inverted micellar solutions in the clear regions of samples 6 and 3. However, the conductivities of these samples are much higher than would be expected for a W/O dispersion even at their highest temperatures. Nmr diffusion studies by Lindman et al.[17,18] on $C_{12}E_4$-hydrocarbon-water systems also indicate the absence of inverted micelles at high temperatures in these regions of the phase diagram.

It can be seen that the conductivity of mixture 3 rises with temperature after the initial sharp fall and the conductivity of mixture 1 rises with temperature throughout the short range of its existence. Mixture 5 also showed increasing conductivity with temperature after the initial steep fall (201 μSm^{-1} at 17 °C to 4.95 μSm^{-1} at 19.9 °C and 7.12 μSm^{-1} at 24.8 °C) during the first few degrees of its existence as an isotropic liquid. It is considered that the steep falls in conductivity always occur when the previous (lower temperature) phase has been lamellar or contained an appreciable lamellar component. During the first few degrees there are probably large, highly structured bicontinuous regions which break down to form bicontinuous regions of lower conductivity. As these regions become 'stable' their conductivity increases with temperature in a normal way until phase separation occurs. In support of this suggestion it is noticeable that the highest values of K on each ternary diagram in the $L_2(\gamma)$-region below 39 °C occur at compositions closest to those of the lamellar phase containing most heptane i.e. on a line from the heptane corner to between 50 and 60% $C_{12}E_4$ in water on the binary axis.

The $L_2(\varepsilon)$ region. It can be seen from Figure 6 that by 13 °C a new isotropic region has appeared which partially overlaps the composition range of the lower temperature $L_1(\delta)$ region. The $L_2(\varepsilon)$ region is not connected to the $L_1(\delta)$ region when heptane is the hydrocarbon component and at intervening temperatures a three phase system containing lamellar phase occurs. When the hydrocarbon is hexadecane two isotropic regions occur in this part of the phase diagram which are connected over a short range of composition and temperature and then the conductivity remains quite high throughout both regions[8].

The $L_2(\varepsilon)$ region is blue translucent at lowest surfactant content; this may be ascribed to large aggregates. Streaming birefringence is observed at the high water content part of the region at lowest temperatures of occurrence; this suggests easily deformable aggregates as in the $L_1(\delta)$ region. As the temperature is increased the $L_2(\varepsilon)$ region joins the $L_2(\gamma)$-region close to the heptane corner and then moves across the diagram towards higher $C_{12}E_4$ contents and finally merges completely with the $L_2(\gamma)$-region. The conductivity behaviour with temperature in this region was discussed previously[8] and it was concluded that mixtures such as number 7 actually inverted to give an oil continuous phase shortly before becoming unstable at higher temperature. In line with the earlier discussion of the $L_2(\gamma)$-region it would now appear that when it first arises the $L_2(\varepsilon)$-region is bicontinuous in structure but the low surfactant content restricts the thermal stability to a relatively small range of temperature. Mixture 11 behaves similarly to mixture 7 and these two mixtures give the lowest conductivities in any region of the diagram (0.3 μSm^{-1}). The fact that conductivity remains high in the interconnecting $L_1(\delta)$ and $L_2(\varepsilon)$ regions of the $C_{12}E_4$-water-hexadecane system suggests that the bicontinuous structure is more stable in this case. Once the $L_2(\varepsilon)$-region has moved across the diagram and joined with the $L_2(\gamma)$-region the conductivities do not fall to such low values since now the higher surfactant content of the samples helps to stabilise the bicontinuous phase structure.

The $L_3(\xi)$ region. The $L_3(\xi)$ region appears as a narrow isolated region initially but it is in fact an extension of the L_3 region in the $C_{12}E_4$-water system with similar high conductivity values. The addition of heptane to binary mixtures containing < 50% $C_{12}E_4$ in water reduces the temperature of occurrence of the L_3 region in an almost linear manner without appreciably altering its narrow temperature range of existence (ca 2 °C). In the ternary diagram the $L_3(\xi)$ region joins the $L_2(\gamma)$-region at higher temperatures. It can be seen from Figure 6 that at 60 °C the joint region forms a narrow salient across the bottom of the ternary diagram and the linearity of this salient suggests that the surfactant-heptane ratio is critical to its existence as a clear phase. Whatever the structure of the surfactant aggregates in this phase its position on the ternary diagram indicates that much more heptane (13%) is solubilized by these aggregates than by the normal micelles of the $L_1(\alpha)$ phase (3%) for a binary mixture of 30% $C_{12}E_4$ in water. Whilst the conductivities in this

region are high, they do not fit Equation (3) unless one allows
'a' to increase as the temperature and/or concentration of $C_{12}E_4$
decreases. This suggests that the stability of the phase is critically dependent on the degree of hydration of the micelles and
possibly also their size.

CONCLUSIONS

Detailed pictures of the phase behaviour of the $C_{12}E_4$ + water
and $C_{12}E_4$ + water + heptane systems over a wide range of composition and temperature have been obtained.

The electrical conductivities of the binary L_1 phase and the
ternary $L_1(\delta)$ phase have been shown to fit equations appropriate
to water continuous emulsion systems. On the other hand the conductivities of certain ternary mixtures of low surfactant content
suggests the presence of a water-continuous microemulsion at low
temperatures which inverts eventually to an oil-continuous microemulsion before finally separating into two phases at higher temperatures. The phase behaviour and associated conductivities in
certain parts of the ternary phase diagram are best explained in
terms of a bicontinuous structure containing no definable aggregates.

ACKNOWLEDGEMENTS

We thank Dr. M.H. Boyle and Dr. R.M. Wood for helpful discussions and the Science and Engineering Research Council for financial
support to one of us (T.A.B.).

REFERENCES

1. S. Friberg, I. Lapczynska, Progr. Colloid and Polymer Sci.,
    56, 16 (1975).
2. T. M. Prince, Editor, "Microemulsions", Academic Press, New York,
    1977.
3. L.E. Scriven, Nature, 263, 123 (1976).
4. Y. Talmon, S. Prager, J. Chem. Phys., 69, 2984 (1978).
5. M. Lagues, R. Ober, C. Taupin, J. Phys. Lett., 39, 487 (1978).
6. B. Lagourette, J. Peyrelasse, C. Boned, M. Clausse, Nature,
    281, 60 (1979).
7. M. Lagues, C. Sauterey, J. Phys. Chem., 84, 3503 (1980).
8. T.A. Bostock, M.P. McDonald, G.J.T. Tiddy, and L. Waring, in
    "Surface Active Agents",   p. 181, Soc. Chem. Ind.,
    Nottingham, 1979.
9. H. Schott, S.K. Han, J. Pharm. Sci., 64, 658 (1975).

10. D.A.G. Bruggeman, Ann. Phys., 24, 636 (1935).
11. R.A. Mackay and R. Agarwal, J. Colloid Interface Sci., 65, 225 (1978).
12. H. Schott, J. Colloid Interface Sci., 24, 193 (1967). H. Yoshioka, ibid, 63, 378 (1978).
13. T.A. Bostock, M.H. Boyle, M.P. McDonald, and R.M. Wood, J. Colloid Interface Sci., 73, 368 (1980).
14. J.C. Lang and R.D. Morgan, J. Chem. Phys., 73, 5849 (1980).
15. T. Hanai, in "Emulsion Science", P. Sherman, Editor, p.379, Academic Press, London, 1968.
16. M.H. Boyle, M.P. McDonald, P. Rossi, and R.M. Wood, in "Microemulsions", I.D. Robb, Editor, p. 103, Plenum Press, New York, 1982.
17. B. Lindman, N. Kamenka, T-M Kathopoulis, B. Brun, and P-G Nilsson, J. Phys. Chem., 84, 2485 (1980).
18. P-G Nilsson and B. Lindman, J. Phys. Chem., 86, 271 (1982).

EXISTENCE OF TRANSPARENT UNSTABLE SOLUTIONS IN THREE AND FOUR COMPONENTS SURFACTANT SYSTEMS

T. Assih, P. Delord and F.C. Larché

LA 233
Université des Sciences et Techniques du Languedoc
34060 - Montpellier Cedex, France

The phase diagram for the system sodium di-2-ethyl-hexylsulfosuccinate (AOT)/decane/water has been determined at 25°C, in the region containing less than 60 wt.% AOT. There exists a large zone where solutions appear at first transparent and homogeneous, but subsequently slowly separate into two isotropic solutions. The phase that contains the smallest amount of water is a micellar solution. X-ray measurements on solutions in the two-phase region before separation, at a molar ratio $[H_2O]/[AOT]$ of 30, indicate the presence of micelles at low volume fractions of water. As the water concentration is increased, they show the simultaneous existence of a second type of particles, with a radius of gyration of the order of 1 nm. The existence of two-phase zones, in which freshly prepared solutions are transparent and homogeneous, has also been found in the systems sodium octylbenzenesulfonate (OBS)/water/pentanol, and OBS/water/pentanol/decane.

INTRODUCTION

The existence of thermodynamically stable transparent micro-emulsions is now well established.[1,2] More recently, Gerbacia, Rosano and collaborators [3,4] have shown conclusively that some clear w/o dispersions are unstable. The method of preparation was an important factor in obtaining these transparent mixtures.

The presence of a slowly separating two-phase zone in the ternary system sodium octylbenzenesulfonate (OBS)–pentanol–water [5] as well as nonreproducible measurements of self-diffusion coeffi-cients by the capillary method [6] in the ternary sodium di-2-ethyl-hexylsulfosuccinate (AOT)–decane–water prompted us to study very carefully the phase diagrams of these systems. Results of these investigations as well as some low-angle X-ray measurements are presented here.

EXPERIMENTAL

The AOT used was a Fluka product of quality "purum". It was further purified following the method of Rogers and Windsor.[7] The

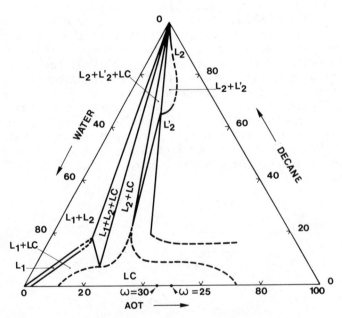

Figure 1. Tentative phase diagram for AOT–water–decane at 25°C.

OBS was synthetized according to Gray et al.[8] and purified by using
the procedure described in reference 9. n-pentanol was obtained
from Merck (quality p.a.) and decane from Fluka (quality purum) and
were used as received. The water was triply distilled.

The components were weighed and mixed in either sealed or
screw-cap glass tubes. They were maintained at 25 ± 0.1°C for three
to four months. Visual observations were made every two weeks. In
most instances the separation of transparent phases was visible to
the experienced eye after 24 to 48 hours. In such cases light microscopy with dark field and phase contrast equipment would clearly show the presence of a coarse emulsion probably produced by the
mixing of the two phases in the capillary tube used for sampling.
The low angle X-ray equipment and the experimental procedure have
been described elsewhere.[10]

RESULTS

AOT-Water-Decane System

The phase diagram, at 25°C, is shown in Figure 1. The most
important result is the presence of a zone noted $L_2 + L_2'$, where two
isotropic phases are in equilibrium. The separation of these phases
is very slow, and the appearance of a visible interface between
them may take several months. However, the presence of two phases
can be seen with a light microscope after only one to two days.

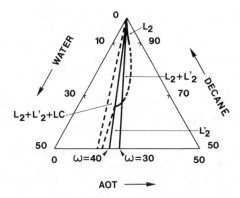

Figure 2. The decane rich corner of the AOT-water-decane system at
19°C.

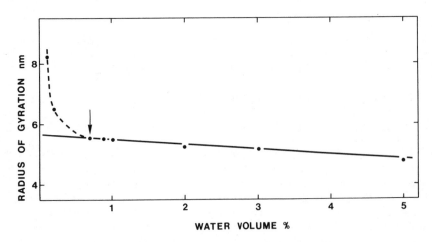

Figure 3. The radius of gyration of the micelles in AOT-water-decane at $[H_2O]/[AOT] = 30$ as a function of the volume fraction of water. The conversion from weight percentage to volume fraction has
been done assuming additivity of the volumes and taking 1000, 730
and 1140 kg m^{-3} as the density of water, decane and AOT, respectively. Below the arrow, the solution is in the $L_2 + L_2'$ zone.

Centrifugation of freshly prepared solutions did not cause a substantial acceleration of the separation. Furthermore once the interface is visible, mechanical agitation cannot reform the transparent system. A coarse unstable emulsion is obtained, that separates in a matter of minutes.

It is interesting to note that when water is added to an AOT-
decane solution, as it is usually done to obtain the limit of stability of the microemulsion region, the solution becomes turbid
when the liquid crystal phase appears. Thus the limit of the three-
phase zone $L_2 + L_2' + LC$ corresponds well to the one obtained by
Huang and Kim at 23°C,[11] and thought to be the line separating a
one-phase from a two-phase region. As already claimed for other
systems by Rosano et al.[4], the optical properties are not always a
good indicator of phase transitions.

At 19°C, the kinetics of phase separation seems to be even
slower than at 25°C, but the corresponding $L_2 + L_2'$ zone is still
present. Its boundary along the $L_2 + L_2' + LC$ region is displaced
towards a higher molar ratio $\omega = [H_2O]/[AOT]$. The diagram is shown
in Figure 2. All the X-ray results, to be discussed below, have
been obtained at this temperature.

We showed previously that at ω = 30, the L_2 phases contain
micelles. Their size increases as the concentration of decane increases.[10] Above ∿ 1 wt.% of water, the solutions are in the two-
phase region. Before separation, small angle X-ray scattering still
indicates the presence of micelles. Their radius of gyration is a
linear function of the water volume fraction (Figure 3). This simply indicates that the interaction between micelles increases, but
their size remains constant. With the three layers model used previously [10] one can calculate the radius of their water core as
5.9 nm. It is important to note that all these results were obtained on freshly prepared solutions; and no discontinuity appears as
the boundary between L_2 and $L_2 + L_2'$ is crossed. At still higher water concentration, ω = 30, and on fresh solutions, the spectra show
a break in a standard Guinier plot (Figure 4). This feature is
usually associated with the existence of two structures [12] (like
spheres with two radii). If one accepts this interpretation, the
radius of gyration of the smallest aggregates would be of the order of 1 nm.

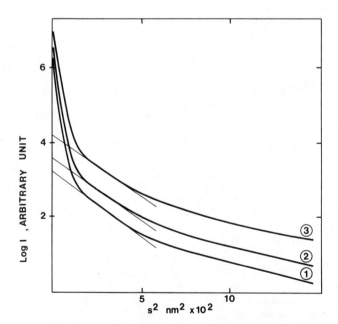

Figure 4. Logarithm of the scattered intensity as a function of s^2.
s = 2 sin θ/λ. λ = 0.154 nm. All the curves are along the line
[H_2O]/[AOT] = 30. Water concentration : curve 1 : 5.3 wt.%;
curve 2 : 9.1 wt.%; curve 3 : 12.8 wt.%.

OBS-Pentanol-Water-Decane System

   A previous study [5] on the ternary OBS-pentanol-water disclo-
sed a region, between the water rich L_1 and the alcohol rich L_2,
where the phenomena described in the preceding section were also
observed. This prompted us to investigate more closely the quaternary OBS-pentanol-water-decane, in the section at constant penta-
nol/OBS (2.1 by weight) ratio. Continuity was previously claimed
between the water rich and decane rich solutions.[13] At 25°C we
found again a region where the solutions are homogeneous and transparent after preparation, but separate slowly on standing. The phase diagram is represented in Figure 5. The boundary between
$(L_1 + L_2)$ and $(L_1 + L_2 + L_2')$ is practically identical to what was
claimed to be the boundary between a water rich and a decane rich
phase.[13] This indicates that as one enters this three-phase zone,
L_2' (oil rich) forms very rapidly, but the subsequent separation of
the water rich phase is a slow process.

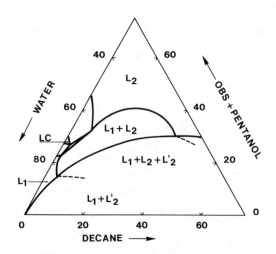

Figure 5. Section at constant ratio pentanol/OBS (2.1 by weight)
and 25°C, of the quaternary OBS-pentanol-water-decane phase diagram.

Centrifugation had no apparent effect on the kinetics of separation, and, as for the previous system, completely separated solutions could not be reverted to homogeneous, transparent solutions by agitation.

DISCUSSION

Transparent mixtures that slowly separate into two isotropic phases have been found in several ternary and quaternary systems, containing an anionic surfactant, water, an alcohol and/or an oil. In all these systems the behaviour of the samples is quite similar: minimal optical effects, time scale for separation of the order of a few months, influence of the method of preparation. In particular, once the phases are completely separated, mechanical action cannot bring back the system to its transparent state. Furthermore extended centrifugation has no effect on the speed of separation.

Rosano et al.[4] suggested that some methods of preparation induce the formation of an unstable microemulsion. Our X-ray results confirm that they are indeed microemulsions, as far as the structure is concerned. The fact that there is no discontinuity on going from the one-phase oil rich region to the two-phase region (on freshly prepared solutions) suggests that these unstable microemulsions are just the metastable extension of the L_2 phase into the two-phase field.

The mechanism proposed by Rosano et al. involve a temporary low interfacial tension sufficient to cause spontaneous emulsification; this was mainly an effect of the cosurfactant. Since we have seen these phenomena in a system without cosurfactant, this mechanism ought to be further looked into.

Several surprising results found in the literature might constitute an indication that such phenomenon is also present in other systems. The case of AOT-water with another oil than decane would be particularly interesting for further investigations. For instance this unsuspected instability might be a reason, as mentioned by Eicke et al.[14], for the observations at high AOT concentration in AOT-water-isooctane. It might also be the reason for the presence of rather large particles together with more standard micelles seen by quasi-elastic light scattering in AOT-water-heptane.[15]

ACKNOWLEDGEMENTS

The authors would like to thank J. Rouvière for the Karl-Fisher determination of water in some phase-separated samples.

REFERENCES

1. S. Friberg and I. Buraczewska, Prog. Colloid and Polymer Sci., 63, 1 (1978).
2. J. Th. G. Overbeek, Faraday Disc. Chem. Soc., 65, 7 (1978).
3. W. E. F. Gerbacia and H. L. Rosano, J. Colloid Interface Sci., 44, 242 (1973).
4. H. L. Rosano, T. Lan, A. Weiss, J. H. Whittam and W. E. F. Gerbacia, J. Phys. Chem. 85, 468 (1981).
5. F. C. Larché, J. L. Dussossoy, J. Marignan and J. Rouvière, (1983) J. Colloid Interface Sci., accepted for publication.
6. N. Kamenka, private communication.
7. J. Rogers and P. A. Windsor, J. Colloid Interface Sci., 30, 247 (1969).
8. F. W. Gray, J. F. Gerecht and I. J. Krems, J. Org. Chem., 20, 511 (1955).
9. N. Kamenka, M. Chorro, H. Fabre, B. Lindman and C. Cabos, Colloid Polymer Sci., 257, 757 (1979).
10. T. Assih, F. C. Larché and P. Delord (1982), J. Colloid Interface Sci., 89, 35 (1982).
11. J. S. Huang and M. W. Kim, NATO Advanced Study Institutes, B73, 809, S. H. Chen, B. Chu and R. Nossal, Ed., Plenum Press, New York. 1981.
12. G. Fournet and A. Guinier, "Small-Angle Scattering of X-Rays", John Wiley, New York, 1955.
13. B. Lindman, N. Kamenka, T. Kathopoulis, B. Brun, P. G. Nilsson, J. Phys. Chem., 84, 2485 (1980).
14. H. F. Eicke and P. Zinsli, J. Colloid Interface Sci. 65, 131, (1978).
15. E. Gulari, B. Bedwell and S. Alkhafaji, J. Colloid Interface Sci., 77, 202 (1980).

THEORY OF PHASE CONTINUITY AND DROP SIZE IN MICROEMULSIONS

II. IMPROVED METHOD FOR DETERMINING INVERSION CONDITIONS

Jyi-feng Jeng and Clarence A. Miller

Department of Chemical Engineering
Rice University
Houston, Texas 77251

The theory of microemulsions developed in a
previous paper has been modified to enable the
"natural" radius of curvature of the drops to be
calculated accurately for large drops. The result is
an improved ability to predict the inversion point
where the natural curvature of the film shifts from an
oil-in-water to a water-in-oil configuration. The
inversion point is of interest because ultralow interfacial tensions between the microemulsion and excess
bulk phases are often found there. Some results are
given for microemulsions containing single-chain
surfactants and short-chain alcohols, a situation for
which only limited results could be obtained with the
earlier theory.

INTRODUCTION

It is well known that thermodynamically stable microemulsions can form in oil-water-surfactant systems. In certain composition ranges the microemulsions consist of tiny drops some 10 nm in diameter of oil in water or water in oil. Surfactant monolayers are believed to cover the surfaces of these drops.

In the first paper of this series[1], hereinafter called Part I, previous theories of microemulsions were reviewed and a new theory was presented. Its novel feature was a lattice model of the hydrocarbon chain region of the surfactant monolayers or films. With this model the amount of oil incorporated into the films and the "natural" radius R^*_0 of the drops could be calculated. The latter is the radius at which the free energy of the films is minimized, irrespective of the free energy of mixing of the drops with the continuous phase and of the energy of interaction between drops. Work must be supplied to the film to bend it from a radius R^*_0 to some other radius. The concept of a natural radius had been used previously by others[2-7], but there was no previous method of estimating its value for oil-containing films. We note that Huh[8] has independently developed a model which also allows R^*_0 to be calculated. His model resembles that of Part I in its general thermodynamic approach but is quite different in various specifics of the model.

The theoretical results of Part I showed that bending effects were of great importance in determining drop radius for microemulsions in equilibrium with an excess oil or with an excess water phase. Predictions of the theory for a system containing a double-chain surfactant, a short-chain alcohol, a straight-chain hydrocarbon, and water were in qualitative agreement with experimental results for a system containing a pure double-chain sulfonate surfactant[9,10]. That is, the theory correctly predicted the direction of change in drop size due to changes in various parameters such as oil and alcohol chain lengths and surfactant-to-alcohol ratio in the film.

While the results given in Part I were thus very encouraging, there were two aspects where improvement was needed. In the first place, the numerical procedure used to solve the equations for the natural radius R^*_0 was of limited value when R^*_0 exceeded about 30 nm. Consequently, it was not possible to accurately predict the conditions at the "inversion point" where R^*_0 becomes infinite. This point is of great interest because of its association with ultralow interfacial tensions, a property of particular importance for application to enhanced oil recovery.

The second difficulty arose when the model was applied to

surfactants having a single, straight hydrocarbon chain. Predictions were again found to be in qualitative agreement with experimental results for the case of an oil-in-water microemulsion in equilibrium with excess oil. But inversion to a water-in-oil microemulsion was not predicted, in sharp contrast to the experimental results.

In this paper we present results obtained with a new procedure for solving the basic equations of the model. Both the above difficulties, which apparently resulted from problems with the numerical solution, are removed. The utility of the basic model is thus considerably enhanced.

THEORY

The basic equations of the theory are given in Part I and are not repeated here except as needed to clarify the improved method of solution. When a microemulsion is in equilibrium with excess oil, the overall free energy of the system must be minimized with respect to two independent variables. In Part I these variables were taken as the number of drops N_d and the number of oil molecules ϕ_o in the microemulsion phase.

The two equilibrium conditions of Part I require knowledge of the partial derivatives $(\partial G_T^*/\partial \phi_o)_{N_d}$ and $(\partial G_T^*/\partial N_d)_{\phi_o}$. In these expressions G_T^* is the free energy of the hydrocarbon chain portion of the surfactant film of a drop. The two partial derivatives were evaluated numerically in Part I since G_T^* is given by a rather lengthy and complicated equation involving the number of possible configurations of surfactant, alcohol, and oil chains within the film. These values of the partial derivatives were then used in the numerical solution of the two equilibrium equations, the final result being equilibrium values of N_d and ϕ_o.

The calculations reported here were made by choosing a different pair of independent variables, the drop radius R_o and the area a_s per surfactant molecule in the film at the position of the surfactant head groups. In this case it is necessary to know $(\partial G_T^*/\partial R_o)_{a_s}$ and $(\partial G_T^*/\partial a_s)_{R_o}$. We have succeeded in developing analytical expressions for these derivatives. As a result, an important source of numerical error in determination of the equilibrium drop size is avoided. The difficulties experienced in Part I for drops with large values of natural radius R_o^* and for single-chain surfactants are thus overcome. Details are given in the Appendix.

J. JENG AND C. A. MILLER

RESULTS

Figure 1 shows variation of the natural radius of curvature R^*_o with surfactant-to-cosurfactant ratio β for a double-chain surfactant with n-propanol as the cosurfactant and n-decane as the oil. Only the portions of the curves below the points P could be obtained by the numerical solution procedure of Part I. With just this information available one can say only that inversion occurs when β is between about 0.5 and about 0.75.

With the present method values of R^*_o in good agreement with those of Part I are found for drop radii below about 30 nm. But, in addition, a solution can be found for values of R^*_o as large as desired, and the inversion point can be located as accurately as necessary. Our calculations show that β at the inversion point is between 0.625 and 0.630. They confirm, moreover, that a single value of β exists where R^*_o becomes infinite. This conclusion is the expected one, but it is by no means obvious from the results of Part I alone.

Also shown in Figure 1 is variation of the actual drop radius R_o with the free energy of mixing effect included for a surfactant concentration in the system of 0.01M. The actual radius R_o is always less than the natural radius R^*_o. The increase in free energy in bending the film from its natural radius to a smaller value is conterbalanced by the simultaneous decrease in free energy of mixing due to an increase in the number of drops. The total interfacial area remains virtually constant during the process.

Again the results are in agreement with those of Part I for small drop sizes (R_o below about 30 nm), and again the present method allows results to be obtained for much larger R_o. A limit is eventually reached where drop size becomes so large that the volume fraction of drops approaches 0.74, the maximum possible for uniform spheres. For the system of Figure 1 these large volume fractions occur only when β is between about 0.61 and 0.64. Under these conditions the available expressions for the free energy of mixing of uniform spheres are of poor accuracy, and the results are of limited value. As the system is near its inversion point, it is likely that the microemulsion does not consist of uniform spheres in any case and hence that some other model is required.

An important advantage of the present method over that used in Part I is that inversion of systems containing single-chain surfactants can be predicted. Figure 2 shows variation of R^*_o with β for a surfactant having a single chain of twelve carbon atoms. As in Figure 1, n-propanol and n-decane are the

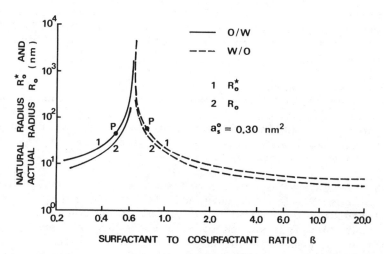

Figure 1.   Natural radius   R* and actual radius R_o for system containing a double-chain surfactant with structure like the pure surfactant Texas 1, n-propanol, n-decane, and water or brine. Values of R_o are for a surfactant concentration of 0.01M.

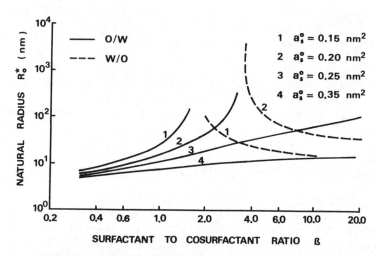

Figure 2.   Natural radius   R* as a function of surfactant-to-cosurfactant ratio   β and surfactant head area a_s^o for a system containing a single-chain surfactant with twelve carbon atoms, n-propanol, n-decane, and water or brine.

cosurfactant and oil. For large values of the actual surfactant head area a_s^o no inversion is seen. With no alcohol present $(\beta \to \infty)$ the surfactant is hydrophilic and forms an oil-in-water microemulsion. Addition of n-propanol makes the surfactant film at the drop surfaces even more hydrophilic and the natural radius R_o^* decreases.

In contrast, when surfactant head area a_s^o is small, the surfactant is hydrophobic and a water-in-oil microemulsion forms in the absence of alcohol. Addition of n-propanol again makes the system more hydrophilic and inversion eventually occurs to an oil-in-water microemulsion. For smaller values of a_s^o more alcohol is required to produce inversion.

When n-hexanol replaces n-propanol as the alcohol, things are quite different. In this case addition of the rather hydrophobic hexanol produces no inversion of the oil-continuous microemulsions which occur for small values of a_s^o, as Figure 3 shows. For somewhat larger a_s^o, addition of hexanol causes inversion from a water-continuous to an oil-continuous microemulsion. Such inversion has been observed in a system contining hexanol, n-decane, brine, and sodium dodecyl sulfate, which, like the surfactant used in our calculations, has twelve carbon atoms[9]. Finally, Figure 3 shows that for very large a_s^o the surfactant is so hydrophilic that even addition of a large quantity of hexanol will not produce inversion.

Also shown in Figure 3 is the actual radius R_o including the free energy of mixing for $a^o = 0.35$ nm^2 (35 A^2). As found previously for the double-chain surfactant R_o is always less than the natural radius R_o^*.

The effect of surfactant concentration on drop radius R_o is shown in Table I. For both oil-in-water and water-in-oil microemulsions R_o increases slightly with increasing surfactant concentration. The same result was found for double-chain surfactants in Part I, and the explanation is the same as given there, namely that the decrease in free energy obtainable by decreasing drop size is less when more drops are initially present.

It should be noted from Figure 3 that the effect of small amounts of hexanol is to increase the drop size in oil-continuous microemulsions and to decrease the drop size in water-continuous microemulsions. Exactly the opposite behavior is found when large amounts of hexanol are present. The reason for this behavior is not entirely clear, but probably stems from the relative magnitudes of alcohol effects on the head and tail regions of the surfactant films.

Replacement of a surfactant molecule by an alcohol molecule

decreases the overall head area of the film since the physical head
size of an alcohol molecule is smaller than the effective head size
(including electrostatic effects) of a surfacant molecule. This
effect alone makes the surfactant film more hydrophobic. But the
shorter alcohol chain length also means a smaller entropy of mixing
in the film when an oil molecule enters the tail region and hence a
smaller area per molecule on the tail side. This effect makes the
film more hydrophilic. Perhaps the latter effect is larger when
little hexanol is present and the former when much hexanol has been
added. Further investigation is in progress using other alcohol
chain lengths to learn more about this effect.

Table I. Effect of Surfactant Concentration c_s on Drop Radius R_o
for a System Containing a Single-Chain Surfactant with Twelve
Carbon Atoms, n-Hexanol, n-Decane, and Water or Brine.

Case 1.   Surfactant head area a_s^o = 0.35 nm^2
           Surfactant-to-cosurfactant ratio β = 1.0
           Natural radius R_o^* = 11.7 nm (oil-in-water)

c_s (mol/l)	R_o (nm)	Volume Fraction of Drops
0.0001	9.3	0.000124
0.001	9.5	0.00127
0.01	9.8	0.0130
0.1	10.0	0.133
0.3	10.2	0.405

Case 2.   Surfactant head area a_s^o = 0.15 nm^2
           Surfactant-to-cosurfactant ratio β = 1.0
           Natural radius R_o^* = 23.6 nm (water-in-oil)

c_s (mol/l)	R_o (nm)	Volume Fraction of Drops
0.0001	15.9	0.000214
0.001	16.8	0.00224
0.01	17.8	0.0233
0.1	18.9	0.244

    The effect of oil chain length is shown in Figure 4. When
decane is replaced by hexadecane as the oil, the system becomes
less hydrophobic, in agreement with many observations of the effect
of oil chain length on microemulsion behavior[11]. Indeed, in the
particular case shown the system does not even invert upon addition
of n-hexanol when the oil is hexadecane.

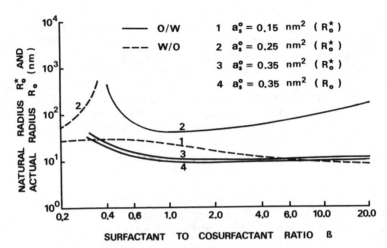

Figure 3.   Natural radius R_o^* as a function of surfactant-to-
cosurfactant ratio β and surfactant head area a_s^o for a system
containing a single-chain surfactant with twelve carbon atoms, n-
hexanol, n-decane, and water or brine.   Values of R_o are for a
surfactant concentration of 0.01M.

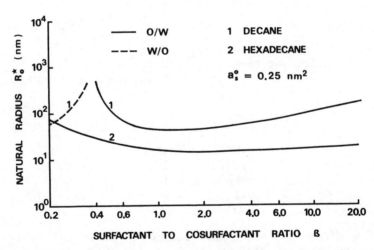

Figure 4.   Natural radius R_o^* as a function of surfactant-to-
cosurfactant ratio   β and oil chain length for a system containing
a single-chain surfactant with twelve carbon atoms, n-hexanol, and
water or brine.

The oil chain length effect can be explained in terms of entropy of mixing effects in the film. The longer-chain oil has a lower entropy of mixing per carbon atom and hence penetrates the surfactant film to a lesser extent than the shorter-chain oil. Less oil in the film leads to smaller drops for water-continuous microemulsions and less tendency to invert to oil-continuous microemulsions, precisely the predictions of Figure 4.

SUMMARY

The improved method of solution of the basic equations of Part I allows the inversion point to be calculated accurately in systems where such an inversion occurs. It also allows the natural radius of curvature to be found near the inversion point. The utility of the previous theory is thus considerably increased.

ACKNOWLEDGMENTS

This work was supported by grants from Gulf Research and Development Company, Amoco Production Company, Exxon Production Research Company, and Shell Development Company. Discussions with S. Mukherjee were useful.

REFERENCES

1. S. Mukherjee, C.A. Miller, and T. Fort, Jr., J. Colloid Interface Sci., 91, 223 (1983).
2. P.A. Winsor, "Solvent Properties of Amphiphilic Compounds," Butterworth, London, 1954.
3. C.L. Murphy, Ph.D. Thesis, University of Minnesota, 1966.
4. M.L. Robbins, in "Micellization, Solubilization, and Microemulsions," Vol. 2, p. 713. Plenum, New York, 1977.
5. W. Helfrich, Z. Naturforsch., 28C, 693 (1973).
6. J.D. Mitchell and B.D. Ninham, J. Chem. Soc. Faraday Trans. II, 77, 601 (1981).
7. A.G. Petrov and A Derzhanski, J. Phys. Suppl., C3, 157 (1976).
8. C. Huh, SPE/DOE Preprint 10728, Presented at Symposium on Enhanced Oil Recovery, Tulsa, April, 1982.
9. W.J. Benton, J. Natoli, S. Qutubuddin, S. Mukherjee, C.A. Miller, and T. Fort, Jr., Soc. Petrol Eng. J., 22, 53 (1982).
10. S. Mukherjee, Ph.D. Thesis, Carnegie-Mellon University, 1981.
11. J.L. Salager, M. Bourrel, R.S. Schechter, and W.H. Wade, Soc. Petrol. Eng. J., 19, 271 (1979).

APPENDIX

We wish to minimize the free energy of a system consisting of an oil-in-water microemulsion (phase I) in equilibrium with an excess oil phase (phase II). Temperature, external pressure, and the total number of surfactant and cosurfactant molecules in the films of all the drops are maintained constant. Independent variables are taken as the area a_s per surfactant molecule in the film and drop radius R_o at the position of the head groups. The basic equilibrium equations are

$$\left(\frac{\partial(G^I + G^{II})}{\partial a_s} \right)_{R_o} = 0 \tag{A-1}$$

$$\left(\frac{\partial(G^I + G^{II})}{\partial R_o} \right)_{a_s} = 0 \tag{A-2}$$

As in Part I, the microemulsion free energy G^I may be written as

$$G^I = \phi_o \mu_o^{II} + N_d G_H^m + N_d G_T^* + N_w \mu_w^I + \Delta G_{mix} \tag{A-3}$$

where ϕ_o is the number of molecules of oil in the microemulsion phase, μ_o^{II} the chemical potential of the bulk oil, N_d the number of drops, G_H^m the free energy of the head region of a drop, N_w and μ_w^{II} the number of water molecules and their chemical potential, ΔG_{min} the free energy of mixing of the drops with water, and G_T^* is given by

$$G_T^* = G_T^m - n_o^m \mu_o^{II} \tag{A-4}$$

Here G_T^m is the free energy of the tail area of a drop and n_o^m is the number of oil molecules in a drop. The free energy of mixing is given by the Percus-Yevick-Carnahan-Starling expression for hard spheres:

$$\Delta G_{min} = N_d kT \left(\ell n\, \eta - 1 + \eta\, \frac{4-3\eta}{(1-\eta)^2} + \ell n\, \frac{V_w}{V_{HS}} \right) \tag{A-5}$$

where η is the volume fraction of drops, V_w the volume of a water molecule, and V_{HS} the volume of a drop. The free energy of the bulk oil phase containing N_o^{II} molecules is given by

$$G^{II} = N_o^{II} \mu_o^{II} \tag{A-6}$$

Substituting (A-3) and (A-6) into (A-1) and (A-2) and noting that N_w and the sum $(\phi_o + N_o^{II})$ are constants, we find that the equilibrium equations become

$$- \Pi_H N_s + \left(\frac{\partial (N_d G_T^*)}{\partial a_s} \right)_{R_o} + \left(\frac{\partial (\Delta G_{mix})}{\partial a_s} \right)_{R_o} = 0 \qquad (A-7)$$

$$\left(\frac{\partial (N_d G_T^*)}{\partial R_o} \right)_{a_s} + \left(\frac{\partial (\Delta G_{mix})}{\partial R_o} \right)_{a_s} = 0 \qquad (A-8)$$

where N_s is the total number of surfactant molecules in the microemulsion and Π_H is the surface pressure contribution of the head groups. No term in Π_H appears in (A-8) because $(N_d G_H^m)$ is independent of R_o at constant a_s.

The most important contribution to G_T^* for present purposes is that due to the configurational entropy of the surfactant, cosurfactant, and oil chains in the film. This contribution is given by $(-kT \ln \omega_T^m)$, where ω_T^m is the number of independent ways of arranging the chains in the tail region. A lattice model was developed in Part I from which an expression for ω_T^m was obtained. The derivation is rather lengthy and will not be repeated here. The chief contribution of the present work is derivation of analytical expressions for the derivatives of ω_T^m. These derivatives were evaluated numerically in Part I. The analytical expressions obtained here are

$$\left(\frac{\partial \ln \omega_T^m}{\partial a_s} \right)_{R_o} = - \left(\frac{\partial n_s^m}{\partial a_s} \right)_{R_o} (1 + \ln n_s^m) - \left(\frac{\partial n_a^m}{\partial a_s} \right)_{R_o} (1 + \ln n_a^m)$$

$$- \sum_{k=1}^{3} \left(\frac{\partial n_o^{(k)}}{\partial a_s} \right)_{R_o} (1 + \ln n_o^{(k)})$$

$$- (r-1) \left(\frac{\partial n_o^m}{\partial a_s} \right)_{R_o} + \left(\frac{\partial (r n_o^{(3)})}{\partial a_s} \right)_{R_o} (1 + \ln(r n_o^{(3)}))$$

$$+ \left(\frac{\partial (t n_o^{(1)} + r n_o^{(2)})}{\partial a_s} \right)_{R_o} (1 + \ln (t n_o^{(1)} + r n_o^{(2)}))$$

$$+ \sum_{i=2}^{I} \left(\frac{\partial W(i)}{\partial a_s} \right)_{R_o} \left[\ln \left(\frac{N_i^m}{W(i)} \right) - 1 \right] + \left(\frac{\partial (n_o^{(1)} (v-1))}{\partial a_s} \right)_{R_o} \ln \left(\frac{\ell z - 1}{N_1^m} \right)$$

$$+ \left(\frac{\partial (n_o^{(1)} t + n_o^{(2)} (r-1))}{\partial a_s} \right)_{R_o} \ln \left(\frac{z-1}{N_T^{(2)}} \right) + (r-1) \left(\frac{\partial n_o^{(3)}}{\partial a_s} \right)_{R_o} \ln \left(\frac{z-1}{N_T^{(3)}} \right)$$

$$+ (\ell n(mz) - 1)\left[\left(\frac{\partial n_s^m}{\partial a_s}\right)_{R_o}(2I - I_1 - 1) + \left(\frac{\partial n_a^m}{\partial a_s}\right)_{R_o}(J-1)\right] \qquad (A-9)$$

$$\left(\frac{\partial \ell n\, \omega_T^m}{\partial R_o}\right)_{a_s} = \left(\frac{\partial N_1^m}{\partial R_o}\right)_{a_s}(1 + \ell n N_1^m) - \left(\frac{\partial n_s^m}{\partial R_o}\right)_{a_s}(1 + \ell n\, n_s^m)$$

$$- \left(\frac{\partial n_a^m}{\partial R_o}\right)_{a_s}(1 + \ell n\, n_a^m) - \sum_{k=1}^{3}\left(\frac{\partial n_o^{(k)}}{\partial R_o}\right)_{a_s}(1 + \ell n\, n_o^{(k)})$$

$$+ \left(\frac{\partial(tn_o^{(1)} + r\, n_o^{(2)})}{\partial R_o}\right)_{a_s}(1 + \ell n(tn_o^{(1)} + rn_o^{(2)})) - (r-1)\left(\frac{\partial n_o^m}{\partial R_o}\right)_{a_s}$$

$$+ \left(\frac{\partial(rn_o^{(3)})}{\partial R_o}\right)_{a_s}(1 + \ell n(rn_o^{(3)})) + \sum_{i=2}^{I}\left(\frac{\partial W(i)}{\partial R_o}\right)_{a_s}\ell n\left(\frac{N_i^m}{W(i)}\right)$$

$$+ \sum_{i=2}^{I} W(i)\left[\frac{1}{N_i^m}\left(\frac{\partial N_i^m}{\partial R_o}\right)_{a_s} - \frac{1}{W(i)}\left(\frac{\partial W(i)}{\partial R_o}\right)_{a_s}\right]$$

$$+ \left(\frac{\partial(n_o^{(1)}(\nu-1))}{\partial R_o}\right)_{a_s}\ell n\left(\frac{\ell z-1}{N_1^m}\right) - \frac{n_o^{(1)}(\nu-1)}{N_1^m}\left(\frac{\partial N_1^m}{\partial R_o}\right)_{a_s} - \frac{n_o^{(1)}t}{N_T^{(2)}}\left(\frac{\partial N_T^{(2)}}{\partial R_o}\right)_a$$

$$+ \left(\frac{\partial(n_o^{(1)}t + n_o^{(2)}(r-1))}{\partial R_o}\right)_{a_s}\ell n\left(\frac{z-1}{N_T^{(2)}}\right) + (r-1)\left(\frac{\partial n_o^{(3)}}{\partial R_o}\right)_{a_s}\ell n\left(\frac{z-1}{N_T^{(3)}}\right)$$

$$- \frac{n_o^{(2)}(r-1)}{N_T^{(2)}}\left(\frac{\partial N_T^{(2)}}{\partial R_o}\right)_{a_s} - \frac{n_o^{(3)}(r-1)}{N_T^{(3)}}\left(\frac{\partial N_T^{(3)}}{\partial R_o}\right)_{a_s}$$

$$+ (\ell n(mz)-1)\left[\left(\frac{\partial n_s^m}{\partial R_o}\right)_{a_s}(2I - I_1 - 1) + \left(\frac{\partial n_a^m}{\partial R_o}\right)_{a_s}(J-1)\right] \qquad (A-10)$$

Part I should be consulted for definitions of the various terms in these equations. Note that W(i) in the above equation is the same as T(i) in Part I.

With (A-5), (A-9), and (A-10) and a suitable expression for Π_H (see Part I), the equilibrium equations (A-7) and (A-8) can be solved numerically for a_s and R_o. This procedure was used to obtain the results presented above.

A similar procedure can be used to develop equations for a water-oil-microemulsion in equilibrium with excess water.

EFFECT OF THE MOLECULAR STRUCTURE OF COMPONENTS ON MICELLAR INTERACTIONS IN MICROEMULSIONS

D. Roux, A.M. Bellocq and P. Bothorel

Centre de Recherche Paul Pascal
Domaine Universitaire
33405 Talence Cedex, France

The effect of the chemical nature of oil, alcohol and surfactant on interactions in W/O microemulsions has been studied by light scattering and photon correlation spectroscopy. The influence of the chain length and of the branching of oil has been investigated. For this purpose the following oils have been used : dodecane, octane, cyclohexane and trimethyl 2-2-4,pentane. The alcohol chain length has also been varied from 5 to 7 carbons. In order to examine the influence of the surfactant structure we have used SDS and α methyl SDS.

All these molecular changes lead to very different behaviors of the interactions. Indeed the second virial coefficients vary over a large range (from positive to very negative values) indicating that interactions in the systems studied change from hard spheres to largely attractive. It seems that the important parameters are the length of alcohol, the molecular volume of oil, and the polar head area of the surfactant. A very simple intermicellar potential is proposed. It accounts for all the light scattering results obtained. This potential is due to van der Waals interactions, and the interpenetration between micelles is taken into account. The contribution of the attractive interactions to the potential is significant only when an interpenetration of two micelles occurs. It is shown that these interactions are proportional to the penetrated volume.

I - INTRODUCTION

The scattering properties of microemulsions were first investigated by Hoar and Schulman[1]. Since that time, these systems have
attracted considerable attention. In some part of the phase diagram
their structure can be pictured as dispersions of water droplets
surrounded by a film of surfactant and alcohol molecules in a continuous medium mainly made of oil. Light scattering techniques have
been extensively applied to these systems, since these methods provide informations on micellar sizes and interactions between droplets[2-5].

Previous results have shown that in water in oil (W/O) microemulsions a large range of interaction forces can be obtained by
varying chemical composition. Modern theories of fluids were used
by several authors to explain the light scattering experimental
results[3,4]. The difficulty in evaluating the interaction potential
for inverted micelles did not permit in most cases a complete analysis of data. Recently two of us have calculated the intermicellar
energy potential between W/O microemulsions[6]. In this calculation
interactions have been evaluated for penetrable particles formed
by spherical aqueous core and concentric spherical layers. Attractive interactions are calculated through integration of semiempi-
rical interatomic potential over the various regions of interacting
micelles. This potential allows us to interpret the effects of micellar size and alcohol chain length on the behavior of the interactions in W/O microemulsions[5].

Our interest in this paper is mainly devoted to investigation
of the effect of the molecular structure of the various components
which form a microemulsion - alcohol, oil and surfactant - on the
interactions between W/O micelles. For this purpose we have studied a large number of W/O microemulsions by static and dynamic
light scattering. Indeed, the static experiments provide access to
the second virial coefficient B of the osmotic pressure which is
directly related to the interaction potential. Moreover, study of
the concentration dependence of the diffusion coefficient D at
moderate concentrations allows us to determine the virial coefficient α of D. Several recent theories relate this virial coefficient α to the interaction potential[7-9]. So in this paper, we have
compared the experimental B and α values with those calculated by
using a simplified interaction potential.

In the following, we will first recall some theoretical results (sec. II), followed by description of the experimental procedure (sec. III) and intensity and diffusion coefficient measurements (sec. IV). In section V we will present an analytical interaction potential between W/O micelles, followed by results of calculation of virial coefficients B and α. In the concluding section

we will discuss the experimental and theoretical results.

II - THEORETICAL BACKGROUND

For unpolarized light, the excess scattering of the particles, assumed of a constant size over that of the continuous phase, is expressed as[10] :

$$I(\theta) = (1+\cos^2\theta) \, K v_m \, \phi \, S(q) \, P(q)$$

where $q = 4\Pi n/\lambda_o \sin \theta/2$ is the scattering wave vector, θ is the scattering angle, v_m is the volume of the micelle, ϕ the micellar volume fraction and

$$K = 2\pi^2 n^2 \left(\frac{dn}{d\phi}\right)^2 (\lambda_o^4)^{-1} \tag{1}$$

with n the refractive index and λ_o the light wavelength in vacuo. $P(q)$ is the intraparticle form factor : we set $P(q) \sim_o 1$ for particles under study since their radius is less than 100 Å. $S(q)$ is the structure factor. In the limit $q \to o$, $S(q)$ is related to the osmotic pressure Π by the compressibility relation[11]

$$S(0) = \frac{k_B T}{v_m} \left(\frac{\partial \Pi}{\partial \phi}\right)^{-1}$$

where k_B is the Boltzmann constant and T the absolute temperature. The relation between compressibility and interaction between particles is not easy. One of the simplest ways to have an idea of the interactions is to develop the osmotic pressure according to the virial formula. Π can be written as a function of different powers of ϕ :

$$\Pi = \frac{\phi k_B T}{v_m} \left(1 + \frac{B}{2} \phi + \frac{C}{3} \phi^2 + \ldots\right)$$

where B and C are virial coefficients. B is directly related to the interaction potential $U(r)$ by :

$$B = - \frac{4\pi}{v_m} \int \left(e^{-\frac{U(r)}{k_B T}} - 1\right) r^2 \, dr$$

In the limit of very small volume fractions, the osmotic pressure can be approximated by

$$\Pi \sim \frac{k_B T}{v_m} \phi \left(1 + \frac{B}{2} \phi\right)$$

From the above equations one can deduce :

$$\frac{\phi}{I} = \frac{1}{K v_m} (1 + B\phi) \tag{2}$$

It appears that the droplet size and second virial coefficient B can be extracted from study of ϕ/I in the low concentration range. In addition, results obtained from more concentrated solutions allow us to determine the variation of the osmotic compressibility.

Modern theories of fluids have been used by several authors to explain the experimental results obtained for microemulsions. One of the main conclusions of these theories is that in dense liquids the spatial structure, which can be represented either by the function of pair distribution $g(r)$ or by the thermodynamic properties, is in a large extent determined by steric repulsions between close particles. The attractive or repulsive effects are treated as perturbation of hard spheres. Calje et al.[3] consider that the intermicellar interaction potential in W/O microemulsions can be expressed as the sum of two terms U_{HS} and U_A. The hard sphere contribution to the osmotic pressure Π_{HS} is described by the equation of state proposed by Carnahan and Starling[12]. Only binary interactions are considered in the perturbation term Π_A. The total osmotic pressure of the solution can be written as $\Pi = \Pi_{HS} + \Pi_A$ with

$$\Pi_{HS} = \frac{k_B T}{v_{HS}} \frac{1 + \phi_{HS} + \phi_{HS}^2 - \phi_{HS}^3}{(1 - \phi_{HS})^3}$$

$$\Pi_A = \frac{k_B T}{v_m} \frac{A}{2} \phi^2$$

$$A = \frac{4\pi}{k_B T} \frac{1}{v_m} \int_{2R_{HS}}^{\infty} U_A(r) \, r^2 dr$$

$U_A(r)$ is the perturbation to the hard sphere potential $U_{HS}(r)$. ϕ_{HS} is the volume fraction of hard spheres of radius R_{HS}. ϕ_{HS} is related to ϕ by $\phi_{HS}/\phi = a$. The second virial coefficient deduced from the whole expression of Π is written as $B' = 8a + A$. In this model, the scattered intensity is :

$$I(\phi) = Kv_m \frac{\phi(1-a\phi)^4}{1+4a\phi+4a^2\phi^2-4a^3\phi^3+a^4\phi^4+A\phi(1-a\phi)^4} \tag{3}$$

The fit of the experimental $I(\phi)$ curve by a least square method allows one to determine three parameters, Kv_m, a and A from which we can deduce the micellar radius and the second virial coefficient B'.

Moreover, the autocorrelation function $g^{(2)}(\tau)$ of the scattered light is given by[13] :

$$g^{(2)}(\tau) = 1 + e^{-2Dq^2\tau}$$

where D is the translational diffusion coefficient and q the wave vector. D is related to osmotic pressure by $D = v_m/f \, \partial\Pi/\partial\phi$ where v_m is the volume of the micelle and f the friction coefficient between micelle and continuous phase.

In the low concentration range, D can be written as[7]

$$D \sim D_o(1 + \alpha\phi) \text{ with } D_o = \frac{k_B T}{6\pi\eta R_H} \qquad (4)$$

η is the viscosity of the continuous phase, R_H the hydrodynamic radius of the micelle. The virial coefficient α is related to that of the osmotic pressure by the equation

$$\alpha = B - \beta$$

β represents the dynamic part which takes into account the volume fraction dependence of the friction coefficient f. Both the static and dynamic contributions of α are related to the interaction potential of particles U(r). The expression for B is well established in the case of rigid spherical particles of radius R_{HS} with a pair interaction potential U(x) where $x = r/R_{HS}$ and r is the distance between the centers of the two particles.

$$B = 8 + \frac{24}{R_{HS}^3} \int_{R_{HS}}^{\infty} \left(1 - e^{-\frac{U(r)}{k_B T}} \right) r^2 \, dr$$

where 8 is the hard sphere contribution.

On the contrary, there is still some discrepancy in the calculations for β presented by different authors[7-9]. However, in the case, where $U(r) = U_{HS}(r) + U_A(r)$ it is possible to write down a relation for β which has the same structure as the preceeding one, that is

$$\beta = \beta_o + \int_{R_{HS}}^{\infty} F(x) \left(1 - e^{-\frac{U(x)}{kT}} \right) dx$$

where β_o is the hard sphere contribution.

A complete treatment has been given by Feldherof[7] who obtained :

$$\beta_o = 6.44$$

and $$F(x) = 12x - 15/(8x^2) + 27/(64x^4) + 75/(64x^5) \tag{6}$$

Others treatments have been proposed by Batchelor[8] and Goldstein and Zimm[9]. As the α values derived from these three calculations are very close, so in the following we will present result obtained with the Feldherof's equation (Equation 6) only.

III – EXPERIMENTAL PART

a – Sample preparation
The samples studied were quaternary mixtures of water, oil, alcohol and surfactant. N-dodecane (D), N-octane (O), isooctane (trimethyl 2-2-4 -pentane) (I), and cyclohexane (C) were used as oil ; 1-pentanol (C5), 1-hexanol (C6), 1-heptanol (C7) as alcohol ; sodium dodecyl sulfate (S) and sodium methyl-1 dodecyl sulfate (M) as surfactant. For the systems containing SDS and dodecane, several microemulsions of water/surfactant ratio (W/S expressed in volume) ranging from 1.74 to 3.50 were studied in order to obtain size variation. All the studied systems are located on the boundary between one and two phase regions. They are designated in an abridged form by the surfactant followed by alcohol and oil ; for example a microemulsion formed with SDS, pentanol and dodecane is designated as S-C5-D. SDS of quality puriss was purchased from Touzart and Matignon ; sodium α-methyl dodecyl sulfate was synthetized in our laboratory according to the method of the ref. 14. The other compounds were Fluka products. The overall composition of the systems studied are given in Table I and reference 5.

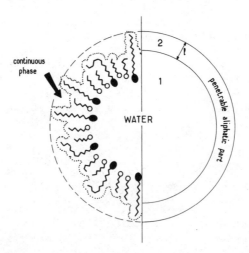

Figure 1. Schematic representation of a W/O microemulsion.

Table I. Volume Compositions of Microemulsions Made with Different Oils, and with α-Methyl-SDS.

	$S-C_6-O$	$S-C_6-C$	$S-C_6-I$	$S-C_6-D$	$M-C_6^1-D$	$M-C_6^2-D$	$M-C_7^1-D$	$M-C_7^2-D$
W/S	2.55	2.55	2.55	2.55	1.75	2.55	2.75	2.55
Water	19.25	21.36	19.33	19.56	13.78	20.51	15.93	20.43
Oil	55.19	55.90	56.46	52.94	65.59	55.47	63.73	55.24
Surfactant	7.51	8.33	7.54	7.66	7.49	8.05	5.80	8.02
Alcohol	18.05	14.41	16.67	19.83	13.14	15.96	14.53	16.31

A schematic description of the W/O studied droplets is given
in Figure 1. The water cores are surrounded by a mixed film of surfactant and alcohol molecules. These micelles are dispersed in a
complex solvent named "continuous phase". The continuous phase
contains primarily oil and alcohol but also a small amount of
water. Analysis of light scattering data in terms of size and interaction requires an extrapolation of the results to zero concentration, therefore it is necessary to use a dilution procedure
which keeps constant both size and composition of micelles. Composition of the continuous and dispersed phases of the microemulsions studied have been determined by using the dilution procedure
described by Graciaa et al.[2]. Validity of such a method has been
checked by Taupin et al.[15] by means of neutron scattering. Indeed
variable contrast gives information about the internal structure
of the object. It has been shown that structure and composition of
the elementary droplet is unchanged up to a water volume fraction
equal to 0.3 in microemulsions where attractive forces are not
very strong.

The volume fraction of the micelles has been defined as :

$$\phi = \frac{V_A^M + V_W^M + V_S}{V}$$

Where V is the total volume, V_S the volume of surfactant, and
V_A^M and V_W^M are respectively the volumes of alcohol and water contained in the micelles. Solutions of volumic micellar fractions
ranging from 0.01 to 0.30 have been prepared. One observes a kinetic effect in systems for which the second virial coefficient is
positive (that corresponds to the less attractive interactions,
Tables II and III). Indeed in these cases, the solution becomes
clear only a few hours after mixing of the components. Compositions of the continuous and dispersed phases are reported in
Tables II and III. They are characterized by the molecular alcohol-oil ratio A^C/oil. It is known that the continuous phase penetrates in the micelle[16] ; usually one assumes that its composition is not changed. With this assumption it is possible to calculate the alcohol volume contained in the micelle : as the
alcohols used are very slightly soluble in water, so we consider
that the alcohol contained in the micelles A^M is only located at
the interface. Hence the composition of the interface is defined
by the molecular alcohol-surfactant ratio (Tables II and III).

For a given alcohol, as the W/S ratio increases the alcohol
concentration in the continuous phase increases whereas in the
interface this concentration decreases. For a given W/S ratio, the
alcohol concentration in the continuous phase depends on the alcohol and oil. The A^C/oil ratio decreases as the alcohol chain

length increases and as the oil chain length decreases. Microemulsions formed with branched surfactants contain less alcohol than
those formed with SDS ; in these systems the decrease of the alcohol concentration is much more marked in the continuous phase
than in the interface. This means that the quantity of surfactant
molecules necessary to solubilize a given amount of oil and water
is less with Me-SDS than with SDS.

b - Method
The various liquid components of the microemulsions (oil,
water and alcohol) were first filtered using fine sintered glass
∿ 1.4 μm) before preparation of the samples and subsequently the
solutions were centrifugated at 5000 rpm for 30 mn. Refractive
indexes were measured using a Pulfrich refractometer. The usual
sin θ correction was made to allow for the angular variation of
the size of the scattering volume. Correction for the solid angle
was also carried out. The angular range studied was $30° < θ < 150°$.

The intensity and the correlation function of the scattered
light were successively measured with a laser beam (Argon ion
laser, Spectra Physics Model 165, $λ_o$ = 5145 A). All the static and
dynamic measurements were made at $21.5 ± 0.5$ C.

Measurements of the viscosities of the continuous phases were
carried out using an improved Oswald-like viscometer.

In the determination of the dilution line, transition from
turbid polyphasic state to transparent one-phase system is visual-
ly observed. However this observation becomes very difficult in
the low concentration range. Moreover we have observed that in some
cases scattered intensity varies largely in the vicinity of the
demixing line. Therefore, all the samples investigated were prepared in the photometer according to the procedure described in the
previous paper[5].

IV - LIGHT SCATTERING RESULTS

We have measured the scattered intensity I_{90} (which is expressed as the Rayleigh ratio in cm^{-1}) and the diffusion coefficient
at θ = 90° for various microemulsions as a function of volume fraction. Figures 2 and 3 show examples of the intensity and diffusion
coefficient variation versus φ for various microemulsions. Variations of both I(φ) and D(φ) are strongly dependent on the alcohol,
the surfactant and the oil. All the I(φ) curves show a maximum for
a certain $φ_{max}$ value. D(φ) curves for most of the microemulsions
studied show a minimum. However for the microemulsions S-C7-D the
translational diffusion coefficient is found to be independent of
φ.

Data analysis shows that the observed differences of intensity are due to the following causes :
  i) variation in size and interactions
 ii) vicinity of a critical point
iii) variation in the increment of refractive index, this latter varies between 0.077 in the pentanol-dodecane system and 0.009 in the hexanol-isooctane microemulsion.

Besides we have measured the angular dependence of the V_v component of the scattered light and of the D coefficient at different volume fractions. It appears that for all the systems studied I and D are independent of the wave vector in the low concentration range ($\phi < 0.04$). For most of the concentrated microemulsions, I and D

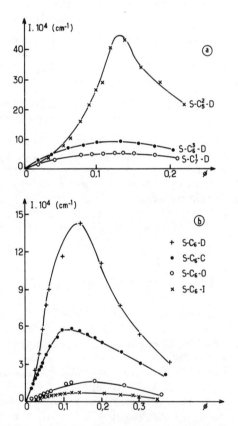

Figure 2. Scattered intensities at $\theta = 90°$ versus the micellar volume fraction. a) influence of alcohol and surfactant ; b) influence of oil.

remain independent of q. However in the case of the S-C5-D and
M-C6-D microemulsions which exhibit a strong variation of I and D
versus ϕ, an angular dependence is observed in the concentration
range around the extremum of I and D. These variations are related
to a critical behavior of these microemulsions[18]. Similar behavior
has been found recently in the three-phase microemulsion systems[19].

Figure 4 shows plots of ϕ/I versus ϕ in the low concentration
range ($\phi < 0.1$) for different microemulsions. Analysis of these
data allows one to derive, by extrapolation at infinite dilution,
(eq. 2) the apparent radius R of the micelle and the second virial
coefficient B (Tables II and III). Besides, for the microemulsions
for which I is independent of q, these two values R and B have also
been determined by analysis of the whole I(ϕ) curve by using equation 3 proposed by Calje et al.[3]. The corresponding values R' and
B' are summarized in Tables II and III. In most cases values of R
and B obtained by the two methods agree within limit of experimental accuracy. However this agreement is only obtained for volume
fractions $\phi < 0.3$. The hydrodynamic radius of the micelle R_H and
the coefficient α are extracted from analysis of the variation of
D versus ϕ. The values obtained are given in Tables II and III.

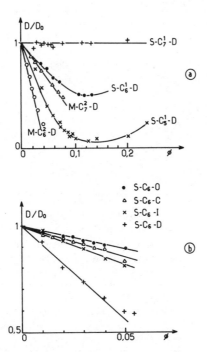

Figure 3. a) influence of alcohol
and surfactant ; b) influence of
oil on diffusion coefficient.

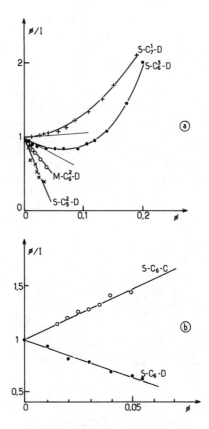

Figure 4. Variation of ϕ/I versus ϕ for various microemulsions
a) influence of alcohol and surfactant ; b) influence
of oil,ϕ/I has been normalized to 1 for $\phi = 0$.

Table II. Effect of Alcohol and W/S Ratio on the Sizes and Virial Coefficients of Microemulsions Formed with SDS and Dodecane. R and B Are Determined by the Extrapolation Method ; R' and B' by the Calje et al' method. a) Volume Ratio of Total Water to Surfactant ; b) Molecular Ratio of Alcohol to SDS in the Micelle ; c) Molecular Ratio of Alcohol to Dodecane in the Continuous Phase.

	$S\text{-}C_5^1\text{-}D$	$S\text{-}C_5^2\text{-}D$	$S\text{-}C_6^1\text{-}D$	$S\text{-}C_6^2\text{-}D$	$S\text{-}C_6^3\text{-}D$	$S\text{-}C_6^4\text{-}D$	$S\text{-}C_7^1\text{-}D$	$S\text{-}C_7^2\text{-}D$	$S\text{-}C_7^3\text{-}D$
W/S [a]	2.55	1.74	1.75	2.32	2.90	3.50	2.55	2.74	3.48
A^M/S [b]	2.99	2.71	2.58	2.8	2.61	1.9	2.69	2.7	2.82
A^C/C_{12} [c]	0.40	0.29	0.24	0.263	0.34	0.454	0.22	0.25	0.34
R (Å)	50 ± 6	62 ± 7	47.6 ± 1	64 ± 4	66 ± 1	70 ± 2	54	62 ± 2	66 ± 3
R' (Å)			48 ± 1	63 ± 5	66 ± 1	74 ± 2	56	61 ± 2	65 ± 3
B	-23 ± 3	-27 ± 3	-0.5 ± 0.5	-4.4 ± 1	-6.1 ± 1	-8.8 ± 2	$+6$	4 ± 2	3 ± 3
B'			-1.5 ± 0.5	-3.8 ± 1	-6 ± 1	-9.2 ± 2	$+5$	3 ± 2	2 ± 3
R_H (Å)	55 ± 10	65 ± 10	52 ± 5	67 ± 10	69 ± 10		61	73 ± 2	75 ± 10
α	-18 ± 3	-21 ± 3	-5.8 ± 1	-9.7 ± 1	-12 ± 1		-1	$\sim 0 \pm 1$	$\sim 0 \pm 1$

Table III. Effect of Oil and Surfactant on the Virial Coefficients.

	S-C_6-O	S-C_6-C	S-C_6-I	S-C_6-D	M-C_6^1-D	M-C_6^2-D	M-C_7^1-D	M-C_7^2-D
W/S [a]	2.55	2.55	2.55	2.55	1.75	2.55	2.55	2.75
A/S [b]	2.45	1.68	2.46	2.77	2.25	2.45	2.4	2.44
A/O [c]	0.306	0.114	0.17	0.36	0.14	0.22	0.16	0.17
R		55.6	68.	62.	54.	75.	61.	57
R'		54.	68.	65.	56.		60.	58.
R_H	61.5	61.	61.	65.		71.	64	
B	~0.	10.	4.	-10.	-13.	-16.	-4.	-9.
B'		4.5	5.	- 5.	- 9.		-3.	-4.
α	- 2.	-3.	-4.	- 8.		-20.	-7.	

Comparison of results presented in Tables II and III shows that the droplet size is mainly dependent on the W/S (water to soap ratio). The radius increases with this ratio. Tables II and III also show that the apparent radius is not very much affected neither by the alcohol nor by oil, but seems to depend on the surfactant. The B and α values vary similarly, they are strongly dependent on the molecular structure of the components. Experimental results obtained for the various alcohols studied in the S-D series exhibit very different behaviors ; the values obtained are largely negative for pentanol, negative for hexanol and positive for heptanol. One concludes from these data that interactions are strongly attractive in microemulsions containing pentanol and much less in hexanol and heptanol systems. Indeed, larger attractive potential leads to a more negative second virial coefficient. Besides, for a given alcohol, an increase of the droplet size seems to induce a decrease in the second virial coefficient. Thus, for example, for the S-C6 series, the B values become more negative as the W/S ratio increases ; these two effects are acompanied by an increase of the alcohol content in the continuous phase. Interactions are also strongly dependent upon the molecular structure of oil. The values obtained are positive for cyclohexane, close to zero for octane and isooctane, and negative for dodecane. Finally our results show that the introduction of a methyl group close to the polar head of SDS leads to a strong decrease in the B and α values.

V - INTERMICELLAR POTENTIAL

Modern theories of liquids have been successfully used by several authors to explain the first light scattering results on inverse microemulsions. In particular Calje et al.[3] have shown that the data obtained can be interpreted by a model of spherical particles in interaction. Interactions in microemulsions can vary over a very large range ; indeed our results show that it is possible to go from very strong attractive interactions to much weaker ones by changing the molecular structure of the microemulsion components (alcohol, oil or surfactant). Then the question is what is the origin of these interactions. Calje et al.[3] have proposed that these interactions are the result of the difference in the molecular compositions between the micelles and the continuous medium. The calculation for homogeneous spheres dispersed in a solvent has been carried out by Hamaker[20]. Application of these calculations to microemulsions leads to a very weak intermicellar potential when realistic values are assumed for the Hamaker constant[3]. Therefore in order to explain this inconsistency Vrij suggested to take into account a possible interpenetration of the micelles. This assumption is supported by the structure of the interface ; indeed the presence of alcohol molecules allows the penetration either of the continuous phase when the micelle is isolated or of the surfactant chain of another micelle during a

collision (Figure 5). A previous calculation has shown that the
most important values of the attractive interactions are obtained
in the overlapping region[6]. Therefore in the following we first
evaluate the interaction potential caused by the interpenetration
of two micelles and in a second step we will examine if such a potential allows satisfactory explanation for the experimental values
of B and α.

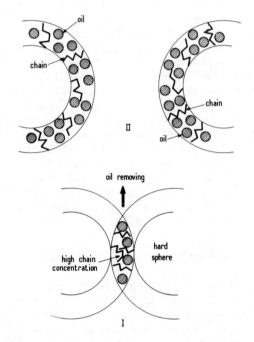

Figure 5. Schematic representation of :
            I   two interpenetrating micelles
            II two isolated micelles.

Calculation of the interaction potential between W/O micelles.

We will consider the case where two micelles overlap ; the
case of non-penetrating micelles can be calculated according to the
Hamaker formula. Following the principle of the Hamaker calculation
one has to estimate the internal energy difference between states I
and II :

- In state I the two micelles are located at a distance r
  (r < 2R).
- In state II the two micelles are located at infinite distance from one another. In this case the oil of the continuous phase replaces the surfactant chains (Figure 5).

The internal energy of each state X (X is I or II) can be written.

$$U^X = \sum_i \int_v \rho_i^X \, dv' \sum_j \int_v U_{ij}(d) \, g_{ij}(d) \, \rho_j^X \, dv$$

Where i and j are relative to the different molecules present, U_{ij} is the interaction potential between two molecules i and j separated by a distance r, $g_{ij}(r)$ is the radial correlation function between these two molecules, and ρ_i the number of molecules i per unit volume. Although calculation of this integral is difficult, an attempt has been recently been made[6]. However a certain number of simplifications allows one to obtain a very simple result.

We only calculate the part of the internal energy relative to the penetration volume ; let $V_o(r)$ be this volume and let consider ρ_i^X as a constant then

$$U^X = V_o(r) \sum_i \rho_i^X \sum_j \rho_j^X \int_{V_o(r)} U_{ij}(d) \, g_{ij}(d) \, dv$$

Consequently one can calculate the interaction potential as the difference between states I and II so :

$$U(r) = V_o(r) \left[\sum_i \rho_i^c \sum_j \rho_j^c \, S_{ij} + \sum_i \rho_i^P \sum_j \rho_j^P \, S_{ij} - 2\sum_i \rho_i^{is} \sum_j \rho_j^{is} S_{ij} \right]$$

$$\tag{7}$$

$$S_{ij} = \int_{V_o(r)} U_{ij}(d) \, g_{ij}(d) \, dv$$

ρ_i^{is} is the density number of the molecules i in the penetrable part of an isolated micelle (state II) ; ρ_i^P is the density number in the overlapped part (state I) and ρ_i^c is the density number in the continuous phase.

If one assumes that the potential $U_{ij}(d)$ is short ranged relative to the size of the volume considered (this potential is typically in $1/d^6$) then one can consider that the integrals S_{ij} are independent of r. If $\rho_i^c = \rho_i^P = \rho_i^{is}$ for all species i $U(r)$ is identically equal to zero. The existence of an interaction potential is due to the difference between the atomic densities in the continuous phase and in the penetrable part of the amphiphilic layer. A more accurate calculation[6] has shown that a small increase of about 1 % of the CH_2 molecular volume can explain such an interaction. If ones supposes that S_{ij} is independent of r the term between brackets in equation 7 can be considered as constant and it is designated $\Delta\rho kT$. Then $U(r)$ is proportional to the volume of penetration $V_o(r)$:

$$U(r) = -V_o(r) \cdot \Delta\rho \cdot kT$$

$\Delta\rho$ is a constant depending only on the compositions of the interface and the continuous phase. $V_o(r)$ can be calculated as a function of the radius of the micelle and the distance r between the micelles.

$$V_o(r) = -\frac{\pi}{6}(2R-r)^2(2R + \frac{r}{2})$$

The presence of alcohol in the micellar interface limits the interpenetration of the micelles ; therefore we assume that penetration occurs over a thickness equal to the difference between the lengths of the entirely stretched alcohol and surfactant molecules $\ell = 1.26(n_S-n_A)$ Å ; n_S and n_A are the numbers of carbon in the alcohol and surfactant molecules ; so in the systems SDS/pentanol, SDS/hexanol, and SDS/heptanol, ℓ equals 8.82, 7.56 and 6.30 Å respectively.

Then one can write :

$$U(r) = 0 \qquad\qquad\qquad\qquad\qquad r > 2R_H$$

$$U(r) = -\Delta\rho\, kT\, \frac{\pi}{6}(2R-r)^2\,(2R + \frac{r}{2}) \qquad 2R_H - \ell < r < 2R_H$$

$$U(r) = +\infty \qquad\qquad\qquad\qquad\quad r < 2R_H - \ell$$

In light of the approximations made in this calculation, this potential is realistic only if $\Delta\rho$ is kept constant for various microemulsions made with the same components. Therefore, $\Delta\rho$ has to be independent of the micellar size. The change of the alcohol chain length is only a geometrical effect, which is included in the ℓ value. Then, $\Delta\rho$ will be not affected by the alcohol chain length. In practice, one looks for the $\Delta\rho$ value which allows to interpret different experiments involving several micellar sizes and alcohols. Then using this $\Delta\rho$ value and the Feldherof formula (eq. 6), we calculate α. In both cases we have carried out numerical integration.

As a first step we have compared the experimental and calculated values for the 9 microemulsions formed with SDS and dodecane. The $\Delta\rho$ value has been found to be equal to $7.1\ 10^{-4}$ Å^{-3}, the experimental and calculated values are given in Table IV ; these results are plotted in Figure 6. An excellent agreement is obtained for both the α and B values. Therefore the proposed potential allows us to interpret very satisfactorily both the static and dynamic results. Besides, the hypothesis that $\Delta\rho$ is constant for a given oil is shown to be valid. In order to interpret results obtained in part II of this work with different oils and also some data published in literature for various systems containing cyclohexane and toluene[4] we have applied the intermicellar potential described above. Figure 7 shows results of the comparison and

Table IV. Comparison of the Experimental (B, B', α) and Calculated
Values (B_{th}, α_{th}) of the Virial Coefficients α and B.

	R_H	B	B'	α	B_{Th}	α_{Th}
$S-C_5^1-D$	55	-23.		-18.	-15.	-19.
$S-C_5^2-D$	65	-27.		-21.	-29.	-33.
$S-C_6^1-D$	52	-0.5	-1.5	-5.8	-1.3	-6.2
$S-C_6^2-D$	67	-4.4	-3.8	-9.7	-5.1	-10.3
$S-C_6^3-D$	69	-6.1	-6.	-12.	-6.	-11.4
$S-C_6^4-D$	76	-8.8	-9.2		-8.6	-13.
$S-C_7^1-D$	61	7.	5.	- 1.	3.5	-1.8
$S-C_7^2-D$	73	4.	3.	$\sim$ 0.	2.5	-2.5
$S-C_7^3-D$	76.	3.	2.	$\sim$ 0.	2.	-3.

Table V. Comparison Between Values of $\Delta\rho$ for Different Oils and
Their Molecular Volume.

	$\Delta\rho$ $\overset{\circ}{A}^{-3}$	Molecular Volume $\overset{\circ}{A}^3$
Dodecane	$7.1\ 10^{-4}$	377.
Octane	$5.5\ 10^{-4}$	270.
Isooctane	$5.5\ 10^{-4}$	273.
Cyclohexane	$4.6\ 10^{-4}$	179.
Toluene	$4.2\ 10^{-4}$	176.

Table V summarizes the values determined for $\Delta\rho$. Let us point out that numerous experimental data for the cyclohexane and dodecane systems are available, this permits one to obtain $\Delta\rho$ with a good accuracy ; however this is not the case for other oils.

The parameter $\Delta\rho$ appears very sensitive to the oil molecule. A very interesting comparison can be made between the values of $\Delta\rho$ and the molecular volume of oil (Table V). These two values are strongly correlated, indeed an increase in the interactions corresponds to an increase of the molecular volumes. We have previously seen that $\Delta\rho$ is sensitive to the difference of densities between the aliphatic layer and the continuous phase. One supposes that this difference is increased with the difficulty for oil to penetrate the aliphatic layer. This difficulty is related to the oil molecular volume.

The preceding results show that the branching of SDS produces two main effects : first a decrease of the amount of alcohol in the interface and secondly an increase of the attractive interactions. One can suppose that the polar head area of the branched surfactant is larger than that of the SDS molecule ; the accuracy of our data is not sufficient to measure this effect. However this assumption is consistent with the decrease of the alcohol content in the interface and with what is generally observed in monolayers. Therefore the micelles formed with the methyl-SDS molecule are probably less compact than those formed with SDS and the strengthening of the attractive forces could result from an increase of the volume of penetration. It is possible to explain this behavior with the interaction potential as an increase of the value of the length of penetration ℓ.

VI - CONCLUSIONS

The results presented in this study clearly show that interaction forces in W/O microemulsions are very sensitive to the molecular structure of the components and to the chemical compositions of the continuous and dispersed phases. Indeed the experimental values of the second virial coefficient B range from -27 to + 10. Our data show that considerable variations of B are realized by changing each component. In order to analyse the intensity data in the low concentration range we have propose an intermicellar potential. In this potential, the most important values of the attractive interactions appear in the overlapping region and results from short interatomic attractive interactions. The attractive energy is proportional to the volume penetrated and to one parameter $\Delta\rho$ which depends only on oil and surfactant. The contribution for the non-penetrated volumes corresponds to the classical Hamaker's contribution, which is negligible in these systems. The behaviors of calculated and experimental virial coefficients are quite similar ; therefore the proposed potential allows us to

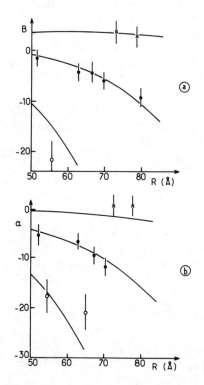

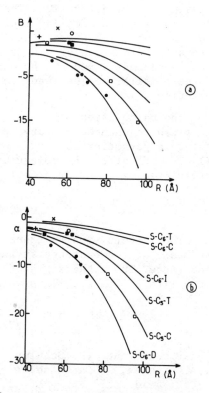

Figure 6. Comparison of the calculated (full line) and experimental B and α values for the microemulsions S-Alcohol-D.
o pentanol, ● hexanol, x heptanol. The value of Δρ is kept equal to 7.1 10⁻⁴ Å⁻³.

Figure 7. Comparison of the calculated (full line) and experimental B and α values for microemulsions made with different oils : ● dodecane-hexanol ; o cyclohexane-hexanol ; □ cyclohexane-pentanol ; ▪ isooctane-hexanol and octane hexanol ; + toluene-pentanol, x toluene-hexanol (from ref. 4).

account for the experimental data obtained in the low concentration range. In particular, the effect of alcohol chain length and micellar size are well represented. The calculation provides an approach for the understanding of the variations of the interactions. Indeed variations in the volume penetrated can be obtained by two different means : either by changing the alcohol chain length or by changing the micellar size. The alcohol effect is the principal one ; for a given micellar radius, as the alcohol chain length decreases, the thickness of the penetrated layer is increased and the interaction is stronger. The second effect relative to the micellar size is secondary and can be explained in the same way. An increase of the micellar radius leads to an increase of the volume of interaction.

The calculation of the potential indicates also that interactions can be changed by varying the $\Delta\rho$ parameter. Our results show that this parameter is very sensitive to the molecular volume of oil. The effect of the molecular structure of the surfactant can be interpreted as a larger penetration of the aliphatic layer.

ACKNOWLEDGEMENTS

The authors wish to thanck Mrs Dubien and Dupart for the synthesis of the branched surfactant and Mrs Maugey for technical assistance. The authors gratefully acknowledge valuable discussions with B. Lemaire. This work was supported by a DGRST contract.

REFERENCES

1. T.P. Hoar and J.H. Schulman, Nature 152, 102 (1943).
2. A. Graciaa, J. Lachaise, A. Martinez, M. Bourrel and C. Chambu, C.R. Acad. Sci. Paris B 282, 547 (1976).
3. A.A. Calje, W.G.M. Agterof and A. Vrij, in "Micellization, Solubilization and Microemulsions" K.L. Mittal, Editor, Vol. 2, p. 779, Plenum Press, New York (1977).
4. A.M. Cazabat and D. Langevin, J. Chem. Phys. 74(6), 3148 (1981).
5. S. Brunetti, D. Roux, A.M. Bellocq, G. Fourche and P. Bothorel, J. Phys. Chem., in press.
6. B. Lemaire, P. Bothorel and D. Roux, Part. I, J. Phys. Chem., in press.
7. B.U. Feldherof, J. Phys. A 11, 929 (1978).
8. G.K. Batchelor, J. Fluid Mech. 74, 1 (1976).
9. B. Goldstein and B.H. Zimm, J. Chem. Phys. 54, 4408 (1971).
10. D.P. Riley and G. Oster, Disc. Faraday Soc. 11, 107 (1951).
11. L.S. Ornstein and F. Zernike, Proc. Acad. Sci. (Amsterdam) 17, 793 (1914) ; Physik Z 19, 134 (1918) ibid 27, 761 (1926).
12. N.F. Carnahan and K.E. Starling, J. Chem. Phys. 51, 635 (1969).

13. B.J. Berne and R. Pecora, in "Dynamic Light Scattering" Wiley, New York (1976).
14. E.E. Drager, K.I. Keim and G.D. Miler, Ind. Eng. Chem., $\underline{36}$, 316 (1944).
15. M. Lagues, R. Ober and C. Taupin, J. Phys. (Paris) Lett. $\underline{39}$, L-487 (1978).
16. M. Dvolaitzky, M. Guyot, M. Lagues, J.P. Lepesant, R. Ober, C. Sauterey and C. Taupin, J. Chem. Phys. 69, 3279 (1978).
17. G.E.A. Aniansson, J. Phys. Chem. $\underline{82}$, 2805 (1978).
18. G. Fourche, A.M. Bellocq, S. Brunetti, J. Colloid Interface Sci., in press.
19. A.M. Bellocq, D. Bourbon, B. Lemanceau and G. Fourche, J. Colloid Interface Sci., in press.
A.M. Cazabat, D. Langevin, J. Meunier, A. Pouchelon, Adv. Colloid Interface Sci., in press.
20. H.C. Hamaker, Physica $\underline{IV}$, 1058 (1937).

THE IMPORTANCE OF THE ALCOHOL CHAIN LENGTH AND THE NATURE OF THE HYDROCARBON FOR THE PROPERTIES OF IONIC MICROEMULSION SYSTEMS

Eva Sjöblom[1] and Ulf Henriksson[2]

[1]The Institute for Surface Chemistry,
P.O. Box 5607, S-114 86 Stockholm, Sweden
[2]Department of Physical Chemistry, The Royal Institute
of Technology, S-100 44 Stockholm, Sweden

The properties of microemulsions stabilized by an
ionic surfactant and an alcohol depend on the alcohol
chain length as well as on the nature of the hydrocarbon. According to the pseudo-phase model (J. Biais *et al.*
J. Dispersion Sci. Technol. 2, 67 (1981) the microemulsion can in many cases be considered as consisting of
an aqueous, a nonaqueous and an interfacial region in
internal equilibrium. The nonaqueous region consists
mainly of alcohol and hydrocarbon, and the thermodynamic
nied by an increased counterion binding and a more
therefore,of relevance. These systems have been studied
by ^{1}H NMR, calorimetry and vapor pressure measurements.
It is found that the interaction between the hydroxyl
group of the alcohol and the π-electrons in aromatic
hydrocarbons reduces the alcohol chemical potential as
well as the degree of self-association. This implies
that as the hydrocarbon content is increased the alcohol is redistributed from the aqueous and interfacial
regions toward the nonaqueous region. The ^{23}Na NMR and
the self-diffusion data indicate that this is accompanied by an increased counterion binding and a more
well-organized hydrophilic/hydrophobic interface offering an effective barrier towards translational diffusion of the water molecules.

INTRODUCTION

The properties of microemulsions containing water, ionic surfactant, alcohol and hydrocarbon depend significantly on the alcohol chain length and on the nature of the hydrocarbon. Figure 1
shows that the existence region of ternary microemulsions drastically decreases when the alcohol chain length is increased. Light
scattering[1], self-diffusion[2] and ^{23}Na NMR[3] show clear differences
between quaternary microemulsions containing aliphatic and aromatic
hydrocarbons. As shown by Biais *et al.*[4] the self-association and
hence the chemical potential of the alcohol is important for the
properties of microemulsions. Therefore, we have found it worthwhile
to investigate the thermodynamic properties of binary hydrocarbon/
alcohol mixtures. In this paper we report results from ^{1}H NMR, calorimetry and vapor pressure measurements, which are discussed in
relation to the properties of microemulsions stabilized by oleate/
pentanol.

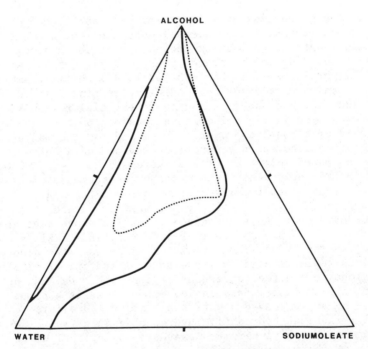

*Figure 1.　Partial phase diagram at 25°C showing the stability
region for ternary microemulsions in the systems
water-sodium oleate-alcohol. The alcohols are
butanol ————— and pentanol .... The phase
diagram is given in weight percent.*

EXPERIMENTAL

Chemicals: Sodium oleate was prepared by titration of sodium ethoxide with oleic acid in ethanol solution[5]. The water was distilled twice. *n*-butanol (Fluka A.G. 99%), *n*-pentanol (Fluka A.G. 99.5%), *n*-decanol (Fluka A.G. 99%), benzene (Merck A.G. 99.5%), cyclohexane (Fluka A.G. 99%) and *n*-decane (Merck A.G. 99%) were dried over anhydrous sodium sulphate.

Methods: The ^{1}H chemical shifts were measured at 200 MHz and 25°C using a Bruker WP-200 spectrometer. The shifts are given to higher frequency relative to TMS in chloroform and have been corrected for the difference in bulk susceptibility between the samples and the reference. The ^{23}Na NMR-spectra were recorded at 27°C on a Bruker CXP-100 spectrometer operating at 23.8 MHz. The self diffusion coefficients were measured at 23°C by the pulsed gradient FT-NMR method[6] on a Jeol FX-100 spectrometer. The alcohol vapor pressures were measured at 20°C by means of a Carlo Erba 4200 gas chromatograph equipped with a head space sampler HS 250. The enthalphy of mixing between alcohol and hydrocarbon were determined at 25°C using a LKB 10700 Batch Microcalorimeter.

RESULTS AND DISCUSSION

Models for the self-association of alcohols: The self association of alcohols in inert solvents is a complex problem since many different molecular species coexist in the solution and many independent equilibrium constants must be used to describe the system adequately. However, experimental data permit only a few equilibrium constants to be determined and it is therefore necessary to adopt some simplified model for the association equilibria. The simplest model for the self-association of alcohols in inert solvents is the so called *1-n*-model in which it is assumed that only two species, the monomers and the *n*-mers are present. If the *n*-mers are cyclic, all hydroxyl protons participate in hydrogen bonds. For the case of association to open chain polymers, one of the hydroxyl protons in the associated species is not hydrogen bonded. Association to cyclic or to open chain *n*-mers thus affects the experimentally measured quantities differently. In solutions of *n*-alcohols in aliphatic hydrocarbons, results from IR-spectroscopy[7], calorimetry[8] and vapor pressure measurements[9] fit the *1-n*-model fairly well and indicate a predominance of tetramers or pentamers. The cyclic tetramer was reported to be slightly more stable than the linear tetramer[7]. A different approach, also with one single equilibrium constant, is the "continous association model"[10]. The equilibrium constant for the reaction

$$A_n + A_1 \rightleftarrows A_{n+1}$$

is assumed to be independent of *n*. This model gives a wide distri-

bution in the degree of self association and at low concentration
the dimers are the dominant associated species[10]. However, several
investigations indicate that in aliphatic hydrocarbons dimers are
present only in small amounts[7,11,12].

The simple models can be extended by introducing more equilibrium constants. For example, Tucker and Becker in their extensive
study of t-butanol in hexadecane found that the so called 1-3-∞-
model best describes the vapor pressure and the ^{1}H NMR data[11]. In
this model it is assumed that no dimers are present, the formation
of trimers is described by a separate equilibrium constant K_3
while the formation of species with higher degree of association
is described by a single equilibrium constant K_∞ in the same way
as in the continous association model.

[1]H Chemical Shifts: The chemical shift of hydroxyl protons
is very sensitive to hydrogen bonding and it has therefore been
frequently used in studies of association phenomena[13]. The exchange of hydroxyl protons between different sites is usually fast
and the observed chemical shift is an average

$$\delta_{obs} = \sum_i p_i \cdot \delta_i \qquad\qquad (1)$$

where δ_i is the chemical shift at site "i" with population p_i.

Figure 2 shows the observed hydroxyl proton chemical shift for
pentanol in three different hydrocarbons. It can be seen directly
from these results that the degree of self association is strongly dependent on the nature of the hydrocarbon and at a given mole
fraction of alcohol increases in the order benzene $\ll$ n-decane $<$
cyclohexane. On the other hand, as shown in Figure 3, the self-
association is practically independent of the alcohol chain length.

The purpose of this work is not to determine which model that
best describes the alcohol self-association but rather to use the
simplest reasonable model to investigate the association behaviour
of alcohols in relation to the properties of ionic microemulsion
systems. In a previous communication[14] we have shown that for pentanol in benzene and decane the 1-n-model and the 1-3-∞-model both
fit the observed hydroxyl chemical shifts for alcohol mole fractions $x_A < 0.25$. Furthermore, the calculated concentration of monomeric alcohol does not differ very much between the two models. In
this work we have used the simple 1-n-model for the interpretation
of the data in Figures 2 and 3. Thus we obtain the following
expressions for the variation in the observed hydroxyl chemical
shift as function of the total mole fraction of alcohol (x_A)

$$\delta_{obs} = (x_1 \delta_1 + n K_n x_1^n \cdot \delta_n)/x_A \tag{2}$$

$$K_n = \frac{x_n}{x_1^n} \tag{3}$$

$$x_A = x_1 + n \cdot x_n \tag{4}$$

where K_n is the equilibrium constant for the formation of a n-mer.

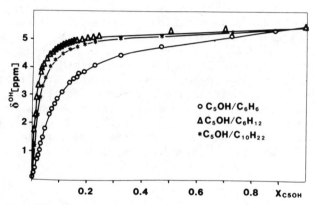

*Figure 2.   The chemical shift of the hydroxyl protons in pentanol
            as function of the total mole fraction of alcohol in
            binary solutions of pentanol/hydrocarbon. The hydrocar-
            bons are benzene (o), decane (*) and cyclohexane (Δ).*

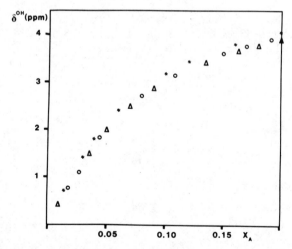

*Figure 3.   The chemical shift of the hydroxyl protons in alcohol
            as function of the total mole fraction of alcohol in
            binary solutions of alcohol/benzene. The alcohols are
            butanol (Δ), pentanol (o) and decanol (*).*

Equation (2) was fitted to the chemical shift data using the
program BMDPAR[15]. The calculations were performed for $x_A < 0.25$
in order to avoid too much variations in the activity factors.
The best values for the parameters δ_1, δ_n and K_n were determined
for different n-values. The results for the n-values giving the
best fit are presented in Table I. It can be seen that the simple
$1-n$-model fits the data quite well. For all three alcohols in
benzene $n = 3$ gives the best fit and the value of K_n is practically independent of the alcohol chain length. For the cyclohexane solutions the best n-value increases from 4 to 6 when passing from butanol to decanol. The concentration of monomeric alcohol calculated from the equilibrium constants in Table I is shown in Figure 4.
It is seen that the monomer concentration in cyclohexane solutions
is very similar for the different alcohols although they give different n-values. The low degree of self association in benzene solutions shows that the difference in free energy between associated and nonassociated alcohol is smaller in this solvent. As pointed out by several investigators [16,17] this indicates a specific
interaction between the alcohol hydroxyl group and the π-electrons
of the benzene ring which lowers the enthalpy of the monomer.

Table 1.   Results from Least Squares Fittings of the 1-n-Model
           to the [1]H Chemical Shift Data.

System	n	δ_1 (ppm)	δ_n (ppm)	K_n	RMSD (ppm)
Butanol/Benzene	3	0.24	5.87	124	0.002
Pentanol/Benzene	3	0.22	5.84	137	0.007
Decanol/Benzene	3	0.28	6.00	140	0.001
Butanol/Cyclohexane	4	0.50	5.40	$4.66 \cdot 10^5$	0.002
Pentanol/Cyclohexane	5	1.61	5.38	$1.08 \cdot 10^7$	0.003
Decanol/Cyclohexane	6	1.89	5.30	$4.37 \cdot 10^8$	0.004
Pentanol/Decane	4	0.62	5.51	$8.16 \cdot 10^4$	0.007

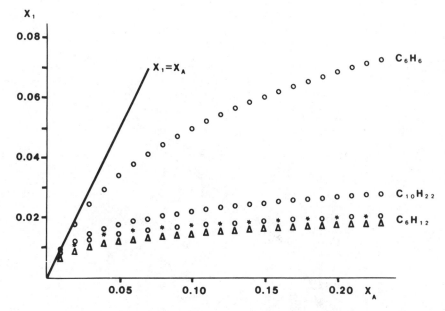

Figure 4.   The mole fraction of monomeric alcohol (x_1) as function
             of the total mole fraction of alcohol (x_A) calculated
             from the equilibrium constants in Table I. The alcohols
             are butanol (Δ), pentanol (o) and decanol (*).

Alcohol Vapor Pressure: Figure 5 shows the measured vapor pressure
of pentanol over benzene and n-decane solutions. At the same
mole fraction, the alcohol vapor pressure and thus the chemical potential is always higher in the decane solutions. These results
clearly reflect the difference in alcohol-hydrocarbon interaction
for aromatic and aliphatic hydrocarbons discussed above. At low
concentrations of alcohol, when the degree of self-association is
negligible, the theory of regular solutions[18] can be applied. For
a regular solution the limiting value of the Henry's law constant
K_H is given by

$$K_H = e^{\frac{\alpha}{RT}} \tag{5}$$

where α is the difference between the alcohol-hydrocarbon and the
average of the alcohol-alcohol and the hydrocarbon-hydrocarbon
interaction energy. From the limiting slope of the alcohol vapor
pressure curves in Figure 5 it is found that the interaction between the hydroxyl group of the alcohol monomer and the π-electrons
of the benzene molecule lowers the free energy of the alcohol by
~4 kJ/mole compared to the decane solutions.

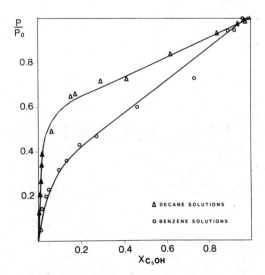

Figure 5. The relative vapor pressure of n-pentanol (P/p_0) as
           function of the total mole fraction of n-pentanol in
           solutions of benzene (o) and decane (Δ).

Calorimetry: The enthalpy of mixing, ΔH_m, between alcohol
and hydrocarbon includes contributions both from the breaking of
hydrogen bonds and from the alcohol-hydrocarbon interaction. As
seen in Figure 6a, ΔH_m is considerably smaller for cyclohexane
than for benzene solutions reflecting the higher degree of self
association in aliphatic hydrocarbons. However, for the lowest
alcohol concentrations where practically all alcohol is present
in the form of monomers, ΔH_m is smaller in the benzene solutions
as shown in Figure 6b. From the slope of the lines the value
5.5 kJ/mole is obtained for the difference in interaction energy
between the hydroxyl group of the alcohol monomers and the two
hydrocarbons, which is in reasonable agreement with the value
obtained from the Henry's law constant.

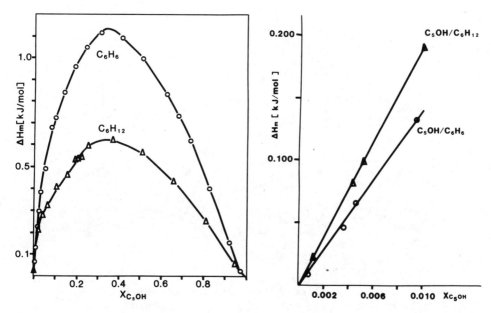

Figure 6 a, b.   The enthalphy of mixing (Δ H$_m$) as function of alcohol mole fraction for the binary systems pentanol/benzene (o) and pentanol/cyclohexane (Δ).

Implications for microemulsions: Microemulsions are thermodynamically stable one phase systems[19]. However, if the concentration of water in a w/o microemulsion is not too low the system can be treated as consisting of three pseudo-phases in equilibrium: an aqueous, a nonaqueous and an interfacial region containing the surfactant and variable amounts of the alcohol[4]. Since the system is in internal equilibrium, the chemical potential of a given component is equal in the different pseudo-phases. This means that a change in the composition in one of the pseudo-phases, *e.g.*, addition of hydrocarbon, affects the distribution of all components in the system.

In ternary microemulsions studied in this work, the nonaqueous pseudo-phase consists mainly of alcohol. Since added hydrocarbon is dissolved in this alcohol-rich region, the alcohol chemical potential is reduced. This causes a redistribution of the alcohol between the pseudo-phases. The thermodynamic properties of binary alcohol/hydrocarbon mixtures are therefore of relevance for the understanding of microemulsion systems. As seen in Figure 5

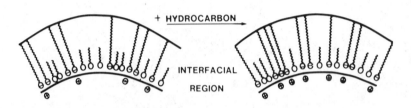

NONAQ. REGION   NONAQ. REGION

+ HYDROCARBON

INTERFACIAL

REGION

AQ. REGION   AQ. REGION

Figure 7.   Schematic illustration showing the division of a
             microemulsion into an aqueous, a nonaqueous and an
             interfacial region.

benzene is much more effective in reducing the pentanol chemical
potential than decane. This is, according to the discussion above,
due to an interaction between the hydroxyl group of the alcohol
and the π-electrons of the benzene ring. As a result the transfer
of alcohol from the aqueous and interfacial regions to the nonaqueous region, as illustrated in Figure 7, is more pronounced
upon addition of benzene. The loss of alcohol from the interfacial
region results in a closer packing of the oleate ions. This will
of course affect the electrostatic interaction and thus the counter
ions. These effects can be studied by means of ^{23}Na NMR[3].

The [23]Na nucleus has spin quantum number 3/2 and relaxes via
the quadrupole mechanism[20]. It has been demonstrated for a number
of amphiphilic systems that the "extreme narrowing" condition,
$T_1 = T_2$, is fullfilled for the counter ions[21]. That this is the
case also in the present system was experimentally verified. The
[23]Na relaxation rate can thus be written

$$R = T_1^{-1} = T_2^{-1} = \frac{2}{5} \pi^2 \left(\frac{e^2 qQ}{\hbar} \right)^2 \cdot \tau_c \qquad (6)$$

where eq is the electric field gradient at the nucleus and eQ is the nuclear quadrupole moment. The quantity e^2qQ/h is the quadrupole coupling constant. For a sodium ion, the electric field gradients are of intermolecular origin and both the magnitude and the direction of the gradient fluctuate due to the molecular motions. τ_c is the correlation time characterizing the time scale of these fluctuations.

Figure 8 shows that addition of hydrocarbon to a water rich microemulsion increases the ^{23}Na relaxation rate. The electrostatically unfavourable closer packing of the oleate ions caused by the redistribution of alcohol can partially be compensated by an increased counter ion binding giving rise to the increased ^{23}Na relaxation rate. This effect is expected to be greater for microemulsions containing benzene which is in agreement with the results in Figure 8.

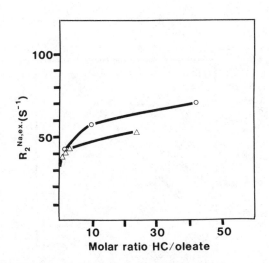

Figure 8. ^{23}Na excess relaxation rates ($R_2^{Na, ex.}$) for quaternary microemusions. The molar ratio water : pentanol : sodium oleate = 44.5 : 10.4 : 1. The hydrocarbon is benzene (o) and decane (Δ).

The redistribution of alcohol also affects the transport properties of the different components in a microemulsion, which have
been studied by means of self-diffusion measurements. The self-diffusion is measured over macroscopic distances and it can thus give
information on whether a component is confined inside a closed aggregate. For a typical reversed micellar system, $i.e.$ an aqueous
core separated from a nonaqueous continous medium by a well-defined
interface, the self-diffusion of the hydrocarbon is expected to be
faster than that of the alcohol which, in turn, is faster than the
water self-diffusion.

With this in mind it is evident from Figure 9 that no typical
reversed micelles with a well-defined interface are present at the
higher water concentrations in the ternary system. The diffusion
coefficient of pentanol is always $\geq$ 70% of the value in pure pentanol, indicating that pentanol is located mainly in the continuous phase. The self diffusion coefficient of the oleate ions can
be estimated to be less than 10^{-10} m^2s^{-1}. Hence, at low water
content the diffusion of the water molecules is comparable to that
of the oleate ions indicating that the water is associated primarily with the surfactant.

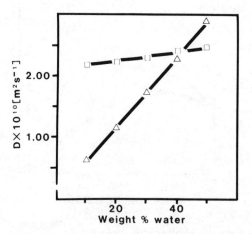

Figure 9.   Self-diffusion coefficients for water ($\triangle$) and pentanol
            ($\square$) in ternary microemulsions with a constant weight
            ratio pentanol : oleate = 68:32.

From the results in Figure 10 it can be concluded that hydrocarbon
and alcohol constitute the continuous phase in the quaternary systems. The increase in the hydrocarbon diffusion is mainly an effect
of the reduced viscosity as the hydrocarbon concentration is increased. The increased diffusion of pentanol is in addition a
result of the reduced degree of self-association but can also be
due to an increased fraction of alcohol in the continuous phase.
The water diffusion is reduced by an order of magnitude and approaches values obtained for typical reversed micellar systems[23].

Considering the distribution of the alcohol as discussed
above, these results confirm the idea that a reduction of the
alcohol content in the interfacial region is accompanied by the
formation of a more well-organized hydrophobic/hydrophilic interface offering an effective barrier towards translational diffusion of the water molecules.

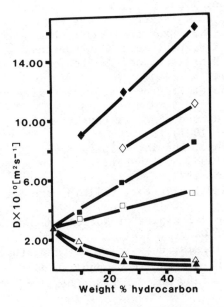

*Figure 10.   Self-diffusion coefficients for water (Δ), pentanol
(□) and hydrocarbon (◇) in quaternary microemulsions with a constant weight ratio pentanol:water:
oleate = 35:49:16. Filled symbols represents microemulsions containing benzene and open symbols microemulsions containing decane.*

ACKNOWLEDGEMENT

This work was financed by The Swedish Board for Technical Development (STU). We wish to express our thank to T. Edberg, A. Jönsson and P. Saris for their careful experimental work.

REFERENCES

1. E. Sjöblom and S. Friberg, J. Colloid Interface Sci., 67, 16 (1978).
2. E. Sjöblom, U. Henriksson and P. Stilbs, in "Proc. of VII Scand. Symp. on Surface Chem.", p. 233. K.S. Birdi, Editor, 1982.
3. E. Sjöblom and U. Henriksson, J. Phys. Chem., 86, 4451 (1982).
4. J. Biais, P. Bothorel, B. Clin and P. Lalanne, J. Dispersion Sci. Techn. 2, 67 (1981).
5. L. Mandell, K. Fontell and P. Ekwall, Acta Techn. Scand., 74, I (1968).
6. P. Stilbs and M.E. Moseley, Chem. Scr., 15, 176 (1980).
7. A.N. Fletcher and C.A. Heller, J. Phys. Chem. 71, 3742 (1967).
8. B.D. Anderson, J.H. Rytting, S. Lindenbaum and T. Higuchi, J. Phys. Chem., 79, 2340 (1975).
9. J.H. Rytting, B.D. Anderson and T. Higuchi, J. Phys. Chem., 82, 2240 (1978).
10. I. Prigoine and R. Defay, "Chemical Thermodynamics", p 426, Longmans, London, 1954.
11. E.E. Tucker and E.D. Becker, J.Phys. Chem., 77, 1783 (1973)
12. H.T. French and R.H. Stokes, J.Phys. Chem., 85, 3349 (1981).
13. J.C. Davis and K.K. Deb, Adv. Magn. Reson., 4, 201 (1970).
14. E. Sjöblom, U. Henriksson and P. Stenius, Finnish Chem. Lett. 6-8, 114 (1982).
15. Biomedical Computer Programs P-series, University of California Press 1979.
16. J, Mullens, I. Hanssens and P. Huyskens, Bull. Soc. Chim. Belges, 79, 539 (1971).
17. E. Nagata, Z. Phys. Chemie, Leipzig, 252, 305 (1973).
18. See e.g. McClelland, B.J. "Statistical Thermodynamics", Chapman and Hall, London, 1973.
19. I. Danielsson and B. Lindman, Colloids and Surfaces, 3, 391 (1981).
20. A. Abragam, "The Principles of Nuclear Magnetism", chapter VIII, Clarendon Press, Oxford, 1961.
21. H. Gustavsson, Thesis, University of Lund, Lund, Sweden, 1978.
22. E. Sjöblom and P. Stilbs (To be published).
23. P. Stilbs and B. Lindman, J. Phys. Chem., 85, 2587 (1981).

TRANSPORT OF SOLUBILIZED SUBSTANCES BY MICROEMULSION DROPLETS

C. Tondre and A. Xenakis

Laboratoire de Chimie Physique Organique
ERA CNRS n°222. Université de Nancy I
B.P. 239 54506 Vandoeuvre-les-Nancy Cedex, France

Biphasic four-component systems (surfactant/co-surfactant/oil/water) were used in order to demonstrate the carrier properties of microemulsion droplets, by studying the transfer rate of probe molecules at liquid-liquid interfaces. The transport of lipophilic substances by o/w microemulsions show that i) the flux is proportional to the initial concentration of the probe to be transported ; ii) the flux is also proportional to the volume fraction of oil in the microemulsions as long as it is less than 0.15 ; iii) at larger values of the volume fraction a dramatic increase of the transfer rate may be attribued to a percolation phenomenon for the first time observed with o/w microemulsions. A model in which the diffusion of the carrier inside the liquid membrane is coupled with a fast solubilization-desolubilization reaction is proposed and its kinetic implications are discussed. Finally an example of transport of hydrophilic solute by w/o microemulsions will be given with possible applications for improving the rate of liquid-liquid metal extraction.

INTRODUCTION

Microemulsion droplets offer the possibility of transporting substances through a medium where they are not soluble or
very poorly soluble. The transport of metabolites or drugs using
the carrier properties of polymolecular assemblies, such as
liposomes or vesicles, is the object of a number of current
studies[1] motivated by the numerous biological and medical possible
applications, but to our knowledge, the ability of a microemulsion
droplet to play the part of a carrier has not yet been clearly
demonstrated.

Besides the possible biological applications, including the
transport of respiratory gases by microemulsions possessing a
perfluorinated core[2], one should also mention the utility of microemulsions in improving the rate of liquid-liquid extraction in
view of metal recovering or separation[3]. Both types of applications are currently under study in our laboratory.

A short preliminary report concerning the transport properties of microemulsions has been published recently[4] and the
present paper will give a full account of the results obtained so
far. On the other hand, some experiments were recently reported
which tend to demonstrate the transport function of micelles[5].

It is important to note that at least two main differences
exist between vesicles and microemulsions : 1) in the case of
vesicles the interior of the globule has the same aqueous nature
as the continuous phase, whereas in the case of microemulsions
one of them is organic whereas the other one is aqueous ; 2) the
dynamic character of the microemulsion globule is much more
accentuated[6] than that of vesicle, whose membrane is constituted
by a bilayer structure.

The microemulsion droplets constituted by a core of oil
surrounded by a layer of tensioactive agents, dispersed in a
continuous aqueous phase, are expected to constitute possible
carriers for a lipophilic substance which could thus be transported from a source organic phase (S) to a receiving organic
phase (R) through the liquid membrane (M) constituted by the o/w
microemulsion. At the opposite end, w/o microemulsions are
expected to allow the transport of water soluble substances
through an organic medium. Examples of both situations will be
given, with emphasis on the former for which the influence of
different parameters on the flux of different transported solutes
will be examined.

The results will be shown to contribute to the understanding of the microemulsion structure and of the kinetics of the
solubilization-desolubilization processes.

EXPERIMENTAL PART

The transport cell used for the major part of the present
work consisted of a U-shaped tube, very comparable to the devices
previously described for ion transport studies using the carrier
properties of antibiotics[7] or macrocyclic compounds[8]. The o/w
microemulsion is put in the bottom of the U-tube and both arms
are then carefully filled with an oil phase in thermodynamic
equilibrium with the microemulsion. (For w/o microemulsion a
device similar in principle is used, but the cell is the reverse
of the previous one, the microemulsion laying on the top of two
aqueous phases). A good agitation and temperature control were
ensured in the three compartments. The solute to be transported
was introduced in the source compartment (S) and the detection of
solute reaching the receiver compartment (R) was carried out by a
UV or visible spectroscopic detection on a continuous basis. A
schematic drawing of the whole set up is shown in Figure 1. The
characteristics of the cells used were the following : the microemulsion compartment (M) was a cylinder of diameter 5 cm and
thickness 1.5 cm where the agitation was achieved by a magnet
with a rotation speed of 120 rpm (this speed was close to the
highest that could be tolerated for keeping a very clean interface
between the two phases) ; the compartments S and R were cylinders

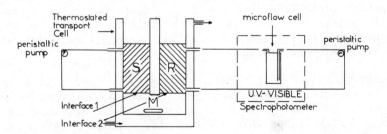

Figure 1. Schematic diagram of the apparatus used for transport
measurements.

of diameter 2 cm and height $\sim$ 5 cm in which a good agitation close
to the interface was insured by peristaltic pumps, whose rotation
speed had practically no influence on the transfer rate of solutes.

It is unfortunately not possible to choose any o/w microemulsion and then superpose an oil phase, because not only the
solute to be transported will migrate from the oil compartment S
to the oil compartment R, but also part of the oil will dissolve
into the microemulsion[9]. For this reason we were led to find
diphasic systems of the Winsor I type (i.e. constituted of one
microemulsion rich in water and one oil-rich phase laying on the
top of it), both phases being perfectly clear and separating
without centrifugation. (Winsor II type systems are of course
needed for transport experiments with w/o microemulsions).

Biphasic systems were prepared from the pseudo-ternary
phase diagram (1-pentanol/sodium-dodecylsulfate (SDS))/n-dodecane/
H2O (with a weight ratio 1-pentanol/SDS = 2), for which various
data are available in the literature (see ref (9)). The phases
being in thermodynamic equilibrium, the small concentration of
solutes employed were assumed not to perturb the established
equilibrium. The composition of the separated phases was characterized by gas chromatography, Karl Fisher and potentiometry in
addition to density measurements. Note that in the separate phases
the ratio 1-pentanol/SDS is no more equal to 2 so that the repre-
sentative points are not in the plane of the pseudo-ternary
diagram. The optical density of the phase R was measured thus
continuously allowing the determination from a calibration curve
of the number of solute molecules crossing the interface with
time.

Blank experiments with water (or water saturated with 1-
pentanol) in place of the microemulsion did not reveal any transport of lipophilic solutes after one day.

For the experiments with w/o microemulsions, the biphasic
system used was obtained from the pseudo-ternary phase diagram
(n-hexanol/tetraethyleneglycol-dodecylether (TEGDE)/n-decane/H2O)
with a weight ratio TEGDE/n-hexanol = 3. From the shape of the
monophasic region determined at 20°C by Friberg *et al*[10], we expected to find Winsor II systems fulfilling the conditions needed for
the present transport studies.

Chemicals used were obtained as follows : n-dodecane, n-
decane, 1-pentanol, 1-hexanol, pyrene, perylene, anthracene and
dicyclohexano-18-crown-6 (Fluka purum or puriss), SDS (Serva,
Heidelberg, W.G.), TEDGE (Nikko Chemicals, Japan). Potassium
picrate was prepared from picric acid (Merck) and recrystallized
three times.

RESULTS

A/ Transport of Lipophilic Solutes by o/w Microemulsions

Different mixtures (referred to as A to F) were prepared
whose initial compositions lie on the bisectrix of the surfactant/
cosurfactant apex of the pseudo-ternary phase diagram (1-pentanol/
SDS)/n-dodecane/H_2O, so that the resulting phases have about comparable volumes. The composition in weight % of the initial mixtures as well as of the separated phases are given in Table I, where
the inferior phase ϕ_i refers to the microemulsion and the superior
phase ϕ_s to the oil phase.

System A was chosen to study the transport of pyrene with
different initial concentrations of pyrene C_p^i in the source compartment S. The plots of the number of moles of pyrene transported versus time through one square centimeter of the second interface are shown in Figure 2. These plots show an initial curvature,
as observed in many other transport studies, and attributed to the
time required to reach a steady-state. The rates of transfer as
given in Table II were obtained from the slopes of the linear part
of the plots after establishment of a steady-state. Figure 3 gives
an idea of how this steady-state is established: the average number of moles of pyrene in compartment M was obtained from the
difference of the total number of moles initially present in S and
the measured number of moles of pyrene both in S and in R versus
time. For this purpose, the same experiment was carried out two
times : the first time was measured the concentration of pyrene in
S and the second time the concentration of pyrene in R.

The influence of the bulkiness of the transported solute
was also examined in the same system A and the results obtained
for the transport of perylene and anthracene are summarized in
Table II in addition to the findings for pyrene.

Pyrene was selected to study the influence of the amount of
tensioactive agents (surfactant + cosurfactant) in the initial
mixture. Table I shows that the higher the amount of 1-pentanol
+ SDS in the initial mixture, the higher the amount of dodecane
incorporated in the microemulsion phase, which is an important
parameter. The flux of pyrene through the second interface is
plotted in Figure 4 versus the volume fraction of dodecane in the
dispersed phase, varying from 8.60 % to 21.09 %. The flux appears
to be proportional to the volume fraction of dodecane up to a
value of ca. 15 % and then increases dramatically.

The partition coefficients k_p of the different solutes
between the two phases have been determined spectrophotometrically.
The solute was then added in the initial mixture before the phase

Table I. - Composition of Initial Mixtures and Separated Phases in Weight % for Winsor I Systems.

	System identification	C	B	F	A	E	D
initial mixture	water	43.8	43	42.2	41.4	40.5	39.5
	n-dodecane	43.8	43	42.2	41.4	40.5	39.5
	1-pentanol	8.26	9.33	10.4	11.47	12.67	14
	SDS	4.14	4.67	5.2	5.73	6.33	7
inferior phase ϕ_i (o/w microemulsion)	water	78.60	74.74	72.35	67.91	62.97	57.25
	n-dodecane	6.82	8.63	9.52	12.25	13.05	17.54
	1-pentanol	7.31	8.56	9.63	10.78	14.41	15.14
	SDS	7.27	8.07	8.50	9.06	9.57	10.07
superior phase ϕ_s (oil phase)	water	0.16	0.24	0.28	0.35	0.40	0.43
	n-dodecane	90.24	89.26	87.71	86.58	89.46	87.84
	1-pentanol	9.42	10.40	11.40	12.51	9.58	11.49
	SDSa	0.18	0.10	0.61	0.56	0.56	0.24

[a]These values were calculated from the difference between the known total amount of SDS in the initial mixture and the amount of SDS measured in the microemulsion phase ϕ_i. They may be overestimated because a very small amount of solutions close to the interface is unavoidably discarded when separating the two phases.

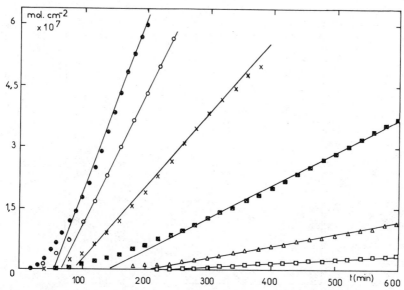

Figure 2. Plots of the number of moles of pyrene transported
versus time through one square centimeter of the second interface
in microemulsion system A. Initial pyrene concentration in S :
$c_p^i = 10^{-2}M$ (⊕), $5\times10^{-3}M$ (O), $2\times10^{-3}M$ (X), $10^{-3}M$ (⊠), $2.5\times10^{-4}M$ (Δ),
$10^{-4}M$ (□).

Table II. Transfer Rates of Solutes in $10^7\times mol/h$ in System A
(cross section = 3.14 cm^2)

initial concentration of solute in S $10^3\times M$	Perylene	Pyrene	Anthracene
0.1		0.18	
0.125		0.25	
0.15	0.34		
0.166		0.33	
0.25	0.53	0.54	0.39
0.4	0.85		
0.5		0.90	0.76
1.0		1.53	
1.25			3.36
1.5		2.07	4.22
2.0		3.51	
5.0		6.14	
10.0		8.66	

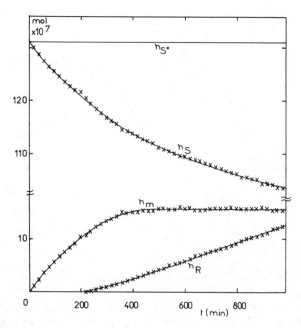

Figure 3. Plots of the number of moles of pyrene in compartments S, M and R versus time (Microemulsion system A with initial pyrene concentration $C_p^i = 5.32 \times 10^{-4}$M)

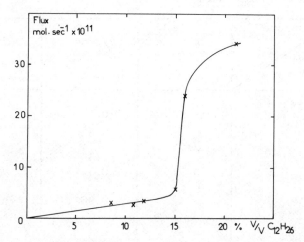

Figure 4. Flux of transported pyrene versus the volume fraction of dodecane in the microemulsion phase.

separation was achieved. The values obtained in system A were 3.7,
2.55 and 7.46 respectively for pyrene, perylene and anthracene.

B/ Transport of Hydrophilic Solute by w/o Microemulsions

An example of the situation reverse to that just described
will be now given with some preliminary results on the influence
of microemulsions on the rate of transfer of solutes in liquid-
liquid extraction. The system used was made of (n-hexanol/tetra-
ethyleneglycol dodecyl ether)/decane/H_2O. The initial composition
as well as the composition of the separated phases are given in
Table III. The transport of salt was performed in this system and
the picrate was chosen because of its convenient light absorption.
A blank experiment with pure decane in place of the w/o microemulsion on the top of an aqueous potassium picrate solution did not
show any transport of picrate after one day.
The efficiency of the above w/o microemulsion for transporting potassium picrate is compared in Figure 5 with the efficiency
of a 10^{-2}M/l solution of dicyclohexano-18-crown-6 in decane. In
the same figure is also shown the transport obtained when the microemulsion itself contains 10^{-2}M/l of dicyclohexano-18-crown-6.

DISCUSSION

We will first discuss the transport of lipophilic substances by o/w microemulsions. Different mathematical treatments have
been proposed in the literature to interpret the transfer of
solutes through liquid-liquid interfaces depending on what is
assumed to be the rate determining process. Rosano at al[11] for
instance attributed the rate-controlling step for the migration of
salts through non-aqueous liquid membranes to the crossing of the
interface and not to the diffusion through the variable concentration layers. On the other hand when a macrocyclic carrier is present in the non-aqueous liquid membrane, a model in which
diffusion is coupled with the chemical reaction leading to the
complex formation between the solute and the carrier is more
appropriate[8]. Exact mathematical solutions are so much complicated[12] that assumptions are usually introduced regarding which one
of the diffusion in the non-stirred layer or of the chemical
reaction is the fastest.

We have considered the transport of lipophilic solute by a
microemulsion droplet in a way similar to that developed to interpret the facilitated transport of ions by crown-ethers[8,12]. The
good mechanical stirring of the membrane compartment M reduced the
diffusion layer thickness 1 on both interfaces to a minimum, so
that the effective membrane has a thickness 2 1 = L. Following a
scheme analogous to that described by Reusch and Cussler[13] and

Table III. Composition of Initial Mixture and Separated Phases in Weight % for Winsor II System.

	water	n-decane	n-hexanol	TEGDE
initial mixture	50	40	2.5	7.5
inferior phase ϕ_i (water phase)	∿ 100	. . . . undetectable		
superior phase ϕ_s (w/o micro-emulsion)	6.46	74.72	4.67	14.01

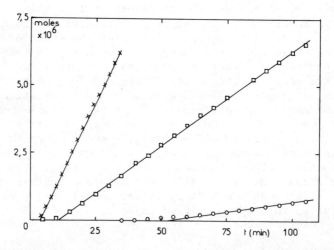

Figure 5. Plots of the number of moles of potassium picrate transported versus time (cross section 3.14 cm²). Initial picrate concentration in S : 5×10^{-3}M ; w/o microemulsion alone (□), solution of dicyclohexano-18-crown-6 in decane alone (O), w/o microemulsion containing dicyclohexano-18-crown-6 (X).

more recently by Lamb et al[14], the following steps can be postulated (with P referring to the lipophilic solute and the circle to
the microemulsion droplet) :

$P_{(S)} \rightleftarrows P_{(continuous\ phase\ of\ M)}$ | interface 1 (a)

$P_{(continuous\ phase\ of\ M)} + \bigcirc \underset{k_2}{\overset{k_1}{\rightleftarrows}} \textcircled{P}$ | membrane   (b)

Diffusion of $\textcircled{P}$ across the membrane | membrane   (c)

$\textcircled{P} \underset{k_1}{\overset{k_2}{\rightleftarrows}} \bigcirc + P_{(continuous\ phase\ of\ M)}$ | membrane   (d)

$P_{(continuous\ phase\ of\ M)} \rightleftarrows P_{(R)}$ | interface 2 (e)

The last step is followed by back diffusion of the empty droplet
across M. Note that the diffusion of free lipophilic solute
through the membrane can be ignored as indicated by the blank
experiments. Steps (a) and (e) are characterized by the same
partition coefficient k of the solute between the oil phase and
the continuous aqueous phase of the microemulsion and not the
microemulsion phase itself : this partition coefficient k is thus
different from k_p for which values have been given previously.
Steps (b) and (d) are characterized by rate constants k_1 and k_2
and an equilibrium constant $K = k_1/k_2$.

The transport of each species across the membrane *at steady-
state* is governed by an equation of mass conservation of the form:

$$D_i d^2[i]/dX^2 = R_i \tag{1}$$

where i designates the three species considered inside the membrane (P, $\bigcirc$, $\textcircled{P}$), X is the coordinate in the direction of
solute transport, D_i is the diffusion coefficient and R_i the rate
of depletion of species i by reaction (b) or (d) within the membrane. These equations are subject to the boundary conditions :

$$c_P^M = k\ c_P^0 \qquad for\ X = 0$$
$$c_P^M = k\ c_P^L \qquad for\ X = L \tag{2}$$

where c_P^M is the solute concentration in the continuous phase of
the membrane and c_P^0 and c_P^L are the concentrations outside the
membrane.

Assuming that step (c) is much slower than all other steps, and that equilibrium between the species involved is always established inside the membrane, it follows that the concentration of droplets containing one solute molecule can be expressed as

$$C_{\left(\boxed{P}\right)}^{M} = \frac{K C_P^M C_D}{1 + K C_P^M} \tag{3}$$

C_D being the total droplet concentration.

Following then the reasoning developed by Ward[15] and Cussler[12], the flux of transported solute can be expressed by

$$F = \frac{D K k C_D}{L} \frac{C_P^O}{1 + K k C_P^O} \tag{4}$$

where D is the diffusion coefficient of a microemulsion droplet.

To obtain this expression, C_P^L was set equal to zero at the beginning of the steady-state (in good agreement with Figure 3) and the transport of free solute was neglected in agreement with blank experiments. In contrast with comparable studies using macrocyclic carrier, C_P^O cannot be taken as the initial concentration C_P^i of solute in compartment S because the concentration in the membrane at steady-state is far from being negligible. The partition coefficient of solute k_p, which has been measured in the biphasic systems allows us to calculate the concentration C_P^O of the solute in the compartment S at the beginning of the steady-state, the membrane being assumed to have reached an equilibrium situation :

$$C_P^O = C_P^i \frac{1}{1 + V/k_p} \tag{5}$$

where $V = V_M/(V_S + V_R)$, ratio of compartments volumes.

As suggested by Cussler[12], a convenient way of checking the validity of equation (4) to describe the experimental results, consists in plotting the reciprocal flux of solute versus the reciprocal of the solute concentration C_P^O. Equation (4) predicts a linear dependence which is indeed observed in Figure 6, with slope $L/DKkC_D$ and intercept L/DC_D, from which the product Kk can be obtained (Note that in the approach of Rosano et al[11], that is if the rate of transport were governed by the transfer across the

interface, the intercept of such a plot should be zero, which is
not consistent with the present results). The intercept is completely determined by the characteristics of the transport cell and
the biphasic system used, and should thus be independent of the
solute used. Figure 6 shows that the results obtained for pyrene
and perylene are in good agreement with this prediction (in the
case of perylene the solubility limit did not allow us to investigate higher concentrations of this solute). The agreement is less
good for anthracene for which an error larger than the expected
one have to be admitted for the points obtained at high concentrations.

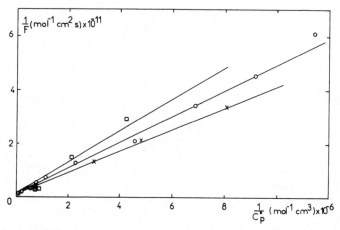

Figure 6. Plots of the reciprocal flux of solute versus the reciprocal of solute concentration at steady-state in source compartment (microemulsion system A) : pyrene (o), perylene (x), anthracene (□).

Nevertheless, the model proposed, which was the simplest
one, seems to reasonably describe our results. The linear dependence observed is an *a posteriori* justification of the simple
assumption that only one solute molecule enters in a microemulsion
droplet and it suggests that the size of the droplets is such that
their concentration was always higher than the solute concentration inside the membrane. If it were not the case, a distribution
of solute molecules amongst the droplets (Poisson's distribution,
for example) should have been considered. Another alternative to
the present model could have been the migration of the transported

solute directly from the oil phase into an oil droplet without
solubilization in water and then a possible transfer by collisions
between droplets. Such a process is very unlikely because of the
electrostatic repulsions between droplets, and it was recently
shown that it can be ruled out in the case of lipid vesicles, the
migration of pyrene taking place mainly *via* the aqueous phase[16].

The values of the product Kk have been determined for the
three solutes, giving 312 $(\pm 50)M^{-1}$, 260 $(\pm 50)M^{-1}$ and 212 $(\pm 50)M^{-1}$
respectively for perylene, pyrene and anthracene (for the last one
the value was obtained by assigning the same intercept as determined for the other two solutes). These values have some kinetic
implications   which will be discussed now. The partition coefficient k can be estimated from solubility values of the solutes in
water and in dodecane or in a similar medium[17], and the equilibrium
constant K can then be determined (k was estimated[17] to be 8×10^{-7},
8.1×10^{-6} and 1.57×10^{-5} respectively for perylene, pyrene and
anthracene resulting in the corresponding K values : $3.9\times10^{8}M^{-1}$;
$3.2\times10^{7}M^{-1}$ and $1.35\times10^{7}M^{-1}$). The forward rate constant k_1 characterizing the entrance of the solute inside a microemulsion droplet
is very likely diffusion-controlled as is the case when ionic
micelles are considered[17]. A comparison of the exit rate of solutes from microemulsion droplets with previously published data on
SDS micelles[17] can thus be performed, using for the diffusion-
controlled rate constant k_1 the same value as used for SDS micelles $(7\times10^{9}M^{-1}s^{-1})$ and correcting it to take into account the
change in the distance of closest approach. The droplet's radius R
can be assumed to be about two to three times the radius of an SDS
micelle. Such a value would be consistent with the above remark
concerning the model (concentration of droplet larger than concentration of solute in the membrane) and in agreement with the few
experiments existing on comparable microemulsions : radius ranging
from 36 to 40 Å have been measured by fluorescence techniques[18]
for the same four-component system with volume fraction of
dodecane from 0.06 to 0.09 whereas values around 75 Å have been
obtained from osmotic pressure measurements[19] in the presence of
salt. With the approximation involved, k_2 would be roughly about
ten times smaller for a microemulsion than for SDS micelles, which
is consistent with Almgren's prediction that $k_2R^2 \simeq$ constant.

Equation (4) also predicts a linear dependence of the flux
F with the droplet concentration C_D. This concentration may be
estimated by :

$$C_D = \emptyset/Nv_D \qquad\qquad (6)$$

where $\emptyset$ is the volume fraction of dodecane in the microemulsion
phase, N Avodagro's number and v_D the volume of the dodecane core
of the droplet. Thus, C_D may be assumed to be roughly proportional·

to $\emptyset$ provided that v_D does not change to much. Figure 4 shows that
the flux is about proportional to the volume fraction of oil up to
~ 0.15. The drastic increase observed for higher values may be
attributed to a percolation phenomenon analogous to that observed
in reverse microemulsion systems[20] and due to the fact that continuous "channels" are likely to be formed from one interface to the
other one. It seems to be the first time that such a behavior is
observed in direct microemulsion systems. Percolation of hard
spheres is known to occur when they fill about 15 % of the space[20],
but it is important to note that the volume fraction considered
above does not include the membrane of tensioactive agents so that
the space really occupied by the droplets (if they are still droplets) is larger than 15 %. This result may also indicate that
when the volume occupied by the hydrocarbon core of the microemulsion phase becomes higher than 15 % of the total volume, a
rearrangement of the structure occurs with the opening of the
droplet membrane and formation of long oil channels bordered by a
membrane of tensioactive agents. The present observation is in
good agreement with previous stopped-flow measurements of the rate
of dissolution of dodecane in o/w microemulsions[9] : the rate of
dissolution becomes too fast to be measured when the oil content
of the microemulsion is larger than 15-18 %, which would indicate
that the system then becomes bicontinuous.

We turn now briefly to the discussion of experiments
carried out with reverse microemulsions. An increase of the transfer rate in liquid-liquid extraction of metal was known to occur
when tensio-active agents were added to the extractive medium, but
it is only recently that the organic phase in the extracting
system of gallium was shown to be a microemulsion[21]. The results
presented in Figure 5 clearly demonstrate the carrying properties
of w/o microemulsions as well as their usefulness in increasing
the transfer rate of ionic solute in liquid-liquid extraction
processes.

The microemulsion used was in fact much more efficient than
a crown-ether for transporting the potassium picrate used here,
but the largest rate of transfer is observed when the crown ether
is contained in the organic phase of the microemulsion. In the
absence of microemulsion the transport of potassium picrate by
crown-ethers of the type of 18-crown-6 is known to be much faster
than the transport of sodium picrate, due to the higher complexation of K^+ compared to Na^+ [8]. Thus, the very small flux observed
in Figure 6 for K^+ picrate transported by the crown-ether alone
would have been even much smaller if Na^+ picrate had been used.

The spectacular effect observed when both the crown-ether
and the water droplets of the microemulsion are present in the
organic phase proves that the method used here can be useful to

optimize the characteristics of the microemulsion in order to
achieve a fast transfer. It is important to note that the effect
is not a simple cumulative effect due to the presence of two types
of carriers. The influence of the characteristics of the microemulsion phase on the rate of transfer in such system is under
investigation in our laboratory.

ACKNOWLEDGMENTS

The authors acknowledge the valuable technical assistance
of Mr. J.L. Vasseur, who has fabricated the transport cells used
in this work.

REFERENCES

1. J.H. Fendler and A. Romero, Life Sci., 20, 1109 (1977).
2. G. Mathis and J.-J. Delpuech, French patent N°8022875 (1980).
3. P. Fourre and D. Bauer, C.R.Acad.Sci.Paris, Ser.II, 292, 1077
    (1981).
4. C. Tondre and A. Xenakis, Colloïd Polymer Sci., 260, 232 (1982).
5. M. Seno, H. Kise, K. Iwamoto, paper presented at the 4th
    Int.Conference on Surface and Colloïd Science, Jerusalem,
    July 1981.
6. J. Lang, A. Djavanbakht and R. Zana, J.Phys.Chem., 84, 1541
    (1980).
7. R. Ashton and L.K. Steinrauf, J.Mol.Biol., 49, 547 (1970).
8. K.H. Wong, K. Yagi and J. Smid, J.Membrane Biol., 18, 379
    (1974).
9. C. Tondre and R. Zana, J.Dispersion Sci.Technol., 1, 179(1980).
10. S. Frieberg, I. Lapczynska and G. Gillbert, J.Colloïd Inter-
    face Sci., 56, 19 (1976).
11. H.L. Rosano, P. Duby and J.H. Schulman, J.Phys.Chem., 65,
    1704 (1961).
12. E.L. Cussler, "Multicomponent Diffusion", Elsevier Scientific
    Publ.Co., New York, 1976.
13. C.F. Reusch and E.L. Cussler, AIChE.J., 19, 736 (1973).
14. J.D. Lamb, J.J. Christensen, S.R. Izatt, K. Bedke, M.S.
    Astin and R.M. Izatt, J.Amer.Chem.Soc., 102, 3399 (1980).
15. W.J. Ward, AIChE.J., 16, 405 (1970).
16. M. Almgren, Chem.Phys.Lett., 71, 539 (1980).
17. M. Almgren, F. Grieser and J.K. Thomas, J.Amer.Chem.Soc.,
    101, 279 (1979).
18. P. Lianos, J. Lang, C. Strazielle and R. Zana, J.Phys.Chem.,
    86, 1019 (1982).
19. A. Graciaa, J. Lachaise, P. Chabrat, L. Letamendia, J. Rouch
    and C. Vaucamps, J.de Physique-Lettres, 39, 235 (1978).
20. M. Lagües, R. Ober and C. Taupin, J.de Physique-Lettres, 39,
    487 (1978).
21. D. Bauer, P. Fourre and J. Lemerle, C.R.Acad.Sci.Paris, Ser.
    II, 292, 1019 (1981).

LIGHT SCATTERING AND VISCOMETRIC INVESTIGATIONS OF INVERSE LATICES

FORMED BY POLYMERIZATION OF ACRYLAMIDE IN WATER-SWOLLEN MICELLES

Y.S. Leong*, S.J. Candau[+] and F. Candau*

* Centre de Recherches sur les Macromolécules,CNRS
6, rue Boussingault, 67083 Strasbourg-Cedex,France
+ Laboratoire d'Acoustique Moléculaire, Université
Louis Pasteur, 4, rue Blaise Pascal
67070 Strasbourg-Cedex, France

Inverse latices of small size (d <500 Å) have been
prepared via polymerization of acrylamide contained
in water-swollen micelles of Aerosol OT in a toluene
medium. Size and interaction parameters of both microemulsions before polymerization and final latices
have been determined using elastic and quasi-elastic
light scattering and viscometry.

INTRODUCTION

Polymerization in microemulsions has recently attracted much
interest because of the great potential of such a process for the
preparation of latices of small size[1-6].

A microemulsion is a thermodynamically stable and transparent system containing surfactant(s), oil and water. It is generally described as a dispersion of spherical droplets of water
(or oil) in a continuous oil (or water) medium , the stabilization
of the droplets being ensured by an emulsifier film[7]. The radius
of the droplets is typically of the order of 50-100 Å. By incorporating a monomer in either the dispersed[1,4-6] or continuous
phase[2,3], polymerization may be achieved. Up to now, most of the
studies have dealt with normal microemulsions leading to hydrophobic polymer latices dispersed in water. The stabilization of a
normal microemulsion usually requires the presence of a cosurfactant (alcohol) which may lead to chain transfer reactions limiting
the molecular weight of the polymer thus formed[3,4].

Recently, two of us (Y.S.L. and F.C.) have reported a first account of the preparation of polyacrylamide latices via a micro-emulsion polymerization process[8]. The small size particles (d < 500 Å) obtained by this method consisted of macromolecules of high molecular weight accompanied by various amounts of water.

As a prelude to a systematic investigation of the properties of the latices, it is necessary to determine the characteristics, and more specifically, the size of the microemulsion droplets in the presence of the monomer. In this paper, we report an investi-gation by means of light scattering and viscometric techniques of the structural properties of the system Aerosol OT/toluene/ water & acrylamide. The elastic and quasi-elastic light scattering measurements in the high dilution limit provide a determination of the size parameter. Furthermore, the analysis of the variation of the scattered intensity with respect to the volume fraction, in the framework of the hard sphere models recently developed, allows us to measure the inter-particle interaction parameters.

The possible change induced by the polymerization process in these parameters is of particular interest as it provides in-formation on the polymerization mechanism. Similar information can also be obtained from a comparison of the concentration dependence of the viscosity of the system before and after poly-merization.

MATERIALS AND TECHNIQUES

Preparation of Samples

Toluene (Merck, Spectroscopy grade) was used as supplied.
Hydrophobic initiator AIBN (azo-bisisobutyronitrile) was recrystallized from absolute ethanol. It was dissolved in toluene (0.23%) and the required quantity weighed in to obtain the desired final concentration with respect to the monomer, for the formation of the microemulsion.

Water was deionized and tri-distilled.
Acrylamide (Merck) was recrystallized twice from chloroform and dried under vacuum.
Aerosol OT (sodium bis-2-ethylhexylsulfosuccinate) supplied by Fluka was purified as described elsewhere[9].

Microemulsions were prepared by dissolving first Aerosol OT (AOT) in toluene and then by adding the mixture of water and acrylamide (AM) in the chosen proportions.

Purified nitrogen was bubbled through the solution for a few minutes to eliminate oxygen in an apparatus similar to that des-cribed elsewhere[10]. Polymerization of acrylamide was achieved

either under ultraviolet irradiation (Philips HPK 125) or thermally.

Table I. Sample Compositions (expressed as percentages by weight).

MICROEMULSIONS

Sample	Toluene	AOT	Water	AM	$(dn/d\phi).10^2$
F_1	72.70	18.20	9.10	0	7.44
C_1	71.30	17.80	8.80	2.10	6.36
A_B	69.20	17.30	10.10	3.40	6.67
N_B	67.80	16.90	10.22	5.08	6.59
E_3	67.40	16.80	10.00	5.80	6.51
C_2	62.40	15.50	20.20	1.90	9.52
L_B	70.10	17.55	10.60	1.75	−
M_B	70.45	17.50	7.65	4.40	−

LATICES

Sample		Toluene	AOT	Water	PAM	$(dn/d\phi).10^2$
T_1	(a)	72.10	17.45	6.95	3.50	5.39
N_A	(b)	67.80	16.90	10.22	5.08	4.93
R_A	(b)	74.90	18.59	2.79	3.72	3.50
A_A	(b)	69.20	17.30	10.10	3.40	6.15
L_A	(b)	70.10	17.55	10.60	1.75	−
M_A	(c)	70.45	17.50	7.65	4.40	5.23

Polymerization carried out thermally at :(a) 50°C ; (c) 45°C ; (b) photo-
chemically (U.V.) at 25°C
The symbols AM,PAM and $(dn/d\phi)$ refer to acrylamide, polyacrylamide
and to the refractive index increment of the solution respectively.

    Table I shows the composition of both microemulsions and lati-
ces. As the continuous phase is believed to consist of pure toluene
with negligible amounts of water, acrylamide and AOT molecules,
the systems can be diluted without significant change in the
droplet size down to a volume fraction of the dispersed phase of
approximately 1%.

Light Scattering Experiments

    A Spectra-Physics argon-ion laser (λ=488nm) was used in con-
junction with a 72 channel clipped digital autocorrelation
(Precision Devices and Systems LTD Malvern) for measuring the
autocorrelation function of the scattered light intensity. The
scattering angle was varied from 20° to 120°C. Intensity correlation data were routinely processed using the method of

cumulants[11] to provide the average decay rate Γ and the variance v. The latter parameter is a measure of the width of the distribution of decay rates and is given by :

$$v = (<\Gamma^2> - <\Gamma>^2)/<\Gamma>^2 \tag{1}$$

where $<\Gamma^2>$ is the second moment of the distribution. The diffusion coefficient D was determined from the average decay rate $<\Gamma>$ according to :

$$<\Gamma> = 2 k^2 D \tag{2}$$

The magnitude of the scattering vector k is given by :

$$k = [4 \pi n \sin (\theta/2)]/\lambda \tag{3}$$

where θ is the scattering angle, λ is the wavelength of the incident light in vacuum and n is the index of refraction of the scattering medium.

The relative scattered intensities were obtained from time averaged rate of photon counts entering the correlator and converted into absolute intensities, using as a reference value the intensity I_B scattered by triply distilled benzene. Only results at $\theta = 90°$ are reported here because the systems investigated did not show any dissymetry.

Refractive index increments were determined with a Brice Phoenix differential refractometer or Abbe refractometer. Their values are listed in Table I.

Viscometry

The viscosity experiments were performed by means of an automatic Gramain-Libeyre[12] viscometer

All the experiments were performed at 25°C.

PHASE DIAGRAM

Hydrocarbon solutions of dialkylsulfosuccinates are able to solubilize significant amount of water, without requiring the addition of a cosurfactant. Ternary Aerosol OT systems have been widely investigated[13-15] and they are known to exhibit phase diagrams with a fairly extensive inverse micellar area L_2. Such an area is schematized in the ternary phase diagram of AOT/Water/ Toluene in Figure 1 (solid line). The dashed line in the figure illustrates the area obtained for a system containing a mixture

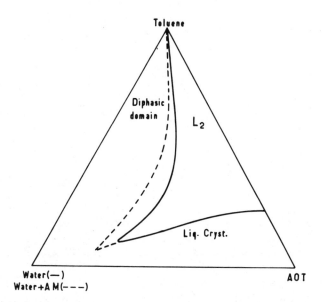

Figure 1. Schematic phase diagram at 20° of the system Aerosol OT/ Toluene/Water (or Water+AM in a weight ratio=1). The dashed line delimits the homogeneous L_2 area. The solid line delimits the same area in the absence of acrylamide (concentrations expressed as percentages by weight).

of water and acrylamide in a 1/1 weight ratio. It is worth noting that the presence of acrylamide increases the extent of the inverse micellar domain and it is likely that the monomer plays here the role of a cosurfactant.

LIGHT SCATTERING RESULTS

Figure 2 shows the volume fraction dependence of the normalized excess of scattering $(I-I_s)/I_s$ (I intensity scattered by the dispersion, I_s that scattered by the solvent) for a microemulsion prior to polymerization. The composition of the system is given in Table I. The curve exhibits a maximum at a volume fraction of the order of 0.13. Such a behavior has been previously observed for other microemulsions[16-18] and was interpreted from a simple model of interaction forces consisting of a hard sphere repulsion and a small attraction.

According to this model, the variation of the excess of scattering with volume fraction can be represented by :

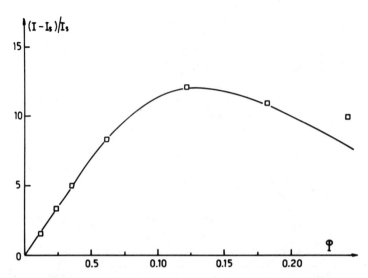

Figure 2. Variation of the normalized scattering excess as a function of the volume fraction of the dispersed phase for the microemulsion C_1. The full line represents the best fit to the data of Equation 4.

$$\left(\frac{\phi}{I-I_s}\right)/\left(\frac{\phi}{(I-I_s)}\right)_{\phi \to o} = \frac{(1+2\ \phi_{HS})^2 - \phi_{HS}^3(4-\ \phi_{HS})}{(1-\phi_{HS})^4} + A\phi \qquad (4)$$

The first term on the right side of Equation (4) derived by Carnahan and Starling[19] arises from the osmotic pressure due to hard sphere repulsion, ϕ_{HS} being the volume fraction occupied by the hard sphere.

The second term introduced by Vrij and coworkers[16-17] derives from the osmotic pressure associated with the attractive forces which are treated as a perturbation. In this term, the coefficient A is given by :

$$A = \frac{4\pi}{k_B T}\ \frac{1}{V}\ \int_{2R_{HS}}^{\infty} V_A(r)\ r^2\ dr \qquad (5)$$

where k_B is the Boltzmann's constant, T the absolute temperature, V the volume of a droplet, R_{HS} the radius of the hard sphere and $V_A(r)$ the attractive part of the interaction potential. This model does not account for attractive interactions between three or more droplets so that it is not expected to hold for high volume fractions.

Our experimental data were fitted with Equation 4 in which we assumed $\phi_{HS}=\phi$, (where $\phi = (V_{AOT}+V_{AM}+V_{H2O})/V_{TOTAL}$), A being the only adjustable parameter. The agreement between calculated curves and experimental data is satisfactory except for the most concentrated samples (cf. the example given in Figure 2). The A values obtained for different systems prior to polymerization are listed in Table II.

Table II. Size and Interaction Parameters from Light Scattering Experiments.

MICROEMULSIONS

Sample	R(Å)	R_H(Å)	B	B'	A	α	α_{cal}
F_1	27±2	30±1	8.4	3	-5.2	-0.5	-0.6
C_1	28±3	30±1	0	0	-8.1	-2.5	-2
A_1^B	35±3	37±2	0	-	-	-4.4	-
N^B	-	50±3	-	-	-	-8.3	-
E_3^B	49±4	57±3	-2.5	-11	-19	-8.5	-7.5
C_2	53±5	65±3	4	-1.5	-9.5	0	-2.7

LATICES

Samples	R(Å)	R_H(Å)	B	α
T_1	125±10	260±10	0	-1.7
N_1^A	140±10	215±5	8	-0.5
R_A^A	65±10	180±5	0	-1
A_A^A	-	160±5	-	0

In the high dilution limit, the concentration dependence of the scattered intensity can be expressed as :

$$\phi/(I-I_s) = K^{-1} \frac{3}{4\pi R^3} \left(\frac{dn}{d\phi}\right)^{-2} (1 + B\phi)$$

(6)

where R is the radius of the particle and

$$K = 2\pi^2 n^2 \cdot \frac{1}{\lambda^4}$$

(7)

The intercept and the initial slope of the curves $\phi/(I-I_s)=f(\phi)$ provide a measurement of R and of the second virial coefficient B/2 of the osmotic pressure. In the hard sphere model of Vrij et al.[16,17] this coefficient should be equal to 8+A, this quantity being quoted in the following as :

$$B' = 8 + A$$

(8)

The values obtained for B, B' and R are reported in Table II.

The volume fraction dependence of the diffusion coefficient is more difficult to assess as the available theories only apply to low values of α where D varies linearly with this parameter according to :

$$D = D_o \ (1 + \alpha \ \phi) \tag{9}$$

For repelling hard spheres, the virial coefficient α is equal to 1,45[20]. This is clearly in disaccord with the experiments which indicate negative or zero values of α (cf. Table II). Models based on several interaction potentials have been derived by Anderson and Reed[21] and Felderhof[22]. Cazabat et al.[18] approximated α from the Felderhof model to be :

$$\alpha_{cal} = 2 + A/2 \tag{10}$$

The values of α_{cal} calculated from Equation 10 are in rather good agreement with the experimental ones, which is surprising considering the crude approximation of the potential.

It must also be noted that some polydispersity could also lead to smaller values of α. This seems unlikely in the present case since the distribution of the particule size is rather narrow as indicated below.

The measurement of the diffusion constant in the high dilution limit provides a determination of the hydrodynamic radius R_H according to :

$$R_H = \lim_{\phi \to o} \frac{k_B T}{6\pi \eta D} \tag{11}$$

The values of R_H are listed in Table II. The polydispersity of the droplets is rather small, as inferred from the values of the variance which are of the order of 0.03.

The general features of the systems after polymerization are qualitatively similar to those of the microemulsions. Figure 3 shows the variation of $\phi I_s/(I-I_s)$ and D as a function of concentration in the high dilution range for the sample T1 (cf. Table I) The corresponding values of R, R_H, B and α are also listed in Table II. it must also be noted that the variance increases after polymerization of acrylamide as it ranges from 0.05 to 0.1, depending on the sample.

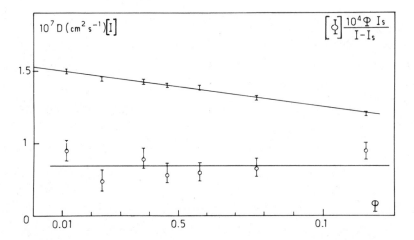

Figure 3. Variation of $\phi.I_s/(I-I_s)$ and D as a function of ϕ for the latex T_1.

VISCOMETRY

Many equations have been proposed to describe the volume fraction dependence of viscosity in dispersed systems[23]. In the range of volume fractions investigated here, the viscosity data are generally well represented by an expansion of the second order on ϕ :

$$\eta = \eta_o (1 + a \phi + b \phi^2) \tag{12}$$

where η and η_o are the absolute viscosities of the dispersion and the solvent respectively, a and b being constants. The coefficient a is equal to 2.5 in the hard sphere model of Einstein. Roscoe[24] proposed a=1.35x2.5 for a system of uniformly sized spheres and a = 2.5 for a polydisperse system.

When the dispersed phase is composed of fluid droplets of viscosity η_i larger than η_o, the flow of fluid inside the droplet lowers the relative viscosity of the dispersion and the coefficient a becomes :

$$a = 2.5 \left(\frac{\eta_i + \frac{2}{5} \eta_o}{\eta_i + \eta_o}\right) \tag{13}$$

Alternatively, some authors fit their data to the well known Mooney equation [25] :

$$\eta = \eta_o \quad \exp \left[a \, \phi / (1 - k \, '\phi) \right] \tag{14}$$

where k' is a hydrodynamic interaction coefficient which depends on the size of the particle. This law was found not to describe adequately our viscosity data contrary to Equation 12 which

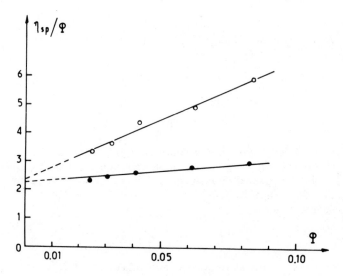

Figure 4.   Variation of η_{sp}/ϕ as a function of ϕ
o Microemulsion M_B
• Latex M_A

provides a good fit except in the high concentration range. This is illustrated in Figure 4 which shows plots of η_{sp}/ϕ ($\eta_{sp} = (\eta/\eta_o) - 1$) as a function of ϕ for a microemulsion containing acrylamide and for the corresponding latex.

The coefficients a and b obtained from the intercept and the slope respectively are listed in Table III.

Table III.   Viscosity Parameters.

Sample		a	b
Microemulsions	L_B	2.5	11.5
	A_B	2.5	36.7
	M_B	2.35	44
Latices	L_A	2.2	7.6
	A_A	2.3	8.2
	M_A	2.25	8.5

DISCUSSION

Inspection of Tables II and III reveals the following features.

Microemulsions prior to Polymerization

1. The hydrodynamic radius is found to be systematically larger than the geometrical one. This is consistent with the observations of previous studies[17,18,26,27] and can be explained by a draining of the toluene through the paraffinic chains of the AOT molecules.

Also the size of the particles increases with the amount of water for a given AM composition[15] and with the AM concentration for a given water content

2. The interactions between particles are described by the coefficients B, α and A. The general features concerning these interactions are quite similar to those observed by Cazabat et al.[18,28] on microemulsions containing alcohol molecules.

Nevertheless specific features seem to arise from the presence of acrylamide. It can be observed in Table II that an increase of AM leads to a strong decrease of the A, B and α coefficients. This expresses an enhancement of the attractive interactions between droplets. As the increase of the particle size is not proportional to the amount of AM(see Table II), it is likely that part of the acrylamide is located at the interface in between the AOT molecules. This should induce some disorganization of the interface and as a consequence enhance the interparticle attractions[17]. On the other hand, the sole increase of the particle size leads to larger attractive forces, as it has been recently shown for microemulsions containing alcohols[29].

3. The viscosity data provide values of the coefficient a in close agreement with the Einstein value of 2.5. This clearly indicates that the droplets behave like hard spheres rather than liquid spheres, their rigidity probably being ensured by the interface.

The values obtained for b confirm the enhancement of the attractive interactions between droplets as the acrylamide content increases.

Latices

1. The polymerization of acrylamide leads to a notable increase of the particle size as determined from quasi-light scattering experiments (cf. Table II). Two possible mechanisms can explain these results.

- The rate of initiation of the monomer is very fast so that the droplets are all nucleated at the same time. The growth of the particles may then simply be explained by their coagulation.
- Only a small number of all droplets are nucleated. The nucleated particles grow, the monomer being supplied from the non-nucleated reservoir droplets either by diffusion through the toluene phase or by collisions between droplets. A mechanism involving diffusion of the monomer though the continuous medium seems more likely as polymerization carried out in heptane, which only dissolves minute amounts of AM, lead to smaller latex particles ($R_H \sim 100$ Å).

2. The particle radius as determined from the intensity of scattered light is found to be much smaller than the hydrodynamic one. This result can be explained from the following considerations. As a consequence of the particle growth, the total interfacial area of the polymer particles is reduced liberating some AOT molecules which would aggregate to form small water-swollen micelles containing a given amount of water. The presence of such micelles has indeed been verified from sedimentation measurements[30]. For a system containing two species of particles, the zero extrapolation concentration of the excess scattering is given by :

$$\lim_{\phi \to 0} (I - I_s) = \frac{4\pi}{3} K \left[\phi_p \left(\frac{dn}{d\phi_p}\right)^2 R_p^3 + \phi_m \left(\frac{dn}{d\phi_m}\right)^2 R_m^3 \right] \tag{15}$$

where subscripts p and m refer to the polymeric particles and AOT micelles respectively. If $R_p \gg R_m$, the first term of the right hand side of Equation 15 is dominant, which explains why the component associated with small micelles is not detected in the correlation function of the scattered light. However, the average value of $\phi (dn/d \phi)^2$ used for the determination of R can be largely overestimated with respect to the effective $\phi_p (dn/d\phi_p)^2$. This would lead to an underestimation of R_p.

Another interesting feature to be noted is the decrease of the attractive interactions after polymerization of the acrylamide inside the droplets. This is evidenced by both the light scattering (cf. Table II) and viscosity interaction parameters (Table III and Figure 4). The presence of two kinds of particles could affect these interaction forces and experiments are in progress to realize a separation of the two species.

REFERENCES

1. C. Schauber, Thesis Mulhouse (France), 1979
2. J.O. Stoffer and T. Bone, J.Dispersion Sci. Tech. 1,No 1,37 (1980)
3. J.O. Stoffer and T. Bone, J.Polymer Sci.Polym.Chem.Ed., 18(8), 2641 (1980)
4. Y.S. Leong, G. Riess and F. Candau, J.Chim.Phys. 73, 279 (1981)
5. S.S. Atik and J.K. Thomas, J.Am.Chem.Soc., 103, 4279 (1981)
6. P.L. Johnson, H.I. Tang and E. Gulari, paper presented at the 183rd ACS Nat.Meeting Las Vegas, (April 1982)
7. See for instance : K.S. Shinoda and S. Friberg, Adv.Colloid Interface Sci., 4, 281 (1975)
8. Y.S. Leong and F. Candau, J.Phys.Chem. (in press)
9. J. Rogers and P.A. Winsor, J.Colloid Interface Sci.,30,247(1969)
10. W.M. Horspool in "Aspects of Organic Photochemistry" p.42 Academic Press, N.Y., 1976
11. See for instant "Photon Correlation and Light Beating Spectroscopy" H.Z. Cummins, Editor, Plenum Press, New York, 1974
12. P. Gramain and J. Libeyre, J.Appl.Polymer Sci., 14, 383 (1970)
13. P. Ekwall, L. Mandell and K. Fontell, J.Colloid Interface Sci. 33, 215 (1970)
14. H.F. Eicke in "Micellization, Solubilization and Microemulsions", K.L. Mittal, Editor, Vol. 2, p. 429 Plenum Pres, New York, (1977)
15. R.A. Day, B.H. Robinson, J.H.R. Clarke and J.V. Doherty, J.Chem.Soc.Faraday Trans., 1, 75, 132 (1979)
16. W.G.M. Agterof, J.A.J. Van Zomeren and A. Vrij, Chem.Phys., Lett., 43, 363 (1976)
17. A.A. Calje, W.G.M. Agterof and A. Vrij in "Micellization, Solubilization and Microemulsions" K.L. Mittal, Editor, Vol. 2, p. 779, Plenum Press, New York, 1977
18. A.M. Cazabat, D. Langevin and A. Pouchelon, J.Colloid Interface Sci., 73, 1 (1980)
19. N.F. Carnahan, and K.E. Starling, J.Chem.Phys., 51, 635 (1969)

20. G.K. Batchelor, J.Fluid.Mech., $\underline{52}$, 245 (1972) ; $\underline{74}$, 1(1976)
21. J.L. Anderson and C.C. Reed, J.Chem.Phys., $\underline{64}$, 3240 (1976)
22. B.U. Felderhof, J.Phys. A.$\underline{11}$, 929 (1978)
23. See for instance : P. Becher, "Emulsions Theory and Practice"
    Reinhold Publishing Corp. pp. 60-85 (1966)
24. R. Roscoe. Brit.J.Appl.Phys., $\underline{3}$, 267 (1952)
25. M. Mooney, J.Colloid Sci., $\underline{6}$, 162 (1951)
26. D.J. Cebula, L. Harding, R.H. Ottewill and P.N. Pusey,
    Colloid Polym.Sci., $\underline{258}$, 973 (1980)
27. D.J. Cebula, R.H. Ottewill, J. Ralston and P.N. Pusey, J.Chem.
    Soc.Faraday Trans., 1, $\underline{77}$, 2585 (1981)
28. A.M. Cazabat and D. Langevin in "Light Scattering in Liquids
    and Macromolecular Solutions", V. Degiorgio, M. Corti and
    M. Giglio, Editors, p. 139, Plenum Press, New York, 1980
29. S. Brunetti, D. Roux, A.M. Bellocq, G. Fourche and P. Bothorel,
    (submitted for publication).
30. Y.S. Leong, G. Pouyet and F. Candau, (Results to be published
    1983).

APPLICATION OF THE ION-EXCHANGE MODEL TO O/W MICROEMULSIONS

Raymond A. Mackay

Department of Chemistry
Drexel University
Philadelphia, PA 19104

The ion-exchange model developed for reactions in aqueous micelles is extended in oil-in-water microemulsions. This model assumes that the concentrations of free and bound counterion and coion are related by means of an ion-exchange constant K_{IE}. The reaction of an oil soluble phosphate ester with fluoride and hydroxide ion in a cationic cetyltrimethylammonium bromide (CTAB) microemulsion is shown to be consistent with this model. A K_{IE} value of 0.1-0.2 is obtained, comparable to values obtained in aqueous CTAB micelles. The model is also applied to the acid-base equilibrium of a neutral aqueous indicator and its cation in an anionic sodium cetyl sulfate (SCS) microemulsion. The model is also consistent with these data, yielding a constant value of K_{IE} over compositions ranging from 30-70% water.

INTRODUCTION

Reactions in oil-in-water microemulsions involving oil soluble substrates and aqueous ions should bear some resemblance to
similar reactions in aqueous micelles. Recently, a psuedophase ion
exchange (IE) model has been applied to reactions between substrates
in micelles and aqueous nucleophiles[1]. The model has been generally successful with only a few partial failures[2]. The binding
of counterions and coions is treated from the point of view of an
ion exchange equilibrium (Equation 1).

$$X_b + Y_f = X_f + Y_b \qquad\qquad (1)$$

where X and Y represent the counterion and coion (e.g. nucleophile),
and the subscripts b and f represent bound and free. Here, the
surface is the region in which the bound ions reside, presumably
the Stern layer.

Thus, the rate constant in micelle relative to that in water
will depend on both the fraction of free counterion (α) and on the
nature of the aqueous (reagent) coion with respect to the micellar
counterion. The other extreme may be called the effective surface
potential (ESP) model. The ESP model is defined here as a limiting
case in which the reaction rate is controlled by the nucleophile
concentration at the surface (Gouy-Chapman), which is determined
by the surface charge. Such a model has also been advanced as a
possible explanation of the apparent failure of the psuedophase
model in a reactive counterion micelle[3]. In other words, the
relative rate constant should depend on the degree of dissociation
of the micelle counterion (α), but not on the identity of the
reactant coion.

In this paper, we will develop a method of applying the ion-
exchange model to microemulsions. The IE and ESP models will be
discussed in terms of two systems for which sufficient data are
available. These systems are the reaction of an oil soluble
phosphate ester with aqueous nucleophile in a cationic microemulsion, and the acid-base equilibrium of a neutral water soluble
indicator and its (exchangeable) cation in an anionic microemulsion.
The ester-nucleophile reaction will be discussed first in order to
establish the basis for the method. The IE model will then be
applied to the experimental data for this system, and then extended
to the indicator equilibrium data.

RESULTS AND DISCUSSION

Phosphate ester - nucleophile reactions. We have previously
investigated the reaction between (oil soluble) phosphate esters,
principally p-nitrophenyldiphenylphosphate (PhP), with aqueous

hydroxide and fluoride ions in both cationic (CTAB) and nonionic (Brij 96) microemulsions[4,5]. These results have been compared with those obtained in simple aqueous micelles. The rate constants and activation enthalpies (ΔH) obtained in aqueous cetyltrimethyl-ammoniom bromide (CTAB) micelles agreed with those reported earlier[6]. Some of the salient features observed were as follows. The activation enthalpies in CTAB microemulsion (μE) were lower than for the corresponding micelles (Table I), and the value of ΔH^+ for both nucleophiles were equal in CTAB micelles but not in μE. Product analysis showed that, in the case of hydroxide, the actual nucleophile was the butoxide ion. In fact, no monoanion product was observed, a 100% yield of the ester being obtained (Equation 2).

$$(C_6H_5O)_2 - \overset{\overset{\displaystyle O}{\displaystyle \|}}{P} - OC_6H_4NO_2 \left\{ \begin{array}{l} \xrightarrow{\quad OH^-\quad} (C_6H_5O)_2 - \overset{\overset{\displaystyle O}{\displaystyle \|}}{P} - O^- \\[2em] \xrightarrow{\quad OR^-\quad} (C_6H_5O)_2 - \overset{\overset{\displaystyle O}{\displaystyle \|}}{P} - O - R \end{array} \right. \qquad (2)$$

However, the equilibrium concentration of butoxide is controlled by the local hydroxide concentration. Therefore, it is still valid to consider hydroxide as the reagent ion since we have also shown that comparisons are best made between charged and uncharged micelles or microdrops rather than between solution

Table I. Comparison of Activation Enthalpies for the Reaction of OH⁻ and F⁻ with PhP in CTAB Micelles and Microemulsion (vide text).

Nucleophile	$\Delta H^{\neq}$ (kcal mol^{-1})	
	Micelle[a]	μE[b]
fluoride	14	10
hydroxide	14	8

a. Aqueous CTAB micelles (concentration 2.5 mM).
b. CTAB/hexadecane/1-butanol microemulsion, 60% water, 10% oil initial (see reference 4).

and organizate or between different organizates[7]. This is an essential point in the application of the IE model to microemulsions, and will be discussed in more detail below. It may be stressed at this point, however, that this will be the case only if the composition and particularly the location of the reaction surface are similar in the ionic and nonionic µE systems being employed for the comparision. We have shown that while a solute may be entirely solubilized in the microdroplet, its relative location can vary, thus affecting the rate constant[7], pK_a[8], etc. By this comparison method, the effective surface potential in CTAB micelles (slightly above the CMC) is determined to be about 130 mV, compared with values of about 30-60 mV for CTAB microemulsion.

Since the nucleophile (Y) is aqueous, its initial actual concentration in the µE should be biven by $[Y]/(1-\phi)$, where [Y] is the overall concentration and ϕ is the volume fraction occupied by the drops (phase volume). Thus, the second order rate constant k_2 calculated by dividing the observed psuedo-first order rate constant by [Y] does not properly reflect the intrinsic (surface) rate constant. We, therefore, have defined a phase volume corrected rate constant $k_{2\phi}$ given by Equation 3.

$$k_2 = k_2(1-\phi) \tag{3}$$

A plot of these rate constants as a function of ϕ for the reaction of PhP with both fluoride and hydroxide is shown in Figure 1. Three features are immediately evident. First the (normalized) results are the same for both nucleophiles. Second $k_{2\phi}$ does not vary with ϕ in the nonionic µE. This is quite reasonable since no major structural changes upon dilution are expected[9]. Third, the value of $K_{2\phi}$ increases with increasing dilution for the cationic µE. In light of the variation of α with ϕ[10,11], this is ascribed to an increased degree of dissociation and thus an increased surface potential with decreasing ϕ.

As mentioned above, all evidence to date indicates that, providing solute locations are similar, the intrinsic properties (rate constants, equilibrium constants, quantum yields) are the same in all microemulsions with similar compositions[7,8,12]. Therefore,we define an effective surface potential (ψ_ϕ) by Equation 4, where e is the electronic charge, k Boltzmann constant, and T the absolute temperature.

$$k_{2\phi}(CTAB)/k_{2\phi}(Brij) = \exp(e\psi_\phi/kT) \tag{4}$$

A plot of ψ_ϕ vs ϕ is shown in Figure 2, along with the corresponding value of α. It may be noted that a value of ψ = 36 mV obtained

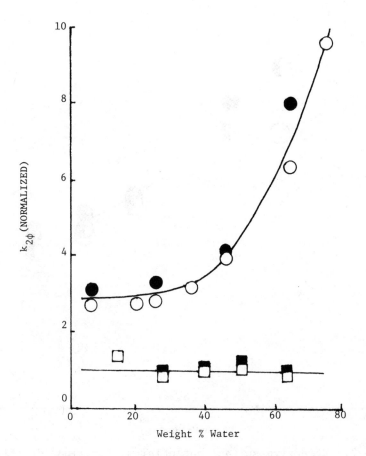

Figure 1.   Phase volume corrected rate constant vs % water (ca. 1 − φ) for the reaction of OH⁻ (open) and of F⁻ (closed) in CTAB (circles) and Brij (squares) microemulsion.

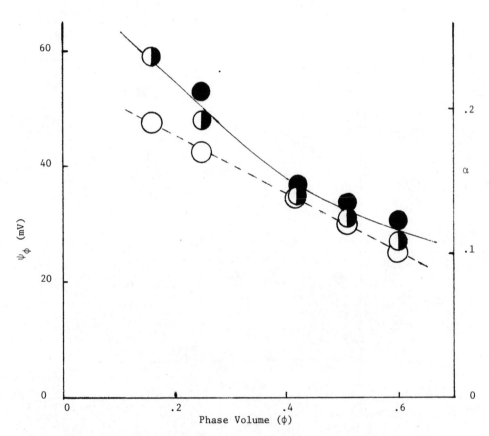

Figure 2.   Effective surface potential ψ_ϕ for fluoride (filled circles) and hydroxide (half-filled circles) and degree of dissociation α (open cricles) in CTAB μE <u>vs</u> phase volume (ϕ)

from an indicator pK measurement in the CTAB microemulsion[8] at $\phi = 0.4$ falls on the same curve.

Now, these results would seem to be in accord with the ESP model since ψ_ϕ is essentially the same for both fluoride and hydroxide ions. However, both of these ions have very similar ion exchange constants against bromide ion[13]. Therefore, we will proceed to apply the IE model to see if these data are also consistent with this model.

Ion-Exchange Model. Neglecting activity coefficients, the ion exchange (K_{IE}) is given by Equation 5, where the concentrations refer to <u>actual</u> (not overall) concentrations, and the subscripts f and b refer to free and bound, respectively.

$$K_{IE} = (X_f)(Y_b)/(X_b)(Y_f) \tag{5}$$

It may then be shown[14] that the ratio of phase volume corrected rate constants in ionic and nonionic (e.g. CTAB and Brij) microemulsions is given by

$$k_{2\phi}ionic/k_{2\phi}nonionic = (r/3s)(1-\phi/\phi)(K_\phi/1+K_\phi)'. \tag{6}$$

where s is the thickness of the reaction layer, r is the drop radius, and $K_\phi = K_{IE}\beta/(1-\beta)$. The fraction of bound counterion $\beta = 1 - \alpha$. If $K_\phi < 1$, and replacing the ratio $k_{2\phi}ionic/k_{2\phi}nonionic$ by $\exp(e\psi_\phi/kT)$, Equation (6) may be written as

$$\psi_\phi = kT/e \ln(1 - \phi/\phi)(\beta/1-\beta) + kT/e \ln(K_{IE} r/3s) \tag{7}$$

Under these conditions, a plot of ψ_ϕ vs. $\ln[(\frac{1-\phi}{\phi})][\beta/(1-\beta)]$ should be a straight line with a slope of 26 mV at 25°C. A plot of data for the reaction of PhP with hydroxide and fluoride, using values of β from the conductivity measurements, is given in Figure 3. The slope is 23 mV, and using a polargraphic radius of 34A°[10,15] and assuming s=2-4A°, $K_{IE} = 0.1-0.2$. This may be compared with values of K_{IE} for exchange between hydroxide and bromide counterions in CTAB micelles of 0.08[1a] and 0.05[1d]. Here, $K_\phi < 1$ (but not <<1), and an iteration of the data yields $K_{IE} = 0.08-0.16$. The important point is that the data also fit the IE model. It should be repeated that, in contrast to the ESP model, the IE predicts that two different nucleophiles should have different values of ψ_ϕ since they have different values of K_{IE}. However, as mentioned above, hydroxide and fluoride exhibit very similar ion-exchange behavior,

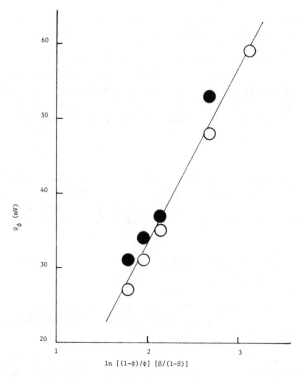

Figure 3. Surface potential (ψ_ϕ) vs. given function of ϕ and β (Equation 7). The open circles are for hydroxide and the closed circles are for fluoride. The slope is 23mV.

and to make a more definitive test another nucleophile with a very different value of k_{IE}, such as cyanide, is needed[1b].

Acid–base equilibrium. It is also possible to apply the IE model to indicator equilibria in the aqueous phase. The compound 1-methyl-4-cyanopyridinium oximate (CPO) has been used as an indicator of environment because of large solvent effects on the wavelength of its intramolecular charge-transfer absorption bands[16]. It undergoes the acid–base equilibrium

$$CPOH^+ = CPO + H^+,\tag{8}$$

with an aqueous pK_a of 4.61.

The CPO is a neutral molecule, and is quite water soluble. This solute can also be solubilized in the oil phase of the μE, depending upon the oil and surfactant. However, as noted above, its location can be very readily determined from its absorption spectrum, and in an anionic sodium cetyl sulfate (SCS) μE it is completely solubilized in the aqueous phase[17]. The CPOH$^+$ cation can exchange with the surfactant cation (Na$^+$), resulting in a change in the apparent acid dissociation constant (K_a) of the indicator. The pK_a was measured in the SCS μE using an oxalate buffer in its mono- and dianion forms to avoid significant buffer uptake by the disperse phase. The pH of the aqueous buffer was used to calculate the pK_a[17]. The ratio of bound (b) and free (f) CPOH$^+$ is given by

$$K_\beta = \frac{[CPOH^+]_b}{[CPOH^+]_f} \tag{9}$$

The value of K_β as a function of ϕ is determined from the pK_a in microemulsion (p$K_a{}^m$) and in water (p$K_a{}^w$) according to Equation (10).

$$pK_a{}^m = pK_a{}^w + \log(1 + K_\beta) \tag{10}$$

Then, since

$$K_{IE} = \frac{(Na^+)_f (CPOH^+)_b}{(NA^+)_b (CPOH^+)_f}, \tag{11}$$

it follows that

$$K_{IE} = K_\beta (1-\beta)/\beta . \tag{12}$$

The values of K_{IE} are given in Table II, and an average value of $K_{IE} = 1.1 \pm 0.2$ is obtained over the range $0.2 \le \phi \le 0.7$. This represents a constant value of K_{IE} since the standard deviation of the average value ($\pm 18\%$) is even less than the estimated uncertainty in an individual value ($\pm 25\%$).

Also shown in Table II are values of K_{IE}. These have been calculated using pH values rather than the pH of the aqueous buffer since there is reason to believe that this may be better representation of the true pH in microemulsions[8]. In any event, the K_{IE} also yield an effectively constant values of 3.0 ± 0.5. These acid-base indicator data are, therefore, consistent with the IE model.

On the other hand, the ESP model predicts that $\ln K$ should be roughly proportional to $\exp(e\psi_\phi/kT)$, which, in turn, should be

Table II. CPO Indicator Equilibrium Measurements in an Anionic
SCS Microemulsion (vide text).

ϕ^a	β^b	K_β^c	K_{IE}^d	$\overline{K_{IE}}^e$
0.70	.93	11.3	0.9	2.1
0.52	.87	7.9	1.2	3.0
0.40	.83	5.6	1.1	3.0
0.35	.82	6.2	1.4	3.7
0.24	.80	4.9	1.2	3.2
0.21	.80	4.2	1.1	2.8

a.   Compositional phase volume for the SCS/1-pentanol/
     mineral oil (21% initial) system (ref. 10).
b.   From conductivity measurements.
c.   Using oxalate buffer (mono and dianion).
d.   Using pH of aqueous buffer. K_{IE} = 1.1 ± 0.2 (ave).
e.   Using pH values (see ref. 8). K_{IE} = 3.0 ± 0.5 (ave).

roughly proportional to α. As α increases (β decreases), K_β should
therefore increase. This is opposite to the observed trend in
Table II.

SUMMARY

     The ion exchange model has been applied to ionic o/w micro-
emulsions. The resulting equations relate the rate constants for
droplet bound substrate with aqueous coion nucleophile in an
ionic and corresponding nonionic system to the phase volume and
degree of counterion dissociation. These equations have been
shown to be consistent with data from a phosphate ester-nucleophile
reaction in cationic microemulsion. However, these reaction data
are also in accord with the ESP model. Since the ion exchange
constants of fluoride and hydroxide are about the same; further
experiments using nucleophiles with different values of the ion-
exchange constant need to be carried out in order to provide a
more definitive test of the model. The IE model has also been
applied to acid-base equilibrium data in an anionic microemulsion.
These data are in accord with the IE model, but directly opposed
to the ESP model.

ACKNOWLEDGEMENT

The author wishes to thank Ms. K. Mazaika for performing the pK_a measurements, and the U.S. Army Research Office for financial support.

REFERENCES

1. See for example (a) H. Chaimovich, J.B.S. Bonilha, M. J.
    Politi and F. H Quina, J. Phys. Chem., 83, 1851 (1979);
    (b) C. A. Bunton, L. S. Romsted and C. Thamavit, J. Am. Chem.
    Soc., 102, 3900 (1980); (c) F. H. Quinn, M. J. Politi, I. M.
    Cuccovia, E. Raumgarten, S. M. Martins-Franchetti and H.
    Chaimovich, J. Phys. Chem., 84, 361 (1980); (d) C. A. Bunton,
    L. S. Romsted and L. Sepulvda, J. Phys. Chem., 84, 2611 (1980),
    and references therein.
2. (a) C. A. Bunton, L. S. Romsted and G. Savelli, J. Am. Chem.
    Soc., 101, 1253 (1979); (b) C. A. Bunton, J. Frankson and L.
    S. Romsted, J. Phys. Chem., 84, 2607 (1980).
3. C. A. Bunton, J. Frankson and L. S. Romsted, J. Phys. Chem.,
    84, 607 (1980).
4. C. Hermansky and R. A. Mackay, in "Solution Chemistry of
    Surfactants", K. L. Mittal, Editor, Vol. 2, pp. 723-730,
    Plenum Press, New York, 1979.
5. R. A. Mackay and C. Hermansky, J. Phys, Chem., 85, 739 (1981).
6. C. A. Bunton and L. Robinson, J. Org. Chem., 34, 773 (1969).
7. R. A. Mackay, Adv. Colloid Interface Sci., 15, 131 (1981).
8. R. A. Mackay, K. Jacobson and J. Tourian, J. Colloid Interface
    Sci., 76, 515 (1980).
9. C. Hermansky and R. A. Mackay, J. Colloid Interface Sci., 73,
    324 (1980).
10. R. A. Mackay, in "Microemulsions", I.D. Robb, Editor, pp.
    Plenum Press, New York, 1982.
11. R. A. Mackay and C. Hermansky, submitted for publication.
12. C. A. Jones, L. E. Weaner and R. A. Mackay, J. Phys. Chem.,
    84, 1495 (1980).
13. J. X. Khym, "Analytical Ion-Exchange Procedures in Chemistry
    and Biology", Prentice-Hall Inc., Englewood Cliffs, NJ, (1974).
14. R. A. Mackay, J. Phys. Chem., in press.
15. R. A. Mackay and R. Agarwal, unpublished results.
16. R. A. Mackay and E. J. Poziomek, J. Am. Chem. Soc., 94, 6107
    (1972).
17. R. A. Mackay, K. Letts, and C. Jones, in "Micellization,
    Solubilization and Microemulsions", K. L. Mittal, Editor,
    Vol. 2, pp. 807-, Plenum Press, NY, 1977; see also reference
    10 for phase map.

Part IX
General Overviews and Other Papers

HLB - A SURVEY

Paul Becher

Paul Becher Associates Ltd.
P. O. Box 7335
Wilmington, DE 19803

A survey of methods of determining Hydrophile-
Lipophile Balance (HLB) is presented. From these, it may
be deduced that HLB is of the nature of a thermodynamic
quantity, with the dimensions of a free energy. No
attempt is made to treat the application of this concept
in a critical way; rather, certain inconsistencies in
the various approaches are indicated (if indirectly),
and some proposals are made for the directions of further
research.

INTRODUCTION

Margarine, invented in 1869 by the Frenchman, Hippolyte Mège-Mouriès, was for many years an inferior product. When heated in the saucepan, it spattered ferociously, and, to avoid serious burns, the act of sauteeing became an exercise in agility. In the 1930's however, it was discovered that certain additives ("anti-spattering agents"), when present in small quantities, effectively inhibited this undesirable phenomenon. It was also noted, in a qualitative way, that the effectiveness of these antispattering agents was a function of their structure, i.e., certain types of molecular structures worked better than others[1]. These anti-spattering agents were, of course, surfactants, and the observations correlating structure and effectiveness constitute the first recognition of Hydrophile-Lipophile Balance.

However, the quantification of this concept had to wait until 1948, 79 years after the invention of margarine, and a quarter-century after the introduction of antispattering agents. In that year, William C. Griffin, working in the laboratories of what was then the Atlas Powder Company, was studying the emulsifying properties of the newly developed ethoxylated nonionic surface-active agents. Griffin noted that in a homologous series of ethoxylates, certain combinations of lipophilic chain-length and ethoxylate chain-length produced optimum oil-in-water emulsions for a given oil phase. In fact, this optimal property was quite simply related to the weight per cent of ethylene oxide in the molecule. These observations were reported in a series of papers[2].

However, for commercial reasons, Griffin initially chose to conceal the simplicity of the relationship, and described a complicated experimental technique for the determination of what he called the Hydrophile-Lipophile Balance, or HLB. Even though he later revealed that the HLB number (for ethoxylates) was simply the weight per cent EO (divided by 5), for some strange reason, the experimental technique has been regarded as a stumbling block in the utilization of the concept. As recently as 1981, a writer states, "Its direct measurement requires a heroic effort and, to a research chemist, enormous samples"[3]. This is quite incorrect.

Empirical quantity or not, it is not incorrect to say that the introduction of HLB revolutionized the formulation of emulsions. I have never attempted to determine the number of references to HLB in the patent literature, but such references must number in the thousands. In a bibliography published in 1974, Becher and Griffin[4] listed more than 300 papers which dealt with HLB in a substantive way, that is to say, represented a significant addition to the theory, application of HLB, or added to the experimental data on it.

The fact is, however, that in the thirty-odd years since the concept was introduced a substantial effort has been expended on developing methods (both experimental and theoretical) for the determination of hydrophile-lipophile balance, and on attempts to raise it from the level of an empirical concept to one having thermodynamic validity. Although this latter effort has not totally succeeded, I argue, here and now, that the HLB number is a thermodynamic quantity, having the dimensions of a free energy. It is, in fact, the free energy associated with positioning the amphiphilic molecule at the oil-water interface.

To begin, I can do not better than to tabulate some of the methods reported for the determination of HLB, in a manner suggested by Müller[5]:

I.   General Methods

     Griffin HLB [2]
     Davies, HLB [6,7]

II. Special Methods

     Spreading Coefficient [8]
     Polarity Index (GLC/TLC) [9-18]
     Polarography [19]
     NMR and Mass-Spectroscopy [20-22]
     Calorimetry [23]
     Dielectric Constant [24-27]
     Partition Coefficient [6, 28-30]
     Heat of (Interfacial) Adsorption [31]
     Solubility Parameter [32-34]
     Partial Molar Volume [35-36]
     Monolayer Properties [37]
     Critical Micelle Concentration [38-39]
     Water Number, Cloud Points [2, 40-43]
     Phase Inversion Temperature [44]
     Interfacial Tension [45]
     Foaming [46]

The references listed above are by no means complete. They do serve, however, as an introduction to the literature on the determination of HLB over the last two decades.

In the remainder of this paper, I shall review, in more or less detail, some of the approaches to HLB (and associated quantities) which will serve to illuminate the statement made above with respect to the relevance of thermodynamics to HLB.

APPROACHES TO HLB

Approach of J. T. Davies

An early attempt to categorize HLB in a fundamental manner came in 1957, from the Cambridge University group in colloid science. Davies[6], in a highly original series of investigations, related the HLB number of a group of emulsifying agents to the coalescence rates of oil and water droplets at the oil-water interface, as well as to the partition coefficient of the emulsifier between the phases in question.

Thus, if the rate of coalescence of an oil droplet at the interface between an oil and an aqueous surfactant solution was defined as $Rate_1$ and that of a water droplet at the same interface as $Rate_2$, then

$$\ln (Rate_2/Rate_1) = 2.2\theta(HLB - 7) \qquad (1)$$

where θ = fraction of interface covered by surfactant (in principle, determined from measurement of interfacial tension).

Similarly, if c_w is the equilibrium concentration of surfactant in the aqueous phase, and c_o that in the oil phase

$$(HLB - 7) = 0.36\ln(c_w/c_o) \qquad (2)$$

Strictly speaking, of course, the ratio of concentrations on the right-hand side of the above equation is not the partition coefficient, since the use of the bulk concentration of the surfactant in the aqueous phase does not take into account the effect on the thermodynamic activity brought about by the organization of some of the surfactant into micelles. Thus, for example, the excellent data of Greenwald et al.[47] for the partition of poly-ethoxylated nonyl phenols between water and iso-octane, when substituted in the above equation, do not yield satisfactory results.

The arguments behind both of the equations are reasonable, however. The latter equation, although numerically unsatisfactory, is of the form showing the relationship between free energy and an equilibrium constant, and thus illuminates the thermodynamic nature of HLB.

In another approach, Davies[6] has attempted to formulate the HLB in terms of surfactant structure by assigning group numbers to various structural elements and combining them according to the equation:

$$HLB = 7 + \sum(\text{hydrophilic group numbers})$$
$$- \sum(\text{lipophilic group numbers}) \quad\quad (3)$$

where the last term on the right-hand side is usually 0.475n, the quantity n being the number of $-CH_2-$ groups in the lipophile. The group numbers defined by Davies are listed in Table 1.

A problem with this ingenious approach arises from the fact that, for polyoxyethylene derivatives, it treats each oxyethylene group as having the same effect on the HLB. This is clearly not the case, as a moment's thought reveals. If one oxyethylene is added to a molecule which already contains 10 such groups, the result is a 10 per cent increase in hydrophilicity. If, on the other hand, one oxyethylene is added to a molecule which already contains 100 such groups, the increase in hydrophilicity is only one per cent. A similar objection obtains to the concept of "linear HLB" introduced by Rimlinger[7]. It is possible, indeed, that a similar objection could be made for the summation for the lipophile. As a practical matter, however, the range of chain-lengths commonly employed is so small that it may not be numerically significant.

Table I. HLB Group Numbers According to Davies (Ref. 6).

Hydrophilic Group Numbers

Group	Number
$-SO_4Na$	38.7
$-COOK$	21.1
$-COONa$	19.1
$-N$ (tertiary amine)	9.4
Ester (sorbitan ring)	6.8
Ester (free)	2.4
$-COOH$	2.1
$-OH$ (free)	1.9
$-O-$	1.3
$-OH$ (sorbitan ring)	0.5

Lipophilic Group Numbers

Group	Number
$-CH-$	-0.475
$-CH_2-$	
CH_3-	
$=CH-$	

Derived Group Numbers

Group	Number
$-(CH_2-CH_2-O)-$	0.33
$-(CH_2-CH-O)-$ $/$ CH_3	-0.15

A more pertinent objection is found in the fact that Equation (3) yields, for a comparitively simple molecule as POE(23) lauryl alcohol (BRIJ 35), a value of the HLB which is in error by a factor of 36 per cent, which the reader may verify by means of the data of Table I.

Approach of Lin et al.

Some of the objections to the approach of Davies are mitigated by the work of Lin and various co-workers[38], which identifies the empirical group numbers of Davies with the free energy of transfer of the surfactant to the interface. While this does not entirely remove all the problems connected with the calculation of HLB for nonionic surfactants, it puts the calculation for ionic surfactants on firmer ground, a factor lacking in the original Griffin formulation.

It follows from the Traube Rule that there should be a linear relationship between the logarithm of the critical micelle concentration c_0 and the number of $-CH_2-$ groups in a lipophilic chain for a fixed hydrophilic head group, i.e.

$$\ln c_0 = a + bn \qquad (4)$$

where n is equal to the number of carbon atoms in the lipophilic moiety.

In the case of polyoxyethylated nonionic surfactants, it follows that a similar relation should govern the case where the lipophile is fixed and the EO-chain length of the hydrophile varied[53]:

$$\ln c_0 = A + B(EO) \qquad (5)$$

where EO is the number of ethylene oxides in the hydrophile. Further, EO is related to the HLB by

$$EO = M_L HLB / 44(20 - HLB) \qquad (6)$$

where M_L is the molecular weight of the lipophilic moiety.

Since the relationship between HLB and EO is hyperbolic, it is clear that if Equation (4) is linear, the same cannot be said for Equation (5). In fact, although Equation (5) is close to linear over a limited range of EO it cannot really be linear for the reasons suggested in connection with the work of Davies, mentioned above.

That this is the case is illustrated, for example, by the data of Crook, Fordyce, and Trebbi[54]. As a result, a combination of Equations (4) and (5) is, in fact, fortuitously linear[53]:

$$\ln c_o = A' + B'HLB \tag{7}$$

Lin and Somasundaran[38] have derived Equation (7) in a fundamental way, based on a consideration of the thermodynamics of micellization derived from the work of Shinoda[55] and Phillips[56]. On thermodynamic grounds it is shown that the critical micelle concentration is given by

$$c_o = ConstXexp[W_e - n\phi_m/kT] \tag{8}$$

where W_e is the electrostatic free energy per molecule, n is the number of carbon atoms per molecule and ϕ is the van der Waals interaction per $-CH_2-$ group between adjacent chains arising from micelle formation, or in other words, the free energy of transfer for the lipophilic chain from an aqueous to a lipophilic environment. In logarithmic form, this equation is, of course, equivalent to Equation (4). Equation (8) can be modified, according to Shinoda[55], to take into account the effect of counter-ions.

The result of these calculations is finally a relation of the form

$$\ln c_o = -[n\phi_m'/(1 + K_g)kT] = K'' \tag{9}$$

where ϕ_m' is the transfer energy per $-CH_2-$ group ($=\phi_m/n$) from the aqueous environment to the micelle, Kg is the ratio of counter-ions to surfactant ions, and K" is a constant characteristic of a given homologous series.

The point of all this is that the transfer free energies may be determined for various molecular groups, and that one therefore has the possibility of ultimately calculating the HLB values through Equation (7). Thus, Lin and Somansundaran identify the right-most term in the Davies Equation (3) with this transfer energy, i.e.,

$$\Sigma(lipophilic\ group\ numbers) = n\phi_m'/2.303kT \tag{10}$$

Accepting Davies empirical value of 0.475 for the $-CH_2-$ group, the transfer energy is seen to be equal to 1.09kT, which is in good agreement with values reported in the literature as far back as Langmuir's paper of 1917[57].

This approach has been most successful in arriving at values for the HLB of ionic surfactants. A number of HLBs determined in this way are shown in Table II.

The resulting HLB-values are in the range of approximately 10-45, in good agreement with the qualitative estimates of Griffin[2].

Table II.   Relation between Critical Micelle Concentration and
HLB (Ref. 38).

Surfactant	c_0 vs. HLB
RCOOK	$\log c_0 = -15.23 + 0.611[\text{HLB}]$
RCOONa	$= -16.33 + 0.718[\text{HLB}]$
RSO_4Na	$= -26.96 + 0.621[\text{HLB}]$
RSO_3Na	$= - 8.28 + 0.510[\text{HLB}]$
$C_nF_{2n+1}COONa$	$= -3.476 + 0.510[\text{HLB}]$
$C_nF_{2n+1}COOK$	$= -15.155 + 0.658[\text{HLB}]$

When there are structural variations, e.g., the presence of
two lipophilic chains or chain-branching, it is necessary to
modify the basic equation by the introduction of an effective
carbon number n_{eff}. However, the results of such calculations
of the effective chain length have been questioned (cf. below,
under the discussion of the approach of Wade et al.).

Although the technique of Lin and co-workers is apparently
effective for anionic surfactants containing short ethylene oxide
chains, it fails in the same way as the fundamental Davies equation
for simple nonionics based on ethylene oxide.

Approach of A. Beerbower and M. W. Hill

A little more than a decade ago, Beerbower and Hill[33] made
a most ingenious approach to the thermodynamics of HLB. Unfortunately, for reasons beyond the control of these workers, the treatment
is somewhat incomplete, and was not given the wider circulation it
deserved. As a result, it has been improperly neglected. I shall
attempt, in some part, to rectify that omission in the following
discussion.

In brief, Beerbower and Hill attempted to combine the so-called
"R-theory" of Winsor[48] with the solubility parametere approach of
Scatchard and of Hildebrand and Scott[49], as amplified by, for
example, Hansen[50].

However, it may be well first to review briefly Winsor's ideas,
as described in his book, Solvent Properties of Amphiphilic Compounds, using the nomenclature of Beerbower and Hill, which differs
somewhat from Winsor. Winsor's theory is a qualitative statement
of emulsion structure and phases in terms of the ratio of molecular
interactions on each side of the interface.

In this treatment, we start with the balance of the cohesive

Table III.   Ten Cohesive Energies in Emulsions (Ref. 33).

C_{OO}	C_{HW}
C_{LL}	C_{HH}
C_{OL}	C_{WW}
C_{OH}	C_{WL}
C_{OW}	C_{HL}

energies which control the mutual solubility of two liquids.   Two
types of cohesive energies are assumed to control this relationship
(nothing in the Winsor theory, however, prohibits the introduction
of other types of energy).   The first of these two energies is the
well-known London nonpolar dispersion (D), and the second is the
sum of all the polar cohesive energies (P) which Winsor ascribed
mainly to hydrogen bonding.   Thus, if we define the cohesive energy
of oil for oil as C_{OO} and that of water for water as C_{WW}, the energies which prevent miscibility and which favor phase separation are

$$C_{OO} = D_{OO} + P_{OO} \qquad\qquad\qquad (11)$$

$$C_{WW} = D_{WW} + P_{WW} \qquad\qquad\qquad (12)$$

and accordingly the cohesive energy between oil and water would be

$$C_{OW} = D_{OW} + P_{OW} \qquad\qquad\qquad (13)$$

Now, C_{WW} is so much greater than either C_{OO} or C_{OW} that separation
and immiscibility is the usual situation.   Indeed, C_{OO} and C_{OW} are
of the same order of magnitude, i.e.

$$C_{WW} > C_{OO} \sim C_{OW} \qquad\qquad\qquad (14)$$

We consider the introduction of a surfactant molecule L-H,
composed of lipophilic (L) and hydrophilic (H) moieties into an
oil/water system.   There result   ten possible cohesive energies,
as shown in Table III.

The relationships giving rise to these energies are illustrated
in Figure 1 from Beerbower and Hill[33].

Clearly, the relative magnitudes of these cohesive energies
control the phase relationships and miscibility in the three-
component system consisting of oil, water, and surface-active
agent.

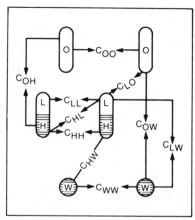

O-Oil Molecule

L-Lipophile   ) Segments Of

H-Hydrophile   ) Emulsifier Molecule

W-Water Molecule

Figure 1. Relation between various cohesive energies in an emulsion (from Ref. 33).

Winsor defines an ideal emulsion as one in which the cohesive energy of the lipophile for water C_{LO} and the hydrophile for water C_{HW} is greater than the other eight energies. When this criterion is met, the surfactant molecule is properly oriented in the oil/water interface, and the magnitude and ratio R of these two quantities determines the nature and stability of the emulsion. Thus, if

$$R = C_{LO}/C_{HW} \qquad\qquad (15)$$

it follows that if R>1 the interface will be concave towards the water phase (owing to the excess of cohesive energy on this side), resulting in a W/O emulsion, and vice versa. When $C_{LO} = C_{HW}$ then R = 1, and this is the inversion point for the emulsion.

Further, when R>>1 then the emulsifier will be solubilized in the oil phase, and when R<<1 the emulsifier will be solubilized in the aqueous phase. It will be recognized that this is nothing more than a quantitative statement of the Bancroft Rule[12].

Thus, it is possible to define a relationship between the HLB system and Winsor's R, as shown in Table IV[33].

Table IV. Relation between HLB and Winsor's R (Ref. 33).

HLB RANGE	R Values	Emulsion Type
16–20	$\ll 1$	No emulsion, solution of emulsifier in water
12–15	<1	Oil-in-water emulsion
8–10	$=1$	Planar, inversion point
4–6	>1	Water-in-oil emulsion
1–4	$\gg 1$	No emulsion, solution of emulsifier in oil

Following Beerbower and Hill, we define these cohesive energies (and hence R) in terms of the concept of solubility parameter, as introduced by Hildebrand and Scott[10]. In applying this theory to mixtures, the necessary assumption is that the change on mixing is negligible, and the force between two unlike molecules is the geometric mean of the forces between the like molecules, i.e., that $C_{AB} = [C_{AA}C_{BB}]^{1/2}$.

The cohesive energy of a pure liquid is calculated most simply from the energy of vaporization, which is related to the (experimentally available) heat of vaporization by the well-known thermodynamic relation

$$\Delta E_v = \Delta H_v = P\Delta V = \Delta H_v - RT \qquad (16)$$

This quantity, divided by the molar volume [V], is the so-called cohesive energy density (CED). This quantity is then the energy required to separate all the molecules in a cubic centimeter to a distance at which they will no longer interact, and is thus a measure of cohesive energy.

As a matter of convenience, Hildebrand defined the square root of the CED as the solubility parameter, with the symbol δ . For example, the solubility parameter of benzene is equal to 9.2, while that of hexane is 7.3, clearly indicating that their solvent properties are quite different, even though they are both hydrocarbons containing six carbon atoms.

This works well enough if we are dealing with nonpolar molecules, where the cohesive energy arises solely from London dispersion forces. If, however, polar or hydrogen-bonding contributions to the cohesion are present (and this is the case in most, if not all, emulsion situations), Hansen[50] and others made the assumption (justified probably only by its success) that these other contributions could be handled in exactly the same way as the dispersion contribution, and that they make an additive contribution to the cohesive energy density

$$\Delta H_v - RT = \Delta E_d + \Delta E_p + \Delta E_h \qquad (17)$$

where the subscripts d, p, and h refer to the dispersion, polar, and hydrogen-bonding contributions to the CED.

From this, it follows that the total solubility parameter may be expressed as

$$\delta^2 = \delta d^2 + \delta p^2 + \delta h^2 \qquad (18)$$

On the basis of extensive experimental data, Beerbower and Hill have determined that one may write a separate solubility parameter for the hydrophilic and lipophilic moieties of the surfactant molecule:

$$\delta H^2 = \delta d^2 + 0.25 \ \delta p^2 + 0.25 \ \delta h^2$$
$$\delta L^2 = \delta d^2 + 0.25 \ \delta p^2 + 0.25 \ \delta h^2 \qquad (19)$$

where the individual partial solubility parameters are determined by the methods described by Hansen and Beerbower[52], and the scaling factor of 0.25 for the polar and hydrogen-bonding contributions was introduced by Hansen, in connection with work on paint systems.

It should be noted that this scaling factor is empirical, but may no doubt be ascribed to effects arising from the local environment of the surfactant molecule, i.e., it orientation at an interface.

This selection of an appropriate emulsifier molecule is now reduced to finding one in which the solubility parameter of the hydrophilic moiety matches that of the aqueous phase, and that of the lipophilic moiety that of the oil phase. If R_o is defined as the cohesive energy ratio corresponding to this perfect match then

$$R_o = [V]_L \ \delta_L^2 / [V]_H \ \delta_H^2 \qquad (20)$$

where $[V]_L$ and $[V]_H$ are the molecular volumes of the lipophilic and hydrophilic moieties, respectively.

Beerbower and Hill now relate this to the emulsifier HLB through the Griffin definition for nonionic, ethylene-oxide based, emulsifiers:

$$HLB = 20M_H/(M_L + M_H)$$

$$= 20[V]_H d_H/([V]_L d_L + [V]_H d_H) \qquad (21)$$

where the quantities M and d are the molecular weights and densities of the hydrophilic and lipophilic moieties, respectively. By a combination of Equations (20) and (21)

$$R_o = [d_H \delta_L^2/d_L \delta_H^2] \times [20/HLB - 1] \qquad (22)$$

Since the three partial solubility parameters are either known or can be calculated, and since for polyoxyethylene compounds, d_H is fairly constant (=1.125) we can write

$$K = R_o \delta_H^2/d_H \qquad (23)$$

Accordingly, from Equation (22),

$$HLB_o = 20 \delta_L^2/(\delta_L^2 + Kd_L) \qquad (24)$$

where HLB_o is the optimum HLB when $\delta_L^2 = \delta_o^2$.

Since it has been shown that the solubility parameter is related to the surface tension[53], it is also possible to calculate the required HLB for a given oil from the relation

$$HLB_o = 20\gamma/(\gamma + 2.49[V]_o 1/3) \qquad (25)$$

Clearly, although it is necessary to introduce empirical elements to this theory in order to obtain calculable results, its basis is in a fundamental and accepted thermodynamic theory of solution. It is interesting to note, in addition, that the author has found Equation (25) to be quite satisfactory in predicting the required HLB for a number of liquids.

Approach of K. Shinoda

As is well known, the solubility properties of both nonionic and ionic surfactants change with temperature, and, in fact, in both cases rather abrupt changes take place at a temperature characteristic of the surfactant, i.e., the cloud point for non-ionics and the Krafft point for anionics[58].

These changes in solubility are reflected in changes in surface activity, and, by consequence, in changes in the hydrophile-lipophile balance. In fact, a nonionic surfactant can undergo such a drastic change in HLB that it may change from an oil-in-water emulsifier to a water-in-oil emulsifier. Shinoda[44] has observed that, for a given emulsion system, this is accompanied by emulsion inversion at a temperature characteristic of the

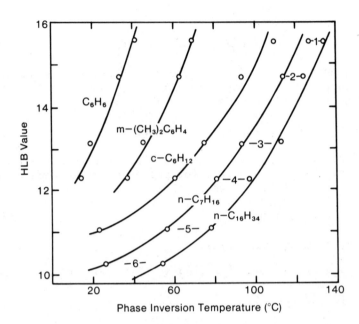

Figure 2.   Relation between HLB and PIT (from Ref. 44)   for a
variety of oil phases.   See original[33] for identification of surfactants.

system.   This temperature he named the Phase Inversion Temperature, or PIT.

It is immediately obvious that at the PIT the emulsifier
system is in the condition where the Winsor R is equal to one,
and the correlation between the type of emulsion and its stability
follows from the considerations developed earlier, and that, by
the same token a correlation between HLB and PIT is to be expected.

This is illustrated in Fig. 2, from Shinoda[44], for a variety
of oil phases, and in Fig. 3 for the system cyclo-hexane/water in
the presence of a variety of nonionic emulsifiers (cf. the original
papers for identification of surfactants employed).   This correlation is so pronounced that Shinoda now prefers the designation
HLB-temperature to PIT.

Determination of the PIT serves to characterize the optimum
HLB of the emulsifier for the particular oil phase in a more
convenient way than the classical stability measurements described
by Griffin[2] (cf. Table V).   Moreover, the most highly-dispersed
emulsions are produced by agitation near the HLB-temperature, with

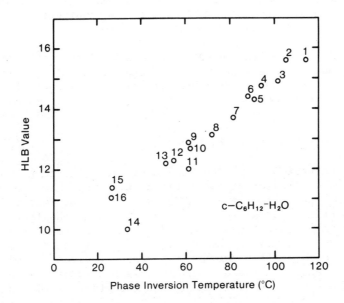

Figure 3.   Relation between HLB and PIT (from Ref. 44) for the system cyclohexane-water in the presence of a variety of nonionic emulsifiers See original reference for identification of surfactants.

Table V.   Comparison between Griffin Required HLB and HLB Estimated from PIT (Ref. 44).

Type of Oil	HLB Numbers (Griffin)	HLB Numbers (PIT)
Mineral Oil (Paraffinic)	10	10
Propane Tetramer	14	12
Kerosene	14	12
Trichlorofluoroethane	14	12.5
Cyclohexane	15	13
Carbon Tetrachloride	16	13.5
Xylene	14	14.5
Toluene	15	15.5
Benzene	15	16.5

subsequent cooling (for o/w emulsions) or heating (for w/o emulsions). If the storage temperature is of the order of 30-40° away
from the HLB-temperature, the stability is maximized. Again, the
relation of this to the Winsor R, and to the considerations of
Beerbower and Hill, will be recognized.

Shinoda and co-workers have also observed that the interfacial
tension between the aqueous and oil phases is at minimum (and quite
low) at the HLB-temperature. The significance of that observation
will be enlarged upon below.

Approach of W. H. Wade, R. S. Schechter, et al.

The observation by Shinoda and co-workers, cited above, with
respect to the behavior of the interfacial tension in the region of
the HLB-temperature is supported by recent work by Wade and others[45],
on the measurement of ultra-low interfacial tensions (LIFT). Wade
has demonstrated that the LIFT behavior of surfactants can be
characterized in terms of the oil phase (in this work, restricted
to hydrocarbons) against which the surfactant gives its minimum
interfacial tension (Fig. 4). It is possible to define an equivalent
alkane carbon number (EACN), such that a particular oil phase may

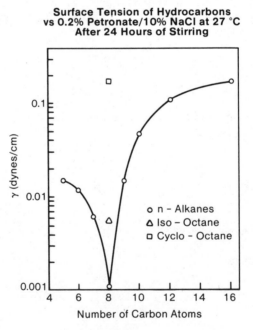

Figure 4. Interfacial tension minimum for a series of hydrocarbons
(from Ref. 45).

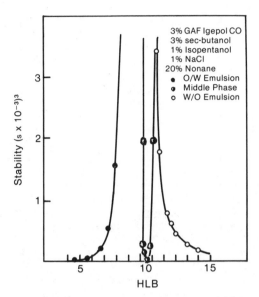

Figure 5. Relationship between emulsion stability and HLB (from Ref. 45). The minimum in stability corresponds to the minimum in interfacial tension.

be assigned an alkane carbon number corresponding to the alkane which has the same optimum properties (in a given system) as the oil phase in question.

For ethoxylated alcohols, it has been found possible to assign an $EACN_{min}$, according to the relationship

$$EACN_{min} = EACN_{min_o} + m(n_o - n) \qquad (26)$$

where $EACN_{min}$ corresponds to the EACN for the minimum interfacial tension of any member of the homologous series having n_o EO groups, and m is $\Delta EACN_{min}/\Delta EO$ for a particular base alcohol. From the definition of HLB it follows that

$$EACN_{min} = EACN_{min_o} + M\{n_o + (0.05HLBxM)/(44 - 2.2HLB)\} \qquad (27)$$

where M is the molecular weight of the base alcohol. A similar relation can be shown to exist for ionic species.

It appears from this work that the minimum in emulsion stability occurs at the minimal interfacial tension, corresponding in effect to HLB-temperature and Winsor R=1 (see Figs. 5 and 6). The significance of these finding must be explored further, but Wade's work serves finally to put to rest the simplistic view that emulsion stability is favored by low interfacial tension. That

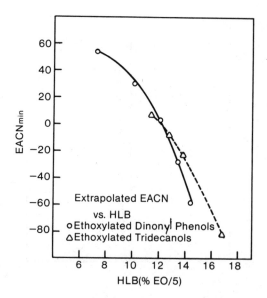

Figure 6.   Relation between $EACN_{min}$ and HLB (from Ref. 45).

this is the case is amply revealed by the relationship between emulsion stability (and HLB) and spreading coefficient[8].

Wade and co-workers have approached the question of the effect of chain-branching from a different point of view of that of Lin et al. Without going into details, these two methods are often inconsistent in the sense that where the method of Lin would predict a lowering of HLB for a particular structure, Wade's method predicts an increase. No resolution of this discrepancy has been made.

Future Approaches

Although we have treated only a few (if the most significant approaches) to HLB, it is clear that there is much to be done. The method of Beerbower and Hill, for example, requires much work. It should be more explicityly wedded to the treatment of Davies and of Lin, and the relation between Wade's EACN and Shinoda's HLB-temperature needs study.

Beyond that, the work of Hansch and co-workers[59] on the relation between the lipophilic character (in fact, the HLB) of drugs and their pharmacological effect, and the work of Drago[60] on the so-called "hard acid-soft acid/hard base-soft base" theory offer possibilities which invite exploration.

CONCLUSIONS

This is a conclusion in which, I hope, there is no conclusion. Evidence has been presented to show that the HLB-number, although derived on empirical grounds, is not an empirical quantity, but has the properties of a free energy. To be sure, there are still large empirical elements in this picture, but perhaps the following two decades will permit a fully thermodynamic representation of the relationship between the structure of the emulsifier and its surface-chemical properties.

I cannot end this paper without mentioning my long-time friend and colleague, William C. Griffin. From Bill I learned a good deal more than just how to calculate an HLB, and I am grateful. To Bill, who I hope is sailing happily on the Chesapeake Bay, this paper is affectionately dedicated.

ACKNOWLEDGEMENT

The financial assistance of INTEVEP, S.A. towards the presentation of this paper is gratefully acknowledged.

REFERENCES

1. W. Clayton, "Theory of Emulsions", 4th ed., p. 127,(1949), The Blakiston Company, Philadelphia, 1943.
2. W. C. Griffin, J. Soc. Cosmetic Chemists, $\underline{1}$, 311 (1949); ibid., $\underline{5}$, 249 (1954); Off. Dig. Federation Paint & Varnish Production Clubs, $\underline{28}$, 466 (1956).
3. R. G. Laughlin, J. Soc. Cosmet. Chem., $\underline{32}$, 371 (1981).
4. P. Becher and W. C. Griffin, in "McCutcheon's Detergents and Emulsifiers", North American Ed., p. 227, Allured Publishing Corp., Ridgewood, NJ, 1974.
5. B. W. M. Müller, Deutsche Apoth.Ztg., $\underline{118(11)}$, 404 (1978).
6. J. T. Davies, in "Proc. Intern. Congr. Surface Activity 2nd", London, Vol. 1, p. 426, Butterworth's, London, 1957.
7. G. Rimlinger, Am. Cosmetics Perfumery, $\underline{82(12)}$, 31 (1967); ibid., $\underline{84(5)}$, 31 (1969); Labo-Pharma Probl. Techn., No. 224, 67 (1973), Relata Tech., (18), 24 (1979); Relata Tech., (19), 45 (1979).
8. S. Ross, E. S. Chun, P. Becher, and J. J. Ranauto, J. Phys. Chem., $\underline{63(10)}$, 1681 (1959); P. Becher, J. Soc. Cosmet. Chem., $\underline{11(6)}$, 325 (1960).
9. V. R. Huebner, Anal. Chem., $\underline{34}$, 488 (1962).

10. P. Becher and R. L. Birkmeier, J. Am. Oil Chem. Soc., 41(3), 169 (1964).

11. I. Fineman, J. Am. Oil Chem. Soc., 46, 296 (1969).

12. G. E. Petrowski and I. R. Vanatta, J. Am. Oil Chem. Soc., 50(8), 284 (1973).

13. M. Leca and J. R. H. Perez, Rev. Roum. Chim., 22(8), 1117 (1977).

14. H. Olano and J. Martinez, Tenside, 12, 334 (1975).

15. V. S. Krivich, G. G. Kochurovskaya and M. Kh. Gluzman, Zhur. Fiz. Khim., 46(4), 973 (1972); V. S. Krivich, M. A. Trunova M. Kh. Gluzman, Zh. Priklad. Khim., 48(6), 1315 (1975); L. P. Bakholdina and V. S. Drivich, Kolloidn. Zhur., 38(6), 1056 (1976.

16. J. Broniarz, J. Szymanowski and M. Rogala, Poznan. Tow. Przyj. Nauk. Pr. Kom. Nauk Podstawowych Stosow., 3(3), 3 (1972); J. Broniarz, J. Szymanowski and M. Wisniewski, Przem. Chem., 51(8), 517 (1973); J. Broniarz, J. Szymanowski and M. Wisniewski, Tluszce, Srodki Piorace, Kosmet., 17(3), 131 (1973); J. Szymanowski, Progr. Colloid Polym. Sci., 63, 96 (1978); J. Szymanowski and K. Prochaska, Fette Seife Anstrichm., 83(5), 172 (1981).

17. F. M. Khutoryanski, D. N. Levchenko, E. N. Malkal'skaya and N. M. Nicolaeva, Neftpererab. Neftkhim., (1), 37 (1981).

18. S. Hayano and T. Asahara, in "Proc. Intern. Congr. Surface Activity 5th", Barcelona, 2, 843 (1968).

19. L. Marszall, Riv. Ital. Essenze, Profumi, Saponi, 57, 113 (1975); Colloid Polymer Sci., 255, 62 (1977).

20. P. A. Crooks, J. H. Collett and R. Withington, Pharm. Acta Helv., 49, 274 (1974).

21. G. Ben-Et and D. Tatarsky, J. Am. Oil Chem. Soc., 49(8), 499 (1972).

22. J. R. Bergueiro, M. Bao and J. J. Casares, Ann. Ouim., 74, 529, 1441 (1978).

23. I. Racz and E. Orban, Acta Pham. Hung., 34(4), 175 (1964); I. Racz and E. Orban, J. Colloid Sci., 20(2), 99 (1965); E. Orban, Tenside, 7(4), 203 (1970).

24. W. G. Gorman and G. D. Hall, J. Pharm. Sci., 52, 442 (1963).

25. A. Rehula, Farm. Obz., 38, 489 (1969).

26. I. Lo, T. LeGras, M. Seiller, M. Choix and F. Puisieuz, Ann. Pharm. Franc., 30(3), 211 (1972).

27. H. Schott, J. Pharm. Sci., 60(4), 648 (1971).

28. G. Szatlmayer, Kolor. Ert., 13(7/8), 160,(1971).

29. P. M. Kruglyakov and A. F. Koretskii, Izv. Sib. Otd. Akad. Nauk SSSR, Ser. Khim. Nauk, (9), 11 (1971); P. M. Kruglyakov, T. V. Mikina and A. F. Koretskii, Izv. Sib. Otd. Akad. Nauk SSSR, Ser. Khim. Nauk, (2), 3 (1974).

30. I. J. Lin and L. Marszall, Progr. Colloid Polym. Sci., 63, 99 (1978).

31. P. M. Kruglyakov and A. F. Koretskii, Dokl. Akad. Nauk SSR, 197, 328 (1971).

32. C. McDonald, Can. J. Pharm. Sci., 5, 81 (1970).

33. A. Beerbower and M. W. Hill, in "McCutcheon's Detergents and Emulsifiers", pp. 223-235, Allured Publishing Corp., Ridgewood, NJ, 1971; Am. Cosmet. Perfumery, 87(6), 85 (1972).

34. R. C. Little, J. Colloid Interface Sci., 65(3), 587 (1978).

35. S. H. Yalkowsky and G. Zografi, J. Pharm. Sci., 61, 793 (1972).

36. L. Marszall, Kolloid Z. Z. Polymere, 251(8), 609 (1973); J. Pharm. Pharmacol., 25, 254 (1973).

37. R. Heusch, Kolloid Z. Z. Polymere, 236(1), 31 (1970); Fette Seifen Anstrichm., 72(11), 969 (1970).

38. I. J. Lin and P. Somasundaran, J. Colloid Interface Sci., 37(4), 731 (1971); I. J. Lin, J. P. Friend, and Y. Zimmels, J. Colloid Interface Sci., 45(2), 378 (1973); I. J. Lin, in "Colloid and Interface Science", M. Kevker, Editor, Vol. 2, p. 431, Academic Press, NY, 1976.

39. H. Schott, J. Pharm. Sci., 58, 1131 (1969).

40. H. Schott, J. Pharm. Sci., 58(12), 1443 (1969).

41. A. T. Florence, F. Madsen and F. Puisieux, J. Pharm. Pharmacol., 27, 385 (1975).

42. J. Broniarz, Tluszce, Srodki Piorace, Kosmet., 16(2), 19 (1972).

43. L. Marszall, Cosmet. Toiletries, 94(9), 29 (1979); Fette, Seifen, Anstrichm., 82(1), 40 (1980); 82(5), 210.

44. K. Shinoda and H. Arai, J. Phys. Chem., 68, 3485 (1964); K. Shinoda, Proc. Intern. Congr. Surface Activity 5th, Barcelona, 3, 275 (1968); K. Shinoda and H. Saito, J. Colloid Interface Sci., 30(2), 258 (1969); K. Shinoda and H. Sagitani, J. Colloid Interface Sci., 64(1), 68 (1978).

45. M. E. Hayes, M. El-Emary, R. S. Schechter and W. H. Wade, J. Colloid Interface Sci., 68(3), 591 (1979); M. Bourrel, A. Graciaa, R. S. Schechter and W. H. Wade, J. Colloid Interface Sci., 72(1), 161 (1979).

46. A. A. Badwan, T. M. Cham, K. C. James and W. J. Pugh, Int. J. Cosmetic Sci., 2, 45 (1980).

47. H. L. Greenwald, E. B. Kice, M. Kenly and J. Kelly, Anal. Chem., 33, 465 (1961).

48. P. Winsor, "Solvent Properties of Amphiphilic Compounds", Butterworth Scientific Publications, London, 1954.

49. J. H. Hildebrand and R. L. Scott, "The Solubility of Non-electrolytes", 3rd Edition, Dover Publications, New York, 1964; J. H. Hildebrand, J. M. Prausnitz and R. L. Scott, "Regular and Related Solutions", Van Nostrand, New York, 1970.

50. C. M. Hansen, J. Paint Technol., 39(505), 105 (1967); 39(511), 500.

51. P. Becher, "Emulsions: Theory and Practice", 2nd Edition, ACS Monograph No. 162, p. 97, Krieger Publishing Corp., Huntington, NY, 1977.

52. C. Hansen and A. Beerbower, in "Kirk-Othmer Encyclopedia of Chemical Technology, Supplement", p. 889, John Wiley, New York, 1977.

53. P. Becher, in "Nonionic Surfactants", M. J. Schick, Editor, pp. 487-492, Marcel Dekker, Inc., New York, 1967.

54. E. H. Crook, D. B. Fordyce and G. F. Trebbi, J. Phys, Chem., 67, 1987 (1963).

55. K. Shinoda, Bull. Chem. Soc. Japan, 26, 101 (1953); J. Phys. Chem., 58, 1136 (1954).

56. J. N. Phillips, Trans. Faraday Soc., 51, 561 (1955).

57. I. Langmuir, J. Am. Chem. Soc., 39, 1848 (1917).

58. K. Shinoda and P. Becher, "Principles of Solutions and Solubility", Chapter 9, Marcel Dekker, Inc., New York, 1978.

59. C. Hansch and W. J. Dunn, III, J. Pharm. Sci., 61(1), 1 (1972); C. Hansch and J. M. Clayton, J. Pharm. Sci., 62(1), 1, (1973).

60. R. S. Drago, G. C. Vogel and T. E. Needham, J. Am. Chem. Soc., 93(23), 6014 (1971); R. S. Drago, J. Chem. Ed., 51(5), 300 (1974).

POLYMERIZATION OF ORGANIZED SURFACTANT ASSEMBLIES

Janos H. Fendler

Department of Chemistry
Clarkson College of Technology
Potsdam, New York 13676

Polymerization of organized surfactant assemblies leads to unique systems which show greatly enhanced stabilities compared to nonpolymeric surfactant aggregates. Available information on the polymerization of micelles, microemulsions, monolayers, multilayers, bilayer lipid membranes, liposomes, and synthetic surfactant vesicles are critically surveyed. Preparation, characterization and applications of the different polymeric systems are discussed. Emphasis is placed on recent developments and future potentials.

INTRODUCTION

Micelles, microemulsions, monolayers, multilayers, and vesicles, collectively referred to as organized assemblies or membrane mimetic agents, are increasingly utilizied as unique reaction media.[1] They solubilize, concentrate, compartmentalize, organize and localize reactants and products; maintain reactant gradients; alter dissociation constants, oxidation and reduction potentials; and affect chemical pathways and rates. These properties render membrane mimetic agents to be potentially useful in various applications such as reactivity control, energy conversion, tertiary oil recovery, molecular recognition and target directed drug delivery.[1]

Early investigations of micelles had been prompted by the expected analogies between micellar and enzymatic catalysis.[2] Micelles, unlike enzymes, are however fluid structures. Substrate solubilization sites are dynamic and there is no equivalent of

1947

"lock and key" interaction bringing about stereospecific recognition. Inherent instabilities of liposomes and synthetic surfactant vesicles are also detrimental. Recognizing the need for stability, controllable rigidity and permeability lead to the development of polymerized membrane mimetic agents. Ideally, polymeric surfactant assemblies should combine the beneficial properties of stable polymers[3] with the fluidities of micelles, microemulsions and vesicles.[2] Available information on the polymerization of micelles,[4-10] microemulsions,[11,12] liquid crystals,[13,14] black (bilayer) lipid membranes or BLMs,[15] monolayers,[16-48] organized multilayer assemblies,[16-48] liposomes[49-65] and vesicles[49-65] will be surveyed in the present overview. The subject demands an interdisciplinary approach. Polymerization of organized assemblies has been investigated by solid state physicists, polymer physicists, polymer chemists, organic chemists, biochemists, biophysicists, and membrane biologists. Inevitably, the approach is diffuse, the literature is scattered, and the same phenomenon is often described by different terminology. Emphasis here will be placed on current developments and future potentials. The explosive growth of this area precludes, of course, an exhaustive treatment.

POLYMERIZATION OF SURFACTANTS IN AQUEOUS MICELLES

The role of surfactants in emulsion polymerization has been recognized for some time.[66] Ample data are also available on the effect of aqueous micelles on the rates of radical and ionic polymerizations.[12] Polymerization of micelles constituting surfactants themselves above their critical micelle concentration, is a different matter. Is the structure of micelle retained, altered or destroyed upon polymerization? Maintaining the micellar structure requires the proper topological alignment of the surfactants during the course of polymerization, the growth of the polymer chain in directions different from volume contraction, the structural similarity of monomer and polymer units and favorable entropies. No attention has been paid to these requirements and the results obtained are often inconclusive (Table I). Spontaneous polymerization of 4-vinylpyridinium salts illustrates this point. Different polymers are formed at low and high substrate concentrations:

$$\text{(1)}$$

Table I. Polymerization of Surfactants in Aqueous Micelles[a]

Surfactant	Polymerization Method	Characterization	Results, Conclusions	Ref.
$H_2C=C(CH_3)CONR(CH_2)_{10}COX$ $R = CH_3, C_2H_5, nC_3H_7, iC_3H_7$ $X = OH, ONa, (OCH_2CH_2)_xOCH_3$	40-120°C for 12-24 h using 1-10% initiator (BuPO, tBuPO AIBN, CuHP, tBuPO $Na_2S_2O_8$) in H_2O	CMC determination by conductivity and dye solubilization, aggregation numbers by light scattering and ultrafiltration, electron microscopy	No clear cut picture on structures of polymerized aggregate	4
4-Vinylpyridinium salt (VP)	Vinylpyridine initiated, N_2	Kinetics of polymerization and product formation determined	Polymerization rate is strongly increased above CMC; at low monomer concentration product is the 1,6-polymer, while at high concentration it is the 1,2-polymer	5,6
$H_2C=C(CH_3)CO_2(CH_2)_n N^+(CH_3)_3 Cl^-$ $n = 2, 3, 6$	60°C, H_2O, $K_2S_2O_8$ initiator	Polymerization kinetics followed (k_p = propagation, k_d = dissociation, k_t = termination rate constants, f = efficiency factor)	$k_p(2fk_d/k_t)$ increases with decreasing monomer concentration: data rationalized in terms of electrostatic repulsion	7
4-Vinylpyridium perchlorate	self polymerization	^{13}C nmr, ir	The microstructure of the 1,2-polymer is unaffected by micellar polymerization	8
p-Styreneundecanoate	?	Sedimentation constant, intrinsic viscosity and partial specific volume of the polymer combined in Scheraga - Mendelkern equation, electron microscopy	Internally crosslinked micelles are formed	9

(continued)

Table I (continued)

Surfactant	Polymerization Method	Characterization	Results, Conclusions	Ref.
$CH_3C(=CH_2)COO$—⟨benzene⟩—O–$(CH_2)_n X$ $n = 3, 7, 13$ $X = -\overset{+}{N}H(CH_3)_2Cl^-,\ -\overset{+}{N}(CH_3)_3Br^-$	?	Kinetics of poly-merization	High polymerization rate above and below the cmc; polymer radical stabilized by electrostatic repulsions in the micelle	10

[a]Abbreviations used: uv = ultraviolet irradiation (to initiate and affect polymerization); ir = infrared spectroscopy; AIBN = azobisisobutyronitrile; BuPO = dibutyl peroxide; tBuPO = ditert-butyl peroxide (initiator)

Concurrently, the rate of polymerization appeared to be greater above 1.0 M substrate concentration than below it.[5,6] These results were initially rationalized by formation of 4-vinylpyridium halide micelles at 1.0 M concentration and by the different polymerization behavior of monomers and micelles.[5,6] Subsequent separate investigations of the competing processes (Equation 1) showed no breaks in plots of overall polymerization rate against 4-vinylpyridinium concentration.[8] Electrostatic interactions were considered to be responsible for the observed kinetic behavior and product distribution.[67] Similar inconclusive results were obtained in the polymerization of methacryloyl oxyalkyl trimethylammonium chloride.[8] The final verdict must await the outcome of more definitive experiments on carefully chosen polymerizable micelles. Such micellar properties as critical micelle concentration, aggregation number, weight averaged molecular weight, hydrodynamic radius, and polydispersity need to be determined prior and subsequent to polymerization. A critical and yet unanswered question is the fate of "polymeric" micelles on dilution.

POLYMERIZATION OF SURFACTANTS IN MICROEMULSIONS, LIQUID CRYSTALS AND BLMs

At present, experiments in these systems are relatively meager (Table II). Polymerization of oil-in-water hexadecyltrimethyl--ammonium bromide, styrene, hexanol, microemulsion has been studied in most detail. The spherical 200 - 400 Å diameter latex particles formed provided two unique sites for the localization of pyrene derivatives; one between the polymer matrix (pyrene) and one at the surface (pyrene sulfonic acid).[11]

Polymerization of some liquid crystals have resulted in retaining the long range order while structural changes occurred in other systems (Table II).[13,14]

Reproducible preparations of bilayer (black) lipid membranes, BLMs, are notoriously difficult. BLMs rarely last longer than a few hours and their composition are generally unknown.[1] The preliminary report on the polymerization of BLMs, formed from $[CH_3(CH_2)_{17}]_2\overset{+}{N}[CH_3][(CH_2)_3NHC=OC(CH_3)=CH_2]Br^-$ (Table II)[15] is most significant. Increasing the BLMs capacitance during polymerization (Figure 1) had been rationalized in terms of thinning the membrane which allows a greater degree of water penetration into the bilayer. Polymeric BLMs were stable up to 50 minutes. In contrast, nonpolymeric BLMs prepared from the same surfactant lasted only for seconds.

Table II. Polymerization of Surfactants in Microemulsions, Liquid Crystals and BLMs [a]

Surfactant	Polymerization Method	Characterization	Results, Conclusions	Ref.
Hexadecyltrimethylammonium bromide : styrene : hexanol : water (1.0g : 1.0g : 0.5g : 50ml) o/w microemulsion	AIBN 60°C, Cs γ-irradiation	Fluorescence spectroscopy, Electron microscopy	Polymeric particles provide two unique sites for location of reacting molecules	11
Sodium dodecyl sulfate : 1-pentanol : styrene (0.2 or 0.5M : 0.6 or 1.0M : 2 vol %) o/w microemulsion	^{60}Co γ irradiation at 4600 rd/min of Ar bubbled samples	Fluorescence spectroscopy	Polymeric particles only provide two solubilization sites for pyrene at high styrene : surfactant ratios	12
Sodium undecanoate liquid crystal (47 - 59 % sodium undecanoate in water)	Ammonium persulfate initiated (0.05M) 50°C, N_2	Optical microscopy, ir, high pressure liquid chromatography	Structural changes from two dimensional hexagonal close packing prior to polymerization to lamellar liquid crystalline structure subsequent to polymerization	13, 14
$H_3C(CH_2)_n$ ... Br^- ... BLM	uv irradiation of methylazobutyrate dissolved in membrane	Capacitance measurement during polymerization	During irradiation capacitance rose from 0.38 $\mu F/cm^2$ to 2 $\mu F/cm^2$; polymerization thins the BLM, stable for 50 minutes	15

[a] See footnote in Table I for abbreviations.

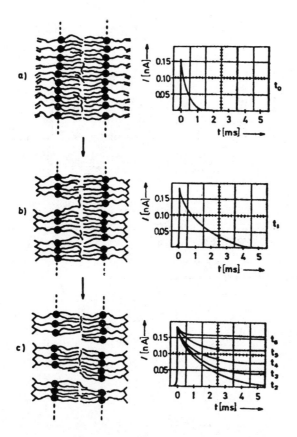

Figure 1.   Schematic representaion of the structural changes in the membrane and course of charging currents (measurement of capacitance) during polymerization: a) monomeric membrane before uv irradiation $(t=t_0)$; b) membrane at the beginning of the polymerization $(t=t_1=0$ min$)$; c) polymeric membrane and change in current across the membrane with time $(t_2=1$ min, $t_6=20$ min$)$. Reproduced with permission from reference 15.

POLYMERIZATION OF SURFACTANTS IN MONOLAYERS AND MULTILAYERS

Polymerization of monolayers and multilayers have been extensively investigated (Table III). Surfactant monolayers can be readily formed in a trough and be successively deposited on a solid support (Figure 2) to form multilayers.[72,73] Polymerization of monolayers has been conveniently studied by following the area contraction in surface pressure - surface area diagrams (Figure 3).

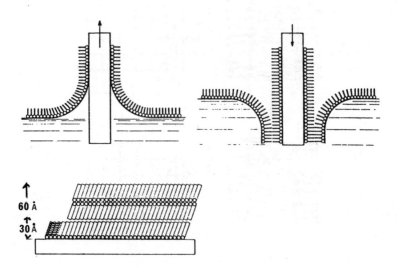

Figure 2.   Schematics of multilayer preparation.   Reproduced with permission from reference 35.

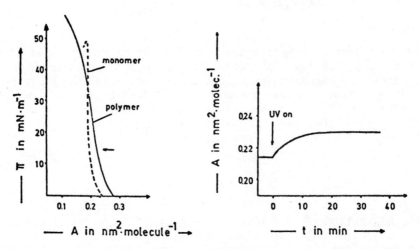

Figure 3.   Surface Pressure (π) – surface area (A) diagrams for monomeric and polymerized octadecanediol (left-hand side).   Changes of surface area as a function of irradiation time (right hand side).   Reproduced with permission from reference 44.

Table III.   Polymerization of Surfactants in Monolayers and Multilayers[a]

Surfactant	Polymerization Method	Characterization	Results, Conclusions	Ref.
Octadecylmethacrylate, Octadecylacrylamide	AIBN + p-octadecylphe- none (sensitizer) + uv (air - H_2O surface)	Surface pressure - area diagrams mass spec.	O_2 containing polymers formed	16
Vinyl + ethyl stearate	Co - 60 radiation (N_2 sat., 100 mono- layer deposited on a plate)	Internal reflection ir, polarized in- ternal reflection ir, x-ray diffrac- tion	Each monolayer is a separate two dimensional reaction site; polymerization in a solid solu- tion of monomer and polymer side chains; monomers can be leached out; orthorhombic side chain packing of vinyl stearate monomer changes to hexagonal packing in polymers	17
RC≡CC≡CR' R = -$(CH_2)_9CH_3$ R' = -$(CH_2)_8COOH$ (tricosa-10,12-diynoic acid)	uv (multilayers of Cd salt of acid on quartz support)	uv - spectra, x-ray	Arrangement of layers remains unaltered	18
Octadecylmethacrylate (ODMA)	uv (monolayer at (N_2 - H_2O Inter- face)	Surface pressure - area diagrams, surface potential measurements, ir, x-ray diffraction	Polymerization in the condens- ed state of monomer (at high surface pressure) is faster than in the expanded state (at low surface interface press- ure); poly-ODMA prepared at interface shows higher tacti- city than that formed in bulk solution	19, 20
Cadmium octadecylfumarate	uv (Y deposited multilayers on a plate)	uv, electron diffraction	Polymerization occurs in each layer separately	21

[a] See footnote in Table I for abbreviation.

(continued)

Table III (continued)

Surfactant	Polymerization Method	Characterization	Results, Conclusions	Ref.
Vinyl stearate, ethyl stearate	uv (multilayer on a plate, irradiation under water at 0°C)	FT-ir, x-ray diffraction, electron diffraction	Order is increasing upon polymerization	22
$H_2C=CH(CH_2)_{21}COOH$ (ω-Tricosenoic acid)	150 - 300 eV electron beam (multilayers on glass, CaF_2, Ge, Si, Al)	ir, selective solubility	Monolayer organization is maintained with molecular reorientation	23
$H_2C=CH(CH_2)_{21}OCOCH=CHCH=CH_2$ (ω-Tricosenyl 2,4-pentadienoate)	uv irradiation (-40°C to 40°C N_2 or O_2, 100 monolayers deposited on a plate)	ir, nmr, thin layer chromatography	Polymerization is lattice controlled, there is a volume contraction	24
$H_3C(CH_2)_nC\equiv CC\equiv C(CH_2)_8COOH$ n = 9, 11, 13	uv irradiation (32 monolayers of Cd salt of the acid on a quartz plate)	Quantum yields and % of conversion determined as a function of time	Quantum yield decreases with increasing conversion time	25
$H_3C(CH_2)_nC\equiv CC\equiv C(CH_2)_8COOH$ n = 4, 8, 9, 11, 13, 15	uv irradiation (monolayers N_2/H_2O interface)	Surface pressure - surface area diagrams	Films polymerized at constant pressure are rigid	26
ω-tricosenoic acid	Electron beam irradiation (multilayers on an aluminum plate)	Scanning electron microscopy, interferencial optical microscopy	The ultrathin (30 - 1000 Å) multilayers provide high resolution resists for electron beam microlithography	27
$H_3C(CH_2)_{15}C\equiv CC\equiv C(CH_2)_8COOH$ (10,12-nonacosadiynoic acid)	uv irradiation (monolayers at N_2 - H_2O interface)	Absorption spectroscopy	Absorbance of monolayer vs polymerization time investigated	28
$H_3C(CH_2)_nC\equiv CC\equiv C(CH_2)_8COOH$ n = 9, 11, 13, 15	uv irradiation (monolayers at N_2 - H_2O interface)	Surface pressure - surface area diagrams, absorption spectroscopy	Polymerization is topochemically controlled	29

Surfactant	Polymerization Method	Characterization	Results, Conclusions	Ref.
Octadecylacrylamide	^{60}Co γ-radiation (multilayers on a glass slide)	FT-ir, x-ray diffraction, electron diffraction	No phase change during polymerization	30
Diacetylenes, $CH_3(CH_2)_{15}C\equiv CC\equiv C(CH_2)_8COOH$	uv irradiation of monolayer	Electron diffraction, electron microscopy	Polymerization occurs in the monomer phase, strong bilayers obtained	31, 32
ω-Tricosenic acid	Electron beam irradiation (multilayers on aluminum plates	ir, scanning electron microscopy	Improved sensitivity to microlithography is related to an increase in the length of the polymer	33
Stearic acid, ω-tricosenic acid	uv and electron beam irradiation (multilayers on aluminum plates)	ir, Auger spectroscopy	3 types of polymerizations: (a) after diffusion polycondensations with bifunctional molecules lead to cross linked polymers (b) polymerization of the double bond gives a linear polymer that is sparingly soluble in cold alcohol but soluble in chloroform (c) cross linked polymers are obtained by the addition of second polymerizable group (acrylate, allyl ester, epoxide, β-vinyl acrylate, vinyl ester). Selective polymerization by near uv light to produce dimers, by far uv to form a linear polymer, which is reticulated by electron beam irradiation	34
$H_2C\equiv CH(CH_2)_{20}COOR$ R = allyl β-vinylacrylic functionalities	Electron beam irradiation (multilayers on aluminum plates)	Contrast optical scanning electron microscopy, esr	Improved microlithography in the 500 – 3000 Å is possible	35, 36

(continued)

Table III (continued)

Surfactant	Polymerization Method	Characterization	Results, Conclusions	Ref.
$H_3C(CH_2)_{12}C{\equiv}CC{\equiv}C(CH_2)_9OR$ $R = H,\ PO_3H,\ PO_3(CH_2)_2NH_3^+,$ $PO_3(CH_2)_2N(CH_3)_2^+$	uv irradiation (monolayers at N_2 – H_2O interface)	Surface pressure – area diagrams absorption spectroscopy	Polymers formed	37
$H_3C(CH_2)_{12}C{\equiv}CC{\equiv}C(CH_2)_8COY(CH_2)_2$ $\diagdown X$ $H_3C(CH_2)_{12}C{\equiv}CC{\equiv}C(CH_2)_8COY(CH_2)_2\diagup$ $Y = 0,\ X = 0;\ Y = 0,\ X = NCH_3;\ Y = NH,\ X = 0;$ $Y = 0,\ X = N^+(CH_3)_2Br^-$				
$CH_3(CH_2)_{12}CH{=}CHCH{=}CH{-}X$ $X = COOH,\ CHO,\ CH_2OH$	uv irradiation	Surface pressure – area diagrams	Polymerization only occurs in monolayers (no polymerization in solution, melts or crystals)	38
$H(CH_2)_nC{\equiv}CC{\equiv}C(CH_2)_8COOX$ $n = 12,\ 14$ or 16 $X = H$ or Li	uv irradiation (monolayers on H_2O/air interface)	Fiber tensile properties tested	Monolayer fibers with unique morphologies are formed	39
$H_3CC{\equiv}CC{\equiv}C(CH_2)_8COOH$ $X = H$ or Li + polyacetylene, I_2 as dopants	uv irradiation (bilayers on a lock-and-key device)	Conductivity, photoconductivity	Doped polymeric bilayers show photoconductivity	40

Surfactant	Polymerization Method	Characterization	Results, Conclusions	Ref.
$CH_2OCO(CH_2)_8C{\equiv}CC{\equiv}C(CH_2)_{12}CH_3$ $\|$ $CHOCO(CH_2)_8C{\equiv}CC{\equiv}C(CH_2)_{12}CH_3$ $\|$ $CH_2O_3{}^-PO(CH_2R$ $R = NH_3{}^+, N(CH_3)_3{}^+$ $CH_2OCO(CH_2)_{16}CH_3$ $\|$ $CHOCO(CH_2)_8C{\equiv}CC{\equiv}C(CH_2)_{12}CH_3$ $\|$ $CH_2O_3{}^-POH$ $CH_2OCOCH{=}CHCH{=}CH(CH_2)_{12}CH_3$ $\|$ $CHOCOCH{=}CHCH{=}CH(CH_2)_{12}CH_3$ $\|$ $CH_2O_3{}^-PO(CH_2)_2N^+(CH_3)_3$	uv irradiation (monolayers at H_2O/air interface)	Surface pressure – area diagrams	Polymers formed from diacetylene-phospholipids are extremely rigid and do not exhibit phase transitions, more flexible systems are formed by the diene-lecithin and by the mixed polymerizable and non-polymerizable lipid systems	41
$CH_3(CH_2)_{12}C{\equiv}CC{\equiv}C(CH_2)_8COO(CH_2)_2{}^+\!\underset{\textstyle N}{}\!^H$ $CH_3(CH_2)_{12}C{\equiv}CC{\equiv}C(CH_2)_8COO(CH_2)_2{}^+N$ $CH_2CH_2SO_3{}^-$	uv irradiation (monolayers at $N_2 - H_2O$ interface)	Surface pressure – area diagrams	No polymerization in the liquid state of monolayers	42

(continued)

Table III (continued)

Surfactant	Polymerization Method	Characterization	Results, Conclusions	Ref.
$H_3C(CH_2)_{12}C\equiv CC\equiv C(CH_2)_9OR$ $R = PO_3H_2$, $PO_2^-(CH_2)_2N^+H_3$, $PO_2^-(CH_2)_2N^+(CH_3)_3$ $H_3C(CH_2)_{12}C\equiv CC\equiv C(CH_2)_8COO(CH_2)_2$	uv irradiation (monolayers at N_2 – H_2O interface)	Surface pressure – area diagrams	Topochemical polymerization	43, 44

$H_3C(CH_2)_{12}C\equiv CC\equiv C(CH_2)_8COX(CH_2)_2$—Y

$H_3C(CH_2)_{12}C\equiv CC\equiv C(CH_2)_8COX(CH_2)_2$

$X = Y = 0;\ X = 0,\ Y = NCH_3;$

$X = NH_3,\ Y = 0;\ X = 0,\ Y = \overset{+}{N}(CH_3)_2Br^-;$

$X = 0,\ Y = N(CH_2)_2SO_3H;\ X = 0,\ Y = NCH_2COOH$

$\begin{array}{l} CH_2OCO(CH_2)_8C\equiv CC\equiv C(CH_2)_{12}CH_3 \\ | \\ CHOCO(CH_2)_8C\equiv CC\equiv C(CH_2)_{12}CH_3 \\ | \\ CH_2O^-_3PO(CH_2)_2\overset{+}{N}(CH_3)_3 \end{array}$

$\begin{array}{l} CH_2OCO(CH_2)_{16}CH_3 \\ | \\ CHOCO(CH_2)_8C\equiv CC\equiv C(CH_2)_{16}CH_3 \\ | \\ CH_2O^-_3POH_2 \end{array}$

Surfactant	Polymerization Method	Characterization	Results, Conclusions	Ref.
$CH_3(CH_2)_nCH(NH_2)CO_2(CH_2)_mCH_3$ $n = 15$, $m = 0$; $n = 15$, $m = 21$; $n = 21$, $m = 0$; $n = 23$, $m = 21$	uv irradiation (monolayers at N_2 – H_2O interface)	Surface pressure – area diagrams,	Polycondensation in monolayers	45
$H_3C(CH_2)_{11}C{\equiv}CC{\equiv}C(CH_2)_8COOH$	uv or Ar-ion laser irradiation (40 layers on a quartz plate)	Absorption and emission spectroscopy	Polymerization of monolayers can only be initiated by uv, excitation of partially polymerized layers by visible light leads to further polymerization	46
$H_3C(CH_2)_{11}C{\equiv}CC{\equiv}C(CH_2)_8{-}COOH$	Cyanine dye sensitized polymerization (40 layers on a quartz plate)	Absorption and emission spectroscopy	Cyanine dyes embedded in multilayers sensitize polymerization; quantum yield is low	47
$CH_2OCO(CH_2)_8C{\equiv}CC{\equiv}C(CH_2)_{11}CH_3$ $CHOCO(CH_2)_8C{\equiv}CC{\equiv}C(CH_2)_{11}CH_3$ $CH_3O_3{}^-PO(CH_2)_2N^+(CH_3)_3$	uv light (multilayers on glass, quartz, Teflon, steel)	Surface pressure – area diagrams, absorption spectroscopy	Stable biomembrane hydrophilic surface is produced	48

Diacetylenes have been the favorite substrates for polymerization [18,25,26,28,31,32,37,39-44,46-48] since the monomer does not absorb in the visible region but, upon polymerization, it turns, first to blue then red (Equation 2). Polydiacetylenes in the solid state are red.

Colorless

λ_{max} = 217,230,242,256 nm

Blue⟶Red

λ_{max} 600 nm 500 nm
ε_{537} nm = 1.5 x 10^4 M^{-1} cm^{-1}
ε_{498} nm = 1.6 x 10^4 M^{-1} cm^{-1}

$$(2)$$

Polymerization can be, therefore, seen by the naked eye. Polymerization of diacetylenes in the monolayers depends on the chain length of the surfactant. Thus, surfactants containing 25 -29 carbon atoms in their chains (n = 12,14, or 16 in $CH_3(CH_2)_{n-1}C\equiv CC\equiv C(CH_2)_8COOH$) readily polymerized at 20°C, while those containing only 23 carbon atoms (n = 10) polymerized only at 0°C, and those containing 21 or fewer carbon atoms (n = 5 or 8) could not be polymerized under any conditions.[26] Further, diacetylenes could be polymerized only in the condensed state (at high surface pressure) when tne monomers are presumably oriented in a topochemically favorable position.[42] Polymerization occurs with retention of lattice and layer structures in the monolayer.[74] Conversely, methacrylate and diene containing surfactants could be polymerized in monolayers both in the liquid crystalline and solid states.[41,55] The greater mobility of the appropriate hydrocarbon segments in these latter systems may be responsible for their ease of polymerization.

Morphologies of polymerized monolayers, bilayers, and multilayers have been examined by optical microscopy between crossed polarizers.[36] Polymerized monolayers are diffcult to deposit on a solid support. They break apart along the crystalline boundaries.[32] Polymeric bilayers, prepared by polymerizing two monolayers successively deposited onto each other, are much more stable. They can cover macroscopic (~0.5 nm diameter) holes and remain stable if kept cold.[32] Polymerization of multilayers, deposited on various solid supports, opens the door to inherently important chemistry and novel applications.

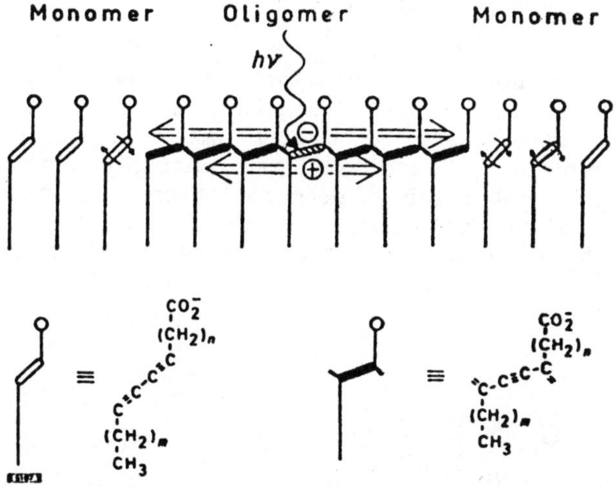

Figure 4.   Proposed diacetylene self-polymerization.   Reproduced with permission from reference 47.

Recent studies established self and cyanine sensitized polymerizations in multilayers. Figure 4 shows the proposed mechanism for the self sensitized polymerization in diacetylene multilayers. Excitation of the oligomers was considered to result in the formation of free carriers whose mobility is comparable to that of semiconductors. This mobile charge carrier then appeared to activate adjacent monomer units and thereby propagate the polymer chain.[47]

POLYMERIZATION OF SURFACTANTS IN VESICLES

Vesicles are smetic mesophases of surfactants containing water between their bilayers.[1]   The term vesicle is used to describe spherical or ellipsoidal single (SUV = small unilamellar vesicle)

or multicompartment (MLV = multilamellar vesicle) closed bilayer
structures, regardless of their chemical compositions. Vesicles
composed of naturally occurring or synthetic phospholipids are
referred to as liposomes.[75-77] In contrast, those formed from
completely synthetic surfactants are designated as surfactant
vesicles.[78]

Vesicles are not only inherently interesting as the most
sophisticated models for the biological membranes but serve for
developing chemistry based on membrane mediated processes. The
need for appropriately functionalized vesicles with long term
stabilities and controllable permeabilities is obvious. Not
unexpectedly, polymeric vesicles have become the subject of
intensive scrutiny in increasing number of research laboratories
around the world.

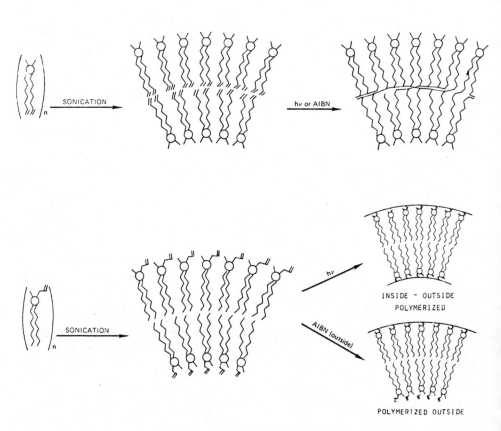

Figure 5. Schematics of polymeric vesicle formation. Reproduced
with permission from reference 62.

Table IV. Polymerization of Surfactants in Vesicles[a]

Surfactant	Polymerization Method	Characterization	Results, Conclusions	Ref.
CH_2OCOR_1 | $CHOCOR_2$ | $CH_2O_3{}^-PO(CH_2)_2N^+(CH_3)_3$ $R_1 = -(CH_2)_{16}CH_3$, $R_2 = -(CH_2)_{11}OCOC(CF_3)=N_2$; $R_1 = -(CH_2)_{16}CH_3$, $R_2 = -(CH_2)_{11}OCOC(CF_3)=N_2$; $R_1 = -(CH_2)_{16}CH_3$, $R_2 = -(CH_2)_{10}O$-[ring] $R_1 = -(CH_2)_{14}CH_3$, $R_2 = -(CH_2)_{10}O$-[ring] $R_1 = -(CH_2)_{14}CH_3$, $R_2 = -(CH_2)_8CH=CHCO(CH_2)_5CH_3$; $R_1 = -(CH_2)_{14}CH_3$, $R_2 = -(CH_2)_7CH=CHCH_2CHN_3(CH_2)_5CH_3$; $R_1 = -(CH_2)_{14}CH_3$, $R_2 = -(CH_2)_8CHN_3(CH_2)_5CH_3$	Sonicated dispersions irradiated by uv light	ir, uv spectroscopy, gel permeation chromatography, analysis of products formed by transesterification, degradation with phospholipase A_2 and C, glpc – mass spec, ^{14}C labelling	Intermolecular crosslinking occured by carbene insertion into the C–H bonds of a second acyl chain	49
$O\big\langle\ {}^{(CH_2)_2NHCO(CH_2)_8C{\equiv}CC{\equiv}CC_{13}H_{27}}_{(CH_2)_2NHCO(CH_2)_8C{\equiv}CC{\equiv}CC_{13}H_{27}}$	?	Electron microscopy	Multicompartment liposomes are polymerized	29

(continued)

Table IV (continued)

Surfactant	Polymerization Method	Characterization	Results, Conclusions	Ref.
$(CH_3)_2\overset{+}{N}$ —$(CH_2)_{15}CH_3$ / —$(CH_2)_{11}OC\text{-}C\text{-}CH_3$ (O, CH_2), Br^-	Sonicated surfactant·heated with AIBN (6h, 80°C)	Electron microscopy, 1H nmr, 14C glucose entrapment, stability to EtOH	Closed polymerized vesicles retained entrapped glucose and remained stable to 25% EtOH	50
$H_3C(CH_2)_{12}C\equiv CC\equiv C(CH_2)_8COO(CH_2)_2\text{-}X$ $H_3C(CH_2)_{12}C\equiv CC\equiv C(CH_2)_8COO(CH_2)_2\text{-}X$ $X = O(CH_2)_2O(CH_2)_2\text{-}O$ $X = \overset{+}{N}(CH_3)_2,\ Br^-$ $X = \overset{+}{N}H(CH_2)_2SO_3^-$ $X = NCH_3$ $CH_2OCO(CH_2)_8C\equiv CC\equiv C(CH_2)_{12}CH_3$ $CHOCO(CH_2)_8C\equiv CC\equiv C(CH_2)_{12}CH_3$ $CH_2O_3^-PO(CH_2)_2N^+(CH_3)_3$ $CH_2OCO(CH_2)_{16}CH_3$ $CHOCO(CH_2)_8C\equiv CC\equiv C(CH_2)_{12}CH_3$ $CH_2O_3^-POH_2$ $H_3C(CH_2)_{12}C\equiv CC\equiv C(CH_2)_8\text{-}R$ $R = COOH$ $R = CH_2OH$ $R = CH_2OPO_3H_2$	Sonicated dispersions (50°C, N_2) irradiated by uv-light	Absorption spectroscopy	Topochemical polymerization of vesicles observed; polymeric vesicles are stable in up to 50% EtOH	51

Surfactant	Polymerization Method	Characterization	Results, Conclusions	Ref.
CH$_2$OCO(CH$_2$)$_8$C≡CC≡C(CH$_2$)$_{11}$CH$_3$ CHOCO(CH$_2$)$_8$C≡CC≡C(CH$_2$)$_{11}$CH$_3$ CH$_2$O$_3^-$PO(CH$_2$)$_2$N$^+$(CH$_3$)$_3$	Sonicated and ether injected surfactants irradiated by uv light	Calorimetry, absorption spectrophotometry	Large unilamellar and multilamellar liposomes polymerized below phase transition temp. but not above; small unilamellar liposomes did not polymerize at any temperature	52
CH$_2$OCO(CH$_2$)$_{14}$CH$_3$ CHOCO(CH$_2$)$_{11}$OCOC(CH$_3$)=CH$_2$ CH$_2$O$_3^-$PO(CH$_2$)$_2$N$^+$(CH$_3$)$_3$	uv irradiation of sonicated lipids, N$_2$	^{1}H nmr, ir, absorption spectroscopy, electron microscopy ^{14}C sucrose entrapment	Polymeric vesicles are more stable than nonpolymeric ones, allow for controllable substrate release	53
CH$_2$OCO(CH$_2$)$_{10}$COC(CH$_3$)=CH$_2$ CHOCO(CH$_2$)$_{10}$COC(CH$_3$)=CH$_2$ CH$_2$O$_3^-$PO(CH$_2$)$_2$N$^+$(CH$_3$)$_3$				
CH$_2$OCO(CH$_2$)$_{14}$CH$_3$ CHOCO(CH$_2$)$_{11}$OCOC(CF$_3$)=N$_2$ CH$_2$O$_3^-$PO(CH$_2$)$_2$N$^+$(CH$_3$)$_3$	Irradiation at 366 nm Hg - Xe lamp, N$_2$, dispersed surfactants	Differential scanning calorimetry, radio thin layer chromatography	Photolysis results in covalently linked dimers which exhibit temperature dependent phase separations	54
CH$_2$OCO(CH$_2$)$_{14}$CH$_3$ CHOCO(CH$_2$)$_{10}$O—C$_6$H$_4$—C(N$_2$) CH$_2$O$_3^-$PO(CH$_2$)$_2$N$^+$(CH$_3$)$_3$	uv irradiation of sonicated dispersions	Absorption spectroscopy, electron micrography	Shape of liposome remains unchanged during polymerization, stability, however, increased	43

H$_3$C(CH$_2$)$_{12}$C≡CC≡C(CH$_2$)$_8$COO(CH$_2$)$_2$N (morpholine ring with C=O)

(continued)

Table IV. (continued)

Surfactant	Polymerization Method	Characterization	Results, Conclusions	Ref.
$H_3C(CH_2)_{12}CH=CHCH=CHCOO$ ─ X (branched structure) $H_3C(CH_2)_{12}CH=CHCH=CHCOO$ X = $-(CH_2)_2N(CH_3)(CH_2)_2-$, X = $-(CH_2)_2N^+(CH_3)_2(CH_2)_2-$, Br^- X = $-CH_2CHCH_2OH$, X = $-CH_2CHCH_2O_3{}^-PO(CH_2)_2N^+(CH_3)_2$ $H_3C(CH_2)_{17}OCO$ 　　　　CH_2 　　　　$*CHNHY$ $H_3C(CH_2)_{17}OCO$ Y = $CH_2=CCH_3CO$, Y = $CO(CH_2)_5NHCOCH_3C=CH_2$	uv irradiation of sonicated dispersions	Electron microscopy, freeze drying, gel filtration	Polymeric vesicles are stable for weeks	55
(disaccharide surfactant structure) CH_2OH ... $O-(CH_2)_8-R$　**1** $C-NH-NH-C-(CH_2)_8-R$ OH　CH_2OH $R = C≡CC≡C(CH_2)_{12}CH_3$	uv irradiation of sonicated dispersions	Absorption spectroscopy	Monomeric and polymerized liposomes prepared from 1 react with Concavalin A by agglutination and precipitation within a short time – an effect not observed with liposomes not bearing saccharide moieties	56

Surfactant	Polymerization Method	Characterization	Results, Conclusions	Ref.
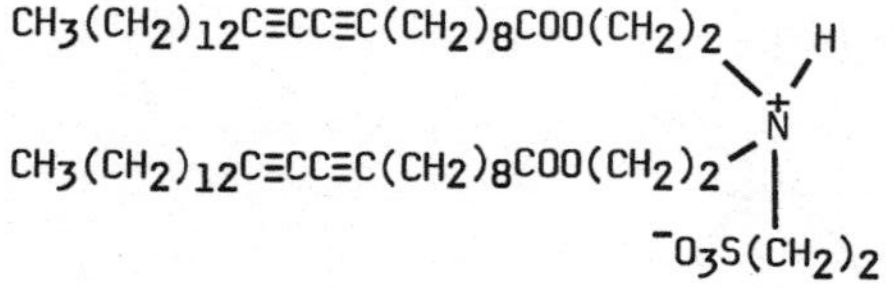 CH$_3$(CH$_2$)$_{12}$C≡CC≡C(CH$_2$)$_8$COO(CH$_2$)$_2$ N$^+$—H, $^-$O$_3$S(CH$_2$)$_2$ CH$_3$(CH$_2$)$_{12}$C≡CC≡C(CH$_2$)$_8$COO(CH$_2$)$_2$—	uv irradiation of sonicated dispersions	Absorption spectroscopy	Liposome formation and polymerization can only be accomplished in alkaline dispersion (pH = 12); polymeric vesicles are stable for months	42
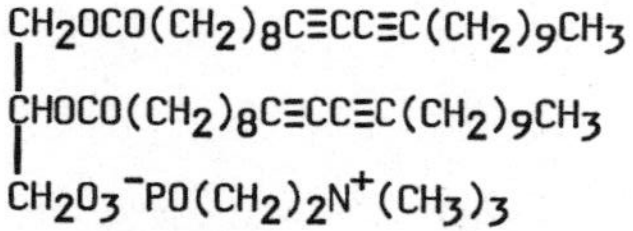 CH$_2$OCO(CH$_2$)$_8$C≡CC≡C(CH$_2$)$_9$CH$_3$ CHOCO(CH$_2$)$_8$C≡CC≡C(CH$_2$)$_9$CH$_3$ CH$_2$O$_3$$^-$PO(CH$_2$)$_2N^+$(CH$_3$)$_3$	uv irradiation of sonicated or alcohol injected dispersions, N$_2$	Electron microscopy, differential scanning calorimetry	Polymerization is facilitated in the crystalline phase below phase transition temperature and inhibited in the lamellar phase	57
CH$_2$OCO(CH$_2$)$_8$C≡CC≡C(CH$_2$)$_9$CH$_3$ CHOCO(CH$_2$)$_8$C≡CC≡C(CH$_2$)$_9$CH$_3$ CH$_2$O$_3$$^-$PO(CH$_2$)$_2N^+$(CH$_3$)$_3$	uv irradiation of Ar flushed samples	Electron microscopy, absorption spectroscopy, photoefficiency of polymerization	Polymerization is topotactic; efficiency depends on the proper alignments of monomer units; polymerization of 2 and 3 ~6,500-fold more efficient than 1	58

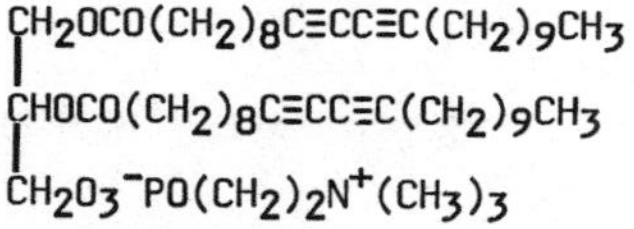

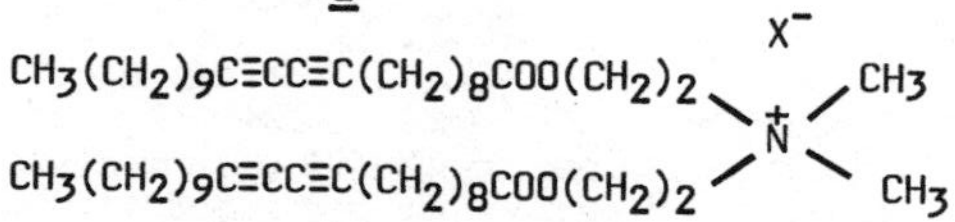

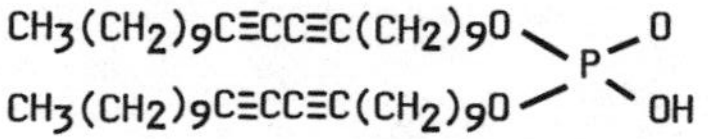

1

CH$_3$(CH$_2$)$_9$C≡CC≡C(CH$_2$)$_8$COO(CH$_2$)$_2$ N$^+$—CH$_3$, X$^-$, CH$_3$
CH$_3$(CH$_2$)$_9$C≡CC≡C(CH$_2$)$_8$COO(CH$_2$)$_2$—

2

CH$_3$(CH$_2$)$_9$C≡CC≡C(CH$_2$)$_9$O—P(=O)—OH
CH$_3$(CH$_2$)$_9$C≡CC≡C(CH$_2$)$_9$O—

3

(continued)

Table IV (continued)

Surfactant	Polymerization Method	Characterization	Results, Conclusions	Ref.
sn-Glycero-3-phosphorylcholins containing the photosensitive ω-[m-(3H-diazirino)-phenoxyl]undecanoyl group in the sn-2 position and a dideuterated palmitic or stearic acid with both deuteriums on specific carbon atoms along the hydrocarbon chain	Vortexed or sonicated dispersions irradiated by uv light	ir, uv, 1H nmr, low resolution electron impact mass spec.	Broad distribution of cross linking along the sn-1 chain; fatty acyl chains are mobile	59
$CH_3(CH_2)_{15}CHNH_2COO(CH_2)_{21}CH_3$ **1** $CH_3(CH_2)_{23}CHNH_2COOCH_3$ **2** $CH_3(CH_2)_{23}CHNH_2COO(CH_2)_{21}CH_3$ **3**	Sonicated dispersion polycondensed	FT-ir, electron microscopy	Vesicles prepared from $\underline{1}$ and $\underline{3}$ precipitate if titrated with 0.02 M NaOH; vesicles prepared from $\underline{2}$ are stable in alkaline solution and undergo uncatalyzed polycondensation - no polycondensation occurs in bulk solution, 2,5-diketopiperidines are formed at higher temperatures	45
$CH_3(CH_2)_{12}C{\equiv}CC{\equiv}C(CH_2)_8COO(CH_2)_2$ $CH_3(CH_2)_{12}C{\equiv}CC{\equiv}C(CH_2)_8COO(CH_2)_2$ $\overset{+}{N}$ with H and $(CH_2)_2SO_3^-$	Irradiation of sonicated dispersions	Absorption spectroscopy, differential scanning calorimeter, electron microscopy	ATP synthetase incorporated into monomeric and polymeric liposomes by incubation; enzyme is reactivated in vesicles; polymerization enhances enzymatic activity; there are "monomeric domains" in enzyme containing polymeric vesicles	60

Surfactant	Polymerization Method	Characterization	Results, Conclusions	Ref.
$CH_2OCO(CH_2)_{11}OCOCH_3=CH_2$ $\|$—$CHOCO(CH_2)_{11}OCOCH_3=CH_2$ $\|$—$CH_2O_3{}^-PO(CH_2)_2N^+(CH_3)_3$	uv irradiation, N_2, of sonicated dispersions	^1H nmr, gel filtration, ir and absorption sepctroscopy, electron microscopy	Polymeric vesicles show enhanced stability and reduced permeability	61
$CH_2OCO(CH_2)_{11}OCOCH_3=CH_2$ $\|$—$CHOCO(CH_2)_{14}CH_3$ $\|$—$CH_2O_3{}^-PO(CH_2)_2N^+(CH_3)_3$				
$CH_2OCO(CH_2)_{14}CH_3$ $\|$—$CHOCO(CH_2)_{11}OCOCH_3=CH_2$ $\|$—$CH_2O_3{}^-PO(CH_2)_2N^+(CH_3)_3$				

(continued)

Table IV (continued)

Polymerization Method	Characterization	Results, Conclusions	Ref.
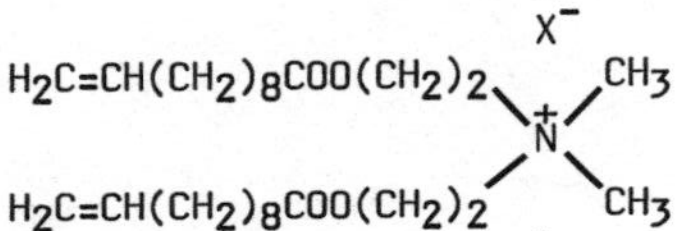 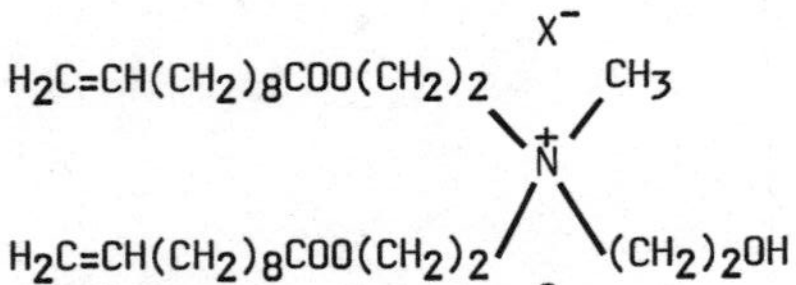 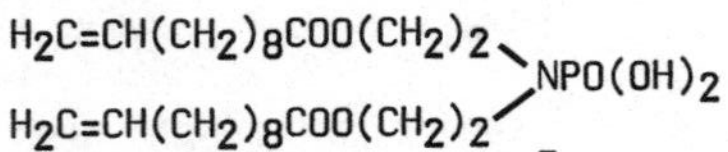 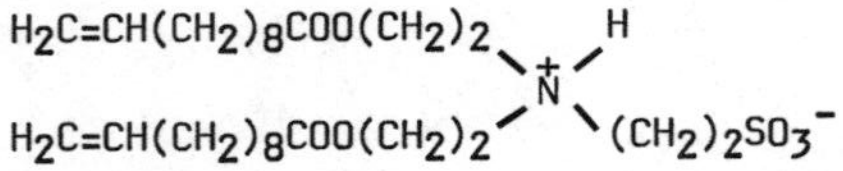 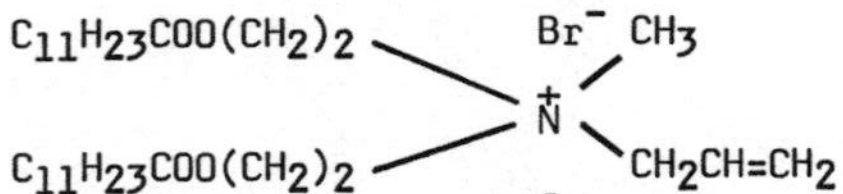 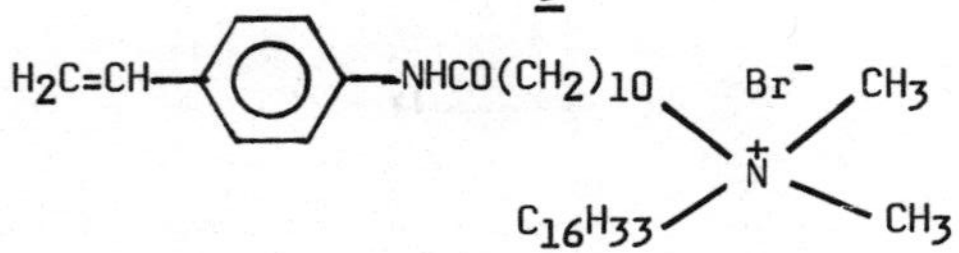 uv irradiation or exposure to AIBN (60°C) of sonicated dispersions	1H nmr absorption and fluoresence spectroscopy, electron microscopy, gel filtration	Polymeric vesicles are considerably more stable than their non-polymeric counterparts; OH⁻ permeability can be controlled by selective polymerization; external exposure of already formed vesicles prepared from 5 to AIBN results in "zipping-up" only the outer surface of the vesicle; exposure to uv irradiation results in the complete loss of double bonds; polymerization rates of 6 are independent of vesicle concentration	62

Surfactant

$H_2C=CH(CH_2)_8COO(CH_2)_2$—$\overset{CH_3}{\underset{CH_3}{\overset{|}{\underset{|}{N^+}}}}$—$X^-$ **1**

$H_2C=CH(CH_2)_8COO(CH_2)_2$—$\overset{CH_3}{\underset{(CH_2)_2OH}{\overset{|}{\underset{|}{N^+}}}}$—$X^-$ **2**

$H_2C=CH(CH_2)_8COO(CH_2)_2$—$\overset{|}{\underset{|}{N^+}}$—$NPO(OH)_2$ **3**

$H_2C=CH(CH_2)_8COO(CH_2)_2$—$\overset{H}{\underset{|}{\overset{|}{N^+}}}$—$(CH_2)_2SO_3^-$ **4**

$C_{11}H_{23}COO(CH_2)_2$—$\overset{Br^-}{\underset{CH_2CH=CH_2}{\overset{CH_3}{\underset{|}{\overset{|}{N^+}}}}}$ **5**

$H_2C=CH$—⬡—$NHCO(CH_2)_{10}$—$\overset{Br^-}{\underset{C_{16}H_{33}}{\overset{CH_3}{\underset{|}{\overset{|}{N^+}}}}}$—$CH_3$ **6**

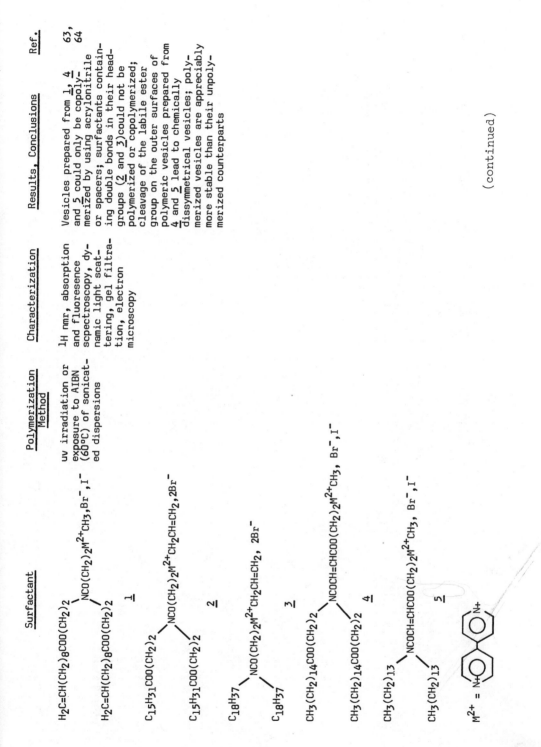

Surfactant	Polymerization Method	Characterization	Results, Conclusions	Ref.
$H_2C=CH(CH_2)_8COO(CH_2)_2$ $\quad$ NCO$(CH_2)_2$M^{2+}CH$_3$,Br$^-$,I$^-$ $H_2C=CH(CH_2)_8COO(CH_2)_2$ **1**	uv irradiation or exposure to AIBN (60°C) of sonicated dispersions	1H nmr, absorption and fluoresence spectroscopy, dynamic light scattering, gel filtration, electron microscopy	Vesicles prepared from **1**, **4** and **5** could only be copolymerized by using acrylonitrile or spacers; surfactants containing double bonds in their headgroups (**2** and **3**)could not be polymerized or copolymerized; cleavage of the labile ester group on the outer surfaces of polymeric vesicles prepared from **4** and **5** lead to chemically dissymmetrical vesicles; polymerized vesicles are appreciably more stable than their unpolymerized counterparts	63, 64
$C_{15}H_{31}COO(CH_2)_2$ $\quad$ NCO$(CH_2)_2$M^{2+}CH$_2$CH=CH$_2$,2Br$^-$ $C_{15}H_{31}COO(CH_2)_2$ **2**				
$C_{18}H_{37}$ $\quad$ NCO$(CH_2)_2$M^{2+}CH$_2$CH=CH$_2$, 2Br$^-$ $C_{18}H_{37}$ **3**				
$CH_3(CH_2)_{14}COO(CH_2)_2$ $\quad$ NCOCH=CHCOO$(CH_2)_2$M^{2+}CH$_3$, Br$^-$,I$^-$ $CH_3(CH_2)_{14}COO(CH_2)_2$ **4**				
$CH_3(CH_2)_{13}$ $\quad$ NCOCH=CHCOO$(CH_2)_2$M^{2+}CH$_3$, Br$^-$,I$^-$ $CH_3(CH_2)_{13}$ **5**				

M^{2+} =

(continued)

Table IV (continued)

Surfactant	Polymerization Method	Characterization	Results, Conclusions	Ref.
$CH_2=CH(CH_2)_8CONH(CH_2)_6$ —$\overset{+}{N}$— CH_3 Br^- CH_3 $CH_3(CH_2)_{15}$ **1**	uv irradiation or exposure to AIBN (60°C) of sonicated dispersions	^1H, ^{13}C nmr, absorption and fluorescence spectroscopy, dynamic light scattering, gel filtration	Increasing sonication times results in the formation of smaller and less polydisperse vesicles; size of vesicles are retained upon polymerization; vesicles containing double bonds in their headgroup can be selectively polymerized (see entry above)	65
$CH_3(CH_2)_{14}COO(CH_2)_2$ —$\overset{+}{N}$— CH_3 Br^- CH_3 $CH_3(CH_2)_{14}COO(CH_2)_2$ $CH_2CH=CH_2$ **2**				
$CH_3(CH_2)_{14}COO(CH_2)_2$ $NCOCH=CHCOOH$ $CH_3(CH_2)_{14}COO(CH_2)_2$ **3**				
$CH_3(CH_2)_{17}$ $NCOCH=CHCOOH$ $CH_3(CH_2)_{17}$ **4**				

aSee footnote in Table I for abbreviations.

Depending on the position of the double bond, vesicles can be polymerized either across their bilayers or across their headgroups (Table IV). Furthermore, vesicles having double bonds in their headgroups can be "zipped-up" either at their inner or their outer surfaces or alternatively be polymerized both at their outer and inner surfaces (Figure 5).[65]   Ultraviolet irradiation of vesicles prepared from 5,[62] 2,[65] 3,[65] and 4[65] (see Table IV for structures under the appropriate entry for the given reference) resulted in the complete loss of vinyl protons (by FT-^{1}H nmr) indicating the crosslinking of both the inner and outer surfaces. Conversely external addition of AIBN to already sonicated vesicles, prepared from the same compounds, and subsequent heating caused incomplete loss of the vinyl protons. Integration of the vinyl signals with respect to CH_3N of the surfactant showed approximately 60% polymerization. This corresponded to "zipping-up" only the external surface of vesicles.

Initial polymerization experiments have been carried out, just like those for monolayers and multilayers, on vesicles prepared from diacetylenes (Table IV).[29, 42, 43, 52, 57, 58] The ease of following the colorless to blue to red color change during polymerization has facilitated these experiments. Electron microscopy (Figure 6) and glucose entrapments provided evidence for maintaining the vesicle structure intact during polymerization.[43,51]

Correct topochemical alignment of surfactants in some, but not all, vesicles is an essential requirement for polymerization. MLV vesicles prepared from a diacetylene containing phospholipid (see structure in Table IV under entry for reference 52) could only be polymerized below the phase transition temperature and SLV could not be polymerized at all.[52] Since diacetylene polymerization is governed by the root mean-square displacement;[79] the closer the reacting moieties are the more readily the polymerization occurs. Vesicles prepared from surfactants containing polymerizable double bonds and methylviologen functionalities in their headgroups (see structure 2 and 3 in Table IV under entry for reference 63 and 64) could not be polymerized under any experimental conditions.[63] Apparently, intermolecular double bonds cannot be brought sufficiently close together in the vesicles for crosslinking. In the absence of bulky and presumably rigid groups viologen polymerization across the surfaces of surfactant vesicles is quite feasible (Table IV). We have recently succeeded in polymerizing vesicles prepared from surfactants which contain a styrene functionality in the headgroups.[80] While vesicles prepared from this surfactant undergo light initiated polymerization with an approximate half life of 50 minutes, there is no measureable polymerization of the same concentration of surfactant in methanol under identical irradiation.[62,80] The importance of topochemical

Figure 6.   Scanning electron micrograph of polymeric vesicles
prepared   from   [H$_3$C(CH$_2$)$_{12}$C $\equiv$ CC $\equiv$ C(CH$_2$)$_8$COO(CH$_2$)$_2$]$_2$N$^+$(CH$_3$)$_2$,Br$^-$.
Reproduced with permission from reference 43.

alignment   can   be   quantitatively   assessed   by   comparing
polymerization efficiencies of diacetylenes in which the triple
bond alignment is varied (see entry under reference 58 in Table
IV).

    The schematics shown for vesicle polymerization in Figure 5
should only be taken as a confession of our ignorance. Surfactants
having photoreactive groups in their hydrocarbon tails may undergo
inter or intramolecular crosslinking within each separate monolayer
in the bilayer vesicle, may crosslink across the bilayer or indeed
react by some combination of all of these processes. Vesicles
prepared from lipids containing photolysable carbene precursors
apparently formed only intermolecular photodimers upon irradiation

(see entries under references 49, 54, and 59 in Table IV). At
present there is only indirect evidence for extended monomer
crosslinking. Careful determination of weight averaged molecular
weights of the polymer units formed following vesicle
polymerization is badly needed!!

All polymeric vesicles show appreciably enhanced stabilities
compared to their nonpolymeric counterparts. They have extensive
shelf lives and do not decompose in most cases, upon the addition
of up to 30% alcohol. Increasing the sonication time results in
smaller and more uniform vesicles. The size of vesicles are
retained upon polymerization. Thus, it is possible to prepare
stable polymeric vesicles with mean hydrodynamic radii ranging from
250 - 2,500 Å.[65] The fluidity of polymeric vesicles depends on
the nature of the surfactants. Polydiacetylene vesicles are rigid
and, in contrast to biological membranes, do not undergo
thermotropic phase transitions.[55] Conversely, polymeric vesicles
prepared from surfactants containing vinyl and styrene groups
showed temperature dependent phase transitions.[62,63,81] Fluidity
of these types of polymeric vesicles (see structure $\underline{2}$ in entry
under reference 65 in Table IV, for example) is also manifest in
their KCl mediated growth. Addition of 3 x 10^{-2} M KCl to both
unpolymeric and polymeric vesicles prepared from $\underline{2}$[65] resulted in
the increase of their hydrodynamic radii from 213 Å to 511 Å
(nonpolymeric) and from 206 Å to 485 Å (polymeric).[65]

Substrate entrapment, retainment and ion permeabilities are
important considerations in designing polymeric vesicles.
Polymeric phospholipid vesicles (see structures in entry under
reference 61 in Table IV) were shown to be completely sealed and to
retain entrapped C-14 labelled glucose better than their
nonpolymeric counterparts.[61] Proton and hydroxide ion
permeabilities have been determined by incorporating pH sensitive
fluorescence probes into the interior of vesicles.[78,82] Hydroxide
ion permeates into polymerized surfactant vesicles with half lives
ranging from 5-20 minutes. Significantly, permeation into
completely polymerized vesicles is appreciably slower than into
vesicles which had only been "stitched-up" at their outer
surfaces.[62]

Extension of the polymeric vesicle concept to
biopolymerization is an interesting recent development. Vesicles
prepared from $\underline{2}$[45] (see structure under reference 45 in Table IV)
underwent spontaneous polycondensation at high pH (Figure 7).
Conversely, competing 2,5-diketopiperidine formation prevails in
homogeneous solution of the same surfactant.[45] Close proximity of
the reacting groups, and apparent hydrophobic environment and high
local pH are the essential requirements for polypeptide formation.
A similar situation has been encountered in the spontaneous alanyl

adenylate polymerization in reversed micelles, formed from sodium di-2-ethyl-hexylsulfosuccinate (Aerosol-OT) and hexadecyltrimethyl- -ammonium bicarbonate, in benzene.[83] In reversed micelles the yield of polypeptides was 94.5% in the $\bar{M}w$ 350 - 3,000 range; in homogeneous solutions only dimers and trimers formed in 30% yield.[83]

Polymerization provides stabilization necessary for creating dissymmetry in vesicles. Domains may be created by copolymerization of two surfactants. Incomplete polymerization could also lead to the formation monomeric domains in the vesicles.[60] Chemical dissymmetry of the inner and outer surfaces of vesicles has been generated in vesicles prepared from surfactants containing labile ester groups adjacent to the polymerizable functionality (see structures 4 and 5 in entry under

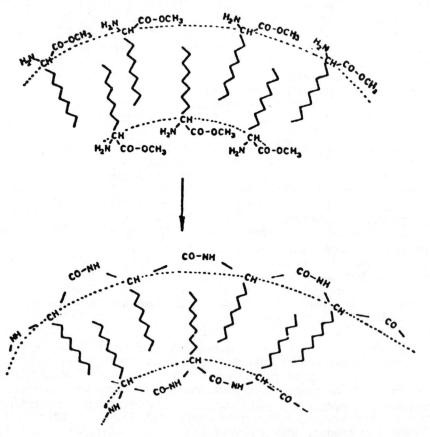

Figure 7.   Schematics for spontaneous polypeptide formation in vesicles. Reproduced with permission from reference 45.

reference 65).   Vesicles prepared from these surfactants could efficiently be copolymerized with acrylonitrile.   Iminodiethanol cleaved only the ester linkages on the outer surface of vesicles; leading to dissymmetrical structures (Figure 8).[64] Dissymmetrical membranes have, of course, vital biochemical functions and dissymmetrical vesicles will have useful applications.

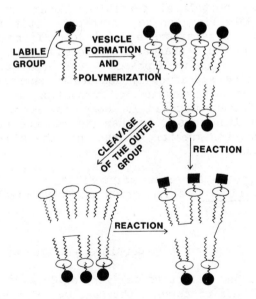

Figure 8. Schematics   for   the   preparation   of   chemically dissymmetrical vesicles.

UTILIZATION AND POTENTIALS OF POLYMERIC ORGANIZED ASSEMBLIES

Beneficial effects of polymeric surfactant aggregates have been utilized in several areas. These, along with future potentials, will be briefly highlighted here.

1. Membrane Modelling

Liposomes, prepared from synthetic phospholipids containing covalently linked carbene or nitrene precursors at selected positions, have been photo-crosslinked in a series of elegant studies by Khorana and his coworkers.[49,54,59,84] This work had aimed at obtaining information on lipid-lipid and lipid-protein interactions in biological membranes.[84] Special care was taken to demonstrate that labelling did not alter the properties of liposomes.[54] Photolysis of liposomes prepared from ω-labelled surfactants (see structures in entry under reference 49 in Table IV) resulted in exclusive crosslinking in the acyl chains of neighboring surfactants. Absence of crosslinking in the headgroup region is in accord with the known structures of liposomes (lack of migration of the terminal hydrophobic groups to the vesicle surface). Crosslinking has been studied for a number of lipids and lipid mixtures as a function of temperature.[54] At temperatures where the solid (gel) and fluid (liquid crystalline) phases coexist, crosslinking preferentially occurred at the fluid phase. Apparently, the photolabel partitions favorably into the fluid phase. Photo-labelling can be, therefore, fruitfully employed for monitoring phase separations and for identifying major components in the solid and fluid phases.[54] Information on the distribution of crosslinking sites in liposomes has been obtained by mass spectroscopy using specifically deuterated surfactants.[49,59] The determined broad distribution of crosslinking sites along the hydrophobic hydrocarbon chain is due to the recognized mobility of acyl chains in liposomes. Further refinements of this technique will provide an even greater insight into the microenvironments of bilayer membranes.

In a different approach polymerization had been induced in Acholeplasma Laidlawii cells grown on polymerizable diacetylenic fatty acid.[52]

2. Target Directed Drug Carriers - Molecular Recognition

The intact and selective delivery of drugs to desired targets is an important aim of current pharmacological research. Although liposomes have been extensively explored as potential drug carriers[76,77,85-87] their inherent instabilities and lack of discrimination have detracted somewhat from their potential. Polymeric vesicles are presumed to overcome these difficulties.[82] Indeed, polymeric phosphatidyl choline vesicles (see entry under reference 53 in Table IV) showed greatly enhanced stabilities and retained glucose longer than their nonpolymeric counterparts.[53] The enhanced activity of polymeric vesicle incorporated ATP-synthetase (see entry under reference 60 in Table IV) is highly relevant.[60] ATP-synthetase consists of a hydrophobic and a hydrophilic portion; the former is incorporated into the bilayer and the latter is exposed into the aqueous environment. The isolated enzyme showed no activity; whereas phospholipids[88] and polymeric surfactant vesicles[60] reactivated the enzyme. Differential scanning calorimetry indicated the formation of "monomeric domains" which provided the site for ATP-synthetase incorporation.[60]

A considerably broader approach has been taken by Ringsdorf who had suggested potential applications of polymeric vesicles in cancer chemotherapy.[89] He envisaged cell specific recognition and tumor cell destruction by membrane destroying agents (lysophospholipid, for example) incorporated into polymeric vesicles. Availability of polymeric vesicles capable of surviving the attack of membrane destroying agents is an essential part of the proposal. Mechanism for the release of the agent would be provided by a trigger mediated opening ("uncorking") of the polymeric vesicle. Alteration of pH, temperature, irradiation and enzymes could trigger the release. Experimental realization of these bold propositions[88] is only a question of time and our ingenuity.

3. Photoconductivity - Microlithography

Electron and energy transfers in organized multilayers have been extensively investigated.[72,73,90-95] Polymerization has provided added dimensions and potentials. A case in point is the development of polymeric multilayers for submicron microlithography used in very large scale integrated (VLSI) micro--circuits.[27,33,35,36] The thickness of the resin and the minimum required electron energy limit resolution to approximately 1.0 nm for conventional systems. Using polymeric multilayers thinner (30 - 1,000 Å), stronger and more uniform films can be prepared which provide resolution as low as 600 Å despite a contrast value of 0.7.[27,33]

Doped monolayers and bilayers can photoconduct current.[40] Stable photoconductors find applications in solar energy conversion[46-98] and photoelectrochemistry.[98]

4. Artificial Photosynthesis - Solar Energy Conversion

In natural photosynthesis the captured visible light is transformed to chemical energy which is then utilized for carbon dioxide reduction. The process is described by the "Z-scheme". Briefly, light is absorbed by two pigment systems: photosytem I, PS I (P700), and photosystem II, PS II (P680). These two systems operate in series; two photons are absorbed for every electron liberated from water. Light induced charge separation in PS II leads to the formation of a stong oxidant, Z^+ (E_O = +0.8V), and a weak reductant, Q^- (E_O = 0.0V). Although the reduction potential of Z^+ is sufficient for water oxidant, evolution of molecular oxygen demands the accumulation of four positive charges. Electron flows from Q^-, via a pool plastoquinones and other carriers, to a weak oxidant, (E_O = +0.4V), generated along with a strong reductant, X^- (E_O = -0.6V) in PS I. This electron flow is coupled to phosphorylation which converts adenosine diphosphate, ADP, and inorganic phosphate to adenine triphosphate, ATP. With the aid of ATP, X^- reduces carbon dioxide to carbohydrate.[100] The precise, yet incompletely understood, arrangements in the thylakoid membrane are responsible for the efficient energy deposition and transmission, for prevention of charge recombination, and for creating a proton gradient essential for photophosphorylation.[101]

In artificial photosynthesis PS I and PS II are substituted by simple sensitizers, S, and electron relays, R. The role of membrane mimetic agents is to perform the functions of the thylakoid membrane. Judicious organization of S and R should bring about favorable energy deposition and transmission and, importantly, prevent back electron transfer between the reduced relay, R^-, and the oxidized sensitizer, S^+.

Ideally (i) the reduced relay (R^-) and the oxidized sensitizer (S^+) should be thermodynamically capable of generating hydrogen and oxygen from water (Equation 3 and 4).

$$2R^- + 2H_2O \longrightarrow 2R + 2OH^- + H_2 \qquad (3)$$

$$(2e^- \text{ reduction})$$

$$4S^+ + 2H_2O \longrightarrow 4S + 4H^+ + O_2 \qquad (4)$$

$$(4e^- \text{ oxidation})$$

and (ii) the process should be cyclic.   In practice, the multielectron steps demand the use of catalysts:

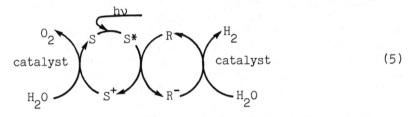

$$\tag{5}$$

and Equation (5) is simplified to two half cells (Equations 6 and 7)

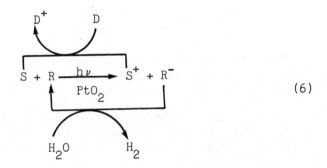

$$\tag{6}$$

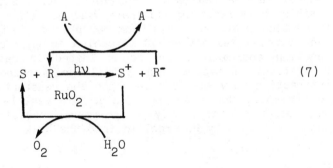

$$\tag{7}$$

which require the use of sacrifical electron donors (D in Equation 6) for H_2 production and sacrifical electron acceptors (A in Equation 7) for O_2 production.   Investigations of sacrifical H_2 generation provided valuable insight into the mechanism of this process. Efficient electron transfer had been demonstrated from a sacrifical donor, ethylenediamine tetraacetate, EDTA, to methylviologen, MV^{2+} via a sensitizer, tris (2,2'-bibyridine)-ruthenium chloride, $Ru(bpy)_3^{2+}$. $Ru(bpy)_3^{2+}$ was attached to the outer surfaces and MV^{2+} was placed on the inner surface of dihexadecyl (DHP) vesicles while EDTA was distributed in the bulk aqueous solution.[102] Initially, electron transfer was considered to occur across the vesicle bilayers.[102] Subsequently, electron transfer was shown to occur on the same outer surface of DHP vesicles, following photosensitized leakage of MV^{2+}.[103] This result as well as the relatively poor long term stability of vesicles demanded alternative approaches. Functionally polymerized chemically dissymmetrical vesicles provide a means to efficient artificial photosynthesis without deleterious instabilities.

5. Reactivity Control

There are four extreme sites of reactant localization in polymeric vesicles. Hydrophobic molecules can be distributed among the hydrocarbon bilayers of the vesicles. Alternatively, they can be anchored by a long chain terminating in a polar headgroup. Polar molecules, particularly those which are electrostatically repelled from the inner surface of the vesicles, may move about relatively freely in the vesicle entrapped water pools or they may be associated with or bound to the inner and outer surfaces of vesicles. Polar molecules can also be anchored to the vesicle surface by a long hydrocarbon tail. A large variety of reactivity control can be realized in polymeric vesicles. Conceivably, the position of a reacting substrate will be different from that of the transition state and that from the product formed in the reaction. Such spacial relocation of molecules as they progress along their reaction coordinates can be exploited in catalyses and product separation. A type of functionally polymerized vesicle - reactant interaction is visualized in which the reactant - an organic ester, for example, would be engulfed in the vesicle. Hydrolysis would then occur in the matrix of polymeric vesicles and the products would subsequently be expelled into the bulk solution.

6. Conclusion

Polymeric organized assemblies are here to stay! Lack of depth in this overview reflects only the infancy of the subject. Practical exploitation of this fascinating area will surely flourish.

ACKNOWLEDGEMENTS

Support of this work by the National Science Foundation and the Army Office of Research is gratefully acknowledged.

REFERENCES

1. J. H. Fendler, "Membrane Mimetic Chemistry", Wiley-Inter--science, New York, 1982.
2. J. H. Fendler and E. J. Fendler, "Catalysis in Micellar and Macromolecular Systems", Academic Press, New York, 1975.
3. A. Blumstein, Editor, "Mesomorphic Order in Polymers and Poly--merization in Liquid Crystalline Media", ACS Symposium Series No. 74, American Chemical Society, Washington, D. C. 1978.
4. V. Kammer and H. G. Elias, Kolloid Z. Polym., 250, 344 (1972).
5. I. Mielke and H. Ringsdorf, Makromol. Chem., 153, 307 (1972).
6. V. Martin, H. Ringsdorf, H. Ritter and W. Sutter, Makromol. Chem., 176, 2029 (1975).
7. H. Ringsdorf and D. Thunig, Makromol. Chem., 178, 2205 (1977).
8. C. M. Paleous and P. Dias, J. Polymer Sci. Polym. Chem., 16, 1495 (1978).
9. W. J. Leonard, cited by H. Ringsdorf in "Polymerization of Organized Systems," H. G. Elies, Editor, Midland Macromolecu--lar Monographs, 3, 187 (1977).
10. K. Dorn, H. Ringsdorf, and W. Siol, Abstract, 3rd Intern. Conference on Surface and Colloid Science, Stockholm (1977).
11. S. S. Atik and J. K. Thomas, J. Am. Chem. Soc., 103, 4279 (1981).
12. P. Lianos, J. Phys. Chem. 86, 1935 (1982).
13. S. E. Friberg, R. Thundathil and J. O. Stoffer, Science 205, 607 (1979).
14. R. Thundathil, J. O. Stoffer and S. E. Firberg, J. Polymer. Sci. Polym. Chem., 18, 2629 (1980).
15. R. Benz, W. Prass and H. Ringsdorf, Angew. Chem. Int. Ed. Eng., 21, 368 (1982).
16. R. Ackermann, O. Inacker and H. Ringsdorf, Kolloid Z. Polym., 249, 1118 (1971).
17. M. Puterman, T. Fort, Jr. and J. B. Lando, J. Colloid Interface Sci., 47, 705 (1974).
18. B. Tieke, G. Wegner, D. Naegele and H. Ringsdorf, Angew. Chem. Int. Ed. Eng., 15, 764 (1976).
19. D. Naegele and H. Ringsdorf, J. Polymer Sci. Polym. Chem., 15, 2821 (1977); B. Tieke, H. J. Graf, G. Wegner, B. Naegele, H. Ringsdorf, A. Banerjie, D. Day, J. B. Lando, Colloid Polymer Sci., 225, 521 (1977).
20. D. Naegele and H. Ringsdorf, in "Polymerization of Organized Systems," H. G. Elias, Editor, Midland Macromolecular Monographs, 3, 79 (1977).

21. D. Naegele, J. B. Lando and H. Ringsdorf, Macromoleculer, 10, 1339 (1977).
22. V. Enkelmann and J. B. Lando, J. Polymer Sci. Polym. Chem., 15, 1843 (1977).
23. A. Barraud, C. Rosilio and A. Ruaudel-Teixier, J. Colloid Interface Sci., 62, 509 (1977).
24. A. Barraud, C. Rosilio and A. Ruaudel-Teixier, Polymer Preprints, 37, 52 (1978).
25. B. Tieke and G. Wegner, Makrom. Chem., 179, 1639 (1978).
26. D. Day and H. Ringsdorf, J. Polymer Sci. Polym. Lett., 16, 205 (1978).
27. A. Barraud, C. Rosilio and A. Ruaudel-Teixier, Solid State Technology, 120 (Aug. 1979).
28. D. Day and H. Ringsdorf, Makromol. Chem. 180, 1059 (1979).
29. D. Day, H. H. Hub and H. Ringsdorf, Israel J. Chem., 18, 325 (1979).
30. A. Banerjie and J. B. Lando, Thin Solid Flims, 68, 67 (1980).
31. D. Day and J. B. Lando, Macromolecules, 13, 1483 (1980).
32. D. Day and J. B. Lando, Macromolecules, 13, 1478 (1980).
33. A. Barraud, C. Rosilio and A. Ruaudel-Teixier, Thin Solid Films, 68, 99 (1980).
34. A. Barraud, C. Rosilio and A. Ruaudel-Teixier, Thin Solid Films, 68, 7 (1980).
35. A. Barraud, C. Rosilio and A. Ruaudel-Teixier, Thin Solid Films, 68, 91 (1980).
36. A. Barraud, Abstract, 151th Meeting of the American Electro-chemical Society, St. Louis, MO, May 13 (1980).
37. H. H. Hub, B. Hupfer and H. Ringsdorf, Org. Coatings and Plastics Preprints, 42, 2 (1980).
38. H. Ringsdorf and H. Schupp, Org. Coatings and Plastics Preprints, 42, 379 (1980).
39. D. Day and J. B. Lando. J. Polymer Sci. Polym. Physics, 19, 165 (1981).
40. D. Day and J. B. Lando, J. App. Polymer Sci., 26, 1605 (1981).
41. B. Hupfer, H. Ringsdorf and H. Schupp, Makromol. Chem. 182, 247 (1981).
42. H. Koch and H. Ringsdorf, Makromol. Chem., 182, 255 (1981).
43. H. H. Hub. B. Hupfer, H. Koch and H. Ringsdorf, J. Macromol. Sci. Chem., A15, 701 (1981).
44. H. Ringsdorf and H. Schupp, J. Macromol. Sci. Chem., A15, 1015 (1981).
45. T. Folda, L. Gros and H. Ringsdorf, Makromol. Chem. Rapid Commun., 3, 167 (1982).
46. C. Bubeck, B. Tieke and G. Wegner, Ber. Bunsengesell. Phys. Chem., 495 (1982).
47. C. Bubeck, B. Tieke and G. Wegner, Ber. Bunsengesell. Phys. Chem., 499 (1982).
48. O. Albrecht, D. S. Johnson, C. Villaverde and D. Chapman, Biochim. Biophys. Acta, 687, 165 (1982).

49. C. Gupta, R. Radhakrishnan, G. E. Gerber, W. L. Olsen, S. Quay and H. G. Khorana, Proc. Natl. Acad. Sci. USA, 76, 2595 (1979).
50. S. C. Regen, B. Czech and A. Singl, J. Am. Chem. Soc., 102, 6638 (1980).
51. H. Hub, B. Hupfer, H. Koch and H. Ringsdorf, Angew. Chem. Int. Ed. Eng., 19, 938, (1980).
52. D. S. Johnson, J. Sanghera, M. Pons and D. Chapman, Biochim. Biophys. Acta, 602, 57 (1980).
53. S. L. Regen, A. Singh, G. Oehme and M. Singh, Biochim. Biophys. Res. Commun., 101, 131 (1981).
54. W. Curatolo, R. Radhakrishnan, C. M. Gupta and H. G. Khorana, Biochemistry, 20, 1374 (1981).
55. A. Akimoto, K. Dorn, L. Gros, H. Ringdorf and H. Schupp, Angew. Chem. Int. Ed. Eng., 20, 90 (1981).
56. H. Bader, H. Ringsdorfand J. Skura, Angew. Chem. Int. Ed. Eng., 20, 91 (1981).
57. D. F. O'Brien, T. H. Whitesides and R. T. Klingbiel, J. Polymer Sci. Polym. Lett., 19, 95 (1981).
58. E. Lopez, D. F. O'Brien and T. H. Whitesides, J. Am. Chem. Soc., 104, 305 (1982).
59. R. Radhakrishnan, C. E. Costello and H. G. Khorana, J. Am. Chem. Soc., 104, 3390 (1982).
60. N. Wagner, K. Dose, H. Koch and H. Ringsdorf, FEBS Lett., 132, 313 (1981).
61. S. L. Regen, A. Singh, G. Oehme and M. Singh, J. Amer, Chem. Soc., 104, 791 (1982).
62. P. Tundo, D. J. Kippenberger, P. L. Klahn, N. E. Prieto, T. C. Jao and J. H. Fendler, J. Am. Chem. Soc., 104, 456 (1982).
63. P. Tundo, D. J. Kippenberger, M. J. Politi, P. Klahn and J. H. Fendler, J. Am. Chem. Soc., 104, 000 (1982).
64. P. Tundo, K. Kurihara, D. J. Kippenberg, M. Politi and J. H. Fendler, Angew. Chem. Int. Ed. Eng., 21, 81 (1982).
65. D. J. Kippenberger, K. Rosenquist, L. Odberg, P. Tundo and J. H. Fendler, J. Am. Chem. Soc., 104, 000 (1982).
66. D. C. Blackley, "Emulsion Polymerization", Applied Science, London (1975).
67. V. Martin, H. Ringsdorf and D. Thunig in "Polymerization of Organized Systems," H. G. Elias, Editor, Midland Macromolecular Monographs, 3, 175 (1977).
68. S. Sadron, Pure Appl. Chem., 4, 347 (1962).
69. Y. Bouligand, P. E. Cladis, L. Liebert and L. Strzelecki, Mol. Liq. Cryst., 25, 233 (1974).
70. L. Strzelecki and L. Liebert, Bull. Soc. Chim. France, 597 (1973).
71. L. Liebert and L. Strzelecki, C. R. Acad. Sci. Ser. C., 276, 647 (1973).

72. H. Kuhn, D. Mobius and H. Bucher, in "Physical Methods for Chemistry," Vol. I, Part 111 B, p. 577, A. Weissburger and B. W. Rossiler, Editors, Wiley-Interscience, New York, 1972.

73. D. Mobius, Acc. Chem. Res., 82, 848 (1981).

74. G. Lieser, B. Tieke and G. Wegner, Thin Solid Films, 68, 77 (1980).

75. D. Papahadjopoulos, Editor, Ann. New York Acad. Sci., 308, (1978).

76. H. K. Kimelberg and E. G. Mayhew, CRC Crit. Rev. Toxicol., 6, 25 (1978).

77. G. Gregoriadis and C. Allison, "Liposomes in Biological Systems," John Wiley, New York, 1980.

78. J. H. Fendler, Acc. Chem. Res., 13, 7 (1980).

79. R. H. Baughman and K. C. Yee, J. Polym. Sci. Macromol. Rev., 13, 219 (1978).

80. L. Guterman and J. H. Fendler, (1982), unpublished results.

81. A. Kusumi, M. Singh, D. S. Tyrrell, G. Oehme, A. Singh, N. K. P. Samuel, J. S. Hyde and S. L. Regen, J. Am. Chem. Soc., (1983) in press.

82. K. Kano and J. H. Fendler, Biochim. Biophys. Acta, 509, 289 (1978).

83. D. W. Armstrong, R. Senguin, C. J. McNeil, R. D. Macfarlane and J. H. Fendler, J. Am. Chem. Soc., 100, 4605 (1978).

84. R. Radhakrisnan, C. M. Gupta, B. Erni, R. J. Robson, W. Curatolo, A. Majumdar, A. H. Ross, Y. Takagaki and H. G. Khorana, Ann. New York Acad. Sci., 346, 165 (1980).

85. J. H. Fendler and A. Romero, Life Sci., 20, 1109 (1977).

86. B. E. Ryman and D. A. Tyrrell, Essays Biochem., 16, 49 (1980).

87. D. A. Tyrrell, T. D. Heath, C. M. Colley and B. E. Ryman, Biochem. Biophys. Acta, 457, 259 (1976).

88. E. Schneider, P. Friedl, V. Schwulera and K. Dose, Europ. J. Biochem., 108, 331 (1980).

89. L. Gross, H. Ringsdorf and H. Schupp, Angew. Chem. Int. Ed. Eng., 20, 305 (1981).

90. D. G. Whitten, Angew. Chem. Int. Ed. Eng., 18, 440 (1979).

91. G. Sprintschnik, H. W. Sprintschnik, P. P. Kirsch and P. Whitten, J. Am. Chem. Soc., 98, 2337 (1976).

92. K. P. Seefeld, D. Mobius and H. Kuhn, Helv. Chim. Acta, 60, 2608 (1977).

93. E. E. Polymeropoulos, D. Mobius and H. Kuhn, J. Chem. Phys., 68, 3918 (1978).

94. E. E. Polymeropoulos, D. Mobius and H. Kuhn, Thin Solid Films, 68, 173 (1980).

95. A. F. Janzen and J. R. Bolton, J. Am. Chem. Soc., 101, 6342 (1979).

96. J. Kiwi, K. Kalyanasundaran and M. Gratzel, "Visible Light Induced Cleavage of Water in Hydrogen and Oxygen into Colloidal Microheterogeneous Systems," Springer Verlag, Heidelberg, 1981.

97. H. Gerischer and J. J. Katz, "Light Induced Charge Separation in Biology and Chemistry," Verlag Chemie, New York, 1979.

98. S. Claesson and M. Engstrom, "Solar Engergy - Photochemical Storage and Conversion", National Swedish Board for Energy Conversion and Storage, Stockholm, 1977.

99. R. W. Murray, Acc. Chem. Res., $\underline{13}$, 135 (1980).

100. Govindjee, "Bioenergetics in Photosynthesis," Academic Press, New York, 1975.

101. P. D. Boyer, B. Chance, L. Eruster, P. Mitchell, E. Racker and E. C. Slater, Ann. Rev. Biochem., $\underline{46}$, 957 (1977).

102. M. S. Tunuli and J. H. Fendler, J. Am. Chem. Soc., $\underline{103}$, 2507 (1981).

103. L. Y. C. Lee, J. K. Hirst, K. Kurihara and M. Politi, J. Am. Chem. Soc., $\underline{104}$, 370 (1983).

LIGHT SCATTERING BY LIQUID SURFACES

D. Langevin, J. Meunier and D. Chatenay

Laboratoire de Spectroscopie Hertzienne de l'E.N.S.

24, rue Lhomond, 75231 Paris Cedex 05, France

In this paper, we intend to summarize the theoretical and experimental work done in the recent years in relation to the scattering of light from liquid surfaces.

This field of research is relatively new, essentially because before the appearance of laser light sources less than 20 years ago, the light scattering experiments were very difficult to perform with the available spectral lamps. Moreover, even with laser sources, the practical difficulties encountered in these experiments are numerous. The number of papers published on the subject is still low compared to the case of the now widely developed bulk light scattering techniques. In particular, most of the time, i.e. far from critical points, the surface light scattered intensity cannot be measured because it cannot be distinguished from the light scattered by the bulk of the liquid phases. The distinction becomes feasible if a spectral analysis is performed : indeed, the characteristic frequencies and dispersion relations are very different for bulk and surface scattering. We then mostly discuss in the following the surface inelastic scattering features.

In the Introduction, a short historical survey is presented together with a brief summary of the experiments performed in the recent years. In the second Section the problem of the scattered intensity is considered. In the third Section, a detailed computation of the spectrum of the scattered light for the very general case of the interface between two liquids in the presen-

ce of a surfactant layer is shown. This calculation
embodies other calculations previously performed in
our laboratory for simpler cases (i.e. free surface
of a liquid) which constitute to our knowledge the
only entirely correct treatment of the problem. Experimental technical details are described in Section IV. Finally, in order to illustrate the whole
discussion, two different types of experiments performed recently in our laboratory are presented : monolayers on water and microemulsions systems.

I. INTRODUCTION

The scattering of light from thermally excited surface waves
was first predicted by V. Schmoluchowsky[1] in 1908. He noted that
thermal motion should constantly distort the surface and give rise
to a certain roughness. In 1913, L. Mandelstam[2] derived the mean
square amplitude of these thermal fluctuations and, by using
Rayleigh's[3] theory for the diffuse reflection by a rough surface,
he calculated the intensity of the light scattered in the plane of
incidence.

In this treatment, the vertical displacement of a given point
of the surface $\vec{r}$ and at a given instant t is written as a sum of
Fourier components :

$$\zeta(\vec{r}, t) = \sum_{\vec{q}} \zeta_{\vec{q}}(t) \, e^{i \, \vec{q} \cdot \vec{r}}$$

Each component $\zeta_{\vec{q}}$ behaves like a sinusoidal diffraction grating.
The zeroth order contributes to the regular reflection, and the
first order to the scattering. The higher orders also give a contribution to the scattering, which is negligible compared to the
first order one, because the roughness amplitude is very small :
$< \zeta^2 >^{1/2} \sim 10$ Å. The light scattered by a particular Fourier component $\zeta_{\vec{q}}$ is found in a well-defined direction simply related to
$\vec{q}$ by :

$$\vec{k}'_{\Sigma} - \vec{k}_{\Sigma} = \pm \vec{q}$$

where $\vec{k}'_{\Sigma}$ and $\vec{k}_{\Sigma}$ are, respectively, the projections of the wave
vectors of the scattered and reflected light on the surface plane.

The creation of a surface vibration mode of wave vector $\vec{q}$ (surface ripplon) requires a certain amount of work against gravity and capillary forces. The equipartition theorem leads to

$$< \zeta_q^2 > = \frac{kT}{\gamma q^2 + (\rho_2 - \rho_1)g} \qquad \text{per unit area} \tag{1}$$

where γ is the surface tension, ρ_1 and ρ_2 the densities of the upper and lower fluids respectively. The gravity term $(\rho_1 - \rho_2)g$ is usually negligible for the wave vectors of interest in light scattering. It follows that the intensity of the scattering, which is proportional to $< \zeta_q^2 >$, becomes very large if the surface tension is small. This happens for instance close to a critical point and gives rise to critical "surface opalescence".

Mandelstam qualitatively verified these predictions, by making visual observations of the light scattered by a carbon disulfide-methanol interface near the critical point. In 1925 Raman and Ramdas[4] studied the polarization and the scattered intensity in more detail. In 1926, A.A. Andronov and M.A. Leontovich,[5] and independently R. Gans[5], derived the general formulas for the scattered light intensity. Further papers on the subject[6] were scarce between 1942 and 1964 . Indeed, the conventional spectral lamps made the light scattering experiments very difficult to perform. A renewal in the research field was associated to the appearance of the Helium-neon lasers. The intensity of these new light sources was much higher. They were also very well defined both in direction, thus improving the accuracy on $\vec{q}$, and in frequency, making possible the frequency analysis of the scattered light.

We have seen that the angular study of the scattered intensity reflects the spatial distributions of the fluctuations ζ . The spectral analysis will reflect their temporal evolution. Indeed, the scattered intensity with wave vector $\vec{k}'$, frequency ω' , is[7]

$$I(\vec{k}' , \omega') = F . P(\vec{q} , \omega) \qquad \omega = \omega' - \omega_o \tag{2}$$

ω_o being the frequency of the incident light. F depends only on the incident beam properties and scattering geometry, P only on the scattering system: P is the space-time Fourier transform of the correlation function of ζ, i.e., $< \zeta(\vec{r}, t) \zeta(\vec{r}', t') >$.

After its creation, a surface ripplon will tend to disappear under the influence of restoring forces (capillarity + gravity) and damping forces due to bulk viscosity in the case of simple fluids. Depending on whether the restoring force is larger or smaller than

the damping force, the surface ripplon will propagate as a damped
oscillation or will be damped exponentially without propagation.
It behaves optically as a moving diffraction grating and the
scattered light is frequency shifted by the Doppler effect. If the
fluid viscosities η_1 and η_2 are small, the frequency of propagation,
ω_q, and the lifetime of surface ripplons, τ_q, are (propagating case):

$$\omega_q = \left(\frac{\gamma}{\rho_1 + \rho_2}\right)^{\frac{1}{2}} q^{\frac{3}{2}} \qquad \frac{1}{\tau_q} = \Delta\omega_q = 2\frac{(\eta_1 + \eta_2)}{\rho_1 + \rho_2}q^2 \qquad (3)$$

and if the fluid viscosities are large, or if the surface tension
γ is very small, their lifetime is (overdamped case)

$$\frac{1}{\tau_q} = \Delta\omega_q = 2\frac{\gamma}{(\eta_1 + \eta_2)}q \qquad (4)$$

As a consequence, the spectrum of the scattered light in the
first case has two Lorenzian components symmetric with respect to
ω_o. They are centered at $\omega_o \pm \omega_q$ and their half width is $\Delta\omega_q$
(Equation (3)). In the second case the spectrum is a single
Lorentzian curve centered at ω_o, of half width $\Delta\omega_q$ (Equation (4)).

This was first noted by M. Papoular[8] who associated the
Mandelstam results with classical hydrodynamics. His predictions
were confirmed in our laboratory[9] by the first experimental
results using light beating spectroscopy. At the same time,
Katyl and Ingard[10] obtained independently similar results
under more difficult experimental conditions: Fabry-Perot
interferometer and large scattering angles. Both the frequency
resolution and the scattered intensity were very low. The method
that we had used (light beating spectroscopy) is particularly well
suited for the surface scattering experiments and was adopted in
all further studies.

In the extreme limits of large or small bulk viscosities,
the analysis of the spectrum is very simple; its components have
Lorentzian shapes whose characteristics (frequency, widths) are
deduced from the roots of the dispersion equation of surface
phonons. But in the intermediate viscosity range, which is
encountered very often experimentally, the spectrum is much more
complex. We have then tried to derive theoretically the exact
spectral shape. In a first approach, the surface was replaced by
a thin elastic membrane having the same vibration modes[11]. This
spectrum was identical to that of an harmonic oscillator. But,
we observed small systematic deviations between this theoretical
and the experimental spectra. A more rigorous derivation of the

theoretical spectrum, taking into account the coupling between
the surface and the bulk fluids was then undertaken[12]. We verified
then that the new theoretical spectrum satisfied the fluctuation
dissipation theorem.

The study of the spectrum of the light scattered by surfaces
of simple fluids allows the determination of the surface tension
γ and of the bulk viscosity η . Many others, often more precise,
methods can be used to measure these quantities. The advantages
of the light scattering technique are:

- the system is at thermal equilibrium;
- the only perturbation to the surface is the light beam
  whose intensity can be lowered if necessary.

The technique can therefore be used when more simple methods
may fail. It can also allow the measurement of different
parameters in more complex systems. The following list will review
briefly the experiments performed in recent years:

surface tension measurements

    critical points   pure fluids CO_2[13]   Xe[14] SF_6[15]

                  binary mixtures   cyclohexane-methanol[16]

    nematic liquid crystals[17]

    nematic-isotropic interface[18]

    lipid bilayer[19]

    multiphase microemulsion systems[20]

other surface forces

    surface elasticity: insoluble monolayers on water[21-24]

    interactions between two surfactant layers in soap films
      (electrostatic, van der Waals forces)[25,26]

bulk viscosity

    nematic liquid crystals (see above)

    polymer solutions[27,28]

    critical point of pure fluids (see above)

surface viscosities

insoluble monolayers on water (see above).

In the following we will restrict the discussion to the problem of adsorbed surfactant layers at the interface between two fluids. We will discuss first the scattered intensity problem (§ II). We will then present the calculation of the spectral shape (§ III), give some details on the experimental technique (§ IV) and present a few recent data obtained in our laboratory (§ V).

II.   INTENSITY SCATTERED BY A LIQUID SURFACE

The surface scattered light intensity I was first calculated by Mandelstam[2], Andronov and Leontovich[4], and Gans[4]. They found that I was proportional to the mean square amplitude of the fluctuations. From Equation (1) it follows that I decreases very rapidly when the scattering angle increases. In practice, the experiments are often restricted to very low scattering angles   : $\Delta\theta \lesssim 1°$. In this domain, the scattered intensity per unit solid angle around the reflected beam is

$$\frac{dI}{d\Omega} = I_o \frac{k_o^4}{4\pi^2} <\zeta_q^2> R \cos^3\theta \quad ; \quad q = k_o \Delta\theta$$

where I_o is the incident light intensity, $\vec{k}_o$ the wave vector of the incident light, θ the angle of reflection, R the reflection coefficient.

Let us estimate the intensity in the case of the free surface of water $\gamma = 72$ dyn/cm $\rho = 1$g/cm^3 . If we use an He-Ne laser ($k_o \sim 10^5$ cm^{-1}), with the incident beam normal to the surface $\theta \cong 0$, and if we detect the scattered light in the direction $\Delta\theta = 10'$,   then

$$\frac{1}{I_o}\frac{dI}{d\Omega} \sim 10^{-3}$$

This value is large compared with the scattering by density fluctuations in the bulk water ($\sim 10^{-5}$ for a scattering volume 1 cm thick). But the surface scattering decreases very rapidly as the scattering angle increases, whereas the bulk scattering remains constant.

The optimum geometry for the experiment will correspond to the maximum of R cos$^3\theta$. The reflection coefficient depends on the ratio between the refractive indices of the two fluids $n = n_i/n_r$,

n_i being the index for the incident beam, n_r for the refracted beam. When $n > 1$, $\frac{1}{I_o}\frac{dI}{d\Omega}$ varies smoothly with θ_i . For $n > 1.2$ the maximum corresponds to normal incidence. When $n < 1$, there is a pronounced maximum for the incidence relative to total internal reflection. The gain in intensity with respect to the case of normal incidence becomes very important for $n \sim 1$.

The above calculations of the scattered intensity were all based on classical continuity conditions for the electromagnetic field from both sides of the distorted surface. When the surface is covered by a monolayer there are other kinds of surface fluctuations that do not produce surface deformations, but scatter light for instance:

- density fluctuations in the surfactant layer;
- orientation fluctuations if the surfactant molecules are elongated and lie in the interface plane (two dimensional nematics).

It was therefore necessary to obtain a new method of deriving the scattered intensity, taking into account the nature of the scattering medium and relating the scattered electric field directly to the molecular properties of the surface layer. This can be achieved by considering that the incident electric field induces a dipole distribution in the medium and calculating the field radiated by these dipoles. This was first done by Ewald and Oseen[29] for uniform molecular distributions in the two fluids and a perfectly flat surface. They rederived in this way the reflection and refraction laws. By carrying the procedure further we extended this calculation to the problem of scattering[30].

The new method gives the same result for the rough surface problem as the previous ones. It also allows the calculation of the intensity scattered by density fluctuations in a monolayer. The ratio of intensities scattered by density fluctuations and surface roughness is independent of the scattering geometry and has the value

$$ r = \frac{\gamma}{\varepsilon} (qd)^2 \left[\frac{\alpha_f}{\alpha}\right]^2 $$

where ε is the monolayer dilational modulus, d its thickness, α_f and α the polarizabilities per unit volume of the film and of the underlying liquid respectively. As qd is very small (typically 10^{-4}), it will be very difficult in practice to observe the density fluctuations, except maybe with dye monolayers (large α_f) or monolayers close to a two dimensional critical point (low ε).

In the following we will restrict our discussion to the surface roughness problem.

III. SPECTRUM OF THERMAL FLUCTUATIONS

Landau and Placzek[31] showed that the average evolution of thermal fluctuations is described by the classical hydrodynamic theory since both the characteristic times and distances are large compared with collision times and molecular lengths. The amplitude of the fluctuations being much smaller than their wavelengths, the hydrodynamic equations can be linearized. As a consequence, the surface phonons of different wave vectors $\vec{q}$ will be statistically independent.

The power spectrum of the fluctuations $P(\vec{q},\omega)$ (Equation (2)) is the Fourier transform of the correlation function:

$$G(\vec{q},t) \;=\; < \zeta_{\vec{q}} (t)\; \zeta_{\vec{q}} (0) >$$

$G(\vec{q},t)$ can be calculated by the statistical method used in Reference 12. An equivalent procedure that we will use here for simplicity is based on the fluctuation dissipation theorem[32]:

$$P(\vec{q},\omega) \;=\; \frac{kT}{\pi\omega}\; \mathrm{Im}\; \chi(\vec{q},\omega)$$

where $\chi(\vec{q},\omega)$ is the response function of the system to an external pressure $p_{ext} e^{i\omega t}$: $\chi(\vec{q},\omega) \;=\; \zeta_{\vec{q}}(\omega)/p_{ext}$, $\zeta_{\vec{q}}(\omega)$ being the time Fourier transform of $\zeta_{\vec{q}}(t)$.

The calculation of the response function will be done by solving the hydrodynamic equations and fixing the proper limiting conditions at the surface. These hydrodynamic equations are very simple in the present problem. The frequencies of the surface phonons are much smaller than the frequencies of sound waves and thermal waves. The fluids can then be treated as incompressible and isothermal. One is left with only two equations[33]

$$\mathrm{div}\; \vec{v} \;=\; 0$$

$$\rho \frac{\partial \vec{v}}{\partial t} \;=\; \eta\, \Delta\vec{v} - \mathrm{grad}\; p$$

where $\vec{v}$ is the fluid velocity, p the pressure. By Fourier transforming these equations with respect to time and space, one obtains, by choosing $OX//\vec{q}$, solutions of the form:

$$v \sim e^{iqx} e^{i\omega t} e^{mz}$$

with $m = \pm q$ and $m = \pm q\sqrt{1 + i\omega\rho/\eta q^2}$

Let us take $z = 0$ for equilibrium position of the surface, and suppose that the upper fluid occupies the half space $z > 0$ and the lower fluid, the half space $z < 0$. The boundary conditions for $z \to \pm \infty$ imply that $\vec{v}$ remains finite. Therefore, the general solution is:

$z > 0$ $\quad v_{z1} = A_1 e^{-qz} + B_1 e^{-m_1 z}$ $\qquad m_1 = q\sqrt{1 + i\omega\rho_1/\eta_1 q^2}$

$$v_{y1} = A_1' e^{-qz} + B_1' e^{-m_1 z}$$

$$v_{x1} = -i A_1 e^{-qz} - i \frac{m_1}{q} B_1 e^{-m_1 z}$$

$$P_1 = i \frac{\omega}{q} \rho_1 A_1 e^{-qz}$$

$z < 0$ $\quad v_{z2} = A_2 e^{qz} + B_2 e^{m_2 z}$ $\qquad m_2 = q\sqrt{1 + i\omega\rho_2/\eta_2 q^2}$

$$v_{y2} = A_2' e^{qz} + B_2' e^{m_2 z}$$

$$v_{x2} = i A_2 e^{qz} + i \frac{m_2}{q} B_2 e^{m_2 z}$$

$$P_2 = -i \frac{\omega}{q} \rho_2 A_2 e^{qz}$$

The limiting conditions at the surface $z = 0$, express the continuity of the velocity components and the balance between the strain tensor components $\sigma_{ij} = \eta(\partial v_i/\partial x_j + \partial v_j/\partial x_i) - p\delta_{ij}$

$$\vec{v}_1(z = 0) = \vec{v}_2(z = 0)$$

$$\sigma_{zi_1}(z = 0) - \sigma_{zi_2}(z = 0) = F_i$$

When no monolayer is present $\vec{F}$ is vertical and is equal to[33]:

$$F_z = \gamma \frac{\partial^2 \zeta}{\partial x^2} = -\frac{\gamma}{i\omega} q^2 v_z(z = 0)$$

In the presence of a surfactant layer γ will be modified and generally lowered[34]:

$$\gamma = \gamma_0 - \pi$$

where γ_0 is the surface tension without surfactant layer and π is the surface pressure in the layer. π depends generally on the surface concentration Γ of the surfactant molecules. Since $v_x(z = 0) \neq 0$, there will be a pressure gradient in the layer and consequently a horizontal surface force[33]:

$$F_x = - \text{grad } \pi = - \frac{\varepsilon}{i\omega} q^2 v_x(z = 0)$$

where $\varepsilon = \Gamma \frac{\partial \pi}{\partial \Gamma}$ is the dilational modulus.

Finally, since $v_y(z = 0) \neq 0$, the layer will also experience shear. The corresponding force is also horizontal and has the value[34]:

$$F_y = - \frac{G}{i\omega} q^2 v_y(z = 0)$$

where G is the shear modulus. It must be noted that ε is the sum of the shear modulus G and of the compressibility modulus K.

Dissipation processes may occur in the surfactant layer and give an entropic contribution to the surface force F. This can be easily taken into account by introducing frequency dependent moduli:

$$\tilde{G} = G + i\omega\mu_s$$

where μ_s is the surface shear viscosity. Typical frequencies in the experiment are 10 kHz and typical moduli are of the order of 10 dyn/cm. This means that viscoelastic behaviour will be observed for surfactant layers of surface viscosities $\mu_s \sim 10^{-3}$ s.p. This is a common order of magnitude observed in many insoluble monolayers[34].

Similarly one can write:

$$\tilde{\varepsilon} = \varepsilon + i\omega\kappa \qquad \kappa = \text{dilational viscosity}$$

$$\tilde{\gamma} = \gamma + i\omega\mu \qquad \mu = \text{transversal viscosity}$$

It must be noted that the meaning of these surface viscosity coefficients is questionable. The hydrodynamics of two dimensional

systems is still poorly understood and the two dimensional surface
viscosities are not usual viscosity coefficients: they depend on
the surface dimensions, on the flow velocity, etc. The surfactant
layers viscosities seem better defined, probably because these
systems are not really two dimensional and are strongly coupled
with the bulk fluids. They are certainly better described as
surface excess properties[35]. It can also be noted that the
surface tension γ is an equilibrium property and is certainly not
frequency dependent like the elastic moduli that are response
functions. However, γ can be apparently frequency dependent if
it reflects the presence of a hidden thermodynamic variable. In
the following we will allow for this possibility and use a complex
surface tension.

It appears now that the motion in the plane XOZ   is
completely decoupled from the motion in the direction OY . The
pure shear motion associated with $v_y(x,y,z,t)$ will not produce
any polarizability variation in the system and therefore will not
be associated with scattering of light. The spectrum $P(q,\omega)$ will
only depend on four film parameters π , ε , κ , and μ . The shear
parameters cannot be measured in the light scattering experiment.

The limiting conditions will then become

$$A_1 + B_1 = A_2 + B_2$$

$$qA_1 + m_1B_1 = - qA_2 - m_2B_2$$

$$\eta_1[2q^2A_1 + (m_1^2+q^2)B_1] - \eta_2[2q^2A_2 + (m_2^2+q^2)B_2] + \frac{\tilde{\varepsilon}}{i\omega} q^2(qA_1+mB_1) = 0$$

$$2\eta_1(qA_1+m_1B_1) + i\frac{\omega}{q}\rho_1 A_1 + 2\eta_2(qA_2+m_2B_2) + i\frac{\omega}{q}\rho_2 A_2 +$$

$$\frac{\tilde{\gamma}}{i\omega} q^2(A_1+B_1) = P_{ext}$$

By setting P_{ext} = 0 one obtains the dispersion equation of
surface phonons:

$$\Delta = q(m_1-q)(m_2-q)\left\{[\tilde{\varepsilon}q^2 + i\omega(\eta_1 m_1 + \eta_1 q + \eta_2 m_2 + \eta_2 q)] \right.$$

$$[\tilde{\gamma}q^2 + i\omega(\eta_1 m_1 + \eta_1 q + \eta_2 m_2 + \eta_2 q) - \frac{\omega^2}{q}(\rho_1+\rho_2)] - \qquad (5)$$

$$\left. [i\omega(\eta_2 m_2 - \eta_2 q - \eta_1 m_1 + \eta_1 q)]^2\right\} = 0$$

and finally from the response function one obtains the spectrum

$$P(q,\omega) = \frac{kT}{\pi\omega} \; \text{Im} \; \frac{i\omega[\eta_1(m_1+q)+\eta_2(m_2+q)]+\tilde{\epsilon}q^2}{\Delta'} \tag{6}$$

where $\Delta' = \Delta / q(m_1-q)(m_2-q)$

The case of a free surface will be obtained by setting $\mu_1 = \rho_1 = 0$. By introducing the reduced parameters

$$\Delta = i\omega\tau_0 \qquad \tau_0 = \frac{\rho_2}{2\eta_2 q^2} \qquad y = \frac{\gamma\rho_2}{4\eta_2^2 q} \qquad \alpha = \frac{\epsilon}{\gamma} \qquad \beta = \frac{\kappa q}{2\eta} \qquad \sigma = \frac{\mu q}{2\eta}$$

the spectrum of reference[36] is reobtained.

The case of the interface between two liquids of the same viscosities and densities is interesting: $\eta_1 = \eta_2 = \eta$, $\rho_1 = \rho_2 = \rho$ $m_1 = m_2 = m$. Indeed

$$P(q,\omega) = \frac{kT}{\pi\omega} \; \text{Im} \; \frac{1}{2i\omega\eta(m+q)-2\omega^2\rho/q+\tilde{\gamma}q^2}$$

The dilational properties no longer affect the spectrum. This is due to the fact that the limit condition for v_x becomes $v_x(z=0) = 0$:   the motion at the surface is purely vertical and therefore not associated with surfactant layer compression or dilatation. The dilational properties will strongly affect the spectrum if the two fluids are very different. This is illustrated in Figure 1 where a comparison between the free surface of water and an oil-water interface is given. In practice the light scattering experiments at liquid-liquid interfaces will not give information on the dilational properties of adsorbed layers.

It must be noted finally that the peak frequency ω_{max} and width $\Delta\omega$ of the spectrum are not simply related to the roots of the dispersion equation $\Delta = 0$, as in Equations (3) and (4) for a simple liquid in the extreme cases of small damping or high damping. Let us illustrate the differences by taking the case of pure water for $y = 100$ $(q = 180 \text{ cm}^{-1})$

$$\omega_q = 0.9842 \sqrt{\frac{\gamma q^3}{\rho}} \qquad\qquad \omega_{max} = 0.9819 \sqrt{\frac{\gamma q^3}{\rho}}$$

$$\Delta\omega_q = 0.842 \; \frac{\eta q^2}{\pi\rho} \qquad\qquad \Delta\omega = 0.80 \; \frac{\eta q^2}{\pi\rho}$$

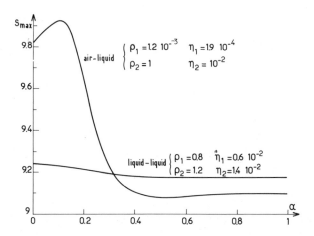

Figure 1. Dependence of the spectrum peak frequency on dilational modulus for the free surface of water (upper curve) and for an oil-water interface (lower curve). CGS units are used.
$$y = \gamma(\rho_1 + \rho_2)/4(\eta_1 + \eta_2)^2 q = 100 \; ; \; s_{max} = \omega_{max}(\rho_1 + \rho_2)/2(\eta_1 + \eta_2)q^2;$$

The difference is of the order of 2‰ for the frequencies and 5% for the width. It can be larger for water covered by monolayers (up to 1-2% and 10-20% respectively). The interpretation of the experimental data must not therefore be limited to the study of the dispersion equation (Equation (5)) and should involve the spectrum $P(q,\omega)$ (Equation (6)).

IV. EXPERIMENTAL TECHNIQUE

A typical experimental set up is shown on Figure 2. A horizontal laser beam is focused onto a diaphragm D_1 (spatial filter). The angle of incidence θ is adjusted by a mirror M_1 for optimal conditions (normal or limit incidence). After the second mirror M_2, the reflected beam is horizontal. The diaphragm D_2 selects the scattered light in a particular direction $\Delta\theta$, D_3 and D_4 reduce the extra light from other sources and the lens L_2 forms the image of the scattering region on the phothocathode of the photomultiplier. The photocurrent is then frequency analysed with a wave analyser, squared and fitted to the theoretical spectrum by a minicomputer. Several remarks can be made

Figure 2. Experimental set up for a surface light scattering experiment.

(a)   Choice of the reflection angle (see § II)

The optimum reflection angle corresponds in all cases to the total limit reflection. It is always interesting to adopt this geometry for liquid-liquid interfaces. For the free surface of liquids the normal reflection in the air is often more practical. The loss of intensity with respect to optimal geometry is in fact compensated by a signal to noise ratio increase because the beam is not subjected to intensity fluctuations caused by small turbulences in the liquid phase.

(b)   Mechanical vibrations

This is the most serious problem of the experiment. Liquid surfaces, especially free liquid surfaces, are very sensitive to mechanical vibrations. Moreover, the scattering angles are very small, thus requiring an excellent angular definition. An anti-vibration table is necessary but may not be sufficient. A very suitable solution was proposed by Hard, Hamnerius and Nilsson[37], and has been used by most authors[20,21,23,24]. They introduced a diffraction grating which gives a well defined "local oscillator" beam. The scattering angle is then determined by the grating spacing and no longer by the distance between the diaphragm D_2 and the reflected beam that could vary largely when the reflected beam oscillates due to vibrations.

The grating must be close to the liquid surface, since the beams scattered by the surface and the grating must be in phase. For thick liquid cells, the best solution is to make the image of the grating onto the liquid surface[38].

When the intensity scattered by the surface is large, for instance when γ is low, the detection can be intermediate between heterodyne and homodyne (optical beating of the scattered light with itself). This can be avoided easily by increasing the local oscillator strength. It must be recalled that the theoretical spectrum $P(q,\omega)$ is only given by Equation (6) when the detection is purely heterodyne. In the intermediate case, the experimental spectrum must be fitted with a much more complicated expression $P(q,\omega) + x\,P(q,2\omega)$, where x is the ratio between the intensity scattered by the surface and the grating respectively.

(c)   Spectrum analysis

The photocurrent can be either analysed with a wave analyser or with a correlator. Although several authors adopted this last solution[23,24], it seems to us that the use of a wave analyser is more convenient for the following two reasons

- the theoretical spectrum is an exact mathematical function
of the frequency and can be fitted directly to the
experimental spectrum. The theoretical correlation
function, which is the Fourier transform of $P(q,\omega)$, can
only be calculated numerically, therefore complicating
very much the fitting procedure;

- the laser sources often have noise components in the 0–100 Hz
range and around 30 kHz. These components are extracted
much more easily from the experimental spectrum (extra-
peaks) than from the correlation function (distorted
asymptote).

(d) Instrumental broadening

In practice one never detects a single well defined surface
phonon of wave vector $\vec{q}$, but a superposition of contributions of
the wave vectors $\vec{q} \pm \Delta\vec{q}$. Therefore the experimental spectrum is
the superposition:

$$P_{exp}(\vec{q},\omega) \ = \ \int f(\vec{q}+d\vec{q}) \ P(\vec{q}+d\vec{q},\omega) \ d\vec{q} \tag{7}$$

where f is the "instrumental" function. If the laser beam is
Gaussian and in the ideal case with no vibrations f is a Gaussian
function with which width is related to the laser beam waist at D_2.
In practice, either with or without grating, f differs slightly
from a Gaussian shape[37,39].

The effect of instrumental broadening is particularly important
when the surface ripplons propagate. Indeed, in the extreme case of
low damping, one has from Equation (3) an instrumental broadening
of the order of:

$$\Delta\omega_q^B \sim \frac{3}{2} \frac{\Delta q}{q} \, \omega_q$$

which can be easily of the same order as the width $\Delta\omega_q$ Equation (4).
The deconvolution of $P(\vec{q},\omega)$ from Equation (7) can be done with a
computer provided f is measured carefully (for instance with a
liquid of very low viscosity for which $\Delta\omega_q^B \gg \Delta\omega_q$, and for which
the shape of f is simply that of P_{exp}).

When damping becomes more pronounced the instrumental broadening
becomes less relevant. In the extreme case of large damping,
$P_{exp}(\vec{q},\omega)$ is rigorously equal to $P(\vec{q},\omega)$. In practice the
instrumental corrections are negligible when the reduced parameter
$y = \gamma(\rho_1+\rho_2)/4(\eta_1+\eta_2)^2 q$ is less than 1.

V.   EXPERIMENTAL RESULTS — INSOLUBLE MONOLAYERS ON WATER

We have studied different kinds of insoluble monolayers;
"liquid condensed" films: stearic acid, polymethylmethacrylate;
"liquid expanded" films: propyl stearate, polyvinyl acetate; and
films showing a phase transition: myristic acid. The surface
pressure of these systems, as measured with a Whilhelmy plate,
are shown in Figure 3 as a function of surface concentration.

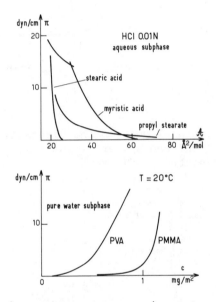

Figure 3.   Surface pressure/area isotherms.

The surface area per molecule cannot be defined for polymer
films since the polymer chains entangle at large surface pressure.
A simple concentration scale has been used.

Two typical scattering spectra are represented in Figure 4,
in order to illustrate the modifications induced by a monolayer
deposited on water.

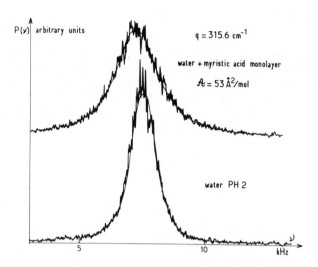

Figure 4.   Typical experimental scattering spectra.

The surface dilational modulus and viscosity values extracted
from the analysis of the spectra after instrumental broadening
correction are represented in Figure 5.

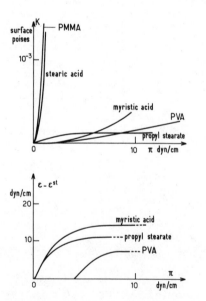

Figure 5.   Surface dilational viscosities κ and moduli ε versus
surface pressure; ε^{st} is the static value deduced from the
slope of the isotherms of Figure 3.

As expected, the "liquid expanded" films have low surface
viscosities and the "liquid condensed" films large ones (as soon
as the surface pressure is above 1 dyn/cm). Below the transition
surface pressure, the myristic acid films are also not very
viscous. When π increases, ε becomes rapidly large and the
spectrum is no longer sensitive to κ or ε variations: these
parameters can no longer be determined.

As expected also, the dilational moduli deduced from the
light scattering measurements at frequencies of the order of 10 kHz
differ from the static moduli deduced from Wilhelmy plate
measurements:

$$\varepsilon^{st} = -\mathcal{A}\frac{\partial \pi}{\partial \mathcal{A}}$$

Although much smaller than κ and $\varepsilon - \varepsilon^{st}$, non-zero μ and $\pi - \pi^{st}$
values were measured in the high surface pressure range of some of
these monolayers: $\mu \lesssim 10^{-4}$ s.p. and $\pi - \pi^{st} \lesssim 1$ dyn/cm. The
experimental accuracy is however of the same order as these numbers.
More experiments are planned with an improved experimental set up
in order to increase the accuracy of the determination of these
parameters. This will help to clarify the existing controversy about
the existence of the transversal viscosity [21-24].
It can be finally noted that the same measurements can be made
on soluble monolayers. The interpretation of the data presents some
differences: ε is not related simply to the surfactant concentration
since the surfactant molecules at the surface are in equilibrium with
surfactant molecules in the bulk ; κ has a large contribution
from exchanges of molecules between bulk and surface[40]. Several
preliminary measurements have been performed with ionic surfactants
aqueous solutions above the critical micellar concentration (cmc),
where large dilational moduli were measured, as
expected. Measurements on copolymer molecules at oil-water interfaces are also in progress.

VI. LOW SURFACE TENSIONS IN OIL-WATER SYSTEMS

When a surfactant is added to oil and water phases in
equilibrium, the surface tension can decrease to very small values.
If γ is smaller than about 10^{-2} dyn/cm, the system can emulsify
spontaneously. If the surfactant concentration is well above the
cmc, the oil (or the water) will solubilize water (or oil), and
form a microemulsion phase[41]. If the surfactant concentration is
not large enough to achieve a complete solubilization of oil and
water, the microemulsion will be in equilibrium with excess water,
excess oil, or both excess oil and water. The surface tensions
between these different phases are usually very low. Their

measurement is therefore particularly easy with the light
scattering technique. As γ is very small, the surface phonons
are overdamped and the spectrum $P(q,\omega)$ is centered at zero
frequency.

The result of the measurements of γ are shown in Figure 6
for a mixture (water + NaCl) = 47%, toluene 47%, sodium dodecyl
sulfate 2%, n-butanol 4%.

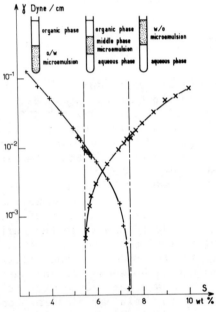

Figure 6. Surface tension measurements in a multiphase
microemulsion system; S is the salinity. The different
phase behaviour at different salinities is shown on the top;
+: oil-microemulsion interfaces; x: water-microemulsion interfaces.

We did not intend to extract from the spectrum the
dilational parameters ε' and κ, since their effect is negligible.
The transversal viscosity could be on the contrary, very easily
deduced from the spectral variations with scattering angle. We
found that if it exists it is beyond the experimental accuracy
($\mu < 10^{-5}$ s.p.) for all the samples.

We have made a detailed comparison of the surface tension
measurements as obtained from the light scattering technique and

from the spinning drop technique, which is the only other method.
suitable for tensions smaller than 10^{-2} dyn/cm. The comparison,
made on a similar system, showed that the two determinations agree
within experimental accuracy[42].

To illustrate the great accuracy of the light scattering
technique, we have plotted in Figure 7 the values of γ versus
those of $\Delta\rho = \rho_2 - \rho_1$, in a logarithmic scale.

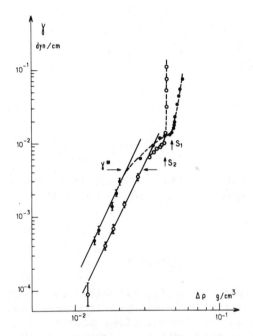

Figure 7. Surface tension versus $\Delta\rho = \rho_2 - \rho_1$ for the water-
microemulsion (●) and oil-microemulsion interfaces (o).

For $\gamma < \gamma^* = 4.5 \ 10^{-3}$ dyn/cm, crossing point of the curves
of Figure 6, γ is proportional to $\Delta\rho^4$, for the two interfaces, as is true
for systems close to a critical consolute point. For $\gamma > \gamma^*$, a
clearly different behaviour is observed in Figure 7. Although
still very low, these surface tension values are no longer due to
the vicinity of a critical point, but rather to surface pressure
effects in the surfactant monolayer. This is proven by the
measurements shown in Figure 8 in which the microemulsion phases
have been diluted and replaced by their continuous phase that
contains very low amounts of surfactant (cmc). The surface
tensions are not changed.

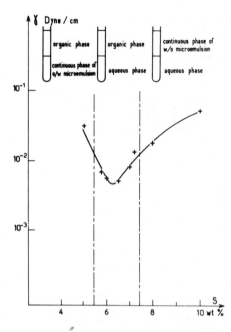

Figure 8. Surface tension measurements in low surfactant
concentration systems. The cmc is about 10^{-5} M .

CONCLUSION

The surface light scattering technique is a very useful tool
for the study of interfacial hydrodynamic properties.   It is
not well developed at the present time, because of some
specific experimental difficulties : these experiments
more sophisticated than the now widely used bulk light scattering
techniques. A further development of the research in the field is
however desirable for the better understanding of many physico-
chemical problems dealing with interfaces: colloidal stability,
coalescence processes in foam or emulsions, can be mentioned
amongst many others.

ACKNOWLEDGEMENTS

We thank J. Joosten (Van't Hoff Laboratory, Utrecht, Holland)
for pointing out to us very interesting problems specific to
surface properties of aqueous surfactant solutions. The preliminary
work on these systems was performed by him during his stay in our
laboratory.

REFERENCES

1. M. Von Schmoluchowski, Ann. Physik, 25, 225 (1908).
2. L. Mandelstam, Ann. Physik, 41, 609 (1913).
3. J. W. Rayleigh, Scientific Papers, Vol. 5, 398 (1912).
4. C.V. Raman and L.A. Ramdas, Proc. Roy. Soc., A108, 561 (1925) and A109, 150, 272 (1925).
5. A.A. Andronov and M.A. Leontovich, Z. Phys., 38, 485 (1926); R. Gans, Ann. Physik, 79, 204 (1926).
6. S. Jagannathan, Proc. Indian Acad. Sci., A1, 115 (1934); F. Barikhanskaya, J. Exptl. Theoret. Phys. USSR, 7, 51 (1937); P.S. Hariharan, Proc. Indian Acad. Sci., A16, 290 (1942).
7. R. Pecora, J. Chem. Phys., 40, 1604 (1963).
8. M. Papoular, J. de Phys., 29, 81 (1968).
9. M.A. Bouchiat, J. Meunier and J. Brossel, C.R. Acad. Sci., 266B, 255 (1968); M.A. Bouchiat and J. Meunier, C.R. Acad. Sci., 266B, 301 (1968); M.A. Bouchiat and J. Meunier, in "Polarization, Matter and Radiation", Presses Universitaires, Paris, 1969.
10. R.H. Katyl and U. Ingard, Phys. Rev. Lett., 20, 248 (1968).
11. J. Meunier, D. Cruchon and M.A. Bouchiat, C.R. Acad. Sci., 268B, 92, 422 (1969).
12. M.A. Bouchiat and J. Meunier, J. de Phys., 32, 561 (1971).
13. J. Meunier, J. de Phys., 30, 933 (1969); M.A. Bouchiat and J. Meunier, Phys. Rev. Lett., 23, 752 (1969); J.C. Herpin and J. Meunier, J. de Phys., 35, 847 (1974). N.B. : In Equation (1), the expression for (S) must be corrected : in the S^2 term, m'(m-1) and m(m'-1) should be replaced by m'(m+1) and m(m'+1).
14. J. Zollweg, G. Hawkins and G. Benedek, Phys. Rev. Lett., 27, 1182 (1971).
15. E.S. Wu and W.W. Webb, J. de Phys., 33, C1-149 (1972); E.S. Wu and W.W. Webb, Phys. Rev., A8, 2070 (1973).
16. J.S. Huang and W.W. Webb, Phys. Rev. Lett., 23, 160 (1969).
17. D. Langevin, J. de Phys., 33, 249 (1972), 36, 745 (1975), 37
18. D. Langevin and M.A. Bouchiat, Mol. Cryst. Liq. Cryst., 22, 317 (1973).
19. E. Grabowski and J.A. Cowen, Biophys. J., 18, 23 (1977).
20. A. Pouchelon, J. Meunier, D. Langevin and A.-M. Cazabat, J. de Phys., 41, L-239 (1980); A. Pouchelon, J. Meunier, D. Langevin, D. Chatenay and A.-M. Cazabat, Chem. Phys. Lett., 76, 277 (1980); A. Pouchelon, D. Chatenay, J. Meunier and D. Langevin, J. Colloid Interface Sci., 82, 418 (1981).
21. S. Hard and H. Lofgren, J. Colloid Interface Sci., 60, 529 (1977).
22. C. Griesmar and D. Langevin, in "Physicochimie des composés ampiphiles", Ed. R. Perron, Editions du CNRS, Paris, 1978;

D. Langevin and C. Griesmar, J. Phys. D, 13, 1189 (1980);
D. Langevin, J. Colloid Interface Sci., 80, 412 (1981).

23. D. Byrne and J.C. Earnshaw, J. Phys. D, 12, 1145 (1979).
24. S. Hard and R.D. Neuman, J. Colloid Interface Sci., 83, 315 (1981).
25. A. Vrij, J. Colloid Sci., 19, 1 (1964);
A. Vrij, Adv. Colloid Interface Sci., 2, 39 (1968);
H.M. Fijnaut and A. Vrij, Nature, 246, 118 (1973);
H.M. Fijnaut and J.G.H. Joosten, J. Chem. Phys., 69, 1022 (1978);
J.G.H. Joosten and H.M. Fijnaut, Chem. Phys. Lett., 60, 483 (1979);
A. Vrij, J.G.H. Joosten and H.M. Fijnaut, Adv. Chem. Phys., 48, 329 (1981).
26. C.Y. Young and N.A. Clark, J. Chem. Phys., 74, 4171 (1981).
27. L. Hammarlund, L. Ilver, I. Lundstrom and D. McQueen, J.C.S. Faraday I, 69, 1023 (1973);
I. Lundstrom and D. McQueen, J.C.S. Faraday I, 70, 2351 (1974).
28. D. Langevin and J. Meunier, in "Photon Correlation Spectroscopy and Velocimetry", pp. 15-16, Ed. H.Z. Cummins and E.R. Pike, Plenum Press, New York, 1977.
29. M. Born and E. Wolf, "Principles of Optics", § 2-4, Pergamon Press, 1959.
30. M.A. Bouchiat and D. Langevin, J. Colloid Interface Sci., 63, 193 (1978).
31. L.D. Landau and G. Placzek, Phys. Z. Sowjetunion, 5, 172 (1934).
32 L.D. Landau and E. Lifchitz, "Statistical Physics", Mir Press, Moscow
33. V.G. Levich, "Physicochemical Hydrodynamics", Prentice Hall. 1962.
34. G. Gaines, "Insoluble Monolayers at Liquid-Gas Interfaces", Interscience, New York, 1966.
35. F.C. Goodrich, Proc. Roy. Soc., A374, 341 (1981).
36. M.A. Bouchiat and D. Langevin, C.R. Acad. Sci., B272, 1422 (1971).
37. S. Hard, S. Hammerius and O. Nilsson, J. Appl. Phys., 47, 2433 (1976).
38. S. Hard and O. Nilsson, Appl. Opt., 18, 3018 (1979).
39. D. Langevin, J.C.S. Faraday Trans. I, 70, 95 (1974).
40. E.M. Lucassen Reynders and J. Lucassen, Adv. Colloid Interface Sci., 2, 347 (1969).
41. P.G. de Gennes and C. Taupin, J. Phys. Chem., 86, 2294 (1982).
42. D. Chatenay, D. Langevin, J. Meunier, A.M. Bellocq, D. Bourbon and P. Lalanne, J. Dispersion Sci. Tech., 3, 245 (1982).

SURFACE CHARGE DENSITY EVALUATION IN MODEL MEMBRANES

C. Stil, J. Caspers, J. Ferreira,
E. Goormaghtigh and J.-M. Ruysschaert

Laboratoire de Chimie Physique des Macromolécules aux
Interfaces-CP 206/2
Université Libre de Bruxelles
1050 Bruxelles, Belgique

A procedure for evaluation of surface charge density and lipid-drug association constants in lipid
monolayers is presented. Lipids are spread at the air-
water interface and the drugs injected into the aqueous
phase. Association constant evaluation is based essentially on the good correlation between the experimental
values of the surface potential and the predictions from
the Gouy-Chapman theory.

INTRODUCTION

Natural membranes bear surface charges due to the presence of ionized lipids and proteins. These fixed charges give rise to an electric field which determines a surface potential. Several membrane phenomena (fusion, permeability, cell-cell adhesiveness) are influenced by this surface potential. For this reason, many procedures have been proposed to define this parameter[1-7]. It is the purpose of this paper to present another approach. The lipid is spread at the air-water interface and a charges species injected into the aqueous phase. Surface potential measurements obtained before and after the injection give direct information about the charge density. Knowledge of this charge density permits evaluation of the association constant between the lipid and the charged species. The interaction between 9-Aminoacridine and lipids was analyzed using this procedure.

MATERIALS AND METHODS

Cardiolipin, phosphatidic acid, phosphatidylinositol, phosphatidylserine, phosphatidylglycerol and 9-Aminoacridine were purchased from Sigma Chemical Co. Phospholipids were spread at the air-water interface from a chloroform solution using an Agla Microsyringe. All chemicals were of analytical grade. Water was triple distilled in presence of permanganate. Buffered solution (Tris-HCl 10^{-4} M, 0,15M NaCl, pH 7,4) were used to prepare the subphases. Experiments were carried out at 25°C. A circular (diam. 2 cm) vibrating inox electrode was employed to measure the surface potential[8-9] Ag electrode was used as reference.

RESULTS

The surface potential associated with a lipid monolayer spread at the air-water interface is given by :

$$\Delta V = \frac{12\pi\ \mu}{A} + \Psi \tag{1}$$

ΔV is the measured surface potential, μ is the vertical contribution to the dipolar moment of the spread molecule (in m Debye), A is the area (A^2) occupied per lipid molecule and Ψ (in mV) is the electrostatic potential.

The Ψ parameter can be evaluated by the Gouy-Chapman theory of the electrical double layer[2,9-11]. At 25°C,

$$\Psi = 50,4\ sh^{-1}\ \frac{136\ \sigma}{c^{1/2}} \tag{2}$$

c is the concentration of univalent electrolyte in the subphase (mole l^{-1}) and σ is the surface charge density (charged group per $\overset{\circ}{A}{}^2$).

When a positively charged drug[12-16] is injected under a lipid monolayer spread at the air-water interface, the following interfacial reaction occurs :

$$D^+ \text{(subphase)} + L^- \text{(monolayer)} \rightleftharpoons DL \text{ (monolayer)} \qquad (3)$$

where D^+ is the drug, L^- the lipid anionic site and DL the complex.

From the association degree β :

$$\beta = \frac{(DL)}{(DL) + (L^-)} \qquad (4)$$

the association constant of reaction (3) can be evaluated :

$$K = \frac{\beta}{1-\beta} \; \frac{1}{(D^+)_s} \qquad (5)$$

where $(D^+)_s$ is the drug concentration D^+ at the interface. $(D^+)_s$ is related to the bulk concentration $(D^+)_\infty$ through a Boltzmann distribution :

$$(D^+)_s = (D^+)_\infty \; \exp(-e\Psi/kT) \qquad (6)$$

e is the electronic charge, k the Boltzmann constant and Ψ the electrostatic potential after complexation. σ is directly related to the association degree.
Indeed,

$$\sigma = \frac{1-\beta}{A} \qquad (7)$$

A is the area occupied per lipid molecule (or per charged group) in the close packed state.

The surface potential associated with a lipid monolayer spread at the air-water interface is given by equation (1).

After injection of a given amount of drug into the subphase, equation (1) becomes :

$$\Delta V' = \frac{12\pi \; \mu'}{A'} + \Psi' \qquad (8)$$

If we assume that the drug-lipid interaction only affects the monolayer charge density, without modification of A and of the dipolar contribution to ΔV, it comes :

$$\Delta V' - \Delta V = \Psi' - \Psi \qquad (9)$$

Ψ is calculated from the initial surface charge density. The knowledge of Ψ and of $(\Delta V' - \Delta V)$ allows one to obtain Ψ' (equation(9)). From Ψ' and equations 5,6,2,7, the association constant is evaluated.

This procedure was used to evaluate the interaction between 9-Aminoacridine (9AA) and lipids. 9AA has been proposed as a probe of the electrical double layer associated with negatively charged biological membranes [5]. It was postulated that when 9AA molecules are in the electrical diffuse layer adjacent to the charged surface their fluorescence was quenched. The quenching could be reversed in a predictable way by adding cations to the suspending medium. From the quenching curves obtained with mono- and divalent ions, Searle and Barber [5] evaluated the surface charge density of liposomes. It was supposed that the probe showed a purely electrostatic interaction with the surface and acts as a diffusible cation. In our measurements, 9-Aminoacridine was injected underneath the lipid monolayer spread at the air-water interface in a close packed state. A constant ΔV modification was obtained 10 minutes after 9AA injection (Figure 1).

Association constants (K/M^{-1}) were calculated from the procedure described before. Figure 2 indicates, for different association constant values, how the initial surface charge density is modified as a function of the drug concentration in the subphase $(10^{-1}$ M NaCl). For high K values $(K = 10^{5}$ M$^{-1})$, it is obvious that the surface charge density is strikingly modified even at low drug concentrations $(10^{-5}$M). Comparison of Figure 2a and 2b shows that the σ decrease will also depend on the initial surface charge density. Indeed, for an initial surface concentration $\sigma_0 = 1,67\ 10^{-3}$ A^{-2} and a drug concentration of 10^{-5}M, the modification is not important for all lipids except cardiolipin (Figure 1). Our results demonstrate that Barber's assumption is valid for chloroplasts thylakoid surface [5]. However, an exact knowledge of the lipid composition ant the absence of high local concentrations of negatively charged lipids would be a prerequisite for an extension of this technique to other charged biological membranes.

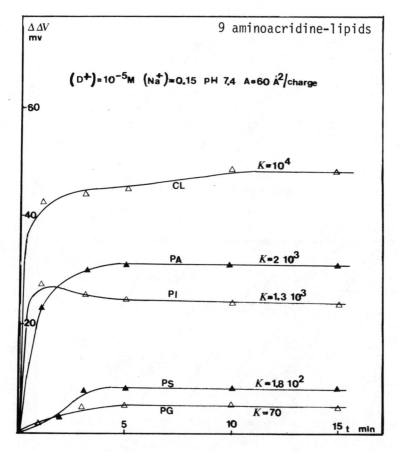

Figure 1. Surface potential variation observed after injection of 9AA (10^{-5}M) in the aqueous subphase. CL: cardiolipin; PA: phosphatidic acid; PI: phosphatidylinositol; PS: phosphatidyl-serine; PG: phosphatidylglycerol. K value are given in M^{-1}.

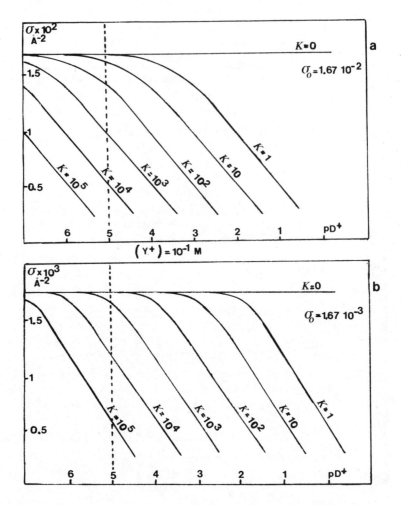

Figure 2.   Evolution of σ as a function of K and of the drug
concentration in the subphase.

a) $\sigma_0 = 1,67\ 10^{-2}\ A^{-2}$

b) $\sigma_0 = 1,67\ 10^{-3}\ A^{-2}$

$PD^+ = - \log(D^+)$, $(Y^+) = (Na^+)$

CONCLUSION

We have described a method for evaluation of surface charge density and of anionic lipid-drug association constants. The method[12-16] has already been applied to several interfacial reactions and could be extended to any kind of lipid-drug interactions. Other model membranes[17,18] have been used to determine surface association constants. However, their use remains limited by the instability of the system[17] (planar lipid bilayers) or by the necessity to use lipids showing a transition temperature in an usual temperature range[18].

ACKNOWLEDGEMENTS

One of us (E.G.) thanks the "Fonds National de la Recherche Scientifique" for financial assistance.

REFERENCES

1. R.C. Mc Donald and A.D. Bangham, J. Membrane Biol., 7, 29 (1972).
2. S. Mc Laughlin, Current Topics Membrane and Transport, 9, 71 (1977).
3. S. Ohki, Physiol. Chem. Phys., 13, 195 (1981).
4. M.S. Fernandez, Biochim. Biophys. Acta, 646, 23 (1981).
5. G.F.W. Searle and J. Barber, Biochim. Biophys. Acta, 502, 309 (1978).
6. S. Ohki and R. Kurland, Biochim. Biophys. Acta, 645, 170 (1981).
7. S. Mc Laughlin, G. Szabo and G. Eisenman, J. Gen. Physiol., 58, 667 (1971).
8. C.D. Kinloch and A.I. Mc Mullen, J. Sci. Instruments, 36, 347 (1959).
9. G.L. Gaines, "Insoluble Monolayers at Liquid-Gas Interfaces", p.188, Interscience, New York, 1966.
10. J.T. Davies and E.K. Rideal, "Interfacial Phenomena", p.56, Academic Press, New York, 1963.
11. A.G. Lee, Biochim. Biophys. Acta, 472, 237 (1977).
12. E. Goormaghtigh, P. Chatelain, J. Caspers and J.M. Ruysschaert, Biochim. Biophys. Acta, 597, 1 (1980).
13. E. Goormaghtigh, P. Chatelain, J. Caspers and J.M. Ruysschaert, Biochim. Pharmacology, 29, 3003 (1980).
14. E. Goormaghtigh, J. Caspers and J.M. Ruysschaert, J. Colloid Interface Sci., 80, 163 (1981).
15. T. Guilmin, E. Goormaghtigh, R. Brasseur, J. Caspers and J.M. Ruysschaert, Biochim. Biophys. Acta, 685, 169 (1982).
16. E. Goormaghtigh, R. Brasseur, M. Vandenbranden, J. Caspers and J.M. Ruysschaert, Bioelectrochem. and Bioenergetics (in press)
17. S. Mac Laughlin and M. Harary, Biochem., 15, 1941 (1976)
18. A.G. Lee, Biochim. Biophys. Acta, 514, 95 (1978)

BREAKDOWN OF THE POISSON-BOLTZMANN APPROXIMATION IN POLYELECTROLYTE SYSTEMS: A MONTE CARLO SIMULATION STUDY

B. Jönsson, P. Linse and T. Åkesson

Physical Chemistry 2
Chemical Centre, POB 740
S-220 07 Lund, Sweden

Håkan Wennerström

Physical Chemistry
University of Stockholm
S-106 91 Stockholm, Sweden

The accuracy and consistency of the Poisson-Boltz-mann approximation in polyelectrolyte systems is investigated using a Monte Carlo simulation technique. It is found that the PB equation gives a good description of the counterion concentration close to the charged aggregate both in the presence of mono- and divalent ions. However, other thermodynamic quantities, such as the osmotic pressure, is less accurate in the PB approximation. In the case of high surface charge density it is in error by ~50% in the presence of monovalent ions, while with divalent ions the PB theory overestimates the osmotic pressure by an order of magnitude. In the PB theory the osmotic pressure attains a constant value, when the surface charge density is increased, while in the MC simulations it becomes zero. These results imply some limitations of the present theory of colloidal stability which is briefly discussed.

INTRODUCTION

In solutions of charged polymers[1], micellar solutions[2], aqueous solutions containing charged membranes[3], lyotropic liquid crystals[4], and in aqueous dispersions of charged latex particles[5], the electrostatic interactions often have a dominant influence on the thermodynamic properties of the system. Due to the long-range character of the interactions, one can, as a first approximation, treat the solvent as a dielectric medium and consider explicitly only the charged species in a theoretical description. For the short range ion-ion interactions, the simplest and most consistent approach is to consider these to be point charges. However, in many theoretical treatments the divergence of the Coulomb potential makes it necessary to assign a radius to the charges and treat them as hard charged spheres – the primitive model of electrolyte solutions. The radius of the hard spheres then becomes a slightly arbitrary parameter in the model and can make a direct comparison between different theoretical approaches difficult. Polyelectrolyte systems, in the broad sense of the word, are characterized as containing highly charged macroions or aggregates as well as mono- and divalent counterions and coions. In contrast to the ordinary electrolyte solutions, there is an asymmetry in the system, in that the pair interactions between the different types of particles are of different orders of magnitude. This circumstance can be used to an advantage, since it gives a motivation for introducing yet another approximation: the cell model[6]. In this model, the total solution is divided into cells, each containing one macroion, counterions, and salt. The cell volume is directly related to the macroion concentration. In this way, one is able to account for effects determined by the small ion-small ion and macroion-small ion correlations, while one has refrained from describing phenomena due to macroion-macroion correlations. Another approximation often used in the electrostatic continuum models for polyelectrolytes is the assumption of a uniformly constant dielectric permittivity, usually the water value, in all regions. This is clearly an approximation and there are some recent results for lecithin-water systems showing that the different dielectric permittivity in the lecithin and water phases gives rise to an additional repulsive force between the neutral lecithin lamellas[7].

In spite of the many simplifications introduced into the model, it is still a very difficult task to evaluate the properties of the model system within an exact statistical mechanical framework. Traditionally, the solution is obtained through analytical approximations, most commonly by solving the Poisson-Boltzmann (PB) equation[8], but also refinements such as the modified PB equation[9] and solutions of the hypernetted chain equation[10-12] have been tried. The PB equation is usually derived as a mean field theory. However, it is also possible to derive the PB equation starting from the expression for the first member of the Yvon hierarchy[13]. The <u>only</u> necessary approximation to make is the neglect of correlation between the mobile ions[14], that is

$$\rho^{(2)}(\overline{r},\overline{r}') = \rho^{(1)}(\overline{r})\rho^{(1)}(\overline{r}') \tag{1}$$

where $\rho^{(n)}$ is the n-particle distribution function. This second derivation may give more insight into the limitations of the PB equation and also how to improve it.

Considering the wide-spread use of the PB equation in different areas of chemistry[8],[17-18] it is an important task to asses its accuracy.   This can be done by performing Monte Carlo (MC) simulations on the systems of interest and this has recently been done covering some aspects of the PB approximation[14-16]. By comparing MC results with those obtained within the PB approximation, one might also obtain a further insight into the approximations made in the PB theory.

THE MONTE CARLO SIMULATION TECHNIQUE

The MC procedure is a general method to solve integrals of high dimensionality, such as the statistical mechanical average of a physical quantity, say the internal energy, that reads

$$\langle U \rangle = \int U(X)\exp(-U(X)/kT)dX/Z \tag{2}$$

where Z is the configurational integral,

$$Z = \int \exp(-U(X)/kT)dX \tag{3}$$

In (2) and (3) X is a shorthand notation for the coordinates of all particles in the system and $U(X)$ is their total interaction energy.

The standard Monte Carlo method would be to pick out one configuration of the counterions at random, calculate the properties of interest for that configuration, and multiply by the appropriate Boltzmann factor. Repeating this procedure many times, adding the terms and normalizing, one will hopefully reach convergence. However, since the integrands in (2) and (3) are rapidly varying functions, very few of the randomly chosen configurations will have any appreciable weight, resulting in a very poor convergence in such a procedure. To circumvent this problem one is forced to use an 'importance sampling' of configurations. This method was introduced in the early fifties by Metropolis et al.[19] based on the generation of a so-called Markov chain. The advantage of importance sampling is that a configuration X_i is chosen according to its statistical weight. This enables (2) to be written as

$$\langle U \rangle = \frac{1}{M} \sum_{i=1}^{M} U(X_i) \tag{4}$$

where M is the total number of configurations. In practice the system is first 'equilibrated' by generating a large number of configurations, so that the total energy or some other relevant property is seen to fluctuate around some stable mean value. Then new configurations are generated and the properties of interest are calculated according to (4). Figure 1 shows a typical behaviour of the total energy for a system as a function of the number of generated configurations. The vertical bars in the figure indicate the configurations on which the calculation of the statistical averages are based (the analysis step).

FREE ENERGY DERIVATIVES IN THE CELL MODEL[20]

A cell of volume V_c contains one fixed macroion of total charge Q and a number N_i of mobile small ions i with charge $z_i e$. A requirement is that the total cell is electroneutral, so that

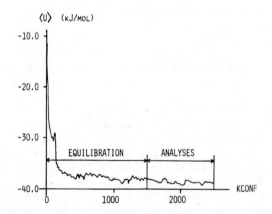

Figure 1. The total internal energy, $\langle U \rangle$, versus the number of configurations (in thousands), KCONF, generated in a MC simulation. The calculation is divided in two parts and the averages are calculated during the second part (analyses).

$$Q + e\Sigma_i z_i N_i = 0 \tag{5}$$

This system would correspond to a solution with the macroion concentration $1/V_c$ and small ion concentration $c_i = N_i/V_c$.

The use of the cell model to take the effects of macroion concentration into account may seem rather crude. There are, however, several important advantages. As the volume of the cell increases, the important infinite dilution limit is obtained. Furthermore, studies using the PB approach indicate that the most important macroion concentration effect is related to the entropy in the counterion distribution.

In the applications considered in this work, three cell geometries have been used: two infinite planes, one infinite plane, and spheres. Furthermore, the macroion geometries are chosen to be identical to those of the cell (see Figure 2). The charge of the macroion is treated as uniformly smeared out so that it has a surface charge density σ.

Within the primitive model, the interaction potential $u_{ij}(r_{ij})$ between two mobile ions is

$$u_{ij}(r_{ij}) = \begin{cases} e^2 z_i z_j / 4\pi \ \varepsilon_r \varepsilon_0 r_{ij}, & r_{ij} \geq R_i + R_j, \\ \infty, & r_{ij} < R_i + R_j, \end{cases} \tag{6}$$

where R_i and R_j are the hard core radius of the ions and $\varepsilon_r \varepsilon_0$ the

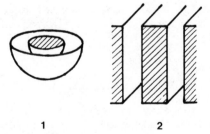

1   2

Figure 2. Schematic picture of two polyelectrolyte geometries. (1) Sphere and (2) parallel surfaces.

dielectric permittivity of the solvent. The macroion-small ion interaction potential $u_{pi}(r_i)$ can be written

$$u_{pi}(\bar{r}_i) = \begin{cases} z_i e\Phi_p(\bar{r}_i), & \bar{r}_i \text{ outside the macroion,} \\ \infty, & \text{otherwise,} \end{cases} \qquad (7)$$

where ϕ_p is the electrostatic potential generated by the macroion.

The configurational partition function

$$Z = \int \dots \int \exp(-U/kT) d\bar{r}_1 \dots d\bar{r}_N \qquad (8)$$

for the N mobile ions can, due to the hard wall interactions, be written

$$Z = \int_{\substack{r_\alpha \\ N}}^{r_\beta} \dots \int_{2N} \int \dots \int \exp[-(\sum_i u_{pi} + \sum_{i>j} u_{ij})/kT] d\bar{r}_1 \dots d\bar{r}_N \qquad (9)$$

where we have used the fact that the cells and the macroions have a simple geometry. Here, r_β specifies the boundary of the cell and r_α is the hard core boundary of the macroion. If the different types of small ions have different hard core radii, the value of r_β and r_α will depend on the particular species. This is a trivial complication and thus will be neglected.

For a classical system, the configurational free energy is

$$A = -kT \ln Z. \qquad (10)$$

In determining the free energy change with variations in a parameter Y, the problematic step is to calculate the derivative $(\partial \ln Z/\partial Y)$. For $Y=r_\beta$, this derivative is related to the osmotic pressure, while if $Y=r_\alpha$, the derivative can give a measure of the free energy change related to changes in the aggregate size and, when $Y=Q$, the process relates to a charging of the macroion, to give some examples.

One Infinite Planar Charged Wall

With the planar geometry and the cell model the ions in the solution are confined to the region $x_\alpha \leq x \leq x_\beta$ by hard walls at x_α and x_β respectively (see Figure 3a). The electrostatic ion-wall interaction is then

$$\sum_i u_{pi} = \sum_i \frac{z_i e\sigma}{2\varepsilon_r \varepsilon_o} (x_i - x_\alpha) \tag{11}$$

As seen from Equation (11), the external field is independent of the location of the charged wall, which implies that also the ion distribution is independent of the exact location of the charge. Consequently, Equation (11) can be split into two parts

$$U_1 = \sum_i \frac{z_i e\sigma}{2\varepsilon_r \varepsilon_o} x_i \tag{12}$$

$$U_2 = - \frac{\sigma^2}{2\varepsilon_r \varepsilon_o} x_\alpha \cdot \text{Area} \tag{13}$$

where only U_1 depends on the mobile ion coordinates. If we denote the mobile ion-ion interaction with U_3, the configurational partition function can be written as

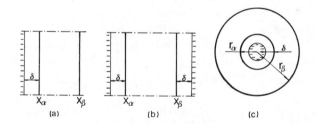

Figure 3. (a) One single charged wall, (b) two parallel charged walls and (c) charged sphere. $x_\alpha(r_\alpha)$ and $x_\beta(r_\beta)$ denote the hard walls. The aggregates are assumed to be negatively charged.

$$Z = \int \ldots \int\limits_{x_\alpha}^{x_\beta} \int\int \ldots \int\limits_{Area}$$

$$\exp[-(U_1+U_3)/kT]d\overline{r}_1 \ldots d\overline{r}_N \exp(-U_2/kT). \tag{14}$$

The derivative of the free energy with respect to the volume $(\partial A/\partial V)$, which is directly related to the osmotic pressure, can be evaluated by changing the positions of either of the two hard walls and

$$P_{osm} = -\frac{\partial A}{\partial V} = -\frac{1}{Area}\frac{\partial A}{\partial x_\beta} = \frac{1}{Area}\frac{\partial A}{\partial x_\alpha}. \tag{15}$$

It is straightforward to determine $(\partial Z/\partial x_\beta)$ from Equation (14) since Z only depends on x_β through the integration limits

$$\frac{\partial Z}{\partial x_\beta} = \rho^{(1)}(x_\beta) \cdot Z \cdot Area, \tag{16}$$

where $\rho^{(1)}$ is the (one) particle density. Equations (15) and (16) show that the osmotic pressure is determined by the ion concentration at the cell boundary

$$P_{osm} = kT\rho^{(1)}(x_\beta). \tag{17}$$

This formula is identical to the one derived by Marcus[8] using the PB approximation.

The effects of changing the constraint at the charged wall is somewhat more complex giving

$$\frac{\partial Z}{\partial x_\alpha} = -\rho^{(1)}(x_\alpha)Z \cdot Area - \frac{\partial U_2}{\partial x_\alpha} \cdot \frac{Z}{kT}. \tag{18}$$

From Equations (15) and (18), the osmotic pressure is

$$P_{osm} = kT\rho^{(1)}(x_\alpha) - \sigma^2/(2\varepsilon_r\varepsilon_o). \tag{19}$$

This relation was previously derived by Henderson et al.[21] using a different approach.

The two independent equations [Equations (17) and (19)] for the osmotic pressure can be combined to give an exact expression for the difference between the densities at the two walls, and if we simplify the notation by using $\rho = o^{(1)}$, we obtain

$$\rho(x_\alpha) - \rho(x_\beta) = \sigma^2/(2\varepsilon_r\varepsilon_o kT). \qquad (20)$$

The derivation above was given for a single ionic species, but the generalization to several kinds of ions is trivial and the formulas remain valid if $\rho(x)$ is replaced by a sum $\Sigma_\gamma\rho_\gamma(x)$ over the different ionic species γ, so that

$$P_{osm} = kT \sum_\gamma \rho_\gamma(x_\beta) = kT \sum_\gamma \rho_\gamma(x_\alpha) - \sigma^2/(2\varepsilon_r\varepsilon_o) \qquad (21)$$

$$\sum_\gamma \rho_\gamma(x_\alpha) - \sum_\gamma \rho_\gamma(x_\beta) = \sigma^2/(2\varepsilon_r\varepsilon_o kT). \qquad (22)$$

In the limit when x_β goes to infinity, Equation (22) is equivalent to the so-called Grahame equation[22] derived within the PB approximation.

Two Infinite Parallel Charged Walls

For two parallel charged plates (see Figure 3b), identically the same procedure can be adopted for calculating the osmotic pressure, as in the previous section. One obtains, from Equation (18),

$$P_{osm} = kT \sum_\gamma \rho_\gamma(x_\alpha) - \sigma^2/(2\varepsilon_r\varepsilon_o) = kT \sum_\gamma \rho_\gamma(x_\beta) - \sigma^2/(2\varepsilon_r\varepsilon_o) \qquad (23)$$

where σ_α and σ_β are the charged densities at the two walls, respectively. In this case, P_{osm} is related to the electrostatic force between the walls, a quantity of the greatest importance in theories of colloidal stability such as the DLVO theory[23]. If the surfaces are immersed in a bulk electrolyte solution with (constant) osmotic pressure p_{osm}^{bulk}, then the force F is

$$F = Area(p_{osm} - p_{osm}^{bulk}). \qquad (24)$$

Charged Sphere in a Spherical Cell

For a charged sphere in a spherical cell (cf. Figures 2 and 3), a commonly used model for micellar solutions, basically the same scheme can be applied as for the planar case. The derivative of Z with respect to the cell radius r_β is simply related to the osmotic pressure and Equation (17) is changed to

$$P_{osm} = kT \sum_{\gamma} \rho_\gamma(r_\beta) \tag{25}$$

However the derivative $(\partial A/\partial r_\alpha)$ does not depend only on the osmotic pressure. In changing the radius of the sphere, the charge density is altered. Thus,

$$\left(\frac{\partial A}{\partial r_\alpha}\right)_N = \left(\frac{\partial A}{\partial r_\alpha}\right)_{N,V} + \left(\frac{\partial A}{\partial V}\right)_{N,r_\alpha} \left(\frac{\partial V}{\partial r_\alpha}\right)_N \tag{26}$$

The unscreened field outside a sphere is independent of the location of the surface of charge, as in the planar case, and the interaction energy can again be divided into three parts as above, where U_2 now is due to the self-energy of the polyion. Repeating the procedure in the derivation of Equation (19), the free energy derivative per unit area is

$$\left(\frac{\partial A}{\partial r_\alpha}\right)_N = kT \sum_{\gamma} \rho_\gamma(r_\alpha) - \frac{\sigma^2}{2\epsilon_r \epsilon_o} \tag{27}$$

and subtracting the osmotic term gives

$$\left(\frac{\partial A}{\partial r_\alpha}\right)_{N,V} = kT \sum_{\gamma} [\rho_\gamma(r_\alpha) - \rho_\gamma(r_\beta)] - \frac{\sigma^2}{2\epsilon_r \epsilon_o} \tag{28}$$

Comparing Equations (21) and (28) one finds that the contact density $\sum_{\gamma} \rho_\gamma(r_\alpha)$ is smaller for a sphere than for a plane with the same charge density and osmotic pressure. The higher the salt content, the smaller the magnitude of $(\partial A/\partial r_\alpha)_{N,V}$ and the two contact values approach one another.

THE ACCURACY OF THE POISSON-BOLTZMANN APPROXIMATION

Contact Concentrations

One important aspect of approximate theories, such as the PB equation, is the ability to preserve the validity of exact relations to as large an extent as possible. It is straightforward to show that Equations (17) and (19) are indeed satisfied by the solutions of the PB equation for a charged wall[20]. However, they are not in general satisfied by the modified (improved) versions of the PB equation. It seems to us that the Equation (19) partly explains the success of the PB equation in predicting the concentration profile close to the charged surface. Figure 4 shows a comparison between the PB and MC concentration profiles between two charged planar

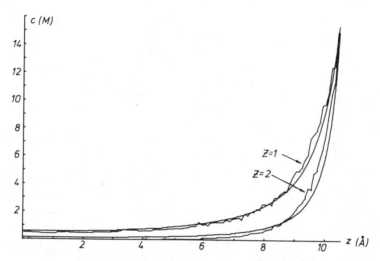

Figure 4. Concentration profiles between two charged walls for mono- and divalent ions. The wall-wall separation is 2b=21Å. The smooth curves are from the PB equation and the   other   are obtained from MC simulations.

walls, without additional salt, a situation common in lamellar liquid crystals. As can be seen, the profiles seem virtually identical. One can note that, for high charge densities and/or low osmotic pressure, $\sigma^2/2\varepsilon_r\varepsilon_o \gg p_{osm}$, and the contact concentration $\sum_\gamma \rho_\gamma(x_\alpha)$ approaches the exact value. Of course the individual concentrations need not be as good. However, preliminary MC simulations with a mixture of mono- and divalent counterions indicate that the individual concentrations are fairly well predicted by the PB theory. The co-ion concentration at the charged surface is too low to be determined in the simulations.

A remarkable feature of Equation (19) is that at high σ, the counterion concentration at the wall is independent of the valency of the ions. This is also reproduced in Figure 4, where it is indeed found that the density at the wall remains unchanged when the counterion charge goes from e to 2e. Similar behaviour is found in micellar solutions[24].

Osmotic Pressure

In the case of a micellar solution, modelled as in Figures 2 and 3, it is straightforward to determine the osmotic pressure via

Equation (25). In Table I, results for two different charges of the
micelle are presented, and a comparison is made with results of the
Poisson-Boltzmann approximation. For monovalent counterions, the PB
equation overestimates the osmotic pressure by 10-40 %. Considering
that this relative error does not vary much over a range of variation of a factor of 400 in the osmotic pressure, we conclude that
the PB equation gives a very useful estimate of p_{osm} in the case of
monovalent ions. As expected from the nature of the PB approximation,
the discrepancy between the MC and PB results are larger for divalent counterions. The PB approximation overestimates p_{osm} by a
factor of 1.5 to 4. The error is strongly dependent on the micelle
concentration. The presence of neutral salt does not seem to alter
these conclusions, although we have only a limited amount of Monte
Carlo data available.

For the important case of two planar charged surfaces, the
osmotic pressure can in principle be evaluated from the contact
density at the wall. However, there are numerical difficulties in-
volved in such an approach, since the constant term $\sigma^2/(2\varepsilon_r\varepsilon_o)$ in
Equation (19) is often substantial, giving the osmotic pressure as
a difference between two large numbers. It is found that the contact density $\rho(x_\alpha)$ can, in general, not be determined with accep-

Table I. The osmotic pressure p_{osm} in micellar solutions calculated
from MC simulations and from the PB equation, for two different
micelle sizes. Only counterions monovalent or divalent, with radii
1A are present in the solution. The listed concentration is the
micellar concentration multiplied by $-Q/e$, i.e., the apparent monomer concentration.

	p_{osm}/RT (mM)					
	Monovalent counterions			Divalent counterions		
c (mM)	MC	PB	PB/MC	MC	PB	PB/MC
$Q=-58e$			$r_\alpha=18$Å			
852	370	450	1.2	32	130	4.1
138	32	39	1.2	4.0	11	2.8
54.5	12	15	1.3			
2.48	0.6	0.8	1.3			
$Q=-12e$			$r_\alpha=10$A			
694	400	550	1.4	68	192	2.8
74.3	38	43	1.1	6.5	13	2.0
38.0	18	22	1.2	3.9	6.8	1.7
1.41	1.1	1.0	1	2.0	3.0	1.5

table accuracy. To circumvent the problem, we have made simulations on a system with only one charged wall, keeping the other uncharged. The uncharged wall can be viewed as inserted in the midplane between the two charged plates in the original system. In this way, the contact density can be evaluated at the uncharged wall with reasonable accuracy and one can note that, within the PB approximation, the two cases are equivalent. Simulations were performed for a single ionic species in the solution. The calculated osmotic pressure at different distances between the hard walls is presented in Table II, together with the corresponding value obtained using the PB approximation. The discrepancies between the MC and PB results are larger than for the micellar system of Table I. This is probably due to the higher ion concentration close to the charged planar wall, which makes ion correlation effects more important. For divalent counterions, there is a factor of ~7 difference in p_{osm} and it is clear that the PB theory fails in this case.

This failure can be illustrated more clearly by displaying the osmotic pressure for a fixed wall-wall separation and a varying surface charge density as in Figure 5. The PB equation[25] shows a monotonic increase of p_{osm} towards an asymptotic value with increasing charge density, while the MC results have a maximum and then go to zero. There are a number of interesting aspects on the comparison in Figure 5: First of all it shows that the PB equation gives the correct answer at low σ. Secondly the erroneous limiting value of p_{osm} in the PB theory is apparently due to the neglect of ion-ion correlation. Thirdly, we can note that the maximum in p_{osm} (actually the repulsive force between the walls) appears at a surface charge density of ~1e/200 Å , a commonly encountered charge density in colloidal systems. Figure 5 has been drawn for divalent counterions, but a similar curve appears for monovalent ions.

Table II. The osmotic pressure p_{osm} in the solution outside a single charged wall. b is the distance between the charged and uncharged walls. See also Fig. 2(b). The statistical uncertainty in the MC result is of the order of a few mM, $\sigma=0.224$ C m^{-2}.

| b | $p_{osm}/RT(M)$ | | | | | |
| | Monovalent counterions | | | Divalent counterions | | |
(Å)	MC	PB	PB/MC	MC	PB[26]	PB/MC
6	1.07	1.60	1.5	0.082	0.493	6
10.5	0.467	0.624	1.3	0.028	0.178	6
15.5	0.207	0.311	1.5	0.014	0.085	6
25.0	0.076	0.128	1.7	0.0045	0.034	8
50.0	0.02	0.034	1.7	0.003	0.009	(3)

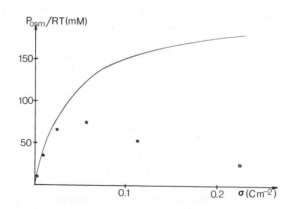

Figure 5. The osmotic pressure as a function of surface charge density. The smooth curve comes from the PB equation and the circles are MC results. The results are for a single charged wall; the distance to the neutral wall is 10.5 Å. Divalent counterions with zero radius.

CONCLUSIONS

The PB approximation seems to work well for monovalent ions even at high surface charge density. This is true for the region close to the charged aggregate, while the osmotic pressure, which is determined by the concentration at the outer boundary, is less accurate. For divalent counterions the situation is somewhat different. The region close to the aggregate is still well described while the osmotic pressure is an order of magnitude wrong even at moderate surface charge density. In fact the PB theory predicts a finite limiting osmotic pressure at high charge density, while in the simulations it goes to zero.

The main object of this work has been to investigate the limitations of the PB approximation. In order to do so we have used a uniformly constant dielectric permittivity and a uniformly smeared out surface charge density on the macroion, although this is by no means necessary in the MC simulations. In fact we have recent indications that the latter approximation is a serious one for low surface charge densities. Using distinct charges on the macroion in the MC simulation, seems to lower the osmotic pressure compared to a uniform charge density.

In this work we have concentrated on the osmotic pressure, due to its importance in the DLVO theory, but further investigations, including other properties, of the PB approximation are clearly desirable in order to find out its limitations.

ACKNOWLEDGEMENTS

As usual we are pleased to acknowledge very interesting discussions with Drs. B. Halle and B. Jönsson.

REFERENCES

1. F. Oosawa, "Polyelectrolytes", Marcel Dekker, New York, 1971.
2. H. Wennerström and B. Lindman, Phys.Rep., $\underline{52}$, 1 (1979).
3. S.A. McLaughlin, in "Current Topics in Membrane Transport", Bronner and Kleinzeller, Editors, vol. 9, p. 71, Academic Press, New York, 1977.
4. V.A. Parsegian, Trans. Faraday Soc., $\underline{62}$, 848, (1966).
5. A. Kose, M. Ozaki, K. Takano, V. Kobayashi and S. Hachisu, J. Colloid Interface Sci. $\underline{44}$, 330 (1973).
6. T.L. Hill, "Statistical Mechanics", Addison-Wesley Reading, Mass. 1960.
7. B. Jönsson and H. Wennerström, J.C.S. Faraday Trans. 2, in press.
8. R.A. Marcus, J.Chem.Phys., $\underline{23}$, 1057, (1955).
9. C.W. Outhwaite, L.B. Bhuiyan and S. Levine, J.Chem.Soc. Faraday Trans., $\underline{276}$, 1388 (1980).
10. C.W. Outhwaite and L.B. Bhuiyan, J.Chem.Soc. Faraday Trans. 2, $\underline{78}$, 775 (1982).
11. D. Henderson, L.Blum and W.R. Smith, Chem.Phys.Lett., $\underline{63}$, 381 (1979).
12. D. Elkoubi, P. Turq and J.-P. Hansen, Chem.Phys.Lett., $\underline{52}$, 493 (1977).
13. R. Balescu, "Equilibrium and Non-equilibrium Statistical Mechanics", Wiley, New York, 1975.
14. B. Jönsson, H. Wennerström and B. Halle, J.Phys.Chem., $\underline{84}$, 2179 (1980).
15. G.M. Torrie and J.P. Valleau, J.Chem.Phys., $\underline{73}$, 5897 (1980).
16. W. Megen and I. Snook, J.Chem.Phys., $\underline{73}$, 4656 (1980).
17. G. Gunnarsson, B. Jönsson and H. Wennerström, J.Phys.Chem., $\underline{84}$, 3114 (1980).
18. B. Jönsson and H. Wennerström, J. Colloid Interface Sci., $\underline{80}$, 482 (1981).
19. N. Metropolis, A.W. Rosenbluth, M.N. Rosenbluth, A.H. Teller and E. Teller, J.Chem.Phys., $\underline{21}$, 1087 (1953).
20. H. Wennerström, B. Jönsson and P. Linse, J.Chem.Phys., $\underline{76}$, 4665 (1982).

21. D. Henderson and L. Blum, J.Chem.Phys., <u>75</u>, 2055 (1981) and
     references therein.
22. D.C. Grahame, Chem.Rev., <u>41</u>, 441 (1947).
23. E.J.W. Verwey and J.Th.G. Overbeek, "Theory of the Stability
     of Lyophobic Colloids", Elsevier, Amsterdam, 1948.
24. P. Linse, G. Gunnarsson and B. Jönsson, J.Phys.Chem., <u>86</u>, 413
     (1982).
25. S. Engström and H. Wennerström, J.Phys.Chem., <u>82</u>, 2711 (1978).
26. This column is incorrectly reproduced in ref. 20.

COLLOIDAL STABILITY OF LIPOSOMES

Lisbeth Rydhag, Kenneth Rosenquist, Per Stenius
and Lars Ödberg

Institute for Surface Chemistry
Box 5607, 114 86 Stockholm, Sweden

In the ternary system of water/dimyristoyl phosphatidyl choline (DMPC)/cetyltrimethyl ammonium bromide
(CTAB), stable bilayer vesicles are easily obtained in the
two-phase region where a swollen lamellar phase is in equilibrium with a monomer solution of DMPC and CTAB in water.
It has earlier been shown that these liposomes are colloidal dispersions which are mainly electrostatically stabilized, i.e. their stability with respect to flocculation
may be predicted from the DLVO theory. An investigation
of negatively charged liposomes prepared from a fractionated commercially available soybean lecithin supports
this assumption.

The question arises whether these colloidal particles can also be sterically stabilized. This is of particular interest with the preparation of concentrated
liposome dispersions in mind. Preliminary results indicate that the incorporation of nonionic amphiphilic molecules with a large hydrophilic moiety in the bilayers
of the liposomes may function as a steric stabilizer of
the liposomes.

INTRODUCTION

It is well known that liposomes (vesicles) can be prepared by dispersion of lamellar liquid crystalline phases formed by swelling lipids in aqueous solution. However, although phase equilibria for typical lipid/water systems that form vesicles are by now known in some detail [1,2], it has generally not been clearly realized that vesicles are formed by dispersing one phase into another phase with which it may be in thermodynamic equilibrium. In earlier papers we have shown that in a model system consisting of a neutral phospholipid (DMPC) and a charged surfactant (cetyl trimethyl ammonium bromide, CTAB) the conditions for the formation of electrostatically stabilized liposomes can be precisely defined if the phase equilibria for the three-component system water/ DMPC/surfactant are known[1-2]. In this paper, we report on similar studies of a phosphatidyl choline of technical quality. The vesicles can be regarded as colloidal dispersions of lamellar liquid crystals in equilibrium with monomeric solutions of the vesicle component in water. The colloidal stability of such dispersions can be predicted from the DLVO theory, i.e. it depends on the surface charge of the liposomes and the charge and concentration of the counter ions [2-4].

The preparation of liposomes in this way implies that long-term stability will be limited to low salt concentrations (in particular, in the presence of bi- or trivalent cations) and low total concentrations of liposomes. In normal dispersion technology this problem is usually overcome by steric stabilization, i.e. by adsorption of high molecular weight water soluble compounds on the particle surface. Since, from a colloidal point of view, liposomes behave as normal lyophobic colloids, we have investigated the possibility to enhance their colloidal stability by adding steric stabilizers. As a model for such a stabilizer a nonionic surfactant with a large hydrophilic group was used.

MATERIALS AND METHODS

Materials. The liposomes were prepared of a highly purified as well as a commercially available phospholipid, i.e. di-myristoyl-L-α-phosphatidyl choline (DMPC), and a fractionated phosphatidyl choline from soybeans (PC$_{SB}$). The DMPC vesicles were charged by addition of the water-soluble surfactant N-hexadecyl-N,N,N-trimethylammonium bromide (CTAB). Details of this system are given earlier[1]. The soybean phosphatidyl choline (PC$_{SB}$) was from Lucas Meyer, Hamburg, containing 96 wt-% PC and 4 wt-% neutral lipids. The fatty acid composition as determined by gas chromatographic analysis[5] was:

C_{16} palmitic acid	14.8 wt-%
C_{18} stearic acid	3.2 "
$C_{18:1}$ oleic acid	13.6 "
$C_{18:2}$ linoleic acid	62.3 "
$C_{18:3}$ linolenic acid	5.2"

Dicetyl phosphate (DCP), 99% from Sigma was used to make the negatively charged vesicles.

A nonionic surfactant with a long hydrophilic chain was used as a steric stabilizer. This surfactant was nonylphenolpoly-ethyleneoxide ether (NP(EO)50) with an average oxyethylene chain length of 50 units (Berol Kemi, Stenungsund, Sweden). The average molecular weight was approximately 2400 and the cmc was 0.6 g/dm^3. The surfactant was used as received from the manufacturer. The salts were all of analytical grade. The water was triply distil-led with a conductivity of about 0.5 μS cm^{-1}.

Phase equilibria. The methods used for determination of the phase equilibria of the DMPC-CTAB-water system at 30°C were repor-ted in detail in ref. 1. They involve careful equilibration and identification of phases by microscopy and X-ray diffraction. These methods were also used to provide a very schematic description of the phase equilibria of the PC$_{SB}$-DCP-water system at 20°C.

Preparation and characterization of unilamellar vesicles.
(a) DMPC/CTAB. Unilamellar vesicles were prepared from samples whose total composition was located in the two-phase region in which lamellar phase and an aqueous monomeric solution of DMPC and CTAB are in equilibrium. As was shown in ref. 1, it is impor-tant that the composition of the lamellar phase is chosen in such a way that the equilibrium concentration of CTAB in the aqueous solution phase is below the cmc.

The total concentration of DMPC + CTAB in each sample was 5 mM or 0.34 wt-%. The dispersions were prepared according to a standardized procedure including probe sonication of a Rapidis 180 equipped with a 3 mm titanium probe (Ultrasonics Ltd., USA) for 30 minutes. The sample volume and the frequency were kept constant. The temperature was 30°C. Particles from the probe were removed by centrifugation at 2.500 x g for 1/2h at 30°C. 4 ml of the original sample volume of 5 ml was recovered and used as the vesicle fraction. Between 97 to 100 wt-% of the total lipid concen-tration was transformed into vesicles as determined by quantita-tive analysis of phosphorous[6]. The average hydrodynamic diameter for the vesicles varied between 40-100 nm with increasing amounts of CTAB[2].

(b) PC$_{SB}$/DCP. In this system vesicles were formed as 5 mM lipid dispersions by 60 min. treatment in a sonication bath (Brasonic 32). Sonication was carried out at 20°C and handling of the samples was made as far as possible in nitrogen athmosphere. Fractionation by size was achieved by ultracentrifugation at 150 000 x g for four hours. In this way a vesicle fraction could be isolated. This fraction contains 95% of the original lipid and its polydispersity is very low as judged by analytical ultracentrifugation (schlieren pattern) or dynamic light scattering (QLS). The z-average hydrodynamic diameter for vesicles prepared at low ionic strength (0.01MNaCl) was 35 nm.

The colloidal stability of the vesicles was characterized by measuring the kinetics of flocculation. The rate of flocculation of the unilamellar vesicle dispersions was measured after addition of electrolyte solutions of mono- and divalent anions. The procedure was similar to the one usually employed for colloidal particles[7]. The change of absorbance with time was determined in a spectrophotometer (Pye Unicam SP-800 Spectrophotometer, Philips, England) after addition of 0.2 ml electrolyte solution to 1.3 ml vesicle fraction in a 5 mm measuring cell. The electrolyte solution was rapidly injected into the measuring cell by means of a thermostated Hamilton syringe (CR 700-200, Whittier, USA). Some of these experiments were later repeated in a Durrum model D-110 stopped flow spectrophotometer equipped with a 20 mm cell. Equal volumes (0.2 ml) were reproducibly mixed within approximately 10 ms by means of a pressure operated actuator. The change of the absorbance of the mixture with time was recorded. All measurements were made at λ = 400 nm and at a temperature of 30°C. For the PC$_{SB}$-DCP-water system all measurements were done in the Durrum spectrometer at 25°C and λ = 450 nm.

Theory of Colloidal Stability

(a) Electrostatically stabilized vesicles. Two criteria were used to characterize the electrostatic stabilization of vesicles. According to the DLVO theory, the critical coagulation concentration of salt, C_c for two particles with Stern potential ϕ_d in a solvent with dielectric constant ε_r is given by[8]

$$C_c = \frac{(3072\pi)^2}{8e^4} \cdot \frac{R^5 T^5 \varepsilon_0^3 \varepsilon_r^3}{F^6} \cdot \frac{B^4}{A_H^2 z^6} \tag{1}$$

where

$$B = \frac{\exp(zF\phi_d/2RT)-1}{\exp(zF\phi_d/2RT)+1} \tag{2}$$

Here, z is the counterion charge, A_H is the Hamaker constant and other symbols have their usual meanings. At low surface potentials,

$$B \approx \frac{zF\phi_d}{4RT} \; ; \; C_c \approx \text{const.} \frac{\phi_d^4}{A_H^2 z^2} \tag{3}$$

and the relationship between the charge of the diffuse layer, σ_d, and ϕ_d is given by

$$\sigma_d = zF(2C_o \varepsilon_o \varepsilon_r / N_A RT)^{\frac{1}{2}} \phi_d \tag{4}$$

Combining (1), (3) and (4) we obtain (at constant T)

$$\log C_c = \frac{1}{3} \cdot \frac{\text{const}}{z^6} + \frac{4}{3} \log \sigma_d \tag{5}$$

If the vesicles are electrostatically stabilized colloids (a) C_c should depend on counterions with different charge according to equation (1) and (b) C_c should depend on the surface charge density of the vesicles according to eq. (5). σ_d can be systematically varied by varying the amount of ionic surfactant in the vesicles.

(b) Sterically stabilized vesicles. The theory of the repulsive interaction by adsorbed polymers (steric stabilization) is based on the Flory-Huggins theory of polymer solutions and has been formulated in detail by, among others, Hesselink et al[9]. The conditions for effective steric stabilization are that the polymer is well anchored at the particle surface, that a complete adsorption layer is formed and that loops or tails of the polymer extend far enough into the solvent to reduce considerably the van der Waals' attraction between the particles. In this investigation we have used a nonionic surfactant with a polyethyleneoxide chain which extends into the aqueous solution and a nonylphenol chain which presumably anchors the stabilizer at the surface.

The theory of steric stabilization shows that the stability should depend, above all, on the polymer/solvent and polymer/polymer interactions and, hence, should not be strongly affected by the addition of salt. In the case that polymer/solvent interactions predominate, the colloid should flocculate on heating (enthalpic stabilization)[10]. In this investigation we have, so far, confirmed the attainment of steric stabilization only by characterizing the coagulation behaviour when salt is added.

(c) Stability ratio. The stability ratio for coagulating the colloid is defined as the ratio between the rate of coagulation without coagulation barrier to the rate in the presence of a coagulation barrier:

$$W = \frac{k_1^{o}}{k_1} \qquad (6)$$

where k_1 and k_1^{o} are the rate constants with and without coagulation barrier, respectively. It was shown by Overbeek et al[11] that for electrostatically stabilized colloids the relationship between W and the electrolyte concentration C_o, is given by

$$\log W = \text{const.} - \frac{16\pi a R^2 T^2 \epsilon_o \epsilon_r B^2}{e \, z^2 F^2} \, \log C_o \qquad (7)$$

where a is the particle radius. For $C_o = C_c$, $\log W = 0$. In this investigation, C_c was determined by measuring W as a function of C_o. W was determined by recording the rate of change of turbidity $(d\tau/dt)$ after addition of salt in a stopped-flow spectrometer. For dilute solutions and at the early stages of coagulation, this rate is proportional to the rate of change of the particle concentration and W may be calculated from

$$W = \frac{(\frac{d\tau}{dt})_o}{(\frac{d\tau}{dt})} \qquad (8)$$

where $(d\tau/dt)_o$ is the maximum rate of change ($W=1$). For a sterically stabilized colloid one expects W to be almost independent of C_o.

RESULTS

The DMPC/CTAB/water system

The phase diagram of this system in the region with more than 90% water is shown in Figure 1. There are two one-phase areas: an aqueous solution L_1 and a lamellar liquid crystalline phase D. Between these phase regions there is a two-phase area in which either D-phase and aqueous solution below the critical micelle concentration (cmc) of CTAB or D-phase and micellar solution are in equilibrium. The salient tip of the D-phase region (which con-

tains so much water that the phase boundary could not be determined accurately because it was difficult to separate D and L_1)
is in equilibrium with L_1 solution at about the cmc of CTAB.

Stable liposomes are formed if D-phase is dispersed in aqueous non-micellar solution L_1 in such a way that the dispersed phase
is in equilibrium with the L_1 solution *below* the cmc of CTAB (i.e.
to the left of the salient region of the D-phase in Figure 1). When
the molar ratio CTAB/DMPC exceeds about 3, the lamellar phase is
in equilibrium with micellar solution. In this case it is no longer possible to form stable liposomes[1].

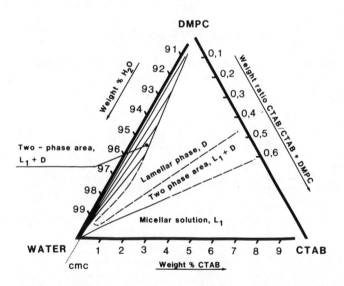

*Figure 1. Phase equilibria in the system H₂O/Dimyristoylphospha-
tidylcholine/cetyltrimethylammonium bromide at 30°C. For CTAB/
DMPC ratios higher than those indicated by the dotted line it was
not possible to separate aqueous solutions from the lamellar phase
by centrifugation. Concentrations are given in weight percent.*

The stable liposomes in all cases are single-lamellar, i.e. they are appropriately called vesicles. The colloidal stability was only investigated for vesicle preparations (for details, see also Methods). There is no reason to assume that multilamellar liposomes are stabilized by other mechanisms.

Vesicle Stability

Flocculation refers to a process in which the aggregated vesicles remain as separate entities in the aggregate while in coagulation the vesicles can coalesce and form one large vesicle. The flocculation process can not be distinguished from the coagulation process with light scattering measurements.

The kinetics of flocculation was investigated for vesicles prepared in the $D + L_1$ region as described above. Flocculation rates were determined by adding NaCl solutions to dispersions of DMPC vesicles containing varying amounts of CTAB. The log W versus log C_0 curves are summarized in Figure 2. Critical coagulation concentrations are easily determined from these plots. Similar curves were also obtained when Na_2SO_4 or sodium citrate (representing bivalent and trivalent counterions, respectively) were added.

The PC_{SB}/DCP/Water System

In this system both components form dispersions in water with very low monomer concentrations. The phase diagram was not investigated

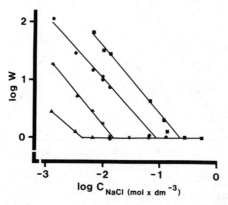

Figure 2. The critical coagulation concentration (ccc) for DMPC-CTAB-vesicles in water was determined for monovalent anions. The logarithm of the stability ratio, W, is given versus the logarithm of the electrolyte concentration. The CTAB concentration of the vesicles is [12]

▲ *0.5 wt-% CTAB*   ● *5.0 wt-% CTAB*
★ *1.0 wt-% "*   ■ *10.0 wt-% "*

in detail. PC_{SB} forms lamellar phase by swelling with water; this phase is able to incorporate up to 15 mol% of DCP. Above 20 mol% DCP phase separation between the components was detected. At a total lipid concentration of 10 weight%, molar ratios DCP/PC_{SB} between 0.04 and 0.1 give rise to a lamellar phase from which no aqueous solution can be separated by ultracentrifugation at 90 000 x g for 2h. This region obviously corresponds to the salient D-phase region in the DMPC/CTAB/water system. Vesicle formation, accordingly, was studied only for lipid composition between 1 and 10 mol% DCP. Lamellar phases within these composition limits will certainly be in equilibrium with non-micellar aqueous solution.

Vesicle Stability

The coagulation kinetics for PC_{SB}-vesicles with 1 and 10 mol% DCP is illustrated as log W/log C_0 plots with Na^+ as counterion in Figure 3 and with Ca^{2+} in Figure 4. C_0 is the counter ion concentration in mol/dm^3 in the final solution. These measurements were done at a final lipid concentration of 1.25 mM. C_c values determined from the break points are summarized in Table 1. The ratios between C_c values for $z = 1$ and 2 are 1:0.08 for vesicles with 1 mol% DCP and 1:0.03 with 10 mol% DCP.

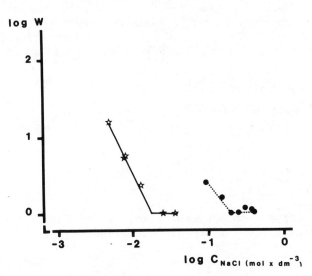

Figure 3. Coagulation kinetics for negatively charged vesicles and monovalent counterion, Na^+. Vesicles were prepared from PC_{SB} and 1 respectively 10 mol% DCP. C_0 values corresponding to inflection points are 0.018 and 0.200 M Na^+. Final lipid concentration is 1.25 mM.

● *PC_{SB} + 10 mol% DCP*

☆ *PC_{SB} + 1 mol% DCP*

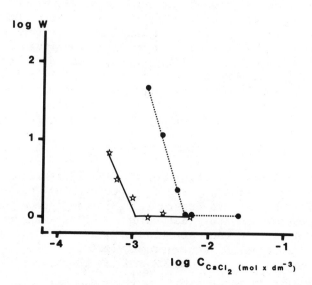

Figure 4.   Coagulation kinetics for negatively charged vesicles and divalent counterion, Ca^{2+}. Vesicles were prepared from PC_{SB} with 1 and 10 mol% DCP. C_c values corresponding to inflection points are 0.0015 and 0.005 M. Final lipid concentration is 1.25 mM.

● PC_{SB} + 10 mol% DCP

☆ PC_{SB} + 1 mol% DCP

Table 1. Comparison between C_c Values for Negatively Charged PC_{SB} Vesicles with Different Counterions.

Mole% DCP	C_c Na$^+$ (mol·dm^{-3})	C_c Ca^{2+} (mol·dm^{-3})	Ratio
1	0.018	0.0015	1:0.08
10	0.200	0.005	1:0.03

Steric Stabilization

The nonionic emulsifier, $NP(EO)_{50}$ was added to vesicles prepared from PC_{SB} with 10 mol% DCP. Several different molar ratios between $NP(EO)_{50}$ and lipid were investigated for destruction of vesicles. No breakdown could be detected by light-scattering measurements. After addition of $NP(EO)_{50}$ the scattered light intensity remained stable at a 10% increased level. Ca^{2+} was added 30 min. after the addition of nonionic surfactant. Intensity curves from stopped flow measurements with 9 mM Ca^{2+} and different molar ratios of the stabilizer to lipid are shown in Figure 5.

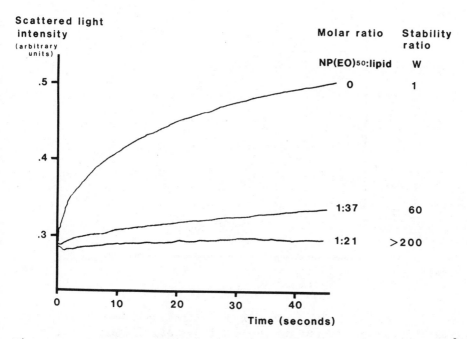

Figure 5. Intensity changes in scattered light after mixing Ca^{2+} ions and negatively charged vesicles with varied amount of $NP(EO)_{50}$. Vesicles were prepared from PC_{SB} and 10 mol% DCP. Final Ca^{2+} concentration was 9mM and final lipid concentration was 1.25mM.

   This counterion concentration causes rapid coagulation of
vesicles without surfactant or after addition of small amounts
of NP(EO)$_{50}$. When NP(EO)$_{50}$ is added at an approximate molar ratio
of 1:21 almost no intensity change is registered. This molar ratio
can be obtained while keeping the NP(EO)$_{50}$ concentration below
the critical micelle concentration in all steps during preparation.

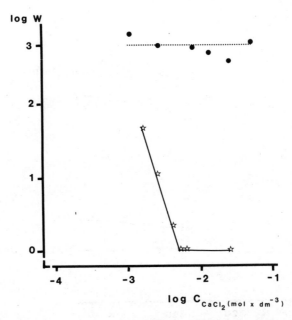

*Figure 6.   Coagulation kinetics for negatively charged vesicles
with NP(EO)$_{50}$ compared to vesicles without stabilizer. Vesicles
prepared from PC$_{SB}$ with 10 mol% DCP and NP(EO)$_{50}$ added at a molar
ratio 1 to 21 lipid. Final lipid concentration 1.25 mM.*

   ● *vesicles with NP(EO)$_{50}$*

   ☆ *vesicles without stabilizer*

Figure 6 shows coagulation kinetics for vesicles prepared with a stabilizer to lipid ratio 1:21 at different Ca^{2+} concentrations. Very small intensity changes were registered at all concentrations investigated up to 50 mM Ca^{2+} which is 10 times above the C_c for vesicles without stabilizer.

DISCUSSION

Electrostatically Stabilized Vesicles

The results shown in Figures 2, 3 and 4 give clear evidence that the vesicles formed by CTAB + DMPC as well as by PC_{SB} + DPC are electrostatically stabilized colloidal particles.

According to equation (3), the ratio between the C_c's for z = 1, 2 and 3 at low potentials should be 1:0.25:0.11. We have previously reported that the C_c's for Cl^-, SO_4^{2-} and Na citrate for vesicles containing low concentrations of CTAB do follow this rule[2].

For vesicles with 10% CTAB we found the ratio 1:0.028:0.0002, which would be more in accordance with equation (1), which at high potentials (for which $B\rightarrow1$) predicts the ratio 1:0.016:0.0014.[2] However, it is well known that the dependency of the C_c's on z often is much higher than expected from estimates of ϕ_d (by measurements of, e.g. the ζ-potential)[13]. An explanation for this discrepancy is that the adsorption of counterions into the Stern layer is strongly dependent on z. It has been confirmed for vesicles of phosphatidyl-serine that the adsorption of counterions to the vesicle surface depends both on the valency (z) and the type of counterion[4].

As shown in Table 1, the stability of PC $_{SB}$/DCP vesicles is much more sensitive to Ca^{2+} ions than predicted by either equation (1) or (3). The Ca^{2+} ions obviously adsorb very strongly on the negative surfaces and decrease the net charge of the vesicles. This has also been found in other investigations[4].

Further evidence that electrostatic stabilization plays an important role in vesicle/vesicle interactions is given in Figure 7. We assume (a) that the charge on the vesicle surface is proportional to the concentration of CTAB in the vesicles: σ_0=const. c (CTAB), and (b) that the diffuse layer charge σ_d is proportional to σ_0. Figure 7 shows a plot of log (C_c) for the vesicles against log c(CTAB). As predicted by equation (5), a straight line of slope 1.34 is obtained for coagulation with NaCl.

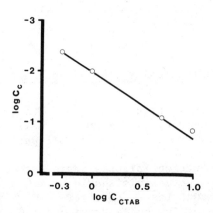

*Figure 7. Critical coagulation concentration for DMPC/CTAB
vesicles as a function of the concentration of CTAB in the
vesicles.*

Flocculation Kinetics

If the vesicles are spherical with radius a and the flocculation is completely diffusion controlled, the rate constant of
flocculation will be given by [14]

$$k_1^{\,o} = \frac{8kT}{3\eta} \tag{9}$$

where η is the viscosity of the solvent. For water at $25^{\circ}C$, $k_1^{\,o} =$
$= 1.4 \cdot 10^{-11} \ cm^3 \cdot s^{-1}$.

From the measured rates of flocculation of the DMPC/CTAB and PCSB/DCP vesicles the rate constants may be calculated[11] from the rate of change of turbidity in the initial state of the reaction using the equation[7].

$$\frac{d\tau}{dt} = k_1 \cdot \frac{2B_o c_o^2}{\rho^2} \tag{10}$$

where τ is the turbidity, measured as absorbance, c_o is the concentration of scattering particles with density ρ and B_o is an optical constant

$$B_o = \frac{32\pi^3 \rho^2 n_o^2}{3\lambda^4} \left(\frac{dn}{dc}\right)^2 \tag{11}$$

For the refraction index increment we use the value 0.162 cm^3/g for DMPC/CTAB vesicles[15] and a measured value of 0.154 cm^3/g for PCSB/DCP vesicles. This gives $B_o = 5.2 \cdot 10^{18}$ cm^{-4} for $\lambda = 400$ nm in the DMPC/CTAB system and $B_o = 3.5 \cdot 10^{18}$ cm^{-4} for $\lambda = 450$ nm in the PCSB/DCP system. For n_o the refraction index of water was used.

Rate constants calculated according to Eq. (10) are given in Tables II and III.

Table II. Rate Constants of Vesicle Coagulation

C_{CTAB} wt-%	Vesicle radius nm	C_e (NaCl) mole/dm³	k_1^o cm³·s⁻¹
0.50	–	$4.0 \cdot 10^{-3}$	$8.1 \cdot 10^{-16}$
1.00	20	$1.3 \cdot 10^{-2}$	$8.1 \cdot 10^{-16}$
5.00	30	$8.9 \cdot 10^{-2}$	$2.0 \cdot 10^{-15}$
10.00	50	$5.0 \cdot 10^{-1}$	$2.3 \cdot 10^{-15}$

The coagulation rate was determined for 5 mM vesicle dispersions of DMPC-CTAB in water at 30°. The optical constant, B_o was $5.2 \cdot 10^{18}$ cm^{-4}. 12

Table III. Rate Constants for Rapid Coagulation of Negatively
Charged PC$_{SB}$ Vesicles with Different Counterions. The final lipid
concentration was 1.25 mM.

Mole% DCP	k_1^o Na$^+$ (cm^3/s)	k_1^o Ca^{2+} (cm^3/s)
1	$3.2 \cdot 10^{-15}$	$1 \cdot 10^{-15}$
10	$5 \cdot 10^{-17}$	$2.7 \cdot 10^{-15}$

The experimentally obtained rate constants are about three
orders of magnitude lower than would be expected if the rate were
completely diffusion controlled. Lansman and Haynes[3] have suggested
that some ordering processes are necessary before the flocculated
vesicles can coagulate (fuse). Nir and Bentz[4,16] have pointed out
that the equation of Smoluchowski or Fuchs[14,17] should be corrected to take into account the hydrodynamic forces due to the finite rate of flow of the solvent in the region between two colliding particles. Spielman[18] has shown that the hydrodynamic correction factor modifies the diffusion coefficients at small distances of separation between the particles. This implies an increase
in the factor W and, hence, would further reduce the rate constant of aggregation of vesicles. Nir and Bentz[4,16] have pointed
out that this explains why Lansman and Haynes[3] have obtained rate
constants two orders of magnitude too low when compared to simple
Smoluchowski coagulation for vesicles of phosphatidic acid and
phosphatidylserine. It should be noted that, in general, experimentally determined rate constants for the coagulation of any
colloid are found to be considerably smaller than predicted by
the Smoluchowski theory[19]. Thus, the results in Tables II and III
are not unexpected and do not contradict that both types of vesicles can be regarded as electrostatically stabilized colloidal particles that coagulate at about the same rate when the electrostatic repulsion is removed.

It should be pointed out that the repulsion forces occuring at close approach of vesicles consisting of phosphatidyl/
choline are not considered in the DLVO theory[22]. It is reasonable
to assume that complicated association processes take place in
order for the vesicles to fuse. Details about the final state of
the flocculated vesicles can, however, not be concluded from
light scattering measurements.

Steric Stabilization

Figure 6 shows very clearly that vesicles can be completely stabilized against coagulation by electrolyte by adsorbing a steric stabilizer on their surface. We have also investigated the stability of the vesicles in the presence of corresponding amounts of polyethyleneoxide ($M_w \approx 1400$). This had no stabilizing effect. Thus, the hydrocarbon chain of the nonionic surfactant obviously anchors at the vesicle surface by adsorption of the nonylphenol chain. We are continuing studies of the effect of this adsorption on vesicle size and stability.

Correlation between Vesicle Stability and Phase Equilibria

We have shown that stable vesicles are formed only in the case where micelles do not occur in the dispersion medium. A reasonable explanation for this is the extremely low solubility of DMPC or PC_{SB} in such aqueous solutions. It was pointed out by Higuchi and Misra[20] that an important cause of the low stability of very small emulsion droplets is the difference in solubility in water of the monomers in droplets with different radii. For two droplets with radii r_a and r_b and monomer solubility C_a and C_b this difference is given by the Kelvin equation

$$\frac{C_a}{C_b} = \exp \frac{2\gamma Vm}{RT} \left(\frac{1}{r_a} - \frac{1}{r_b}\right) \tag{11}$$

If the monomers are slightly soluble in the surrounding solution this difference will result in a net diffusion of material from small droplets to larger droplets in a fairly short time. The rate of decrease of the radius of a droplet with radius r_a in the presence of a droplet with radius r_b is given by

$$\frac{dr_b}{dt} = -D \cdot \frac{2\gamma M}{RT\rho 2} \cdot \frac{c_o}{r_b^2} \cdot \frac{N_a(r_a - r_b)}{N_b r_b + N_a r_a} \tag{12}$$

where D is the monomer diffusion coefficient in the external phase, M and ρ are the molecular weight and density of the material in the droplet, N_a and N_b are droplet concentrations and c_o is the equilibrium solubility at a planar surface.

The solubility of DMPC in water or submicellar aqueous surfactant solution is of the order 10^{-10} mol/dm^3 [21]. Assuming $D \sim 10^{-6}$ cm^2/s, $T = 300$ K, $N_a = N_b$, $r_a = 40$ nm, $r_b = 20$ nm, $M = 700$ g/mol, $\rho = 1$ g/cm^3 and $\gamma = 5$ mJ/m^2 we find that dr_b/dt is about $2 \cdot 10^{-5}$ nm/s, i.e. diffusion is completely negligible.

In micellar solution, however, the solubility is of the order of
10 g/dm^3 (see Figure 1) which increases the rate of change to about
10^3 nm/s, implying that complete breakdown will take place within
a few seconds. Thus, vesicles will remain stable below the cmc as
long as the repulsion (steric or electrostatic) is sufficiently
large while the presence of micelles will cause immediate breakdown. The electrostatic charge introduced by adding surfactant
will cause the lamellae to swell, make dispersion easier and electrostatically stabilize the vesicles against coagulation. The
small amounts of charged lipid required to achieve this may explain why difficulties have been experienced to produce liposomes
with reproducible properties from commercial lipids.

ACKNOWLEDGEMENTS

This work was supported by the Swedish Board for Technical
Development. Skillful assistance by Thomas Gabrán is gratefully
acknowledged.

REFERENCES

1. L. Rydhag and T. Gabrán, Chem. Phys. Lipids, 30, 309 (1982).
2. L. Rydhag, P. Stenius and L. Ödberg, J. Colloid Interface Sci.,
   86, 274 (1982).
3. J. F. Lansman and D. H. Haynes, Biochim, Biophys. Acta, 394,
   335 (1975).
4. S. Nir and J. Bentz, J. Colloid Interface Sci., 65, 399 (1978).
5. R. L. Glass, R. Jeness and H. A. Trookin, J. Dairy Sci., 48,
   1106 (1965).
6. A. Biverstedt and L. Rydhag, in "Proceedings from 9th Scand.
   Lipid Symp., Visby, Sweden, 1977", R. Marcuse, Editor, Lipid-
   forum, Gothenburg.
7. G. Oster, J. Colloid Interface Sci., 2, 29 (1947).
8. J. Th.G. Overbeek, in "Colloid Science", H. R. Kruyt, Editor,
   Chap. 8, Elsevier, Amsterdam, 1952.
9. F. Th. Hesselink, A. Vrij and J. Th. G. Overbeek, J. Phys.
   Chem., 75, 2094 (1971).
10. D. H. Napper, Kolloid-Z., 234, 1149 (1969).
11. H. Reerink and J. Th. G. Overbeek, Disc Faraday Soc., 18, 74
    (1954).
12. L. Rydhag, Ph.D. Thesis, Lund University, Lund, 1982.
13. J. Th. G. Overbeek, Pure Appl. Chem., 52, 1 (1980).
14. M. von Smoluchowski, Z. Physik. Chem., 92, 129 (1917).
15. P. N. Yi and R. C. MacDonald, Chem. Phys. Lipids, 11, 114 (1973).
16. S. Nir, J. Bentz and A. R. Portis, in "Bioelectrochemistry",
    M. Blank, Editor, Adv. Chem. Series, No.188,75,American Chemical
    Society, Washington, D.C., 1980.
17. N. Fuchs, Z. Phys., 89, 736 (1934).
18. L. A. Spielman, J. Colloid Interface Sci., 33, 4 (1970).

19. P. C. Hiemenz, "Principles of Colloid and Surface Chemistry", Marcel Dekker, New York, 1977.
20. W. I. Higuchi and J. Misra, J. Pharm. Sci., 51, 459 (1962).
21. C. Tanford, "The Hydrophobic Effect", Ch. 12, Wiley, New York, 1973.
22. D. M. Le Neveu, R. P. Rand and V. A. Parsegian, Nature, 259, 601 (1976).

FAST DYNAMIC PHENOMENA IN VESICLES OF PHOSPHOLIPIDS

DURING PHASE TRANSITIONS

V. Eck and J. F. Holzwarth

Fritz-Haber-Institut der Max-Planck-Gesellschaft

Faradayweg 4-6, D-1000 Berlin 33, West Germany

A laser-temperature-jump method has been used to study relaxation phenomena in unilamellar vesicles of pure phospholipids. We report five well separated and time resolved relaxation signals between 4×10^{-9} s and 10^{-2} s. By choosing turbidity as the detection parameter, it was possible to distinguish between cooperative and noncooperative processes within the overall phase transition of the membrane. Using this approach it was not necessary to introduce probes. Two noncooperative phenomena occur in the ns-time regime and are due to kink formation[1] associated with an increase of the optical density of the membrane followed by a fast membrane expansion. The three signals in the μs- to ms-time scale show clear, but individually different, cooperativity. Their relaxation signals are caused by a decrease in the optical density of the membrane. The two slowest relaxations could be interpreted using a cluster model derived from a two-dimensional Ising-model[2] for nucleation and growth of clusters. In the end, we present a dynamic model of the molecular processes representing the five individual relaxation times observed.

INTRODUCTION

Vesicles prepared from phospholipids such as dimyristoyl-1-α-lecithin (DML), dipalmitoyl-1-α-lecithin (DPL) or dipalmitoyl-1-α-phosphatidylserine (DPS) are the simplest cell-like aggregates which contain only molecules of natural cell-membranes. The concentration-temperature phase diagram of aqueous phospholipid solutions is shown for DML as an example in Figure 1. The general chemical structure is given in Figure 2. The vesicles of phospholipids have been shown to mimic biological membranes to a certain extent: One of the interesting properties of these aggregates is the thermotropic phase transition which has also been demonstrated in the membranes of living cells[2,3,4]. In the temperature range of the main-phase transition, the fluidity of the bilayer is changed dramatically as the system goes from a so-called crystalline to a liquid crystalline state (Figure 6,7,8). This has a drastic effect on living cells[5], which seem to grow only above the phase transition of their plasma membranes. Other properties associated with biological membranes like transport mechanisms or protein and enzyme activity seem to depend on the state of the membrane lipids.

To investigate this behavior, many equilibrium data have been published but only few kinetic measurements can be found in the literature[1,2,9]. This is partly due to the difficulty of finding a suitable relaxation technique which provides high time resolution without unwanted physical effects disturbing the observation channel. The rapid thermal perturbation produced by a Joule heating T-jump certainly causes leakage or even blast of the bilayer because of the very strong electric field-jump of ~ 20 kV cm^{-1} inevitably preceding the temperature jump. The unwanted perturbation caused by this effect becomes the limiting factor for the possible time resolution of any observed relaxation phenomena. Pressure jump techniques with optical detection can only be used for dynamic processes slower than 200 µs[6]. Our laser-T-jump[6] avoids these difficulties and makes it possible for the first time to cover the complete time-range from 10^{-9} up to 10^{-2} s, using the same technique and providing the same very high time resolution independent of additives.

In this paper we present a complex relaxation spectrum for the main phase-transition of phospholipid-vesicles prepared from DML and DPL. Our data confirm the relaxation signals in the 10^{-4} -10^{-2} s time range, already observed by Joule heating T-jump and pressure-jump experiments[7,8] and in the ns time range where ultrasonic and NMR methods were used[9,10]. In addition to these signals we could examine two further relaxation processes in the 10^{-7} s and 10^{-5} s ranges which are inaccessible to other methods.

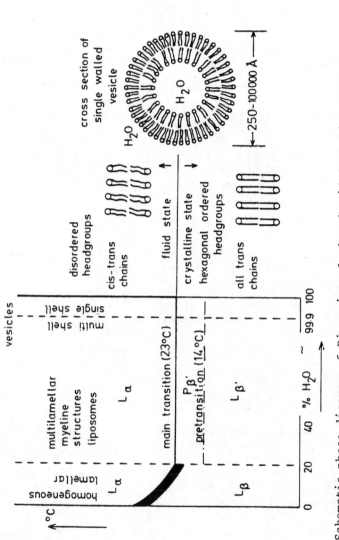

Figure 1. Schematic phase diagram[28] of Di-myristoyl-phosphatidyl-choline (DML) as suggested by X-ray and optical birefringence[28]. The black area is a region of coexistence, β indicates the crystalline phase while α stands for the liquid phase. The main transition temperature $T_m = 23°C$, and the pretransition temperature $T_{pm} = 14°C$.

MATERIALS AND METHODS

Chemicals - The phospholipids DML and DPL (L stands for lecithin, identical with phosphatidycholine in Figure 2) and Tris(hy-
droxy-methyl)aminomethane "Tris" were obtained from Fluka, Switzerland; DPS was purchased from Serva, Germany. Thin-layer chromatography on Merck Kieselgel 60 using a chloroform-methanol-water
(65:35:4 by volume) mixture as solvent showed no impurity for
the three lipids so they were used without further purification.
All chemicals were of purissimum or suprapure grade, only NaN_3 was
of analytical grade.

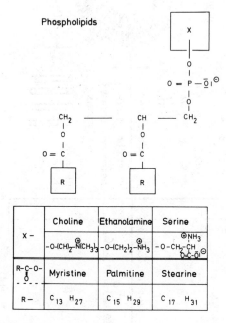

Figure 2. Summary of chemical structure of phospholipids.

Preparation of unilamellar vesicles - The vesicles were prepared according to the injection method developed by Batzri and
Korn[11] and further modified by Kremer et al.[12]. For vesicles of
DML and DPL with a diameter of 100 nm, a solution of 30 µM
lipid in 1 ml absolute ethanol was injected very slowly into
10 ml buffer at pH 7.5, at a temperature well above the main phase-
transition of about $23°C$ for DML and $41°C$ for DPL respectively,
and then dialysed against pure buffer for at least 8 h. The buffer contained 0.1 M NaCl, 10^{-2} M Tris-HCl[13a] and 10^{-3} M NaN_3. By
this method a very homogeneous size distribution was obtained as
shown by gel filtration using Sepharose 4B and electron-micrographs (Figure 3).

The same method was used to prepare vesicles of DPS. In this case 10 μM DPS were dissolved in 3 ml absolute ethanol and the solution was concentrated to 1 ml by heating it up to 65-70°C. This hot solution was injected into 10 ml buffer at the same temperature. The phase transition of DPS occurs around 60°C, and was shown to be dependent on the pH value of the buffer[14]. So two buffers were used, one with the pH of the isoelectric point of DPS around pH 5.9 which consisted of 0.1 M NaCl, 9.2 mM NaH$_2$PO$_4$, 0.8 mM Na$_2$HPO$_4$ and 1 mM NaN$_3$[13b], the other as described above giving pH 7.5. The dialysed solution was gel filtered on a Sephadex G-50 M column to remove smaller aggregates or monomers and afterwards it was gel filtered on a Sepharose 4B column. The elute contained a small amount of very large vesicles with a diameter of 150 nm but most of the vesicles showed a diameter of about 100 nm as shown in the electron-micrograph in Figure 3. Vesicles of DML with impregnated cholesterol were prepared in a similar way as pure DML-vesicles. 1 ml of an ethanolic solution containing 24 mM DML and 6 mM cholesterol was injected into 10 ml Tris-HCl buffer described above. The solution was then dialysed against pure buffer at 25°C for at least 8 h. The homogeneity of this preparation was tested by taking electron-micrographs. An example is shown in Figure 3.

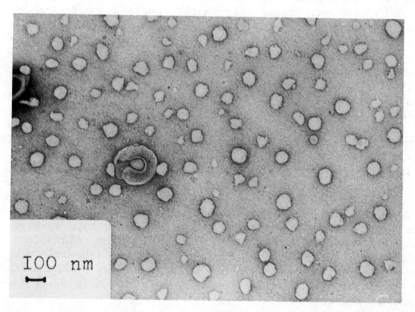

Figure 3. Electron-micrograph of vesicles of DML +20% cholesterol, negatively stained with uranylacetate.

Laser-Temperature-Jump - For laser action, an ultraviolet-flash induced dissociation reaction is used. Perfluoroisoiodopropane (i-C_3F_7I) dissociates into C_3F_7-radicals and excited I* atoms. The wavelength of the laser is determined by the I* ($^2P_{1/2}$) to I($^2P_{3/2}$) transition emitting photons at a wavelength of 1.315 µm. The detailed set-up has been described elsewhere [6,15,16]. The laser was used in various modes. For signals slower than 5 µs, the oscillator mode was used. In this mode the laser emits approximately 2 J in 2.4 µs at 1.315 µm. In the free run mode, a whole pulse sequence is amplified after self-mode locking. The pulse-train length is approximately 80 ns. Finally in the mode-locked version, a single pulse is selected out of the emitted pulse train, amplified and passed through a quartz cuvette containing the aqueous solution to be investigated; this results in a laser pulse halfwidth between 100 ps and 4 ns. The laser light energy is absorbed by rotational-vibrational states of H_2O [17]. The energy equilibration causes a temperature jump of about 1°C within an effective volume of 40-100 µl depending from the quartz cuvettes used. The cuvettes were thermostatically controlled with an accuracy of $\pm$ 0.1°C.

The laser-T-jumps were repeated about every 5 min, the time necessary to allow the sample solution to reach its thermal equilibrium.

Light perpendicular to the laser light beam from a xenon arc lamp (XBO 150 W) was passed through the sample and detected by a very fast low-noise photomultiplier [18] with a rise-time of 0.6 ns which operates only with 5 dynodes and therefore requires a high light power. In the mode-locked version the lamp was pulsed by discharging a capacitor bench of 60-80 V within 500 µs. The light was detected by a photomultiplier tube (RCA 1 P 28) after passing through a filter before and a monochromator after the sample. In both cases, the mode locked or the free run version of the laser, the relaxation signal was monitored through an active probe (Tektronix P 6201, bandwidth 900 MHz). The signal was stored in a Tektronix transient digitizer 7912 AD equipped with plug-ins 7A13 or 7A22 in the case of the slower version of the laser, and a time base 7B92. The overall bandwidth was greater than 100 MHz (rise-time 2.6 ns) or 1 MHz, (rise-time 267 ns). The transient recorder was connected to a calculator (Hewlett-Packard 9845 B) via an IEEE 488-bus system. This arrangement allows one to average several experiments at each temperature and to calculate the relaxation times and amplitudes using a computer-program. A schematic diagram of the experimental arrangement is shown in Figure 4. The emitted laser intensity against time from the "free-run" mode of the laser is shown in Figure 5.

IODINE - LASER TEMPERATURE - JUMP

10^0 s $\geqq$ Timeresolution $\geqq$ 10^{-10} s

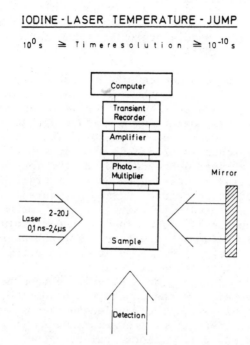

Figure 4. Experimental arrangement.

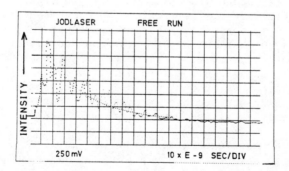

Figure 5. Iodine laser emission with a decay time of 70 ns ("free run").

RESULTS

Equilibrium measurements – All measurements which we report in this paper were performed using only unilamellar vesicles of pure phospholipids. The only exception were experiments with a DML-cholesterol mixture. Special care was taken for a very narrow size distribution (see MATERIALS).

The temperature dependence of the turbidity measured as an absorbance which is related to the optical density of the phospholipid membrane is shown in Figure 6 for vesicles from pure DML, in Figure 7 for vesicles containing a mixture of DML with cholesterol, and in Figure 8 for vesicles of pure DPL. The sharp decrease of the turbidity at temperatures around 23.5°C for DML (Figure 6) and 40.5°C for DPL (Figure 8) are due to the thermotropic transitions from the crystalline or solid-like to the fluid-like state of the membrane. The schematic phase diagram in Figure 1 shows the principle changes associated with this so-called "main phase transition"; the temperature T_m is defined as the midpoint of the normalized parameter θ temperature curve. For DPL a similar diagram exists which differs only in the pretransition temperature T_{pm} = 33°C and the main transition temperature T_m = 40°C. Because of the nature of turbidity (light scattering phenomena) its absolute value not only depends on the concentration of the vesicles but also on the particle size, the wavelength of observation (see Figure 8), and the density of the membrane. As this transition goes from a very rigid two dimensional hexagonal lattice of the phospholipid headgroups and a strict all-trans conformation of the hydrocarbon chains to a random distribution of the headgroups on the surface of the vesicle-membranes and a cis-trans (gauche) conformation of the hydrocarbon chains[19,20], it can be expected that there are several contributions to the overall dynamic behavior representing the main phase transition. The equilibrium measurements for pure unilamellar vesicles in Figure 6 and Figure 8 are reversible in their temperature dependence and show only a single narrow main phase transition temperature range. If cholesterol is added to vesicles of DML a pronounced hysteresis appears in the temperature range of T_m (Figure 7). The reason for this is an irreversible change in the incorporation of cholesterol into the double layer of the phospholipids. This phenomenon is of great interest to biologists because it offers the possibility for the vesicles to memorize if the temperature was raised or lowered.

The static measurements described in Figures 6,7 and 8 were used not only to define the main phase transition temperature T_m but also to monitor any change in the size-distribution of the samples used for our laser-temperature-jump experiments.

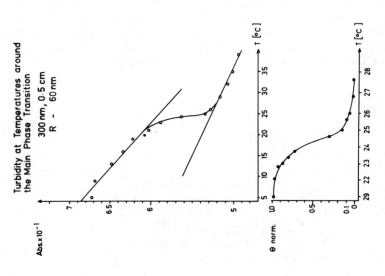

Figure 6. Equilibrium phase transition curve of vesicles of DML measured by the turbidity change at 300 nm.

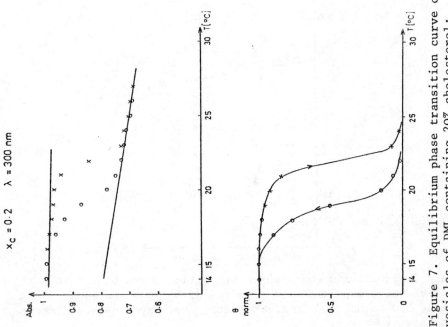

Figure 7. Equilibrium phase transition curve of vesicles of DML containing 20% cholesterol; at 300 nm turbidity shows a hysteresis.

Different slopes of the almost linear dependence of the turbidity above and below T_m can be interpreted as a pretransition
range T_{pm} below T_m (greater slope) with a higher ΔH because of
higher order-changes and very small changes in the order parameter
(smaller slope) well above T_m. The molecular reasons for these effects especially the so-called pretransition at T_{pm} are still unknown; these will not be discussed in this paper.

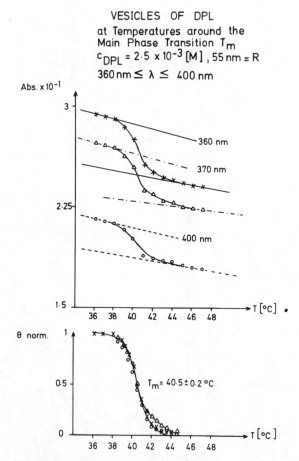

Figure 8. Equilibrium main phase transition curve of single
shell vesicles from DPL with a radius of 55 nm measured at
360 nm, 370 nm and 400 nm demonstrating the wavelength dependence
of turbidity.

Kinetic measurements - In a previous publication[1a] we reported
a very fast relaxation signal caused by a laser-T-jump of 1.3 ns
with a relaxation time of 4 ns, called τ_1. The very important experimental observation about this signal was that it showed an increasing turbidity, in contrast to the equilibrium measurements,
around T_m. In static experiments turbidity always decreases with
increasing temperature. Furthermore the amplitude as well as the
relaxation time showed an extremely weak cooperativity (less than
20% change) which could not be distinguished from the uncertainty
of these very difficult measurements.

We have identified this very fast relaxation time τ_1 with the
formation of rotational isomers (kinks or gauches) in the hydrocarbon chains of the phospholipid-monomers forming single shelled
vesicles. DML and DPL showed no differences. Recent experiments
using an NMR technique confirmed our results as well as the interpretation[10].

In the following part we concentrate on relaxation phenomena
occuring in the time range from 10 ns to 10 ms at temperatures
around T_m.

The experimentally achieved turbidity time signals shown in
Figure 9a covers the range between 10 ns and 2 μs. The first very
fast increase of turbidity is not time resolved (τ_1 = 4 ns discussed before) because the temperature jump applied here had a
rise time of ∿70 ns (Figure 5 "free run mode"). The second relaxation signal with opposite sign (decreasing turbidity) is well
time resolved. We could not detect a marked cooperativity neither
for the relaxation time nor its amplitude. But we found different
relaxation times for vesicles of different phospholipids with a
radius of ∿50 nm: DPL, τ_2 = 420 ns; DML, τ_2 = 325 ns and DPS,
τ_2 = 87 ns. The accuracy of the results is + 15%. No change in τ_2
with the radius of the vesicles, outside the accuracy of the
measurements was observed. (DML radius ∿25 nm, τ_2 = 313 ns).

Figure 9b covers the time range from 2 μs to 10 ms. The first
two very fast signals τ_1 and τ_2 are not seen at the time resolution used here because 2 μs would only correspond to 1 point on
the target of the 7912 AD transient digitizer. Three single relaxation times τ_3, τ_4 and τ_5 can be separated from Figure 9b (during
the actual measurements we used three different sweep rates to
achieve the best time-window for each signal). In Figure 10 we
have summarized the temperature dependence of the relaxation times
τ_3, τ_4 and τ_5 together with the corresponding amplitudes for small
vesicles of DML in the temperature range around T_m. The signals
3, 4 and 5 show strong cooperativity at the midpoint of the main
phase transition, in the relaxation times τ as well as in their relaxation amplitudes A. For signals 4 and 5 the cooperativity seems
higher than for signal 3.

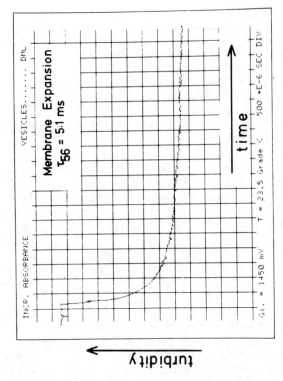

Figure 9b. Relaxation trace of single shell vesicles of DML in the time range from 2 μs to 10 ms.

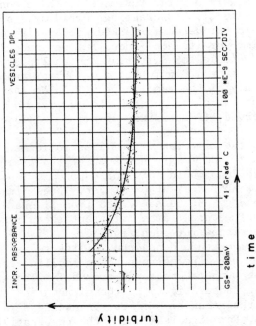

Figure 9a. Relaxation trace of single shell vesicles of DPL in the time range from 10 ns to 2 μs showing two signals of opposite sign.

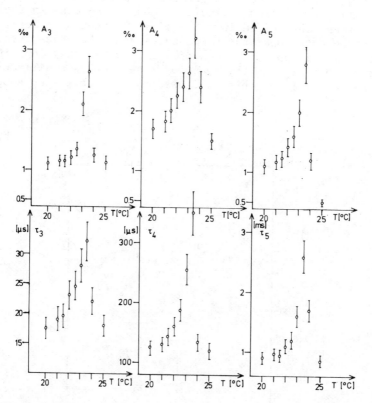

Figure 10. Temperature dependence of the relaxation times τ, and their amplitudes A of signals 3, 4 and 5 for single shell vesicles of DML with a radius of 30 nm demonstrating cooperativity at the main phase transition temperature of 23.5°C.

If we increase the radius R of the vesicles from 30 nm to
60 nm the differences in cooperativity between τ_3 and τ_4 or τ_5 are
more pronounced. This is demonstrated in Figure 11, here log $1/\tau$
is plotted against $1/T$. The strongest cooperativity is observed for
τ_4 followed by τ_5, with τ_3 showing the smallest change. τ_3 for
larger vesicles of DML ($R \sim 60$ nm) is faster than for smaller ones.
This may be due to the stronger curvature of the smaller vesicles
causing a more rigid packing of their hydrocarbon chain ends in the
bilayer of the membrane. The greater cooperativity of τ_4 and τ_5
for the larger vesicles is indicated by an increase in their relaxation times. The change in slope of the Arrhenius plots in Figure
11 at different temperatures indicates that the phase transition
is not symmetric on both sides of T_m. Below T_m the slope is smaller
than that above T_m and the reaction rate becomes smaller with increasing temperature; this is caused by the increasing cooperativi-

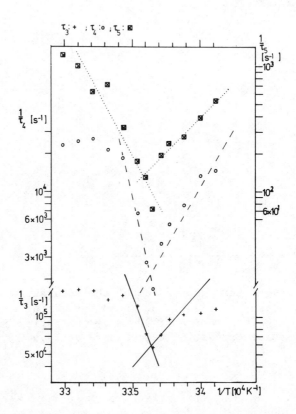

Figure 11. Dependence of the logarithms of $1/\tau$ on the reciprocal
temperature for the signals 3,4 and 5 from vesicles of DML with a
radius of 60 nm around their main phase transition temperature
(T_m) of 23.5°C.

ty. After τ has reached its maximum at T_m the slope of the Arrhenius plot changes its sign and the reaction rate increases with increasing temperature. At values of T much higher than T_m, all three relaxation signals show a very weak temperature dependence.

In Figure 12 we plot the temperature dependence of the relaxation times and amplitudes for the incorporation of cholesterol (20% in moles) into vesicles of DML. The corresponding equilibrium measurements are given in Figure 7.

Incorporation of cholesterol leads to a wide broadening of the melting curve and causes a wide temperature range where cooperativity occurs. The relaxation time τ_3 is faster than for pure vesicles and reaches relaxation times at temperatures 10°C above T_m which are as fast as the temperature jump (2.4 μs) applied during these experiments. Faster T-jumps indicated a relaxation time of approximately 500 ns for temperatures higher than 27°C.

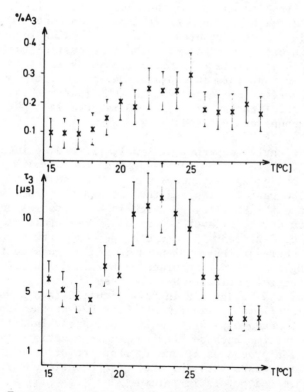

Figure 12. Temperature dependence of the relaxation time τ_3 and its amplitude A_3 for single shell vesicles of DML containing 20% of cholesterol, radius $\sim$40 nm, showing cooperativity.

DISCUSSION

The final aim of kinetic measurements always is to design a molecular model for the dynamic phenomena observed. In the case of our measurements we observed five discrete relaxation times between 1 ns and 10 ms. This covers seven orders of magnitude, therefore it was necessary to summarize all dynamic processes which are known to occur in phospholipid (PL) bilayers of vesicles. Figure 13 shows such a summary in the time range from 10^{-13} s to 10^4 s. The two fastest signals τ_1 and τ_2 we have observed are in this figure together with their molecular interpretation. For the three slower signals τ_3, τ_4 and τ_5 no schematic molecular models are given, because these are developed in the following discussion. Cooperativity used in this paper means that a molecule which undergoes dynamic changes facilitates the same process for all neighboring molecules: the stronger the cooperativity, the more molecules are influenced in the direction given by the first one. The fastest kinetic process with a relaxation time of 4 ns has already been attributed to the formation of kinks[1] and will not be discussed further. The second noncooperative relaxation time is clearly caused by a very fast first expansion of the phospholipid bilayer compensating the increase of the optical density and the decrease of volume occupied from PL molecules and caused by kink formation. Its molecular interpretation seems to be a lateral diffusion process of single phospholipid (PL) molecules which is associated with headgroup motion. This interpretation is confirmed by the rate-decreasing effect of longer chains (DML faster than DPL) and the pronounced acceleration of τ_2 if DPS is used in a pH-range in which the headgroups are charged.[2]

The first clearly cooperative signal, τ_3, can be interpreted as the motion of whole chains forming gauche[3] conformations and changing their headgroup distances. This is backed by the observed dependence on the vesicle size and the influence of cholesterol incorporated into the PL-bilayer[20] which is known to reside in the hydrophobic part of the bilayer[20] and should therefore uncouple the chain ends of different PL-molecules belonging to different sides of the PL-bilayer (PL-monolayers inside or outside the vesicles forming the bilayer-membrane). This leads to less cooperativity – faster relaxation times – as seen in Figure 12. The participation of the headgroups in this signal although weak was concluded from preliminary experiments with DPS at the isoelectric point. Uncharged headgroups with smaller distances lead to a slower relaxation time because of stronger interaction. The two slowest relaxation processes τ_4 and τ_5 show the strongest cooperativity which increases with the size of the vesicles. These signals were observed before in T-jump[7,22,25] as well as in pressure jump-experiments[1b,8]

Relaxation Time s	Dynamic Process	Technique	Schematic Model
$\sim 10^{-13}$	Vibrations Rotations of Single Bonds	IR	
$\sim 10^{-10} \underline{\ } 10^{-13}$	Complex Vibrations and Rotations of Single Bonds and Chain-Parts	Dynamic Neutron Scattering	
$\sim 10^{-10}$	Orientation of Segments of Chains	NMR Order Parameter	
$\sim 10^{-9} - 10^{-8}$	Rotional - Isomers (Kinks)	Laser-T-Jump Fluorescence - Polarization Life-time	
4×10^{-9}	Kinks not coop.	Laser-T-Jump	
$\sim 10^{-7} - 10^{-6}$	Lateral Diffusion of Single PL-Molecu-les„hopping"	ESR Laser-T-Jump	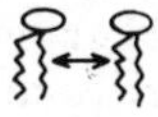
10^{-5}	cooperative	Laser-T-Jump	
$\sim 2 \times 10^{-4}$	very weak $T > T_{MP}$ strongly coop.	Laser-T-Jump	
$\sim 10^{-3} - 10^{-2}$	H_2O Diffusion through PL-BL	Swelling Fluorescence Stopped-Flow	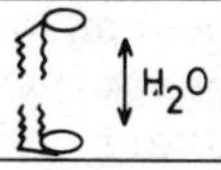 H_2O
$\sim 3 \times 10^{-3}$	strongly coop.	Laser-T-Jump	
$\sim 10^{+1} - 10^{+4}$	Transversal Diffusion of Single PL-Molecu-les„flip-flop"	ESR	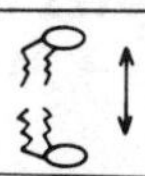
$\sim 10^4$	Exchange of PL between different Vesicles	Change of Phasetransition Temperature	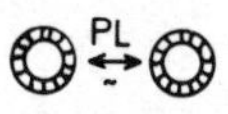 PL

Figure 13. Summary of dynamic phenomena in bilayers of phospholipids in the time range from 10^{-13} s to 10^4 s.

Kanehisa and Tsong[2] interpreted these signals using a cluster
model. We agree with their interpretation in so far as τ_4 can be
associated with the concerted formation of gauche conformations accompanied by further disordering of the headgroups in areas around
molecules of higher order (clusters). Increasing cooperativity
with increasing size of the vesicles (less curvature of the bilayer)
allows for more molecules being involved in this process. The
slowest relaxation time τ_5 might be associated with the complete
melting of the ordered clusters, formed in the time range of τ_4,
leaving the membrane in a completely disordered state with respect
to the headgroups as well as to the conformation of the hydrocarbon
chains which now carry kinks and gauche conformations. The volume
changes caused by the kink formation and the resultant expansion
of the membrane is schematically shown in Ref. 1a. The sum of these
changes result in a total $\Delta V = 25.6$ cm^3/mol which agrees very well
with the equilibrium values given in the literature[23,24]

CONCLUSION

The dynamic phenomena associated with the main phase transition in unilamellar vesicles of phospholipids give rise to five
well separated relaxation times. This result justifies the interpretation of the signals in terms of different molecular processes.
The formation of kinks followed by a fast first membrane expansion,
which produces more space for cis-trans isomers in the hydrocarbon chains and starts to disorder the headgroups, are the two
fastest non-cooperative dynamic events. In a first cooperative
process whole molecules are moving together with their neighbours
forming gauche conformations in areas near kinks. In the second
strongly cooperative process, the disordered areas grow in such a
way that clusters containing molecules of higher order surrounded
by disordered molecules are dissolved along their borderline.
In the slowest but also cooperative process, the ordered clusters are disappearing leaving the PL-membrane in a completely
disordered state. This is schematically demonstrated in Figure 14.

ACKNOWLEDGEMENT

We would like to thank Mrs. I. Henschke from the Freie Uni-
versität Berlin, Inst. Anorg. Chemie, for preparing the electron-
micrographs, Mrs. M. Rokosch for drawing the figures and the
Deutsche Forschungsgemeinschaft for a research grant.

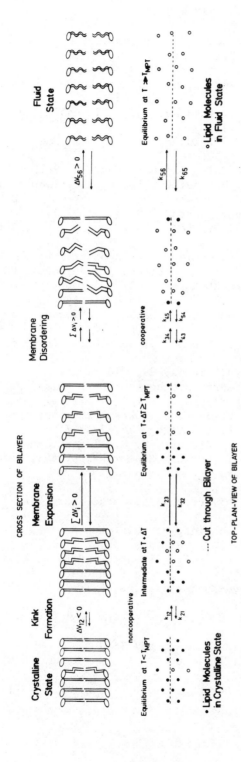

Figure 14. Schematic molecular model of the dynamic changes associated with the five descrete relaxation signals measured with the Iodine laser T-jump in the temperature range of the main phase transition of single shell vesicles of phospholipids. ● molecules in a hexagonal crystalline state; o molecules in a disordered fluid state; ◯ headgroups; // all trans hydrocarbon double chains; ⌇ kinks; ⟨⟨ gauches; and ⟨⟨ completely disordered chains.

REFERENCES

1a) B. Gruenewald, W. Frisch and J. F. Holzwarth, Biochem.
    Biophys. Acta, 641, 311 (1981).

  b) J. F. Holzwarth, W. Frisch and B. Gruenewald, in "Microemul-
    sions", I. D. Robb, Editor, p. 185, Plenum Pub. Corp., New
    York, 1982.

  c) T. Sano, J. Tanaka, T. Yasunaga and Y. Toyoshima, J. Phys.
    Chem., 86, 3013 (1982),

2) M. I. Kanehisa and T. Y. Tsong, J. Am. Chem. Soc., 100, 424
    (1978).

3) P. Overath and H. Träuble, Biochemistry, 12, 2625 (1973).

4a) J. C. Reinert and J. M. Steim, Science, 168, 1580 (1970).

  b) D. L. Melchior, H. J. Morowitz, J. M. Sturtevant and T. Y.
    Tsong, Biochem. Biophys. Acta, 219, 114 (1970).

5) D. L. Melchior and J. M. Steim, Ann. Rev. Biophys. Bioeng.,
    5, 205 (1976).

6) J. F. Holzwarth, in "Techniques and Applications of Fast Re-
    actions in Solution", W. J. Gettins, E. Wyn-Jones, Editors,
    p. 47, Reidel, Dordrecht-Boston, 1979.

7) T. Y. Tsong and M. I. Kanehisa, Biochemistry, 16, 2674 (1977).

8) B. Gruenewald, A. Blume and F. Watanabe, Biochem. Biophys.
    Acta, 597, 41 (1980).

9) R. C. Gamble and P. R. Schimmel, Proc. Natl. Acad. Sci. USA,
    75, 3011 (1978).

10) O. Edholm, Chem. Phys. Lipids, 29, 213 (1981).

11) S. Batzri and E. D. Korn, Biochem. Biophys. Acta, 298, 1015
    (1973).

12) J. M. H. Kremer, M. W. J. v. d. Esker, C. Pathmamanocharan
    and P. H. Wiersema, Biochemistry, 16, 3932 (1977).

13a) R. M. C. Dawson, D. C. Elliott, W. H. Elliott, and K. M. Jones,
    Editors, "Data for Biochemical Research", p. 494, Oxford,
    Clarendon Press, 1969.

  b) ibid., p. 489.

14) H. Träuble and H. Eibl, Proc. Natl. Acad. Sci. USA, 71, 214
    (1974).

15) J. F. Holzwarth, A. Schmidt, H. Wolff and R. Volk, J. Phys.
    Chem., 81, 2300 (1977).

16) W. Frisch, A. Schmidt, J. F. Holzwarth and R. Volk, in
    "Techniques and Applications of Fast Reactions in Solution",
    W. Gettins and E. Wyn-Jones, Editors, Reidel, Dordrecht-
    Boston, p. 61, 1979.

17) D. M. Goodall and R. C. Greenhow, Chem. Phys. Lett., 9, 583
    (1971).

18) G. Beck, Rev. Sci. Instr., 47, 537 (1976).

19) D. Chapman, R. M. Williams and B. D. Ladbrooke, Chem. Phys.
    Lipids, 1, 445 (1967).

20) L. Powers and P. S. Pershan, Biophys. J., 20, 137 (1977).

21) E. Bicknell-Brown and K. G. Brown, Biochem. Biophys. Res. Commun., 94, 638 (1980).
22) T. Y. Tsong, Proc. Natl. Acad. Sci. USA, 71, 2684 (1974).
23) N. I. Liu and R. L. Kay, Biochemistry, 16, 3484 (1977).
24) K. R. Srinivasan, R. L. Kay and J. F. Nagle, Biochemistry, 13, 3494 (1974).
25) B. Gruenewald, Biochem. Biophys. Acta, 647, 71 (1982).

DYNAMIC LIGHT SCATTERING STUDY OF DMPC VESICLES COAGULATION

AROUND THE PHASE TRANSITION OF THE ALIPHATIC CHAINS

D. Sornette and N. Ostrowsky

Laboratoire de Physique de la Matière Condensée
(CNRS : LA 190), Université de Nice
Parc Valrose, 06034 Nice Cedex, France

Monodisperse suspensions of DMPC vesicles can be obtained by sonication and ultracentrifugation above the phase transition temperature $T_\emptyset$ of the chains, forming small single spherical bilayers of a few hundreds Angströms in diameter, and stable at low concentration.

Upon rapid cooling to below $T_\emptyset$, a coagulation phenomenon is triggered, which we have followed experimentally as a function of time by measuring the intensity as well as the spectrum of the light scattered by the suspension. From these data, we can extract the intensity distribution of the scattering particles, which in turn yields their size distribution and thus the time evolution of the coagulation process for different fixed temperatures T below $T_\emptyset$.

The data can be interpreted quantitatively with a simple model which assumes that below $T_\emptyset$ vesicles undergo a diffusion controlled aggregation, partially reversible for the smallest ones (i.e. activation controlled) and which eventually leads to fusion between vesicles. The physical justification of this model emphasizes the importance of the short range repulsive forces between two bilayers which strongly depend on their physical state (liquid- or gel-like), and therefore on the size of the vesicles and temperature.

INTRODUCTION

Due to their amphiphilic nature, phospholipid molecules spontaneously form organized structures when put into water with their polar heads shielding the aliphatic chains from the water.

At low lipid concentration, small single lamellar structures[1], called vesicles, can be formed using sonication, at a temperature above $T_\emptyset$, the temperature of the first order chain melting transition.

The stability of these structures when cooled below $T_\emptyset$ can provide useful information on the nature, the cooperativity as well as the size and curvature dependence of the liquid-gel chain transition. Several experiments have been performed on phospholipid vesicles of different carbon chain lengths[2-6]. In this paper we present a study of the kinetics of the size growth of sonicated dimyristoyl phosphatidylcholine (DMPC) vesicles when they are rapidly cooled at a fixed temperature T , below $T_\emptyset$[7] . We shall first summarize (§I) the experimental procedure and the method of data analysis we have derived[8]. We then argue (§II) that the observed growth of the vesicles is caused by a fusion phenomenon which occurs via a reversible aggregation step. Based on a discussion of vesicles interactions, a phenomenological model for the coagulation kinetics will be introduced and compared to the experimental results.

I. EXPERIMENTAL METHOD AND RESULTS - DATA ANALYSIS

The time evolution of the size distribution was studied by the nonperturbative light scattering technique, measuring both the total scattered intensity and its spectrum with a standard heterodyne technique. The correlogram thus obtained is simply the first order correlation function of the scattered electric field which,for a distribution of particles sizes, has the form :

$$g(t) = \int_0^\infty \exp(-D(R)q^2) t \quad G(R) \quad dR \qquad (1)$$

where $G(R)dR = n(R)I(R)dR$ is the intensity scattered by $n(R)dR$ particles having a radius in the interval $[R , R+dR]$. The scattering fluctuation wave vector q is given by :

$$q = (4\pi / \lambda) \sin \theta /2$$

where θ is the scattering angle, and λ the wavelength of the incident light in the medium. For spherical particles in dilute solution, as is the case in our study, the translationnal diffusion coefficient is given by :

$$D(R) = kT/(6 \pi \eta R)$$

where η is taken to be equal to the viscosity of water.

The data analysis procedure consists of inverting Equation (1) in order to get the distribution function G(R) from the measured correlation function g(t). It is well known [10] that this problem is ill-conditionned : a small noise in the data g(t) may indeed greatly alter the recovery of the distribution function G(R). In order to alleviate this problem, we have derived the so-called "Exponential Sampling Method" [8] which has the advantage of quantifying the maximum amount of information one can extract from the experimental correlation function for a given noise level. However, it can be shown that the problem of determining the cumulative distribution function :

$$F(R) = \int_0^R G(R') \, dR' \tag{2}$$

while imposing the physical constraint that F(R) is non-decreasing, which amounts to saying that G(R) is non-negative, is well-conditionned. We have actually experimentally verified that F(R) is much less sensitive to noise that G(R) and since the cumulative distribution function gives a more practical and reliable representation of the size distribution for comparison between experimental and theoretical data, we present our results in this form. Figure 1 gives examples of the correspondance between G(R) and F(R).

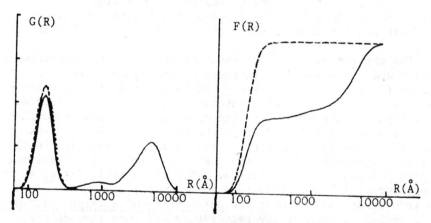

Figure 1. Examples of the correspondence between the intensity distribution function G(R) and its cumulative form F(R). Dashed line : unimodal distribution function. Solid line : large size distribution function typical of the ones obtained in our experiments.

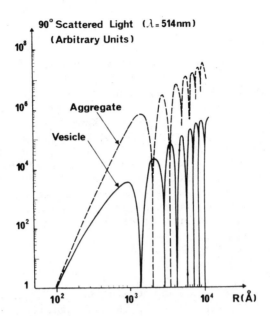

Figure 2. 90° scattered intensity by vesicles, i.e. hollow spheres (solid line), and by roughly spherical aggregates made out of small vesicles, i.e. roughly equivalent to full spheres (dashed lines).

Note from Equation (1) that the comparison between the experimental results G(R) and any model one can build up for n(R) needs the computation of I(R), the intensity scattered by a single particle of mean radius R. This function depends in a crucial way on the shape and inner constitution of the particles, as shown in Figure (2).

The experimental results obtained for the total scattered intensity and the cumulative intensity distributions are shown in Figures (3) and (4) respectively and have the following characteristics. The total scattered intensity, measured at a fixed (90°) angle rapidly increases after the cooling of the sample, with a time scale strongly dependent on the temperature T of the experiment. The lower the temperature, the faster the increase in the scattered intensity. As for the cumulative intensity distribution, it also follows a time evolution, which is markedly faster as the temperature T in this study is further down below $T_\emptyset$. The results do show however the persistence of a large fraction of small vesicles, even at low temperature and long time delays. These features are in qualitative agreement with the previous results on the longer chains vesicles[2,3,6] which have not been accounted for theoretically. We shall outline in the next Section the model we have been working on, based on a discussion of the vesicles interactions around the phase transition of the phospholipids[12,13].

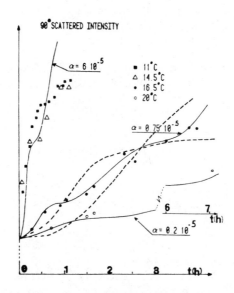

Figure 3. Time evolution of the total scattered intensity at 90°, for vesicle's solutions at very low lipid concentration (10^{-5}M/l). The solid lines represent the theoretical predictions from the model described in §II, for different temperatures (i.e. different parameters α). The time parameter $t_c \simeq 0.6$s was fixed in all the experiments with the 16.5°C data. The α values found by fitting the intensity data for each temperature T were used to fit the data in Figure (4). The dashed lines around the 16.5°C data represent the fits obtained with the aggregation model for different values of α, optimizing in each case the time parameter t_c.

II. MODEL FOR THE COAGULATION KINETICS

The purpose of this Section is to present a simple phenomeno-logical model which explains the observed coagulation kinetics of the vesicules cooled at a fixed temperature T below T_ϕ.

We consider an initial population of small identical vesicles (monomers) at low concentration N_o such that the average distance between vesicles is two orders of magnitude greater than their radii. The fact that the actual initial population is not mono-sized is discussed elsewhere [14], but does not change appreciably the predictions of the model. We shall call i-mer a particle for-med by the coagulation of i monomers.

A particle of radius R suspended in such low concentration undergoes most of the time a Brownian motion well described by the

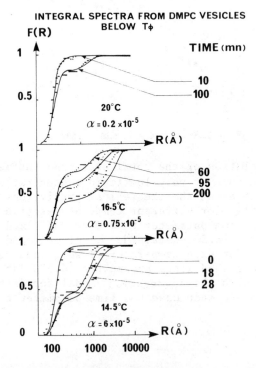

Figure 4. Cumulative intensity distribution obtained for different temperatures as a function of time. The parameters α and t_c being determined by the intensity measurement, no adjustable parameters were used to fit the data (broken lines) on the theoretically pre-dicted curves (dotted lines).

free diffusion coefficient D(R). When the distance between the two particles becomes small a variety of phenomena may be triggered. We make the hypothesis that the observed coagulation process can be described by a mechanism involving essentially two steps, discussed below.

II.1 Diffusion Controlled Approach

It can be shown[14] that the hydrodynamic repulsive force (which physically comes from the fact that two approaching particles must push apart the solvent from their collision path) almost balance the attractive van der Waals interaction, thus giving a negligible correction to the collision rate calculated for pure Brownian particles. For a collision to become effective and lead to an aggregate requires that the two approaching particles have a sufficient amount of initial kinetic energy for the aggregation to prevail against the short range repulsive forces. The close approach of two particles is thus essentially hindered by the tail of the short range repulsive interaction which appears when the distance between the bilayers is around 50 Å or less[15]. At the present stage of our knowledge, we shall assume this tail to be practically independent of temperature and of the particles sizes(*) . This hypothesis leads to a temperature and size independent mean time of aggregation, i.e. effective collision time :

$$t_c = t_o \exp(V_c/kT) \quad \text{with} \quad t_o = 3\pi\eta / (8kTN_o)$$

where t_o is the mean time between collisions for pure Brownian particles (on the order of 3ms in our experiments), and V_c is the height of the potential barrier that must be overcome by the two aggregating particles.

Note that above the phase transition, when the bilayer elasticity increases, an additionnal important repulsive force called the undulatory force greatly enhances the barrier V_c. As a result, the probability for vesicles aggregation above T_ϕ is markedly reduced.

II.2 Time Evolution of an Aggregate

Once an aggregate is formed, it will eventually either break up (§II.2a) or undergo a fusion phenomenon (§II.2b). We shall examine these two possibilities in turn.

(*) Although we emphasize below that this short range repulsive force strongly depends on the structure of the bilayer, one may argue that this is no longer true when the separation between the bilayers becomes large compared to the average distance between adjacent phospholipids.

II.2a <u>The stability of an aggregate</u>. It depends on the gain
in free energy realized from the close contact of two vesicles as
a result of a positive balance between the van der Waals attractive force[16] and the short range repulsive forces of two bilayers
carrying arrays of permanent dipole moments. The present theoretical comprehension of these short range repulsive forces[17-19] leads
to the general statement that these forces, which favour disaggre-
gration, are much weaker and break down at smaller distances when
the bilayers are more into the gel state. As the width and temperature of the phase transition is strongly dependent on the size
of the vesicles[20], we can infer the following qualitative features.
At a given temperature, the repulsive forces will be most effective for the smallest vesicles which are not yet completely into
the gel state. In a similar way, and for a given vesicles'size,
these forces become weaker as the temperature is lowered further.
As a result, aggregates composed of two small vesicles will be
very unstable compared to aggregates formed of two larger ones. In
addition, the stability of all aggregates will be enhanced as the
temperature T is lowered. We thus introduce the probability per
unit time $1/t(i,j)$ that an aggregate (i) (j) formed of an i-mer
and a j-mer breaks up, and postulate the following dependence of
$t(i,j)$ on i and j :

$$1/t(i,j) = 1/t_a \qquad \text{if i or j} = 1$$

$$= 0 \qquad \text{if i and j} \neq 1$$

This sharp simplifying size dependence reproduces qualitatively
the instability of small vesicles'aggregates and the greater, quasi
infinite, stability of the larger ones. Note that a very strong
size dependence has also been found for the thermodynamic properties of the gel-liquid phase transition of vesicles. For example
the phase transition temperature increases by 4 degrees as the
vesicles'radii vary from 120 Å to 350 Å (20). As we mentioned
above, we expect this time constant t_a to be strongly temperature
dependent, increasing markedly as the temperature T is lowered.

II.2b <u>The fusion phenomenon</u>. It is taken phenomenologically
into account by defining the rate coefficient for the process involving the actual fusion of the bilayers of an i-mer and a j-mer ,
which were already in the aggregate (i)(j) form. The potential
barrier V_f between the aggregated state and the fused state may
depend on i and j, but not necessarily in a crucial way as the
life-time of an aggregate does. For the sake of simplicity and as
we lack theoretical knowledge on this point, we shall assume this

rate coefficient to be the same for all aggregates (i)(j), and call it $1/t_f$. Note however that our experimental results suggest that this is no longer true for very large i's and j's (i+j$\gtrsim$1000) as we shall mention in the discussion below.

II.3 The Kinetics Equation

With these simplifying assumptions, we may now write down the equations governing the temporal evolution of the concentration of the i-mers :

$$N_o \frac{dn_i^*}{dt} = \frac{n_{i-1}n_1}{t_c} - N_o n_i^* \left(\frac{1}{t_f} + \frac{1}{t_a}\right) \qquad (3)$$

$$N_o \frac{dn_i}{dt} = \frac{1}{2} \sum_{j=1}^{i-1} \frac{n_j n_{i-j}}{t_c} + \frac{N_o n_i^*}{t_f} + \frac{N_o n_{i+1}^*}{t_a} - n_i \sum_{j=1}^{\infty} \frac{n_j}{t_c} \qquad (4)$$

where n_i is the concentration of an i-mer and n_i^* the concentration of an aggregate (i-1)(1) formed with an (i-1)-mer and a monomer.

Note that this set of equations neglects the concentration of aggregates composed of more than one monomer. This is easily justified since as shall be seen from our experimental results $t_a \ll t_c$. This means that an aggregate (i)(1) does not survive long enough for a second monomer to aggregate on top.

Equation (3) shows how an aggregate (i-1)(1) is formed by the "effective" collision of an (i-1)-mer and a monomer, and has an effective life-time :

$$1/t_m = 1/t_a + 1/t_f$$

reflecting that it can disappear either by disaggregation or by fusion.

Equation (4) implies that an i-mer is formed by all the effective collisions between a j-mer and an (i-j)-mer (where both j and i-j are greater than 1) since they all lead to stable aggregates which sooner or later will end up as a fused i-mer. It may also be formed by the fusion of the aggregate (i-1)(1), with the probability $1/t_f$, as well as by the breaking of the aggregate (i)(1), with the probability $1/t_a$. The disappearance of the i-mer will occur upon any effective collision with all particles in suspension in the solution. In this last term, we shall neglect the presence of unstable aggregates, since we always have $n_i^* \ll n_i$.

Equasions (3) and (4) are valid for all i-mers, except for the monomers which obey the following simpler kinetic equation :

$$N_o \frac{dn_1}{dt} = - \frac{n_1 \sum_{j=1}^{\infty} n_j}{t_c} + \frac{N_o (2n_2^* + \sum_{j=3}^{\infty} n_j^*)}{t_a} \tag{5}$$

In order to solve this system of equations, we made the commonly used hypothesis of pseudo-steady state for the very unstable aggregate n_i^* , which implies that :

$$n_i^* = \frac{t_m n_{i-1} n_1}{t_c N_o} \tag{6}$$

We then get a set of equations which only depend on two adjustable parameters :
 - the effective collision time t_c, which is supposed to be temperature independent,
 - a dimensionless parameter $\alpha = t_m/t_f \approx t_a/t_f$, which, since t_a and possibly t_f depend on the microscopic structure of the bilayer, is expected to be temperature dependent.

This set of equations was solved numerically by incrementing the time, and following step by step the concentration of all i-mers (for i $\leqslant$ 1000)[14]. For larger i's, an asymptotical analytical solution was used. From these results, we computed the time dependent intensity distribution function G(R) and its cumulative form F(R), as well as the total 90° scattered intensity.

Comparison with the experimental data for several fixed temperatures T below T_ϕ are shown in Figures (3) and (4). Let us emphasize that the fits presented in these curves depend only on the two parameters t_c and α , defined above. These two parameters could easily be extrated from the total intensity measurements, leaving thus no adjustable parameters for the intensity distribution fits. We thus find a good agreement between the experimental data and the model discussed above.

The temperature dependence of the parameter α , which increases roughly by a factor of 30 when T goes from 20°C to 14.5°C, was to be expected from the above discussion. As the temperature is lowered, the bilayers go more into the gel state, thus weakening the short range repulsive forces and favouring the formation of stable aggregates, leading eventually to fusion with a characteristic time t_f. Our experiments only allow us to determine, in addition to $t_c \approx 0.6s$ the ratio $\alpha = t_a/t_f$. Nevertheless, the theoretical lines shown in Figures (3) and (4) were computed assuming that practically all the i-mers were in a <u>fused state</u>, i.e. that the size of an i-mer was proportional to the square-root of i and that it scattered light according to the solid line of Figure (2). This implies that t_f is at most of the order of a minute, which is much shorter than the fusion time found for C_{16} chains[2-3]. Note that with the condition $t_f \lesssim 60s$ and $\alpha \approx 10^{-5}$ we find for t_a an upper limit of the order of $10^{-3}s$ which is consistent with the approximation used in the resolution of the system of Equations (3) and (5).

A detailed comparison of the data with predictions computed with the opposite assumption (the i-mers still in the aggregated state) has been worked out but does not give a consistent fit as can be seen from the dashed curves on Figure (3). This must be expected since a simple argument shows from Equation (4) that the radius of the largest particles existing in the suspension at time t obeys the power law :

$$R_{max} \propto t \qquad \text{for the fusion case}$$

$$R_{max} \propto t^{2/3} \qquad \text{for the aggregation case}$$

This indicates that at long times ($t \gg t_c$) the predicted distribution function would not extend towards large enough sizes to be able to account for the experimental data, as we have checked on the intensity distribution curves.

We must however mention that at still longer times, the predicted kinetics seems to move away from the experimental data. In particular, the total scattered intensity does not increase as fast as expected and seems to reach a plateau. Similarly, the time evolution of the intensity distribution lags behind the theoretical prediction. This suggests that the fusion phenomenon becomes less probable for very large vesicles, as has been observed for giant vesicles[21]. It can be shown indeed that for large i's the intensity scattered by an aggregate may be smaller than that scattered by the corresponding vesicle (i.e. containing the same amount of matter[22]).

In conclusion, let us emphasize again the fact that the results presented above give a new and indirect experimental support for the theoretical prediction that short range repulsive forces between zwitterionic bilayers strongly depend on their microscopic structure, and in particular on the ordering of the polar heads on each surface . This conclusion is indeed supported by the fact that the rate of size growth changes faster in the pretransition temperature range (17-13°C) than at the main transition ($\sim$ 21°C). This should encourage more theoretical as well as experimental work in order to further elucidate the correct sources and characteristics of these repulsive forces. In particular, it would be of interest to weigh the comparative importance of the water orientational mechanism[17] to the hydration forces concepts.

REFERENCES

1. C. Huang, Biochem., 8, 344 (1969)
2. S.E. Schullery, C.F. Schmidt, P.L. Felgner, T.W. Tillack and T.E. Thompson, Biochem. 19, 3919 (1980)
3. C.F. Schmidt, D. Lichtenberg and T.E. Thompson, Biochem. 20, 4792 (1981)
4. J. Suurkuusk, B.R. Lentz, Y. Barenholz, R.L. Biltonen and T.E. Thompson, Biochem, 15, 1393 (1976)
5. H.L. Kantor, S. Mabrey, J.H. Prestegard, and J.M. Sturtevant, Biochem. Biophys. Acta 466, 402 (1977)
6. A.L. Larrabee, Biochem. 18, 3321 (1979)
7. D. Sornette, C. Hesse-Bezot and N. Ostrowsky, Biochimie 63 955 (1981)
8. N. Ostrowsky, D. Sornette, P. Parker and E.R. Pike, Optica Acta 28, 8, 1059 (1981)
9. For a recent review of the light scattering technique and its application, see for example :
   "Scattering Techniques Applied to Supramolecular and Non-Equilibrium Systems", S.H. Chen, B. Chu and R. Nossal, Editors, Plenum Press, New-York, 1981
10. J. Mc Whirter and E.R. Pike, J. Phys. A, 11, 1729 (1978)
11. A.V. Goncarskii and A.G. Jagola, Soviet Math. Dokl. 10, 1 (1969)
12. J.N. Israelachvili and B.W. Ninham, J. Colloid Interface Sci. 58, 1 (1977)
13. B.W. Ninham, J. Phys. Chem. 84, 1423 (1980)
14. D. Sornette and N. Ostrowsky, to be published
15. V.A. Parsegian, N. Fuller and R.P. Rand, Proc. 1, Natl. Acad. Sci. USA 76, 6, 2750 (1979)
16. For a review of Van der Waals interactions and their relevance to vesicles' systems, see for example : S. Nir, Prog. Surf.Sci. 8, 1, 1 (1976)
17. S. Marcelja and N. Radic, Chem. Phys. Lett., 42, 1 (1976)
18. B. Jönsson, private communication.
19. W. Helfrich, Z. Naturforst., 33a, 305 (1978)
20. D. Lichtenberg, E. Freire, C.F. Schmidt, Y. Barenholz, P.L. Felgner and T.E. Thompson, Biochem., 20, 3462 (1981)
21. W. Helfrich, private communication (1981)
22. D. Sornette, unpublished results (1982)

THE EFFECT OF GINSENG SAPONINS ON BIOCHEMICAL REACTIONS

Chung No Joo

Department of Biochemistry
College of Science, Yonsei University
Seoul 120, Korea

Saponins are plant glycosides which have long been known to lower the surface tension of water and therefore their aqueous solutions froth readily. These studies of saponins from various sources, however, show that they behave differently in biological systems. Some are toxic but others are not; some are hemolytic while the others are not. It has been reported that the saponins from Korean ginseng roots (Panax gineseng C. A. Meyer) are mainly triterpenoidal dammarane glycosides which showed no hemolytic activity but were good solubilizers of nonpolar lipids such as triglycerides and cholesterol. Moderate amounts of the saponin were found to stimulate several enzyme reactions in vitro so far examined in this laboratory suggesting that the surface activity of the saponin might give rise to better condition for the reactions to proceed. The saponin was also found to stimulate intermediary metabolism in the animal body. The protective effect of the saponin from alcohol intoxication in liver and its preventive effect against aortic atheroma formation in rabbits fed cholesterol for a long time were demonstrated. Clinical application of the saponin in patients with acute viral (B type) hepatitis showed that the saponin has the preventive effect against the development of the disease. It seems likely that the effect of ginseng saponins might be explained, at least partly, by their suface activity. Ideal amounts of physiologically nontoxic saponins might play a significant role in metabolism.

INTRODUCTION

Saponins, some of which cause hemolysis, are plant glycosides which lower the surface tension of water and therefore their aqueous solutions froth readily. It is easily understood from their structure that they are amphipathic having both the hydrophobic sapogenin part and the hydrophilic sugar part in the molecule. Therefore they disperse lipid in aqueous medium.

The action of saponins from various sources, however, differs from each other in biological system. Some are toxic but the others are not; some are hemolytic while the others are protective against hemolysis. It has been reported that the saponins of Korean Panax ginseng C. A. Meyer roots are mainly triterpenoidal dammarane glycosides, and they have no hemolytic activity but rather are protective against hemolysis.[1]

Korean ginseng belongs to the family of Araliaceae and the genus of Panax, and is the root of a perennial plant grown under a shade. It was first scientifically named Panax schinseng Nees by a German botanist in 1833, and later Panax ginseng C. A. Meyer by a Russian botanist in 1843. The latter name which has been generally used, in which 'Pan' means all and 'axos' cure, originated from the Greek. Ginseng was derived from the Chinese name jen-sheng, implying the herb whose roots resemble the human body.

Since World War II, studies of chemical constituents of ginseng have kept pace with the newly developed analytical tools. In surveying many ginseng studies, ginseng is known to contain useful medicinals for human body; among them saponins have been believed to be the most active components of ginseng.

It has been demonstrated that ginseng has a wide range of pharmacological properties, including antifatigue and antistress actions, mild normalizing effects on blood pressure and carbohydrate metabolism, suggesting central nervous system stimulatory properties and an effect on macromolecular synthesis in the liver.[2,3] Brekhman and Dardymov[4] in 1969 mentioned in a review article that the basic effect of ginseng is due to its capacity to increase nonspecific resistance of the organism to various untoward influences.

During the past two decades, physiological and biochemical approaches to elucidate the mechanism of ginseng effect on the body have been intensively made. Oura and Hiai found that aqueous extracts of ginseng stimulated the synthesis of rat liver nuclear RNA in vivo.[5]

They compared the effects of 18 well known tonic medicinals such as radix Rebmanniae (China), Fracus Lycil (China), rhizoma Chidii (Hokkaido), Panax ginseng (Kumsan, Korea) on the incorporation of labelled orotic acid into rat liver nuclear RNA and found that panax ginseng (Kumsan, Korea) was most effective. Their further studies suggested that ginseng saponin might be an active component. They also found that ginseng could stimulate carbohydrate metabolism in the liver and could increase the lipid content of adipose tissue. They considered that the action of ginseng had special features and suggested that ginseng saponin was a kind of metabolic regulator or hormone-like substance.

Joo and his coworkers investigated the solubilizing effect of ginseng saponin by determining the critical micellar concentration (CMC) of the saponin. They found that the CMC of the saponin aqueous solution was about 2.0%, however, when the saponin was present together with cholesterol, the CMC of the saponin was lowered to as little as 0.1%. Also the CMC of Na-Cholate (5mM) was lowered below 1.0mM in the presence of 0.1% saponin. When the saponin was added to chicken intestinal lumen fluids, the intestinal lumen lipids were found to be dispersed effectively.[9,10] Observations of the effect of ginseng saponin on serum lipoproteins, pancreatic lipase and cholesterol esterase supported the idea that saponin might act as an excellent lipid solubilizer.[6-9]

The authors further investigated the effect of saponin on various enzymes such as mitochondrial dehydrogenases and found that moderate amounts of the saponin stimulated the enzyme reactions. They have considered that the mode of the saponin action is not specific and probably the surface activity of the ginseng saponin might play a singnificant role in enzyme catalyzed reactions.[10-17]

They also investigated the distribution of radiolabelled lipids in livers of the rats administered ginseng saponin prior to acetate-1,2-C[14] injection intraperitoneally and found that the saponin stimulated the lipid metabolism and transport in the animal body.[18]

This paper describes the general action of the saponin on enzyme reactions in vitro and on intermediary metabolism in the animal body. The effect of ginseng saponin on alcohol detoxication and some preventive effect of the saponin against aortic atheroma formation in rabbits fed cholesterol for a prolonged time are discussed.

In connection with the results obtained from experiments using animals, a clinical application of the saponin to patients suffering from viral (B type) hepatitis was attempted to understand the liver protective action of the saponin in the human body.

EXPERIMENTAL

Fifteeng of ginseng saponin mixture was obtained from 300 g of powdered Korean white ginseng (Keumsan, 4 years, 50 pieces/300 g) according to the modified procedure described elsewhere.[13] The chromatogram of the saponin showed that it contained several saponins with R_f values of 0.66, 0.59, 0.50, 0.43, 0.33, 0.25 on silica gel thin plate using chloroform-methanol-water (14:6:1, v/v) as a developing solvent. It appeared that the saponin with R_f value of 0.59 was the most abundant, the saponins with R_f values of 0.43, 0.33 were less abundant, and the saponins with R_f values of 0.66, 0.50, 0.25 were the least. The above saponin mixture was used without further purification in this study.

To observe the effect of ginseng saponin on the absorption of water insoluble vitamins such as vit. A and vit. E, Wistar rats were fed ordinary diet, which contained crude protein above 19.6%; crude cellulose below 7.0%; crude ash below 9.0%; Ca below 0.6%; P below 0.4%; crude fats above 3.0%, total digestable nutrient above 73.0% until required. After overnight starvation with free access to water, 12mg of the corresponding vitamin and 50mg of saponin were dissolved in 2ml of 50% ethanol and fed orally by stomach tubing. To control animals, 12mg of the corresponding vitamin dissolved in 2ml of 50% ethanol were fed orally similarly as the test animals. The levels of the vitamins in blood serum, liver and kndney were then monitored at timed intervals. Vitamin A was determined according to Carr-Price[25] and vit. E was assayed by Emmeire-Engel reaction.[20]

Observation on the saponin effect on fat absorption in cannulated rats, in which the secretion of bile and pancreatic juice into the small intestine was interrupted, was made as follows. Wistar rats were fasted for 24 hr, and fed 0.5ml of corn oil and 0.05mg of ginseng saponin per rat orally by stomach tubing. They were then anesthesized by intraperitoneal injection of 0.4ml of pentothal sodium per rat and were cannulated according to the method of Bloom et al.[21] The control group was treated similarly but no ginseng saponin was given. They were killed at specified times (30 min.-9 hrs.) and blood was taken from the heart. Bloods were placed in centrifuge tubes without anticoagulant and chilled in a refrigerator for 2 hours. The clot was sedimented by centrifugation and the supernatants were taken for analysis.

The effect of surface active substances such as Na-taurocholate and Triton X-500 on several enzymes was observed in order to compare with that of ginseng saponin. The reaction mixture (3.0ml) for the assay of alcohol dehydrogenase contained (final conc.) 48mM glycine buffer (pH 9.6), 0.4mM NAD^+, 3mM ethanol, surface active substance and 0.4ml of enzyme solution. The assay mixture (3.0ml)

of glutamate dehydrogenase contained 18mM phosphate buffer (pH 7.6),
0.25 mM NAD, 5 mM Na-glutamate, surface active substance and 0.2 ml
of enzyme solution. The reaction mixture of succinate dehydrogenase
assay contained 17 mM phosphate buffer (pH 7.4), 6.7 mM Na-malate,
2.8×10^{-5} M dichlorophenol-indophenol (DICPIP), 0.67 mM KCN, surface active substance and 0.3 ml of rat hepatic mitochondrial
preparation.

Examination of the saponin effect on the reaction catalyzed
by regulatory enzymes in the presence of effectors was made. The
reaction mixture (3.0 ml) of glutamate dehydorgenase assay contained 0.133 mM NADH(reduced), 0.02 M phosphate buffer (pH 7.7), 1.33 mM
α-ketoglutarate(-Na), 0.016 M NH_4Cl, effector, saponin and 0.2 ml of
enzyme solution. For the assay of succinate dehydrogenase, the
reaction mixture contained 0.017 M phosphate buffer (pH 7.4), 0.01 M
succinate, 0.28×10^{-4} M DICPIP, 6.7×10^{-3} M malate, 6.7×10^{-4}
KCN, malonate and 0.3 ml chicken hepatic mitochondrial preparation.
The effect of the saponin on NAD^+ specific isocitrate dehydorgenase
in the presence of adenine nucleotide such as AMP, ADP and ATP was
studied. The reaction mixture (4.0 ml) contained (final concentration) 1.25×10^{-2} M PIPES buffer (pH 6.5), 5×10^{-4} M NAD^+,
2.5×10^{-3} M $MgSO_4$, 1.27×10^{-5} M DICPIP, 1.25×10^{-3} M threos-Ds(+)
-isocitrate, 5×10^{-4} M nucleotide, 1×10^{-2}% saponin and 0.5 ml pig
cardiac mitochondrial preparation.

The effect of ginseng saponin on glucose oxidation by E. coli
(will type 3110) was carried out as follows. E. coli on nutrient
agar slant was inoculated on nutrient agar plate and incubated at
$37^\circ C$ for 18 hours and the procedure was repeated 3 - 5 times. The
grown colony was suspended in 50 ml of culture medium, and incubated at $37^\circ C$ for three days and harvested. The precipitated cells
were washed twice with saline and suspended in 50 ml of the above
culture medium and 5 ml of each was kept in a sterile capped tube
at $4^\circ C$. 0.2 ml of the above E. coli was allowed to stand at room
temperature for 30 minutes and suspended in 10 ml of the above
culture medium and incubated at $37^\circ C$ for 2 hours and used to
observe glucose oxidation by the orgaism. The rate of glucose
oxidation by E. coli (w-3110) cells in the presence of ginseng
saponin was determined manometrically.

The rate of pyruvate oxidation by rat hepatic mitochondrial
preparation was determined manometrically. Rat hepatic mitochon-
dria were prepared as described elsewhere.[14] The test volume was
3.0 ml. The test samples containing 0.4 ml of 0.3 M phosphate
buffer (pH 7.4), 0.3 ml of saponin solution (various concentrations),
0.2 ml of 0.001 M cytochrome C and 0.5 ml of mitochondrial preparation were placed in the main vessel and 0.1 ml of 0.1 M Na-fumarate,
0.1 ml of 0.1 M Na-pyruvate, 0.2 ml of 0.02 M $MgSO_4$, 0.2 ml of
0.01 M ATP, 0.8 ml of 0.5 M sucrose and 0.2 ml of 0.03 M NAD^+ in

0.2 M nicotinamide were placed in the side arm. 0.2 ml of 20% KOH
with 1 cm x 3 cm filter paper was placed in the center well of the
vessel.

The manometer was preincubated at 37°C for 10 min, and the
reaction initiated by adding the contents of the side arm to the
main vessel and the height of manometer was read at suitable intervals (10 min. 30 min.). Oxygen uptake was calculated according to
the equation, $\mu O_2 = KO_2 \times h$ (where μO_2 : oxygen uptake, KO_2 : flask
constant for oxygen and h is the reading). The flask constant (KO_2)
could be obtained using the formula, $KO_2 = V_g \times 273/T + (V_f \cdot \alpha) / Po$
(where Po : 1 atm. corresponding to 10,000 mm height of Brodie solution used, V_f: volume of fluid, V_g: volume of gas phase, α: Bunsen
constant of oxygen and T was 310°K). Further experiments using
radioactive Na-pyruvate-3-^{14}C were carried out to measure the CO_2
absorbed in KOH of the center well of the manometric vessel.

The effect of ginseng saponin an phospholipid biosynthesis
using mitochondrial and microsomal fractions of rat (Sprague Dawley)
liver was studied. The reaction mixture (1 ml) contained 50 mM
Tris-HCl buffer (pH 7.0), 200 µM CoA, 100 µM sodium palmitate,
2 mM potassium cyanide (KCN), 5 mM $MgCl_2$, 1.5 mM DL-glycerol-phosphate
containing (U-^{14}C) sn-glycerol-3-phosphate (0.04 µCi), 10 % ginseng
saponin and 0.4 ml of rat hepatic mitochondrial or microsomal preparations. At the termination of the reaction by adding methanol,
the lipids were extracted and the phospholipids were separated by
TLC. The corresponding fractions in the plate were scraped and the
radioactivity counted.

The effect of ginseng saponin in rat (Sprague Dawley) hepatic
microsomal ethanol oxidizing system (MEOX) was observed in vitro
as well as in vivo. The reaction mixture of the in vitro assay
(total volume : 3 ml) contained 1.0 mM Na_2-EDTA, 5.0 mM $MgCl_2$, 0.1 M
phosphate buffer (pH 7.4), 1 mM sodium azide, 2 mM pyrazole, 20 mM
nicotinamide, 5 mM glucose-6-phosphate, 1 mM $NADP^+$, 2 units of
glucose-6-phosphate dehydrogenase, various concentrations of ginseng saponin and 1 ml of rat hepatic microsomal preparation. The
above reaction mixture was placed in the main vessel of the Warburg
manometric flask, and in the center well, 0.6 ml of 0.16 M phosphate buffer (composition 0.16 M phosphate buffer pH 7.0, 0.015 M
semicarbazide hydrochloride). Following 10 minutes preincubation
at 37°C, ethanol (0.5 µCi) was added to be 50 mM in the assay mixture and the incubation was continued for 15 min. At the termination of the reaction by TCA solution (5% TCA, 45 mM thiourea) in
the side arm, the mixture stood overnight and the radioactivity of
acetaldehyde absorbed by semicarbazide was measured. The effect of
ginseng saponin on ethanol oxidation in pyrazole fed rats which
were fed with ginseng saponin (2 mg/day/rat) for 8 weeks prior to

1 ml of 25% ethanol containing 1 - [14]C ethanol (2 µCi) administration by stomach tubing was observed.

Observation of the radioactivities in blood serum of prolonged cholesterol fed with and without the ginseng saponin prior to cholesterol-4-^{14}C administration were made on timed interval (2-20 hr.). Rabbits (giant) were divided into three groups, each of which consisted of two male rabbits (initial wt. 2.0-2.2 kg). 500 mg of cholesterol, 3 g of corn oil, 100 mg of Na-cholate and 10 mg of ginseng saponin preparation were administered for 4 weeks to the 1st group (test group). 2nd group (control group) was administered exactly the same regimen as test group but no ginseng saponin. 3rd group (normal group) was fed normal diet only. After 24 hour starvation, 50 mg of cholesterol containing cholesterol-4-[14]C (0.77 µCi) were fed orally and the radioactivity of the blood serum taken from ear vein every 2 hr. up to 20 hr. was traced.[26]

For the observation of the radioactivity of cholesterol-1-^{14}C in blood serum of rabbits fed cholesterol chronically after 4 days following the cholesterol-4-^{14}C administration, three groups (test, control and normal), each consisting of two rabbits, were fed the corresponding cholesterol diet in addition to ordinary diet as described above. This was followed by two doses of cholesterol-4-^{14}C administration at 24 hr. interval after the first 2 days with a continuous feeding of cholesterol diet with saponin (test) and / without saponin (control) for 4 days. The blood was taken directly from the heart and the sera were prepared as usual, the radioactivity (^{14}C) and cholesterol were assayed.

Among the patients with liver diseases admitted to Hanyang University Hospital, Yongdeungpo City Hospital and Haesung Hospital, Seoul, Korea, patients with acute viral (B-type) hepatitis were selected based on clinical features and pathological tests. Radioimmunoassay were done to distinguish B-type viral hepatitis using Abbott laboratory's AUSTRIA II-125 system (HB-AG) AVSAB system (HB-Ag) and CORAB system (HBc Ag). Patients were divided into control and ginseng administered groups and were treated in the usual way. In addition to the ordinary medicinal, 5 g panax ginseng powder (Keumsan white ginseng root powder)/day was given orally to the ginseng treated group. Blood tests were made everyweek to follow the liver function. It included protein measurement,[27] thymol turbidity test,[28] bilirubin (total and conjugated)[29] measurment, cholesterol level determination,[30] transaminase (serum-GOT serum-GPT)[31] and alkaline phosphatase activity measurement.[32]

Radioactivies were measured using a Beckman liquid scintillation spectrophotometer model LS 3150T or a Packard model 3320 Tricarb liquid scintillation spectrophotometer.

RESULTS AND DISCUSSION

The Effect of the Saponin on Lipid Absorption

When water insoluble vitamins such as vit. A and vit. E were administered with ginseng saponin by stomach tubing, the vit. A conc. in blood serum of test group was twice as much as that of control, but the vit. A content started to decrease after 3 hours. The vit. A content of the liver of test animals was also about 1.5 times that of control and its value returned to the normal value three hours later. The vit. E content of the blood and liver of test animals were 1.4-1.5 times as much as that of control at 40 minutes after the feeding. These results suggested that saponin stimulated the absorption of water insoluble vit. A and E. On the other hand, the levels of the above vitamins in the kindneys of test animals were significantly higher, suggesting that the ginseng saponin might also affect the transport of the water insoluble vitamins (Table 1).

Table 1. The effect of ginseng saponin on the absorption of vit. A and E in Wistar rats. 12 mg vit. A-acetate or vit. E(α-tocophe-rol) and 50 mg ginseng saponin in 2 ml of 50% ethanol were fed by stomach tube.

Vitamin	Tissue	Group	Time (min)				
			0	40	60	120	180
	Serum	cont.	1.0	1.5	2.3	3.7	2.2
	(ug/ml)	test	1.1	2.9	6.0	7.6	1.7
Vit. A	Liver	cont.	209.5	216.8	286.4	339.4	212.8
	(ug/rat)	test	214.1	280.4	310.9	510.4	231.3
	Kindney	cont.	-	-	3.4	3.4	8.9
	(ug/rat)	test	-	-	5.8	15.2	15.8
	Serum	cont.	-	5.6	6.2	6.7	7.7
	(ug/ml)	test	-	7.7	7.6	7.2	10.0
Vit. E	Liver	cont.	-	80.2	114.0	160.5	217.9
	(ug/rat))	test	-	117.1	114.9	173.5	203.8
	Kindney	cont.	-	-	3.5	-	3.5
	(ug/rat)	test	-	-	8.9	20.1	18.1

When Wistar rats were fed fats either with ginseng saponin (0.05 mg/rat) or without the saponin, the blood serum lipids of both groups increased gradually with time and no appreciable difference in the absorption pattern between the two groups could be observed. However, in cannulated rats, in which the secretion of bile and pancreatic juice into the small intestine was impaired, the absorption of fats was greatly stimulated in the saponin-fed group suggesting that even the small amount of ginseng saponin greatly favoured lipid absorption in the absence or on the deficiency of bile and pancreatic juice (Figure 1 and Figure 2).

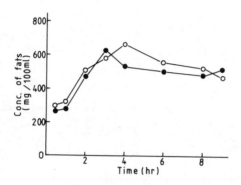

Figure 1. Blood serum lipid levels of rats fed with ginseng saponin and control rats at intervals after the fat feeding. O : Control, ● : Test.

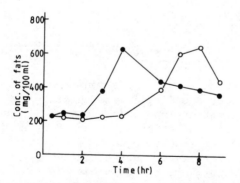

Figure 2. Blood serum lipid levels of cannulated rats fed with ginseng saponin and control rats at timed intervals after fat feeding. O : Control, ● : Test.

The effect of Ginseng Saponin on Enzyme Catalyzed Reactions

It has been observed during the past several years that adequate amounts of saponin stimulated enzyme reactions. As shown in Figure 3, the dual effect, stimulation at the moderate but inhibition at excess amounts of the saponin suggests that the enzyme stimulation might be not brought about by direct binding of the saponin to a specific site of specific enzymes, but rather by some common property as the surface activity of the saponin.

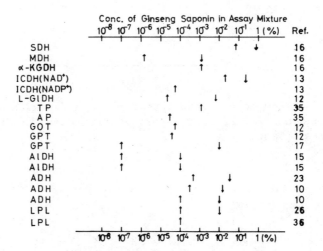

Figure 3. The effect of ginseng saponin on enzyme catalysed reactions. The corresponding enzyme reaction rate reached a maximum at the concentrations shown by the mark (↑) and was inhibited when the saponin concentration was over those shown by the mark (↓). Abbreviation: Succinate dehydrogenase (SDH), malate dehydrogenase (MDH), α-Ketoglutarate dehydrogenase (α-KGDH), isocitrate dehydrogenase (ICDH), L-glutamate dehydrogenase (GlDH), glutamate-oxaloacetate transaminase (AlDH), alcohol dehydrogenase (ADH), lipoprotein lipase (LPL), alkaline phosphatase (AP), tryptophan pyrrolase (TP). Numbers in parentheses denote references.

The actions of surface active materials such as Na-taurocholate and Triton X-100 on alcohol dehydrogenase, L-glutamate dehydrogenase and succinate dehydrogenase were compared with that of ginseng saponin. As shown in Figure 4, they all showed a similar action pattern on the above enzyme reactions.

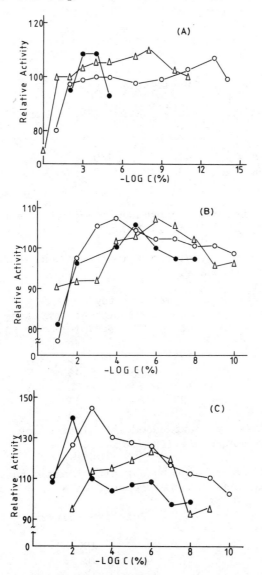

Figure 4. The effect of Na-taurocholate (o), Triton X-100 (Δ) and ginseng saponin (●) on alcohol dehydrogenase (A), L-glutamate dehydrogenase (B) and succinate dehydrogenase (C).

Examination of the saponin effect on the reactions catalyzed by regulatory enzymes such as glutamate dehydrogenase, isocitrate dehydrogenase and succinate dehydrogenase showed that the saponin was unable to repress the inhibitory action of the inhibitors on the above enzymes. This suggested that the saponin would not compete with the inhibitors (Table II, III & IV).

Table II. The effect of ginseng saponin on bovine hepatic glutamate dehydrogenase in the presence of effector.

conc. of effector	Enzyme activity*				
	conc. of saponin				
	control	1.7×10^{-5}%	8.3×10^{-5}%	1.7×10^{-4}%	8.3×10^{-4}%
no effector	100	112	124	105	103
ADP 5×10^{-5}M	326	426	450	460	485
GTP 5×10^{-5}M	41	38	2	26	38
GTP 5×10^{-5}M ADP 5×10^{-5}M	118	85	97	94	82

* The enzyme activities are expressed as relative activities assuming the activity of control is 100.

Table III. The effect of ginseng saponin on succinate dehydrogenase in the presence of malonate.

conc. of saponin in assay mixture	Enzyme activity*		
	conc. of malonate in assay mixture		
	0	1.67×10^{-3}%	1.67×10^{-2}%
0	100	44	26
2.5×10^{-2} %	122	43	25
5.8×10^{-2} %	132	46	29
8.3×10^{-2} %	137	50	34
16.7×10^{-2}%	87	48	32
23.4×10^{-2}%	–	48	32

* The enzyme activies are expressed as relative activities assuming the activity of control is 100.

Table IV. The effect of ginseng saponin on pig cardiac NAD^+ specific isocitrate dehydrogenase in the presence of AMP, ADP and ATP.

Assay Condition	Relative acitvity
Control	100
AMP	100
ADP	125
ADP+ATP	93
ADP+saponin(1×10^{-2}%)	150
ADP+ATP+saponin	93

* The enzyme activities are expressed as relative activities assuming the activity of control is 100.

Kinetic data of various enzyme reactions show that the Michaelis constant (Km) of the enzyme for substrate was not exceptionally lowered so as far examined in the presence of moderate amounts of saponin, suggesting again that the saponin acts nonspecifically on various enzyme catalyzed reactions.

We know little about the mechanism of how the saponin works. However, it is expected from the above results that moderate amounts of saponins may bring about a slight change in the aqueous environment of the enzymes, resulting in a change of enzyme conformation, which would be in favour of the reactions being accelerated.

The Effect of Ginseng Saponin on Intermediary Metabolism

It was found in this laboratory that when E. coli (ADO1) cells were grown in nutrient rich medium such as BHI broth, no effects of ginseng saponin were observed. However, when E. coli were grown in basic medium, the ginseng saponin (10^{-3}% - 10^{-2}%) stimulated the growth of E. coli. Analysis showed that when the concentration of ginseng saponin was 10^{-3}% - 10^{-2}%, the synthesis of lipid, protein, nucleic acid and one enzyme (GOT) of the cell were stimulated. At lower or higher concentrations of saponin, no significant effect was observed.

Glucose oxidation by E. coli (w 3110) was examined manometrically. It was found that when the concentration of the saponin was 10^{-3}%, oxygen uptake was 1.4-1.6 times that of control, but when the saponin concentration was 10^{-1}%, oxygen uptake was greatly lowered as shown in Table V. This suggested again that moderate amounts of ginseng saponin stimulated glucose oxidation.

Table V. The effect of ginseng saponin on glucose oxidation by E. coli (w 3110). The test soln. (3.0 ml) contained (final concentration) phosphate buffer (pH 7.4) 4×10^{-2}M, glucose 3.3×10^{-2}M, fumarate 3.3×10^{-3}M, MgSO$_4$ 1.3×10^{-3}M, ATP 6.7×10^{-4}M, NAD$^+$ 2×10^{-3}M, cytochrome C 6.7×10^{-5}M, E. coli preparation 0.5 ml and various concentrations of ginseng saponin.

| incubation time (min) | Oxygen Uptake (µℓ) | | | |
	conc. of ginseng saponin 0	10^{-5}	10^{-3}	10^{-1}
30	23.1	27.7	33.3	6.7
60	27.7	38.8	45.1	8.4

The effect of ginseng saponin on pyruvate oxidation by a rat
hepatic mitochondrial preparation was observed in vitro by measuring oxygen uptake using the Warburg manometric apparatus. As shown
in Table VI, the rate of pyruvate oxidation increased gradually as
the concentration of saponin in the reaction mixture increased and
the maximum oxidation occurred at the concentration of saponin of
1×10^{-4}%. When the saponin concentration increased further, however,
it was found that the oxidation was rather inhibited.

Table VI.   The effect of ginseng saponin on pyruvate oxidation
by a rat hepatic mitochondrial proparation. The test volume was
3.0 ml. It contained (final concentrations): phosphate buffer
(pH 7.4) 4×10^{-2}M, sucrose 1.3×10^{-1}M, cytochrome C 6.7×10^{-5}M,
mitochondrial preparation 0.5 ml and various concentrations of
ginseng saponin. Oxygen uptake was measured at the intervals of
20 and 30 minutes incubation at 37°C. Values are the mean value of
three determinations in $\mu\ell$.

conc. of ginseng saponin (%) incubation time (min)	Oxygen Uptake ($\mu\ell$)							
	0	10^{-7}	10^{-6}	10^{-5}	10^{-4}	10^{-3}	10^{-2}	10^{-1}
20		31.8	33.2	40.8	47.3	55.4	27.6	21.1	4.0
30		45.4	44.9	46.0	55.0	68.1	41.1	27.8	3.1

In an experiment using radioactive Na-pyruvate-3-^{14}C as
substrate, the CO_2 production was indirectly measured by the radioactivity of CO_2 absorbed in KOH of the center well of the manometric
vessel at 20 min. and 30 minutes incubation at 37°C. As shown
in Table VII, the percent isotope recovered in CO_2 of the test
(20 min. incubation) was as much as 2.2 times that of the control
supporting again that the pyruvate oxidation was greatly stimulated
by the saponin under the condition described in Table VII.

Table VII. The radioactivity of CO_2 absorbed in KOH of the center
well of Warburg manometric vessel during 20 min. and 30 minutes
incubation at 37°C. The reaction mixture contained 0.3 M phosphate
buffer (pH 7.4) 0.4 ml, 0.1 M fumarate 0.1 ml, 0.1 M pyruvate 0.1ml,
Na-pyruvate-3-^{14}C 0.25 μCi, 0.02 M $MgSO_4$ 0.2 ml, 0.01 M ATP, 0.2 ml,
0.03 M NAD$^+$ in 0.2 M nicotinamide 0.2 ml, 0.5 M sucrose, 0.8 ml
0.001 M cytochrome C 0.2 ml, mitochondrial preparation 0.5 ml and
10^{-3}% saponin 0.3 ml (final concentration : 1×10^{-4}%).

Incubation time (min)	20 min.		30 min.	
	Control	Test	Control	Test
Total radioactivity (DPM)	52,150	117,196	97,489	141,128
% isotope recovered	9.66	21.73	18.08	26.7
Relative acitivity (T/C)	1	2.23	1	1.45

It has been found that ginseng saponins stimulate lipid meta-bolism in the animal body.[18] It was realized from the recent study[22] in this laboratory that the saponin stimulated phospholipid biosynthesis in the liver of rats, which were administered with ginseng saponin for 10 days (5 mg of ginseng saponin/day/rat) prior to intraperitoneal injection of $H_3{}^{32}PO_4$ at timed intervals. In both, test and control group, the highest radioactivity was reached after 7 hours of isotope injection. At that time the specific radioactivity was as much as 4 times higher in the liver of ginseng administered group than in that of control rats, suggesting that the saponin stimulated greatly the incorporation of H_3PO_4 into the hepatic phospholipids of this animal.[22,34]

Using rat liver homogenate and its mitochondrial and microsomal fractions, it was observed that the saponin stimulated the phospho-lipid biosynthesis (Table VIII and Table IX).

Table VIII . Incorporation of $U-^{14}C$-sn-Glycerol-3 phosphate into various lipids by rat hepatic microsomal preparation. The incubation time was 60 minutes and the values are mean values of three tests.

Group	Total	PS+S	PC	PE+PA	CHOL.	TG
	Radioactivity incorporated (CPM)					
Control	2945	282	272	389	116	95
Test	3030	449	465	474	131	90
Ratio of (T/C)	1.03	1.59	1.70	1.22	1.13	0.95

Table IX. Incorporation of $(U-^{14}C)$-sn-glycerol-3-phosphate into various lipids by rat hepatic mitochondrial preparation. The values are mean values of three tests. Incubation time was 30 minutes.

Group	Total	PS+S	PC	PE+PA	CHOL.	TG
	Radioactivity incorporated (CPM)					
Control	2588	320	414	284	109	91
Test	2948	367	495	662	99	81
Ratio of (T/C)	1.14	1.15	1.20	2.33	0.91	0.90

Abbreviation: PS(phosphatidyl serine). S(sphingomyelin). PC (phosphatidyl choline), PE(phosphatidyl ethanolamine), PA (phosphatidic acid), Chol(cholesterol), TG(triglyceride)

The effect of Ginseng Saponin on Alcohol Metabolism

Previous work[15,23] showed that the ginseng saponin stimulates alcohol dehydrogenase and aldehyde dehydrogenase resulting in a rapid oxidation of ethanol in the body. It was also observed[23]

that the hepatocytes of rats which had free access to 12% ethanol
and normal diet for 14 days showed severe injury such as vacuolic
degeneration, glycogen deposition and fatty degeneration. Those
rats which had free access to 12% ethanol and 2% ginseng extract
containing diet showed no severe injury suggesting that the ginseng
extract protected the liver from alcohol intoxication.

     In addition to liver alcohol dehydrogenase, a hepatic micro-
somal ethanol oxidizing system (MEOS), and especially its capacity
to increase in activity after ethanol feeding, have been
reported.[37,38] Although the MEOS in vitro has been challenged
partly because the rate of ethanol oxidation accounts for only a
small fraction of ethanol metabolism in vivo, the MEOS may explain
various effects of ethanol, including proliferation of hepatic
smooth endoplasmic reticulum, induction of other hepatic microsomal
drug-detoxifying enzymes, and the metabolic tolerance to ethanol
which develops in alcoholics.[39-43]

     Examination of the effect of ginseng saponin on rat hepatic
MEOS showed that the oxidation of ethanol was significantly
acclerated in the presence of the saponin as shown in Table X & XI.

Table X. The effect of ginseng saponin on rat hepatic MEOS in
vitro.

Concentration of saponin (%)	0	10^{-10}	10^{-8}	10^{-6}	10^{-4}	10^{-2}
Aldehyde formed (CPM)	1013	2647	3357	3963	2026	1595
Ratio of Test/Control	1.0	2.6	3.3	3.9	2.0	1.6

Table XI. The effect of ginseng saponin on ethanol oxidation in
pyrazole fed rats which were administered ginseng saponin
(2 mg/day/rat) for 8 weeks prios to 1 ml of 25% ethanol containing
1-^{14}C ethanol (2 µCi) by stomach tube.

Radioactivity (CPM)

	Fraction	Aldehyde	non-volatile fraction
Liver (CPM/g)	Control	163	383
	Test	185	416
	Ratio of T/C	1.13	1.09
Blood serum (CPM/ml)	Control	677	166
	Test	525	158
	Ratio of T/C	0.78	0.95

The Preventive Effect of Korean Ginseng Saponins on Aortic Atheroma Formation in Rabbits fed Cholesterol Chronically.

It has been observed microscopically that saponin administration lowered significantly the increased total lipid and cholesterol levels in blood sera caused by continuous cholesterol feeding up to 4 weeks feeding, with a longer feeding of cholesterol diet for more than 4 weeks, no appreciable effect of the saponin could be seen.[26]

Analysis of the radioactivity of the blood serum lipid fraction of rabbits fed with cholesterol with saponin (test group) and without saponin (control) for 4 weeks prior to cholesterol-4-^{14}C administration at timed intervals (2-20 hrs) showed that the highest radioactivity of test, control and normal group was observed at 6-8 hours, 16-18 hours and around 14 hours respectively after the isotope administration. The radioactivity seemed to disappear gradually thereafter in all the above three groups. The radioactivities of blood serum lipid fractions of rabbits fed with cholesterol with saponin (test), without saponin (control) for 4 weeks followed by two doses of cholesterol-4-^{14}C at 24 hour intervals showed that the counts of the test group were only two-thirds (617 cpm/ml) that (927 cpm/ml) of the control group.

No occurrence of atheroma in the aorta, coronary and renal arteries was observed microscopically in the saponin-chloesterol fed rabbits up to four weeks feeding, while in the cholesterol fed rabbits without the saponin, atheroma in the ascending aorta appeared after 2 weeks feeding. It suggested that the saponin prevented the formation of atheroma in aortic tissue.

The ratio of specific radioactivity of cholesterol in aortic tissue to blood serum cholesterol of the rabbits fed with a high cholesterol diet containing (4-^{14}C) cholesterol with and/without the saponin for 4 weeks showed that the elevation of cholesterol influx rate from blood to aortic tissue by prolonged high cholesterol diet was significantly diminished by simultaeous feeding of ginseng saponins (Table XII).

Table XII. Penetration of (4-^{14}C)-cholesterol into aortic tissue.

	Normal	Control	Test
Radioact. of Blood Serum(CPM/ml)	286	727	659
Radioact. of Aortic Tissue(CPM/g)	614	481	443
Sp. Radioact. of Cholesterol			
in blood serum(CPM/mg)	403	85	119
in Aortic tissue(CPM/mg)	212	152	157
Aorta/serum ratio	0.53	1.79	1.32

In view of the presence of cholesterol in the atherosclerotic
plaques of the aorta, it seems likely that a casual relationship
between the disease and an excess of the substance exists.
Arterial tissue is perhaps capable of synthesizing cholesterol in
situ, and the cholesterol deposits may originate from either
endogeneous or dietary cholesterol. In the absence of conclusive
evidence, it must be assumed that atherosclerosis is probably due
to some abnormality in lipid metabolism, perhaps in the handling of
cholesterol. It seemed that ginseng saponin, might stimulate the
enzymes relating to the metabolism of lipid (including cholesterol),
particularly to cholesterol transport, resulting in the delay of a
rise of the cholesteral level in blood, consequently, the prevention
of atheroma formation in such tissue as the aorta.

Clinical Observation on the Preventive Effect of Panax Ginseng
against Liver Cirrhosis and its Early Phase Treatment

The effect of ginseng administration on the patients with acute
viral (B type) hepatitis has been observed. We found that the
albumin/globulin ratio of the ginseng administered patients improved
4 weeks after admission while that of control group has not been
improved at that time, suggesting that the ginseng might be effective
in improving the protein metabolism. The thymol turbidity test
again gave a similar result.

Recovery from the disorder of bilirubin metabolism was also
accelerated in the treated group compared with the control. The
increased bilirubin value of the former returned to the normal value
2 weeks after admission while that of the latter reached normal
4-5 weeks after admission. However no significant difference in the
bilirubin level between ginseng-treated and non-treated groups could
be observed.

Cholesterol metabolism was also stimulated in the ginseng-
administered group. The lowered cholesterol level of the ginseng
group returned to normal 3-4 weeks after admission, while that of
latter reached to normal 5-6 weeks after admission. The raised
S-GOT and S-GPT levels of the ginseng-treated group returned to the
normal 3-4 weeks after admission, while those of control group
returned to normal in 5 weeks after admission suggesting that the
ginseng improved impaired liver function.

From the above results, it seems that ginseng might stimulate
the improvement of the disturbance of liver function, particularly
in the early phase of the development of acute liver disease, suggesting that panax ginseng might play a significant role in preventing
the disease from developing to a chronic disorder.

ACKNOWLEDGEMENTS

Support of this work by Korean Traders Scholarship Foundation, Korea Ginseng and Tobacco Research Institute and Korean Ginseng Products Co. Ltd. is gratefully acknowledged.

REFERENCES

1. T. Namba, M. Yoshizaki, T. Tominori, K. Kobachi and J. Hase, in "Proceeding of 6th Symp. of Oriental Drugs", p88, Toyama University, Toyama, Japan (1972)
2. C. P. Li and R. C. Li, American J. Chinese Med., 1, 249 (1969)
3. H. Oura, S. Hiai, S. Nabetani, H. Nakagawa, Y. Kurata and N. Sasaki, Plant Medica, 28, 76 (1975)
4. I. I. Brekhman and I. V. Dardymov, Ann. Rev. Pharmacol., 9, 419 (1969)
5. H. Oura and S. Hiai, in "Proceedings of International Ginseng Symposium, Korean Ginseng Res. Inst., Seoul (1974)
6. C. N. Joo, H. S. Yoo, S. J. Lee and H. S. Lee, Korean Biochem. J. 6, 177 (1973)
7. C. N. Joo, S. J. Lee, S. H. Cho and M. H. Son, Korean Biochem. J. 6, 185 (1973)
8. C. N. Joo and S. J. Lee, Korean Biochem. J. 10, 59 (1977)
9. C. N. Joo, Yonsei Nonchong, Yonsei Univeristy, Seoul, 10, 487 (1972)
10. C. N. Joo, R. S. Choe, N. P. Chunh, S. J. Lee, and O. H. Kim, Korean Biochem. J. 7, 75 (1974)
11. C. N. Joo, B. H. Yoon, S. J. Lee and J. H. Han, Korean biochem. J. 7, 231 (1974)
12. C. N. Joo, J. H. Oh and S. J. No, Korean Biochem. J. 9, 53 (1976)
13. C. N. Joo and J. H. Han, Korean Biochem. J. 9, 237 (1976)
14. C. N. Joo and T. Y. Kim, Korean Biochem. J. 10, 13 (1976)
15. C. N. Joo, J. D. Koo, D. S. Kim and S. J. Lee, Korean Biochem. J. 10, 109 (1977)
16. C. N. Joo and J. H. Han, Korean Biochem. J. 9, 43 (1976)
17. C. J. Park, J. H. Koo and C. N. Joo, Korean Biochem. J. 4, 168 (1978)
18. C. N. Joo, H. B. Lee and D. S. Kim, Korean Biochem. J. 10, 71 (1977)
19. C. N. Joo and S. J. Lee, Korean Biochem. J. 10, 59 (1977)
20. P. P. Hawk, B. L. Oser and W. H. Summerson, in "Practical Physiological Chemistry" (13th Ed), p1272 Churchill, London, (1963)
21. B. Bloom, I. L. Chaikff, W. O. Reinhardt, C. Entenman and W. G. Dauben, J. Biol. Chem., 198, 1. (1953)
22. S. O. Lee, J. H. Koo and C. N. Joo, Korean Biochem. J. 14, 161 (1981)

23. C. N. Joo, J. H. Koo and B. H. Kang, Korean Biochem. J. 12, 81 (1980)

24. C. N. Joo, D. S. Kim and J. H. Koo, Korean Biochem. J. 13, 51 (1980)

25. D. T. Plummer, in "Introduction to Practical Biochemistry", McGraw-Hill, (1971)

26. C. N. Joo, J. H. Koo and T. H. Baik, Korean Biochem. J. 13, 63 (1980)

27. T. E. Weichsellaum, Amer. J. Clin. Path. 7, 40 (1946)

28. R. E. Shauk and C. W. Hoagland, J. Biol. Chem. 162, 133 (1946)

29. H. T. Malley and K. A. Evelyn, J. Biol. Chem. 119, (1937)

30. S. Pearson, Analyst. Chem. 25, 813 (1953)

31. S. Reitman and S. Frankel, Amer. J. Clin. Path. 28, 56 (1957)

32. E. J. King and A. R. Armstrong, Canad. Med. Ass. J. 31, 376 (1934)

33. C. N. Joo, Y. D. Cho, J. H. Koo, C. W. Kim and S. J. Lee, Korean Biochem. J. 13, 1 (1980)

34. T. K. Park, C. N. Joo and K. S. Cho, Korean Biochem. J. 14, 337 (1981)

35. N. P. Chung, Private Communication (1982)

36. C. N. Joo and Y. J. Cho, Korean Biochem. J. in press (1982)

37. O. A. Iseri, L. S. Gottlieb and C. S. Lieber, Fed. Proc. 23, 579 (1964)

38. E. Rubin, F. Hutterer and C. S. Lieber, Science, 159, 1469 (1968)

39. N. Grunnet, B. Quistorff, and H. I. D. Thieden, Eur. J. Biochem. 40, 275 (1973)

40. C. S. Lieber and L. M. DeCarli, J. Biol. Chem. 245, 2505 (1970)

41. C. S. Lieber and L. M. DeCarli, J. Pharmacol. Exp. Ther. 181, 279 (1972)

42. R. Teschke, Y. Hasumura and C. S. Lieber, J. Biol. Chem. 250, 7397 (1975)

43. R. Teschke, Y. Hasumura and C. S. Lieber, Arch. Biochem. Biophys. 175, 635 (1976)

DETERMINATION OF VERY LOW LIQUID-LIQUID INTERFACIAL TENSIONS

FROM THE SHAPES OF AXISYMMETRIC MENISCI

Y. Rotenberg[1], S. Schürch[2], J.F. Boyce[2]
and A.W. Neumann[1]

[1] Department of Mechanical Engineering
    University of Toronto
    Toronto, Ontario, Canada, M5S 1A4

[2] Department of Biophysics
    University of Western Ontario
    London, Ontario, Canada, N6A 3K7

We have recently developed a user-oriented scheme to determine liquid-fluid interfacial tensions and contact angles from the shapes of axisymmetric menisci, i.e. from sessile as well as pendant drops. The input information consists, besides gravity constant and density difference between the two fluid phases, only of a set of coordinate points of the drop profile.

Sessile and pendant drops were trans - illuminated by means of fibre optics and photographs of the drops were taken with a Nikon SMZ microscope equipped with polarizing filters. From enlarged photographs, approximately 35 points along the drop profile were generated with a Talos 600 series digitizer.

The interfacial tension of water/n-butanol was studied for which a literature value of the interfacial tension exists (1.8 ergs/cm^2). Five sessile drops, ranging in diameter from 1.5 to 4.5 mm yielded an interfacial tension of 1.78 ± 0.04 ergs/cm^2. Measurement with three pendant drops yielded a value of 1.73 ± 0.03 ergs/cm^2. Measurements were also performed with aqueous solutions of sodium dodecyl sufate in contact with a solution of cholesterol in di-n-butyl phthalate. Interfacial tensions in the range of 10^{-3} to 10^{-4} ergs/cm^2 were recorded.

INTRODUCTION

Methods for calculating interfacial tension from the dimensions of axisymmetric pendant and sessile drops have existed since the late 19th century. Most of these methods pertain to a solution of the Laplace equation of capillarity:

$$\gamma J = \Delta P \tag{1}$$

where γ is the interfacial tension, J is the first (mean) curvature of the interface and ΔP is the pressure difference across the interface. Bashforth and Adams[1] were the first to present a numerical solution to this differential Equation (1) and various others have since improved and extended the range of application of this method.[2,3,4,5] Most of these approaches place critical emphasis on the location of specific features on the meridian curve, such as the drop apex and the equator. Procedures for extracting this information were developed,[6,7] but the fundamental problem of describing the drop's meridian curve with two or three critical points still existed. Maze and Burnet[8] attempted to overcome this difficulty by analyzing the entire meridian curve of the sessile drop but their solution was not general enough and its implementation often proved to be quite difficult.

In light of the difficulties experienced using such methods, a user-oriented scheme to determine interfacial tensions and contact angles from the shapes of both pendant and sessile drops[9] was developed recently. The computational procedure constructs an objective function which expresses the error between the physically observed curve and a theoretical Laplace curve. This function is minimized by using incremental loading in conjunction with the Newton-Raphson iterative procedure. Details of the mathematics and the computational scheme may be found elsewhere.[9] The results reported in this investigation demonstrate the value of such a computational procedure for calculating very low interfacial tensions in a simple and convenient way.

MATERIALS AND METHODS

For the system butanol/water sessile and pendant drops were trans-illuminated with a 150 W fibre light source. The incident light was polarized, and then dispersed with a frosted glass plate placed behind the testing apparatus as shown in Figure 1. Droplets were formed inside a small (1 cm x 10 cm) glass sided test chamber filled with water and photographed through a Nikon SMZ stereomicroscope. A second polarizing filter, mounted onto the microscope, enhanced contrast and an auxiliary 2X objective

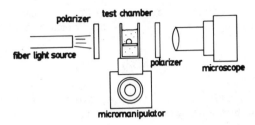

Figure 1. Schematic diagram of apparatus.

increased the total magnification to 120X. The chamber rested on a tilting plate (Newport Research, Fountain Valley, California) attached to a Leitz micromanipulator.

A sodium dodecyl sulfate (SDS)/cholesterol system was used to produce ultra-low interfacial tensions. The test droplets were a mixture of dibutyl/dioctyl phthalate (Eastman Chemicals) 1:1 by volume, with cholesterol dissolved in them. Such low interfacial tensions as those produced by this system present special optical difficulties. These problems were overcome using a horizontally positioned Nikon Optishot microscope equipped with Nomarski differential interference optics. A stage micromanipulator held a specially constructed chamber 0.5 cm wide and 6.0 cm long. The sessile drops were placed onto a plate that could be tilted slightly thus allowing the base and contact points of the drop to be seen clearly. Substrates were chosen to produce contact angles as large as possible. Glass slides coated with FC 721, a lightly fluorinated acrylic polymer (3 M Corporation), were found to be quite satisfactory for this use.

Pendant drops were formed on squared tips of stainless steel hypodermic needles of various diameters from 18 Gage to 25 Gage, mounted onto a Gilmont ultra precision micrometer syringe. This micrometer syringe was attached to a Brinkman micromanipulator which permitted easy adjustment of the drop for photography. The densities of the fluids were determined by weighing a stainless steel ball of known diameter in air and in the liquid with a Mettler analytical balance.

From enlarged photographs, approximately 35 points along the drop profile were generated with a Talos 600 series digitizer (resolution $\approx$ 0.003 cm). The size calibration of these images was done by measuring, from the photographs, the diameter of the stainless steel needles. These coordinates

along with pertinent physical data (density difference and local
acceleration of gravity) were used as the computer input.

RESULTS

In order to illustrate the approach, the interfacial
tension of water/n-butanol was determined. The literature

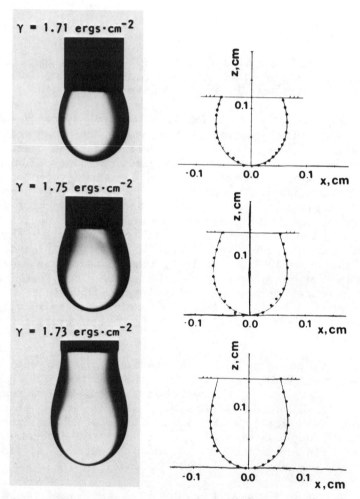

Figure 2. Pendant drops of water in butanol and their profiles
as fitted by the computer. γ is the computed interfacial
tension.

value[10] for this system is 1.8 ergs/cm^2. Three pendant drops of water in butanol of different sizes, but all suspended from the same capillary, are shown on the left hand side of Figure 2. The coordinate points taken from these profiles are shown in the right hand side of Figure 2, together with the Laplace curves fitted by the algorithm[9] through these points. The interfacial tension obtained as the parameter for the best fit of a Laplace curve to the digitized image is given on the left hand side for

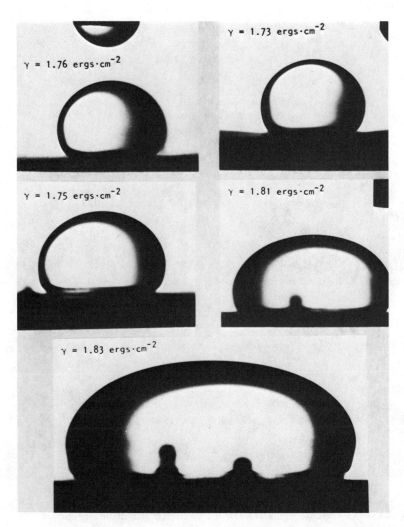

Figure 3.   Sessile drops of water in butanol. γ is the computed interfacial tension.   The tip of a pipette shown in the top, left figure is 1.05 mm in diameter.

each of the three pendant drops.   The statistical mean of these values is 1.73 ± 0.03 ergs/cm^2.

Similarly, a series of five sessile drops of water in butanol is shown in Figure 3, together with the interfacial tension obtained in each case.   These drops are all shown at the same magnification.   They range from 1.5 mm to 4.5 mm in diameter.   The mean of the surface tension values obtained is 1.78 ± 0.04 ergs/cm^2, in agreement with the value obtained from pendant drops and consistent with the literature value.

Next, the approach was applied to a system of very low interfacial tension, i.e. an aqueous solution of 0.005 M sodium

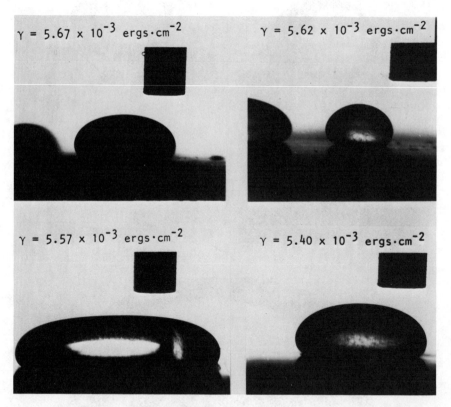

Figure 4.   Sessile drops of a mixture of dibutyl/dioctyl phthalate (1:1) with cholesterol added in an aqueous solution of 0.005 M sodium dodecyl sulfate (SDS). γ is the computed interfacial tension.   The diameter of the pipette is 0.304 mm in all four cases.

dodecyl sulfate (SDS) and a 1:1 mixture of dibutyl/dioctyl
phthalate to which a small amount of cholesterol had been added.
Four drops of the aqueous phase in the organic liquid are shown
in Figure 4. The diameter of the pipette (the same one in all
cases) is 0.304 mm and provides a scale for the size of the
drops. These four drops are part of a series of ten drops which
were all evaluated. The mean value of the interfacial tension
for the whole series is $(5.45 \pm 0.17) \cdot 10^{-3}$ ergs/cm^2, where the
error represents the 95% confidence limit.

Two drops of even lower interfacial tension are shown in
Figure 5. The system is the same as in Figure 4, except that
the cholesterol concentration was increased to 0.031 M. The
pipette tip is 0.040 mm in diameter and the diameter of the
droplets is approximately 0.3 mm. The interfacial tension
obtained for each drop is indicated in Figure 5.

DISCUSSION

The comparison with literature values as well as between
the results obtained for sessile and pendant drops,
respectively, indicates that the method works independently of
drop size and shape (Figures 2 and 3). The chance of error in
the interfacial tension determination has been reduced, because
this method does not place any emphasis on one or two critical
measurements of drop height and diameter, as do most other
solutions of the Laplace equation. Compared with Maze and
Burnet's technique,[8] the method used here will handle pendant
drops in addition to sessile drops and requires only a minimum

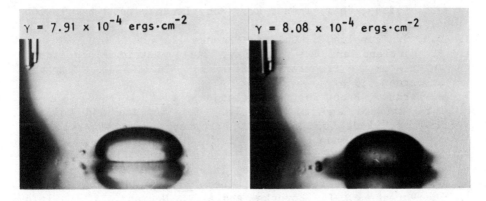

Figure 5. Two sessile drops of the SDS/cholesterol system. In
this case, the concentration of cholesterol in the dibutyl/
dioctyl phthalate mixture is 0.031 M. γ is the computed
interfacial tension. The diameter of the pipette is 0.040 mm.

of input data, i.e. the density difference between bulk phases, the local acceleration of gravity and a set of coordinates on the drop profile. No estimates of interfacial tension, apex location, radius of curvature, etc. are required.

The experimenter has a choice between pendant and sessile drops. For higher values of interfacial tension (e.g. butanol/water) both approaches are easily implemented. However, for very low interfacial tensions ($<10^{-2}$ ergs/cm^2) with substantial differences between the densities of the bulk phases, the sessile drop is far faster and easier to use. For example, systems with values in the range of 10^{-3} ergs/cm^2 such as the SDS/cholesterol system require pipette tips of 10 μm or less for the pendant drop method. Once formed, pendant drops in this range are somewhat unstable and fall off easily, making measurement quite difficult. To determine such low interfacial tensions from sessile drops (Figures 4 and 5) is relatively simple; only a good optical system for resolving small droplets (phase contrast or differential interference) and a proper choice of substrate producing a large contact angle and hence producing significant distortion of the drop are required. This ease of implementation will be appreciated by those familiar with the experimental techniques of spinning drop[11] and variable density gradients.[12]

REFERENCES

1. F. Bashforth and J.C. Adams, "An Attempt to Test the Theories of Capillary Action," Cambridge University Press, Cambridge, 1892.
2. J.F. Padday, Phil. Trans. Roy. Soc. London, Ser. A 269, 265 (1971).
3. D.N. Staicopolus, J. Colloid Science 17, 439 (1962).
4. S. Hartland and R.W. Hartley, "Axisymmetric Fluid-Liquid Interfaces," Elsevier Scientific Publishing Company, Amsterdam, 1976.
5. S. Fordham, Proc. Roy. Soc. (London), 194A, 1 (1948).
6. D.S. Ambwani and T. Fort Jr., in "Surface and Colloid Science," R.J. Good and R.R. Stromberg, Editors, Vol. 11, pp. 93-119, Plenum Press, New York, 1979.
7. J.F. Padday, in "Surface and Colloid Science", E. Matijevic, Editor, Vol. 1, pp. 151, Wiley Interscience, New York, 1969.
8. C. Maze and G. Burnet, Surface Sci. 24, 335 (1971).
9. Y. Rotenberg, L. Boruvka and A.W. Neumann, J. Colloid Interface Sci., in press.
10. A.W. Adamson, "Physical Chemistry of Surfaces," 3rd Ed., John Wiley and Sons, New York, 1976.
11. H.M. Princen, I.Y.Z. Zia and S.G. Mason, J. Colloid Interface Sci. 23, 99 (1967).
12. J. Lucassen, J. Colloid Interface Sci. 70, 355 (1979).

MECHANISM OF USING OXYETHYLATED ANIONIC SURFACTANT TO

INCREASE ELECTROLYTE TOLERANCE OF PETROLEUM SULFONATE

Ying-Chech Chiu

Department of Chemistry
Chung-Yuan Christian University
Chung-Li, 320, Taiwan
The Republic of China

Petroleum sulfonate, being effective in lowering oil-water interfacial tension in low salinity water, is used as a primary surfactant in enhanced oil recovery. In saline formation water, oxyethylated sulfate is added to increase the electrolyte tolerance of the aqueous sulfonate solution for flooding low temperature reservoir. The mechanism of increasing electrolyte tolerance was investigated by using specific ion electrode to study the ion-surfactant interaction and using light scattering to study the surfactant aggregate size. It was found that the oxyethylated sulfate cannot prevent the chemical reaction between sulfonate and cations. Only weak interaction was detected between oxyethylate groups and cations. Large dispersion effect on petroleum sulfonate aggregates was observed when oxyethylated sulfate was added to the solution. It was concluded that complex formation with oxyethylated sulfate is not the mechanism of increasing the electrolyte tolerance of petroleum sulfonate. The proposed mechanism works via decreasing the size of the resulting mixed aggregates. A delicate balance of surfactant structures must be maintained in order that the mixed surfactant system can achieve low interfacial tension and high electrolyte tolerance.

INTRODUCTION

This paper discusses some fundamental study concerning tertiary oil recovery. After secondary oil recovery of petroleum re-
servior due to high interfacial tension at the oil-brine interface.[1] To lower the interfacial tension, petroleum sulfonate solution has been used under low electrolyte condition.[1] In the
presence of saline formation water, oxyethylated sulfate has been
added to increase the electrolyte tolerance of the sulfoante system at low temperature.[2] In such a mixed surfactant system,
petroleum sulfonate is referred to as the primary surfactant and
oxyethylated sulfate as the cosurfactant. At high reservoir temperature and field scale time, the rate of hydrolytic decomposition
of oxyethylated sulfate becomes significant and cosurfactants with
stable structures are more desirable. Concentrated effort has been
developed to finding thermally stable cosurfactnats. Although some
stable cosurfactnats have been reported,[3-15] only Dowfax is
commercially avaiable.

In order to find new cosurfactants and to predict the per-
formance of a mixed surfactant system. It is necessary to understand the mechanism of increasing sulfonate electrolyte tolerance
by oxyethylated sulfate. Publications concerning this particular
subject cannot be found in the literature. A general speculation
has been that the oxyethylated sulfate may work through a complexing mechanism with metal cations. In another case, no reaction
between Neodol 25-3S and calcium ions was assumed in salinity
requirement calculations given by Glover et al.[16]

It is the purpose of this paper to explain the function of
Neodol 25-3S in a petroleum sulfonate solution. The experiments
were carried out using specific ion electrodes to study the interaction between metal ions and different surfactants. Since the
specific ion electrode measurement determines only the ionic activity of a specific ion in solution, it eliminates the confusion
of counting the neutral species formed with the ion as compared to
ordinary chemical analysis. This approach is useful in predicting
ionic equilibria of electrolytes in surfactant solutions. Some
light scattering measurements were also made to study the change
of particle size.

EXPERIMENTAL

Specific Ion Electrode Measurement

The basic equipment consists of an Orion Model 801 digital PH/mv meter, a Beckman 39278 sodium ion electrode, an Orion model 92-20 calcium electrode and a saturated calomel electrode as reference. The experiments were performed by adding NaCl or $CaCl_2$ solutions to the solutions of interest and recording the potential at each addition at 24 ^{O}C. Unless the whole range is indicated, all concentrations shown in the figures are initial concentrations of the solutions.

Light Scattering and Cloud Point

The light scattering was measured by using a Fluorometer (Turner Associate, Model 110) with a primary filter and a secondary filter of 405 mμ. In order to adjust the output of the scattered light within the measured range, the intensity of the scattered light was cut down to 10%. The measurement was made at 24°C.

The cloud point was determined by observing the turbidity of the nonionic solution during one heating and cooling cycle with a strong light. The finding was checked by using the Fluorometer when visual observation was uncertain.

Chemicals and Analysis

All NEODOL samples were commercial products from Shell Chemical Company. Siponate DS-10 and Abex 707 were commercial products of Alcolac, Inc. Dowfax surfactants were supplied by the Dow Chemical Company. Cheelox BF-13 was a commercial sample of EDTA produced by GAF, Inc. The narrow range nonionic surfactants were manufactured by Nikko Chemical Company, Japan and had a purity of 96-99%.

Active ingredient contents of some commercial surfactants were determined by the hyamine titration according to the detailed procedures given by Egham Industrial Chemical Laboratory (Determination of Anionic Detergents, Research Report EICH-60, Egham, England).

RESULT AND DISCUSSION

The Requirement of a Cosurfactant in Tertiary Oil Recovery

In this paper, the word cosurfactant refers to surfactant molecules which can be used with the primary surfactants (usually, petroleum sulfonates) in saline reservoir brine and which give reasonably good oil recovery. According to this definition, a cosurfactant must : (1) have high electrolyte tolerance by itself, (2) be able to increase electrolyte tolerance of a primary surfactant mixed with it, and (3) be able to recover oil after mixing it with the primary surfactant. A surfactant may have high electrolyte tolerance by itself, but it may not increase the electrolyte tolerance of the other. The resulting system may have high electrolyte tolerance but may not be able to recover oil. Therefore, it usually takes a long series of tests before one can select a cosurfactant which satisfies those three conditions.

Based on the discussion in a previous paper[17] that a surfactant system must build up large size aggregates to achieve recovery activity and that the addition of NEODOL 25-3S to petroleum sulfonate results in a system which has smaller aggregate size, one may deduce two alternate conditions for a cosurfactant: (1) a cosurfactant must be able to reduce size of the surfactant aggregate under the same electrolyte condition to prevent coagulation of the surfactant and (2) a cosurfactant must not reduce the aggregate size of the resulting system to a value below the required size for achieving recovery activity. To satisfy these two conditions simultaneously, it requires a delicate balance in the cosurfactant structure.

General Properties of Oxyethylated Anionic Surfactant

NEODOL 25-3S has a structural formula $C_{12-15}H_{(24-30)+1}$ $(OCH_2CH_2)_3 \cdot OSO_3Na$. It is a commercial product of Shell Chemical Company. This type of surfactant is sometimes referred to as anionic nonionic surfactant or oxyethylated anionic surfactant. Although it has been widely used in industry[18], it has been studied much less than the simple anionic or the nonionic surfactants. Having the characteristics of both nonionic and anionic surfactants[19-26], the oxyethylated anionic surfactants do not follow the rules of micellization given by Klevens.[27] Weil, et al.[19] found that increasing the oxyethyl group number to three decreased the CMC of the parent alkyl sulfate, four oxyethyl groups reversed this effect; no such reversal was found by Tokiwa and Ohki.[20] Barry and Wilson[26] found that up to two oxyethyl groups decreased the CMC of cation in surfactants while a third oxyethyl group reversed the effect. Weil, et al.[19] concluded that while the oxyethyl groups increase the hydrophilic nature of the surfactant, the CMC was reduced because it was the total length of the surfactant (other than its ionic head) rather than only its hydrocarbon portion which determined its CMC. However, Shinoda and Hirai[24] stressed that the hydrated water and counter ions around the oxyethyl groups depress the electrical potential of the ionic heads and cause the CMC decrease. In general, rules applied to nonionic or anionic surfactants do not always apply to the oxyethylated anionic surfactants.

NEODOL 25-3S molecule has the combined structure of oxyethylated alcohol and alkyl sulfate and has very high electrolyte tolerance. Generally, alkyl sulfate surfactants have medium tolerance toward monovalent cations but low tolerance toward multivalent cations. The high multivalent cation tolerance of NEODOL 25-3S has been considered to be related to the complex formation of the oxyethyl groups with the metal ions. This kind of complex was first reported by Doscher, Myers and Atkins[28] who isolated a crystalline material having approximately 3 $CaCl_2$, 15 H_2O, 1 Renex. Renex is a polyglycol ester of mixed fatty and rosin acids. The complex was precipitated by adding 30% or more $CaCl_2$ to a 5% solution of nonionic surfactant. Following Doscher, et al.[28], Schott[29] attributed the salting-in of nonionic surfactants to the complexing of oxyethlene groups with the metal ions. This work gave rise to the speculation that similar complexing could play a role with NEODOL 25-3S. In order to examine this mechanism, we performed specific ion electrode measurements.

NEODOL-Metal Ion Interaction

Ordinary petroleum sulfonate of molecular weight around 430 separates out of water in a Na^+ concentration of 0.4 M or in a Ca^{++} concentration of 0.02 M at room temperature. In this paper, the word separation refers to the formation of aggregates sufficiently large to cause coagulation. The addition of NEODOL 25-3S increases the concentration of Na^+ and Ca^{++} in a petroleum sulfonate solution depending on the ratio of NEODOL to sulfonate. At room temperature, enough NEODOL is usually added to the sulfonate solution to prevent the separation of sulfonate below 1 M Na^+ or 0.1 M Ca^{++}. Therefore, most of our study was carried out in these concentration ranges.

Figure 1 shows sodium ion electrode measurements of the cosurfactant solutions. Theoretically, the potential of a specific ion electrode follows the Nernst equation :

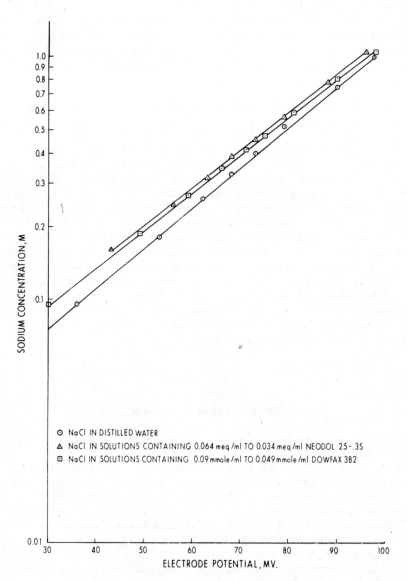

Figure 1. Potential of Na$^+$ as a function of Na$^+$ concentration in cosurfactant solutions.

$$E_{obs} = E^o - \frac{RT}{nF} \ln a_{ion}$$

E_{obs} = potential observed for any given activity of ion.

E^o = potential in the standard state

R = universal gas constant in joules

T = absolute temperature in degree Kelvin

n = number of electrons transferred in the reversible reaction

F = Faraday constant

$\ln a_{ion}$ = natural logarithm of the activity of free ion in solution.

Three straight lines were obtained in Figure 1, representing Na^+ potential in distilled water, in NEODOL 25-3S and Dowfax 3B2 solutions. The concentration of cosurfactant is about two to three times of that usually used in the sulfonate formulation. The sodium concentration represents the added NaCl to the solution. The three straight lines have a slope very near to the Nernst equation (59 mv per 10-fold change in concentration). This means that the added sodium ions are all free in solution. No complexing effect was detected between the Na^+ and the cosurfactants. It should be mentioned here that, for some unknown reason, the electrode potential may change when one takes the electrode out of the solution. Therefore, one should not do this during the run and the potential values between two runs can not be compared accurately with each other. Nevertheless, this behavior does not prevent us from detecting the interaction between the cosurfactant and the ions, since any interaction will not result in a linear response of ion potential with respect to ion concentration with a slope predicted by the Nernst equation.

Figure 2 shows calcium ion potential as measured by the calcium electrode as a function of added $CaCl_2$ concentration. The straight line represents the response of the calcium electrode in distilled water. The slope of the line corresponds to what is predicted by the Nernst equation, 29 mv per 10-fold change in concentration. In the presence of NEODOL and Dowfax, the potential of the electrode increases slowly at the beginning as $CaCl_2$ is added but it increases almost as predicted by the Nernst equation at the end. The early part of the curves clearly indicates that Ca^{++} ion is reacting with the cosurfactants. When the reaction is finished, the response is Nernstian again. The strong chelating effect of EDTA with Ca^{++} is presented here for comparison. The bonding is so strong that nearly all the added Ca^{++} goes into chelation (potential does not increase) and the completion of the reaction is indicated by a clear break in the curve. Comparision of curves indicates that the interaction between the cosurfactant and Ca^{++} is much weaker. The difference in the initial potential

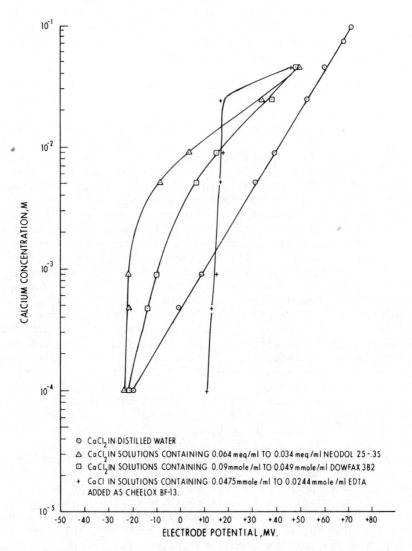

Figure 2. Potential of Ca++ as a function of Ca++ concentration in cosurfactant solutions.

between the EDTA curve and the others can be attributed to the difference in pH of the solutions. Cheelox BF-13 is an alkline solution, whereas the other solutions are neutral.

Figures 3 and 4 show the calcium ion potential as $CaCl_2$ is added to nonionic surfactants. The narrow range oxyethylates are high purity samples from Nikko Chemical Company with 96-98% purity at the specified oxyethylene number (E_3, E_4, etc.). In this paper CxEy means a molecule having x carbon in the hydrocarbon chain and y oxyethylene groups. These oxythylates have dodecyl hydrocarbon chain. The broad range oxyethylated alcohol are commercial samples from Shell Chemical Company. Both figures show very similar patterns, namely, a very weak interaction between the oxyethylene groups and the calcium ion. The reaction is so weak that only slight curvature is seen at the beginning and the later Nernstian response (a slope of 29±3mv is within experimental error considered as Nernstian) is resumed. The reaction does not seem to be enhanced by the increment of oxyethylene number in the $C_{12}E_8$ samples as compared to $C_{12}E_4$ or $C_{12}E_3$ samples.

Comparing the NEODOL curve in Figure 2 with Figures 3 and 4, one can see that the interaction between Ca^{++} and NEODOL 25-3S is much stronger than the interaction between Ca^{++} and the oxyethylated alcohol. Obviously, the difference is due to the presence of an anionic group in the NEODOL 25-3S molecule. It is reasonable to consider at this state that the interaction between Ca^{++} and NEODOL 25-3S (such as the curve shown in Figure 2) is a combination of two reactions: (1) a strong interaction between Ca^{++} and the sulfate group with a reaction constant K_1 and (2) a weak interaction between Ca^{++} and the oxyethylene groups with a reaction constant K_2. Doscher et al.[28] Proposed that the ether oxygen of the oxyethylated nonionic surfactants forms a complex with the hydrated calcium ion.

Sulfonate-Calcium Interaction in the Presence of NEODOL 25-3S

In order to examine the effect of NEODOL 25-3S on the sulfonate-calcium interaction, we made some calcium electrode measurement to determine whether the calcium-NEODOL interaction would decrease the sulfonate-calcium interaction or not. Figure 5 shows two titration curves of NEODOL 25-3S with calcium. The response of the electrode was first checked by adding $CaCl_2$ to distilled water. Nerstian response was shown in line AB. Then NEODOL 25-3S was added, and the electrode potential was observed to drop to point C. As more $CaCl_2$ was added to the solution, the electrode potential increased slowly until it reached point D. Then it increased rapidly. D is the point which shows the largest change of potential per unit change of concentration and is taken as the end point of the titration. Through the course of the titration, the NEODOL solution remained clear to the eye.

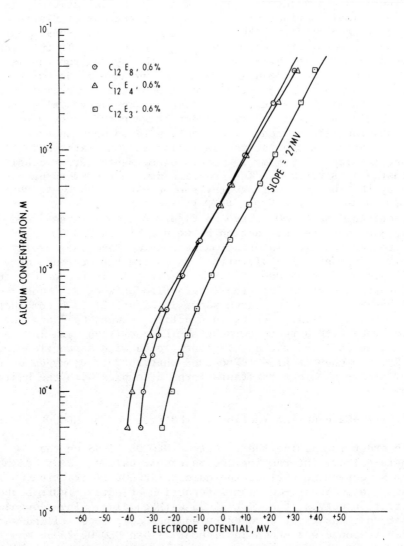

Figure 3. Interaction of calcium with narrow range alcohol oxyethylates.

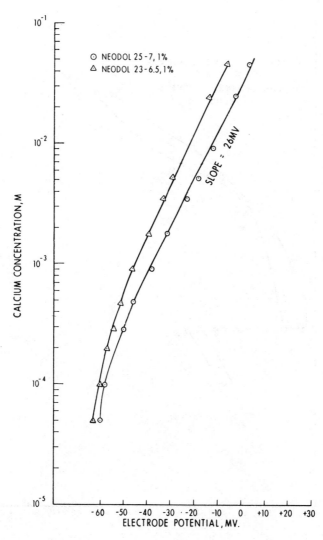

Figure 4. Interaction of calcium with broad range alcohol oxyethylates.

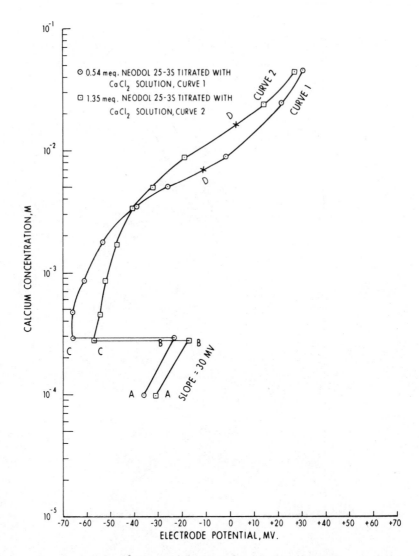

Figure 5. Calcium potential as a function of calcium concentration in the presence of NEODOL 25-3S.

Figure 6 shows a titration curve of sulfonate with $CaCl_2$. Siponate DS-10 (commercial sample of sodium dodecylbenzene sulfonate) was chosen because it forms colorless solution in which changes are easily observed. The electrode response was first checked as shown in line AB. Then Siponate was added to the solution and the potential dropped to point C. As more $CaCl_2$ solution was added, the potential increased only slightly. The solution became increasingly turbid and the potential fluctuated within 1 mv. As the total calcium concentration approached 0.0275 M, the solution became very turbid; precipitation occurred and the potential became stable. Further addition of $CaCl_2$ passing point D caused a steady increase of the potential. D is taken as the stoichiometric end point of the titration.

Figure 7 shows the titration curve of Siponate with $CaCl_2$ in the presence of NEODOL 25-3S. The experiment was done by first checking the electrode response as shown in line AB. When siponate and NEODOL were added to the solution, the potential dropped to point C. The titration with calcium was continued until it passed the end point D. During this titration, the turbidity of the solution increased more slowly than the titration of Siponate alone. Precipitation did not occur until calcium concentration reached 0.1M. Furthermore, the precipitate of Siponate in the presence of NEODOL is much finer than the precipitate without the NEODOL.

Table 1 shows some stoichiometric calculation based on the end points taken from Figure 5-7. This calculation should be considered approximate since the end point in Figure 5 is not a clear break due to the nature of the two interactions with calcium. However, the end point D was taken consistently in every case. A material balance of the calcium taken under the same criterion is probably valid in the calculation. This calculation shows that when NEODOL and Siponate were added together, each component reacted with the same amount of calcium as if they were in separate solutions. Therefore, one may conclude that the addition of NEODOL 25-3S does not decrease the interaction of sulfonate and calcium but does increase the tolerance of sulfonate to precipitation by calcium (compare the value of 0.0275M to 0.1M of calcium in the absence and presence of 0.0096M NEODOL 25-3S, as mentioned above, to cause precipitation of the sulfonate).

Up to this point, we may conclude that the increased electrolyte tolerance (to separation or precipitation) of sulfoante by adding NEODOL 25-3S is not due to complex formation of Na^+ or Ca^{++} by NEODOL because: (1) the increased electrolyte tolerance is far more than the stoichiometric amount of NEODOL added to the solution, (2) no complex effect was found between NEODOL 25-3S and Na^+ in the Na^+ concentration range of interest (Figure 1) and (3) sulfoante reacted with calcium to form neutral species in the

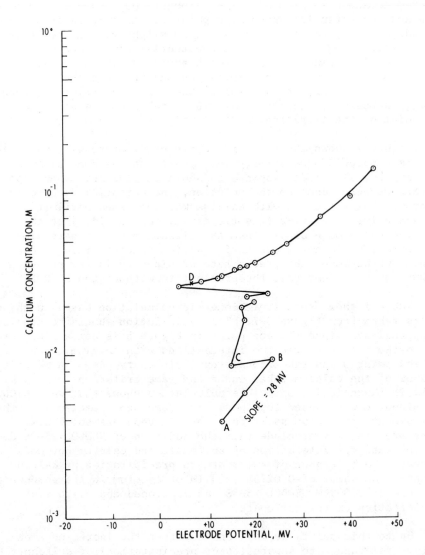

Figure 6. Calcium potential as a function of calcium concentration in the presence of Siponate.

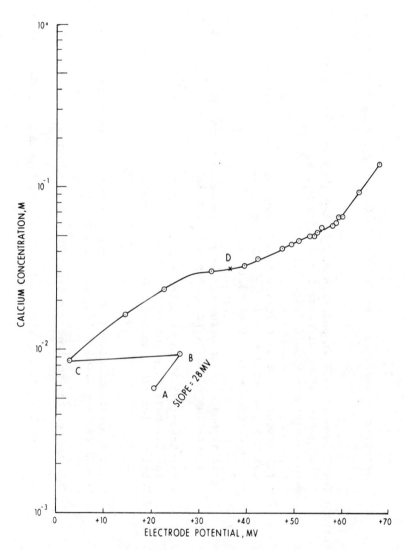

Figure 7. Calcium potential as a function of calcium concentration in the presence of Siponate and NEODOL 25-3S.

Table 1. Stoichiometric Calculation from End Points Given by Figure 5 Through 7.

Figure 5

Curve 1   Total Neodol 25-3S in solution = 0.54 meq (hyamine titration)
           Total Calcium added at point D = 0.174 m moles
           Neodol/Calcium = 0.54/0.174 = 3.1

Curve 2   Total Neodol 25-3S in solution = 1.35 meq (hyamine titration)
           Total Calcium added at point D = 0.44 m moles
           Neodol/Calcium = 1.35/0.44 = 3.1

Figure 6   Total Siponate in solution = 1.68 meq (hyamine titration)
            Total Calcium added at point D = 0.70 m moles

Figure 7   Total Neodol 25-3S in solution = 0.405 meq
            Total Siponate in solution = 1.68 meq
            Total Calcium added at point D = 0.85 m moles

            Under separate condition (from Figures 5 and 6) Calcium reacted to
            Neodol 25-3S = 0.405 X 1/3 = 0.135 m moles

            Calcium reacted to 1.68 meq Siponate = 0.70 m moles.

            Total Calcium reacted with Neodol 25-3S and Siponate = 0.84 m moles.

presence of NEODOL. This effect of NEODOL 25-3S to prevent sulfonate precipitation could be due to one or both of these reasons: (1) a salting in effect of metal ions complexing with the oxyethyl groups in the mixed micelles formed by sulfonate and NEODOL 25-3S and (2) a dispersion effect of NEODOL 25-3S to the calcium sulfonate or sodium sulfonate.

Salting-in of Oxyethylates by Metal Ions

Schott[29] has examined the cloud point of Triton X-100 (containing 10 oxyethylene) and Carbowax Compound 20M (containing 400 oxyethylene). He concluded that H^+, Pb^{++}, Cd^{++}, Mg^{++}, Ni^{++}, Al^{+++}, Ca^{++} and Li^+ ions salt in the polyoxyethylated compounds (raising the cloud point). Only Na^+, K^+ and NH_4^+ ions salt out these compounds. Schott has attributed the salting-in of these compounds to complexation of the ether oxygen with the metal ions. Some different results and explanation on this subject have also been given by Kuriyama[30] and Nishikido and Matuura.[31] In order to examine this effect in compounds containing low oxyethylene, light scattering measurements were performed with narrow range oxyethylated alcohols.

Figures 8-9 show the effect of different salts on the solubility of the narrow range oxyethylated alcohols. The scattered light was measured at 90° to the incident light. The intensity of the scattered light reflects the size of the aggregate. Figures 8-9 seem to show that for each oxyethylated compound a similar pattern appears irrespective of the type of electrolyte added to the solution. In the concentration range we studied, little effect was found on $C_{12}E_6$ by adding electrolyte. A gradual increase of scattered light was found in $C_{12}E_3$ solutions while a slight decrease of scattered light was found in $C_{12}E_4$ solution as electrolyte concentration was increased. The increase of scattered light in $C_{12}E_3$ solutions appeared to be caused by a salting out effect as evidenced by the fact that the solutions containing 1.0M NaCl precipitated after aging for two weeks. The effect on the $C_{12}E_4$ solutions could also be a slight salting out. The CMC values of $C_{12}E_3$-$C_{12}E_6$ are in the range of 0.001-0.004%[32,33] and the cloud points are 10°C ($C_{12}E_4$) and 55°C ($C_{12}E_6$)[32,33]. The cloud point of $C_{12}E_3$ is too low to be measured by our present setup. The addition of electrolyte to $C_{12}E_3$ at 24°C causes the large aggregates first to increase in size and then to coagulate and precipitate. It may also cause the large micelles formed above the cloud point in $C_{12}E_4$ solutions to dehydrate and shrink in size. The reason to believe this is that the CMC of this compound is decreased by adding CaCl$_2$ at 24°C.[33] In summary, we have not seen a significant salting in effect in these low oxyethylene compounds.

If we look at Schott's results[29] again, we may come to the conclusion that the salting-in of nonionic surfactants by metal

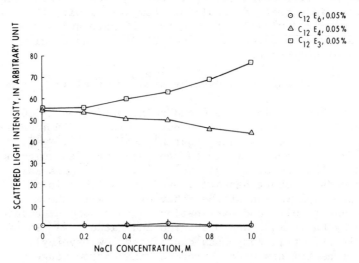

Figure 8. Effect of NaCl on solubility of alcohol oxyethlyates

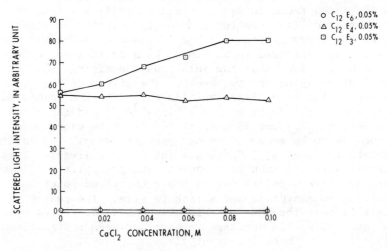

Figure 9. Effect of CaCl$_2$ on solubility of oxyethylated alcohol.

ions may be induced by the complexation, but the effect of raising the cloud point is not proportional to the amount of complexing agent in the solution. For instance, the cloud point of a 2% Triton X-100 can be increased continuously as the concentration of the metal salts is increased even above 2 or 3 molar. A 2% Triton X-100 contains only 0.29 M of the oxyethylene (taken 680 as the molecular weight[28] of Triton X-100). After the reaction of all of the oxyethelene in Triton X-100 with the metal ion, further increase of the cloud point is probably due to change of water structure by the metal salts.[34]

Dispersion Effect of the oxyethylated surfactants

Schönfeldt[35] studied the effectiveness of lime soap dispersants and found that surfactants containing oxyethyl groups, either the nonionic or the anionic surfactants are very effective lime soap dispersants. With these surfactants, usually less than 6% (based on the soap) is required to disperse all the calcium soaps formed in 534 ppm hardness. To produce the same effect would require 40 times this amount of sodium dodecylbenzene-sulfoante or 25 times this amount of sodium lauryl sulfate (that is more lime soap dispersant than soap). If 3 oxyethyl groups are built in the lauryl sulfate, only 1/30 of this surfactant is required to produce the same effect as sodium lauryl sulfate.

The tremendous effectiveness of the oxyethyl group is probably due to the very weak interaction between the oxyethyl groups and calcium (such as those illustrated by Figures 3-4). As a result, the molecule remains hydrated and has good dispersing power. On the other hand, the alkyl sulfate or the alkylaryl sulfonate molecules react strongly with calcium. After all the calcium is reacted, the excess surfactant can now act as dispersant for all the calcium salts formed in the solution. That is why it requires so much of the anionic surfactants to perform the same job. Another example is shown in the Abex-calcium interaction.

Abex-Calcium Interaction

Abex 707 is a dodecyl diphenyl disulfonate produced by Alcohac Chemical Corporation. It has a similar basic structure as Dowfax except the ether oxygen between the two benzene rings is missing. Figure 10 shows the interaction between Abex and calcium. Very strong interaction occurred when $CaCl_2$ was added to the Abex solution. The reaction caused precipitation of the surfactant at high surfactant concentration and interferred with the experiment. Therefore, the measurements were made at 0.1% and 0.3% Abex solutions. Due to the very strong interaction, end point (D) is detected easily. It is interesting to see that without the ether oxygen, the Abex surfactant has little tolerance to calcium.

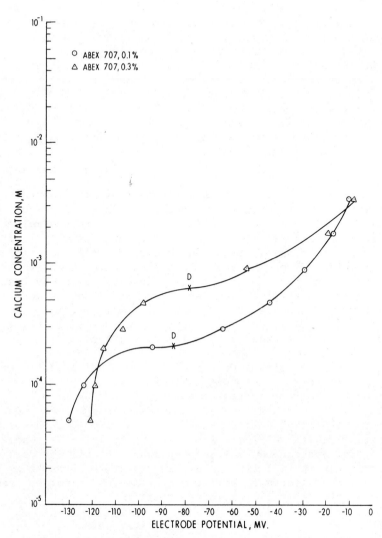

Figure 10. Abex and calcium interaction as measured by a calcium electrode.

Since we do not have enough information about the Abex sample, no attempt has been made to calculate the stoichiometric ratio of calcium to Abex at the end point. With the ether oxygen, the Dowfax molecule shows high calcium tolerance.

Mechanism of NEODOL 25-3S as a Cosurfactant

From the discussion presented above. it appears that the capability of NEODOL 25-3S to increase electrolyte tolerance of sulfo-

nate toward precipitation depends largely on the dispersion power
of the NEODOL 25-3S molecule. The dispersion effect of NEODOL
25-3S on petroleum sulfonate has been observed during our daily
practice of surfactant dissolution. When a petroleum sulfonate
separates out of the solution containing a large amount of electrolyte (2M NaCl or so), the addition of NEODOL 25-3S always brings it
back into solution. The mechanism of such dispersion is probably
via the formation of mixed surfactant aggregates. Stirton et al[36]
have given a schematic representation of the action of a lime soap
dispersing agent through mixed micelle formation. Additional evidence that suggests mixed aggregate formation in a surfactant-
cosurfactant system is given in Reference 17. Furthermore, the
mixed aggregates of surfactant and cosurfactant have smaller sizes
than the aggregates of the surfactant under the same salinity
condition.[17]

According to the mechanism of Stirton, et al,[36] the molecular
structure of an effective lime soap dispersant is characterized by
a bulky hydrophilic polar end which does not precipitate with
calcium. In the NEODOL 25-3S molecule, the combination of the
oxyethylene groups and the sulfate group constitutes a bulky hydrophilic polar end which does not precipitate with calcium. Barry
and Wilson[26] found a significant increase of the area per molecule
of the sodium dodecyl sulfate when one and two oxyethyl groups are
inserted in the molecule. They have also found that the presence
of the oxyethylene group increases the degree of dissociation of
the sodium ion[26] and decreases the aggregation number of the
micelles.[25] Hato and Shinoda[23] measured the Kraft points of sodium and calcium dodecylpoly (oxyethylene) sulfates (with 1-3 oxyethylene groups) and found that the Kraft points were depressed
with an increase in oxyethylene chain length. When the calcium
salt of dodecyl sulfate is mixed with the calcium salt of the
dodecylpoly (oxyethylene) sulfates, the Kraft points of the resulting systems were depressed in a fashion similar to the freezing
point depression of binary mixtures of ordinary substances.[23]

From our specific ion electrode measurements, we found that
at the end point of the NEODOL 25-3S titration with calcium, there
are three NEODOL sulfate molecules per calcium ion in the solution
(Figure 5 and Table 1). Since the interaction of calcium with the
oxyethylene groups is weak (Figures 3 and 4), one could assume that
when the calcium concentration is limited, most of the calcium
reacts with the sulfate group. In order to form a neutral compound, each calcium must take up two NEODOL sulfate molecules. In
effect, one-third of the NEODOL sulfate in the solution is ionized
and the solution is perfectly clear. The micelles in this system
are probably charged negatively, have high degree of ionization
and low aggregation number compared with the parent alkyl sulfates
(no oxyethylene groups).

The presence of the oxyethylene groups in the NEODOL 25-3S provides solubility for the molecule when a large amount of calcium has reacted with the sulfate group and increases the ionization of the metal salts over that of the parent alkyl sulfates. (This also implies that the oxyethylated derivatives would have higher solubility in high electrolyte concentrations). When NEODOL 25-3S is mixed with petroleum sulfonates, mixed micelles (or aggregates) are formed. The mixed micelles have properties intermediate between NEODOL 25-3S and the petroleum sulfonate depending on the proportion of these two components. Since NEODOL 25-3S has high solubility in high electrolyte concentrations the solubility of the sulfonate-NEODOL system is also increased in high electrolyte concentrations.

The three oxyethylene groups (average number) provides optimum effect on increasing solubility of the parent alkyl sulfate of NEODOL 25-3S for practical utility. More oxyethylene should increase dispersion power of the molecule but may decrease the sulfate aggregate size to below the requirement for maintaining sufficient recovery activity. The presence of an anionic group in a cosurfactant molecule is also important. It increase the counter ion requirement[17] because of its large ionization constant so that the surfactant-cosurfactant system will become active at high electrolyte concentration. At higher temperatures, the anionic group is essential for a cosurfactant. Many nonionic surfactants have low cloud points and cannot be used as cosurfactants at high temperatures. Even those with high oxyethylene content and high cloud point seem to lose the dispersing power for petroleum sulfonate at high temperatures and high calcium concentrations.

NEODOL 25-3S works well with petroleum sulfonate of molecular weight around 430.[2] In this surfactant-cosurfactant system a delicate balance between all groups in the mixed micelles may have been obtained so that this system would have the micelles neither too large to be flocculated by high electrolyte concentration nor too small to lose recovery activity. When the sulfonate molecular weight or structure is substantially changed, the NEODOL structure must be adjusted to maintain the delicate balance.

REFERENCES

1. SPE Symposium on Improved Oil Recovery, Tulsa, Oklahoma, 1972 and thereafter.
2. J. Reisberg and J. B. Lawson, U. S. Patent 3,508,612, (1970).
3. D. L. Dauben and H. R. Froning, J. Pet. Tech. 23,614 (1971).
4. J. Maddox Jr., J. F. Tate and R. D. Shupe, U. S. Patent 3,916,994, (1975).
5. R. D. Shupe, J. Maddox Jr. and J. F. Tate, Patent 3,916,995, (1975).
6. R. D. Shupe, J. Maddox Jr. and J. F. Tate, U. S. Patent 3,916,996, (1975).

7. R. D. Shupe, J. Maddox Jr. and J. F. Tate, U. S. Patent 3,945,493, (1976).

8. J. F. Tate, J. Maddox Jr. and R. D. Shupe, U. S. Patent 3,946,813, (1976).

9. W. W. Gale, R. K. Saunder and T. L. Ashcraft Jr., U. S. Patent 3,997,471, (1976).

10. R. F. Farmer, J. B. Lawson and W. M. Sawyer, U. S. Patent 3,943,160, (1976).

11. V. K. Bansal and D. O. Shah, Soc. Pet. Eng. J. 18,167 (1978).

12. V. K. Bansal and D. O. Shah, J. Colloid Interface Sci., 65, 451 (1978).

13. V. K. Bansal and D. O. Shah, J. Amer. Oil Chem. Soc., 55, 367 (1978).

14. Y. C. Chiu and H. J. Hill, U. S. Patent 3,945,437, (1976).

15. Y. C. Chiu and H. J. Hill, U. S. Patent 4,013,569, (1977).

16. C. J. Glover, M. C. Puerto, J. M. Maerker and E. L. Sanduik, paper presented at the Symposium on Improved methods for Oil Recovery of the SPE-AIME, paper No. 7053, April, 1978.

17. Y. C. Chiu, in "Solution Behavior of Surfactants : Theoretical and Applied Aspects", K. L. Mittal and E. J. Fendler, Editors, Vol 2, pp.1415-1440, Plenum Press, New York, 1982.

18. W. M. Linfield, Editor,"Anionic Surfactants", pp. 136-217, Marcel Dekker, Inc. New York, 1976.

19. J. K. Weil, R. G. Bistline and A. J. Stirton, J. Phys. Chem., 62, 1083 (1958).

20. F. Tokiwa and K. Ohki, J. Phys. Chem., 71, 1343 (1967).

21. F. Tokiwa, J. Phys. Chem., 72, 1214 (1968).

22. D. Attwood, Kolloid-Z. Z. Polym. 232, 788 (1969).

23. M. Hato and K. Shinoda, J. Phys. Chem., 77, 378 (1973).

24. K. Shinoda and T. Hirai, J. Phys. Chem., 81, 1842 (1977).

25. B. W. Barry and R. Wilson, Colloid Polymer Sci., 256, 44 (1978).

26. B. W. Barry and R. Wilson, Colloid Polymer Sci., 256, 251 (1978).

27. H. B. Klevens, J. Amer. Oil Chem. Soc., 30, 74 (1953).

28. T. M. Doscher, G. E. Myers, and D. C. Atkins, J. Colloid Sci., 6, 233 (1951).

29. H. Schott, J. Colloid Interface Sci., 43, 150 (1973).

30. K. Kuriyama, Colloid Z.u.Z. Polym. 181, 144 (1962).

31. N. Nishikido and R. Matuura, Bull. Chem. Soc., Japan, 50, 1690 (1960).

32. H. L. Benson and Y. C. Chiu, paper presented at 70th Annual Meeting of Amer. Oil Chemists' Soc., (May 1979).

33. H. L. Benson and Y. C. Chiu, Technical Bulletin SC:443-80,Shell Chem. Co., Houston, Texas, U. S. A. 1980.

34. J. L. Kavanau,"Water and Solute-Water Interaction," Holden-Day, Inc., 1964.

35. N. Schönfeldt, J. Amer. Oil Chem. Soc., 45, 80 (1968).

36. A. J. Stirton, F. D. Smith and J. K. Weil, J. Amer. Oil Chem. Soc., 42, 114 (1965).

LOCAL ANESTHETIC-MEMBRANE INTERACTION: A SPIN LABEL STUDY OF

PHENOMENA THAT DEPEND ON ANESTHETIC'S CHARGE

Shirley Schreier, Wilson A. Frezzatti, Jr., Pedro S. Araujo, and Iolanda M. Cuccovia

Departmento de Bioquímica, Institute of Chemistry University of São Paulo, C.P. 20780, São Paulo, Brazil

The electron paramagnetic resonance (EPR) spectra of spin probes were examined in order to obtain information about the interaction between the local anesthetic tetracaine and egg phosphatidyl choline membranes. We found that (1) at low pH, where the charged form of tetracaine predominates, a small degree of lipid organization is observed; the opposite is true in the high pH region; (2) the plot of an empirical parameter that reflects the degree of lipid organization as a function of pH yields a curve whose mid point occurs at lower pH than the pK of the anesthetic in aqueous solution. This suggests that the EPR spectra are sensitive to changes in apparent pK due to partitioning of the anesthetic in the membrane; (3) increasing the amount of uncharged tetracaine beyond a given concentration, a saturation level is attained where cluster formation is detected, as indicated by a second spectral component; (4) the charged form of tetracaine forms micelles capable of disrupting phospholipid membranes (Fernandez, Biochim. Biophys. Acta 646, 27 (1981)); it was suggested that the phospholipid was incorporated into mixed micelles. Spectral changes observed for a spin-labeled phospholipid provide further evidence for this incorporation.

INTRODUCTION

The mechanism of action of the tertiary amine local anesthetics is thought to involve binding at the sodium channel in nerve cells[1]. However, those compounds have been seen to interact with lipid membranes in a variety of ways[2-6], and these interactions have been proposed to play a role in the overall mechanism of anesthesia. In particular, it has been proposed that the uncharged form of these tertiary amines is necessary to cross the membrane and that the protonated form, regenerated in the cell cytoplasm, binds to a receptor site in the sodium channel located near the inside surface of the plasma membrane[1].

Recent work, making use of magnetic resonance techniques[7-12], has provided insight into the molecular details of the interaction between local anesthetics and lipid membranes. Employing several deuterated derivatives of tetracaine (Figure 1), Boulanger and coworkers[11] demonstrated that the binding of the local anesthetic to egg phosphatidyl choline membranes can be described by multiple equilibria, and that the uncharged form of tetracaine binds to a site located more deeply in the membrane than that for the protonated form. The latter was seen to be in fast exchange with a population of tetracaine in the aqueous phase. When deuterated lipids were used[12], the results corroborated the previous findings in that large effects on the nuclear magnetic resonance spectra (NMR) of the deuterated head group were seen in the presence of the cationic form of tetracaine. The results resembled those obtained when trivalent ions were added to a phospholipid suspension[13]. In contrast, much smaller changes were detected at the head group when the pH was raised to yield the uncharged form of the anesthetic. As expected, the effects caused by this form were more pronounced in the acyl chain region[12].

The interest in examining whether local anesthetics would be involved in lipid polymorphism phenomena led Hornby and Cullis[14] to investigate the effect of these compounds on the bilayer to hexagonal II phase transition of egg phosphatidyl ethanolamine. It was found that the anesthetics are capable of stabilizing the bilayer phase of the phospholipid.

In addition, Fernandez[15] found that the charged form of tetracaine is capable of forming micelles with a critical micelle concentration (cmc) of ca. 0.07M. She found that the tetracaine micelles could disrupt unilamellar vesicles of egg phosphatidyl choline and suggested that the phospholipid was incorporated in the anesthetic micelle[16].

In this paper, we report on the increased lipid organization caused by the charged form of tetracaine, and on the opposite effect induced by the uncharged form. In addition, the EPR spectra of spin probes seem sensitive to the change of anesthetic apparent

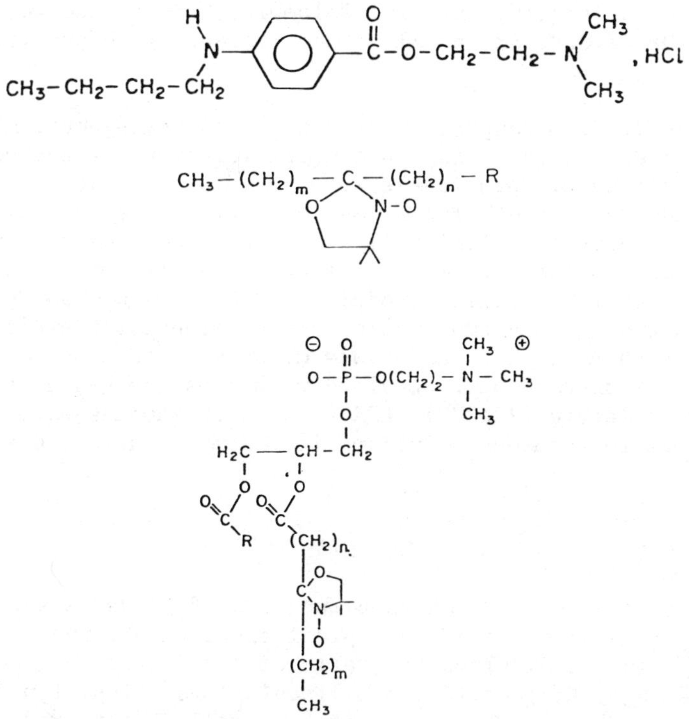

Figure 1. Top: tetracaine hydrochloride. Middle: Probe I,
R = $-\underset{\underset{OCH_3}{|}}{C}=O$, n=3,m=12,stearic acid methyl ester containing the nitroxide
moiety at C(5); probe II. R = $-\underset{\underset{OH}{|}}{C}=O$,   n=3, m=12, stearic acid con
taining the nitroxide moiety at C(5). Bottom: Probe III, spin-
labeled derivative of dipalmitoyl phosphatidyl choline containing
the nitroxide moiety at C(5) of the <u>sn</u>-2 acyl chain.

pK in the presence of membrane. The spectra are also indicative of
cluster formation, in the membrane, by the uncharged form of tetracaine. Finally, we provide spectroscopic evidence for the incorporation of phospholipid into micelles of tetracaine.

MATERIALS AND METHODS

Tetracaine hydrochloride, from Sigma Chemical Co., St. Louis,
Mo., was a generous gift of D. I.C.P. Smith, National Research
Council of Canada. Egg phosphatidyl choline came from Lipid
Products, South Nutfield, England and was also extracted in our
laboratory according to Nielsen[17]. The spin probes used in this
work are shown in Figure 1. Labels I and II were purchased from
Syva, Palo Alto, Calif. Spin label III was synthesized by P. Laks,

Simon Frazer University, British Columbia, Canada, and was also
a gift of Dr. I.C.P. Smith. All other reagents were analytical
grade.

Borate-citrate-phosphate buffer was used throughout. Membranes
were prepared by evaporating chloroform solutions of the components
under wet nitrogen. Spin labels were added at a ratio of 1:100
label:lipid (mole:mole). The films were dried under vacuum for no
less than two hours. Multilamellar membranes were dispersed in a
Vortex mixer. Unilamellar membranes were prepared by sonification
with a Braunsonic sonifier, (Model 1510) at 100W nominal power
for 30 minutes, with a two minutes on-one minute off cycle, in an
ice-water bath under N_2. The pH was measured with a Metrohm E512
pH meter. EPR spectra were obtained with a Varian E-4 spectrometer
at room temperature (22±2°C). EPR spectra of samples were taken in
quartz cells for aqueous solutions from James Scanlon, Costa Mesa,
CA.

RESULTS

The effect of tetracaine as a function of pH was examined
with probe I. Figure 2 shows a typical spectrum obtained in this
study and Figure 3 displays the ratio of the heights of the low
field line (h_{+1}) to the mid field line (h_o) as a function of pH.
At low pH, the h_{+1}/h_o ratio undergoes a small, but reproducible
decrease. Between pH 6.5-9.0, the ratio goes through and increase
and then levels off. The increase in h_{+1}/h_o at high pH depends on
anesthetic concentration. Values of h_{+1}/h_o were calculated for
spectra of samples containing variable tetracaine concentrations
at pH 10. The differences Δ between these values and those obtained in the absence of anesthetic are given in Table I. It is
seen that Δ increases up to ca. 5mM tetracaine. For higher anesthetic concentrations, the h_{+1}/h_o ratios become smaller (smaller
Δ), and can reach values below that of the control. Figure 4
illustrates the results obtained for 10mM anesthetic. The decrease
in h_{+1}/h_o correlates with the appearance of a second component in
the spectra of I (Figure 5). The low field shoulder that corresponds to this component is indicated by an arrow in Figure 5.

The new spectral component is associated with a new phase
being probed by the spin label. In the pH region above the pK
of tetracaine (8.5, ref. 18, and our own determination), the
anesthetic precipitates at the concentrations used in this study.
In order to check whether the additional spectrum was due to label
extracted from the membrane into some aggregate of uncharged tetracaine, we made a film of I on the walls of a tube and added to it
tetracaine at pH 10.0, in the absence of membrane. This system
was not able to extract the label, yielding no detectable EPR
spectrum.

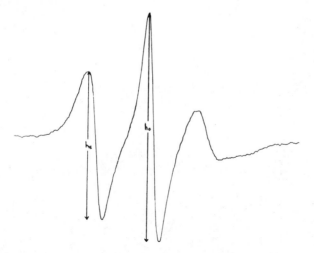

Figure 2. EPR spectrum of probe I in a 5.2mM multilamellar dispersion of egg phosphatidyl choline, pH 7.0. The figure shows how the heights of the low field (h_{+1}) and mid field (h_o) resonances were measured in order to calculate the h_{+1}/h_o ratio.

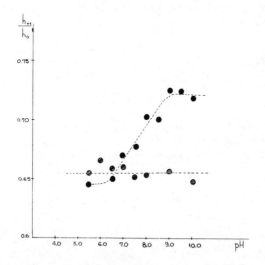

Figure 3. h_{+1}/h_o ratio as a function of pH for 5.2mM phosphatidyl choline membranes in the absence (○) and in the presence (●) of 2.3mM tetracaine.

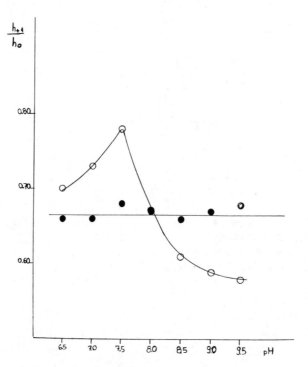

Figure 4. h_{+1}/h_0 ratio as a function of pH for 5.2mM egg phosphatidyl choline membranes in the absence (●) and in the presence
(○) of 10mM tetracaine.

Table I. Difference (Δ) between h_{+1}/h_0 Ratios for Spectra of
Probe I in 5.2mM Egg Phosphatidyl Choline Samples Containing
Variable Tetracaine (TTC) Concentrations and in Controls (pH
10.0).

[TTC] (mM)	0.68	1.27	1.85	2.44	3.92	4.69	5.44	6.95
Δ	0.035	0.040	0.056	0.063	0.083	0.079	0.067	0.033

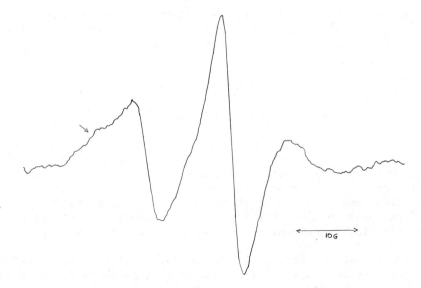

Figure 5. EPR spectrum of probe I in 5.2mM egg phosphatidyl choline membranes in the presence of 20mM tetracaine, pH 9.5. The arrow indicates the low field shoulder due to the appearance of a second spectral component.

Next, we examined the formation of micelles by tetracaine, and their effect upon lipid structure. We repeated the previous experiment (i.e., an attempt to solubilize a film of label, this time II, with a solution of anesthetic) with various concentrations of tetracaine at pH 6.0. Spectra characteristic of the probe in an aggregate were not obtained until a concentration of anesthetic above its reported cmc (0.07M[15]) was used. Above this concentration, the spectra of II were similar to those yielded by other micellar systems[19].

Labels II and III are known to yield similar spectra in lipid bilayers. The spectra of II in micelles are well known[19], and have been shown to differ greatly from those in membranes. In order to examine the effect of charged tetracaine on the structure of egg phosphatidyl choline membranes, the spectra of the phospholipid probe III in these membranes were obtained as a function of anesthetic concentration at pH 6.0. In the absence of anesthetic (Figure 6A) and below the cmc of tetracaine, the spectra of III are characteristic of a lipid bilayer. A drastic change occurs

above the cmc of tetracaine and the spectra become typical of the
label in a micelle (Figure 6B). The results were the same whether
the membranes were unilamellar or multilamellar.

DISCUSSION

Effect of Anesthetic on Lipid Organization

The results in Figure 3 provide information about a variety
of events taking place during the interaction between tetracaine
and egg phosphatidyl choline membranes. Since the partition
coefficients, K_p, for protonated and unprotonated tetracaine are
very   different   (K_p=22 at pH 5.5 and K_p=660 at pH 9.5)[11] , it
should be borne in mind that for a given aqueous concentration of
the anesthetic, the amount in the membrane will vary with pH.

Figure 3 shows that, although the effect of protonated tetracaine seems to be small, there seems to be a definite decrease in
the h_{+1}/h_o ratio. Probe I is an ester, thus lacking a group that
would strongly anchor it to the membrane surface. As a consequence,
its spectra lack the inner and outer extrema that allow the calculation of an order parameter[20]. Instead, the spectral features
are indicative of a considerable degree of motional freedom, and,
although the probe undoubtedly undergoes a certain degree of ordering, it has been considered acceptable, for comparison purposes,
to assume the motion of I as pseudo-isotropic and to measure the
h_{+1}/h_o ratio as an indication of that motional freedom. The
decrease in h_{+1}/h_o in the presence of tetracaine at low pH is thus
associated with a small decrease of motional freedom in the membrane.
This is in agreement with previous results found by deuterium
NMR[11,12].

When the anesthetic is present mainly in the uncharged form,
the h_{+1}/h_o ratio increases, suggesting that this species is solubilized in the acyl chain region of the membrane causing an increase in the freedom of motion in this region. This again is in
agreement with deuterium NMR results[12] and with previous spin label
studies[8]. Increasing the anesthetic concentration increases the
effect (Table I) due to a greater partitioning of tetracaine in
the membrane.

Effect of Membrane on Anesthetic Apparent pK

A striking effect concerns the pH value where the mid-point
in the h_{+1}/h_o versus pH curve occurs (Figure 3). This value is
lower than the pK for the anesthetic in aqueous phase (8.5). Titration studies[21] indicate a measurable decrease of anesthetic pK in
the presence of membrane. This is in agreement with the prediction

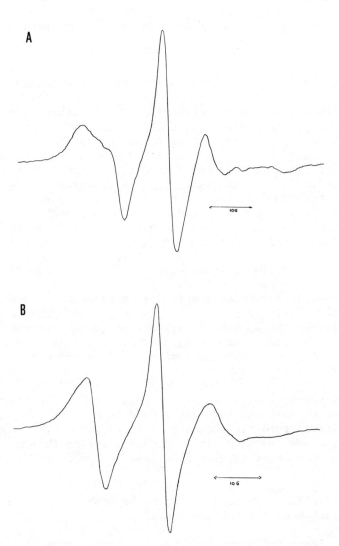

Figure 6. A. EPR spectrum of probe III in 5.2mM egg phosphatidyl choline membranes, pH 6.0. B. Same as A, in the presence of 80mM tetracaine.

by Lee[22] and in disagreement with his experimental findings. Our results suggest that the mid-point in the inflection between pH 6.5 and 9.0 in Figure 3 corresponds to the theoretically predicted and experimentally demonstrated (titration studies[21]) decrease of tetracaine pK in the presence of egg phosphatidyl choline membranes, as a result of the different partitioning of the charged and uncharged forms of the anesthetic.

Clustering of Uncharged Tetracaine in the Membranes

At high pH, as the concentration of anesthetic increases, the h_{+1}/h_o ratio increases too, till a maximum value is reached. For higher concentrations, the ratio decreases (Table I and Figure 4). In this case, the decrease is not due to a decrease in motional freedom, but rather to the appearance of the additional spectral component clearly seen in Figure 5. The decrease in h_{+1}/h_o is greater as the concentration of anesthetic increases. We interpret this result in the following manner: at high enough tetracaine concentration, the membrane become saturated with anesthetic, and additional molecules segregate into clusters, giving rise to a second spectrum. We cannot at present establish whether the clusters also contain phospholipid.

The cluster formation observed in this work is interesting in view of the existing theories of anesthesia that propose a fundamental role for the anesthetic solubility in the membrane[23]. Our results suggest that a more complex picture should be envisioned, i.e., that different membrane regions could contain different anesthetic concentrations.

The appearance of a second component in the EPR spectra of spin probes in membranes has often been interpreted as being due to immobilized lipid. In the case of tetracaine, by virtue of the molecule's shape and size, it is highly unlikely that the additional spectral component (Figure 5) is due to immobilized lipid.

Micellar Tetracaine Disrupts the Membrane Forming Lipid-Anesthetic Mixed Micelles

Fernandez[15] has shown that the protonated form of tetracaine forms micellar aggregates. From light scattering measurements and phosphorus analysis of filtrates, she showed that micellar tetracaine was capable of disrupting unilamellar vesicles of phosphatidyl choline at an anesthetic:lipid ratio of 100:1. It was suggested that the lipid was incorporated into the anesthetic micelles[16].

Our results in Figure 6 confirm that indeed this is the fate of the lipid. Our data represent a more direct evidence of the event, since the EPR spectrum of a spin-labeled phospholipid clearly indicates that the molecule is in a micellar aggregate when protonated tetracaine is present at concentrations above its cmc. In addition, we found that the formation of mixed micelles takes place at higher lipid:anesthetic molar ratios (up to 1:14, in this work) and that the detergent action of tetracaine does not depend on whether the membrane is uni- or multi-lamellar.

In conclusion, we have shown that charged tetracaine increases the degree of lipid organization, while the opposite takes place in presence of uncharged form. We have found that the EPR spectra can monitor the change in anesthetic apparent pK in the presence of membrane. Moreover, we have shown that the uncharged form of tetracaine has the ability to cluster in the membrane. Finally, we have provided spectroscopic evidence for membrane disruption and lipid incorporation in micelles formed by the protonated anesthetic.

Hornby and Cullis[14] have shown that local anesthetics influence the bilayer to hexagonal II phase transition, and Menashe et al.[24] have suggested that tetracaine can promote fusion of small unilamellar vesicles. Thus, the anesthetic has the ability to interfere with the state of lipid organization in a variety of ways. These effects should be taken into account in further studies of anesthetic-membrane interaction and in considerations of possible steps in the mechanism of action of anesthetics.

ACKNOWLEDGEMENTS

We thank Dr. H. Chaimovich for helpful discussions and Mr. Wilson R. Toselli for excellent technical work.

REFERENCES

1. G. Strichartz, Anesthesiol., 45, 421 (1976).
2. D. Papahadjopoulos, Biochim. Biophys. Acta, 265, 169 (1972).
3. D. Papahadjopoulos, K. Jacobson, G. Poste and G. Shepherd, Biochim. Biophys. Acta, 394, 504 (1975).
4. S. McLaughlin, in "Molecular Mechanisms of Anesthesia", B. R. Fink, Editor, Vol. 1, p. 193, Raven Press, New York, 1975.
5. M. B. Feinstein, S. M. Fernandez and R. I. Sha'afi, Biochim. Biophys. Acta, 413, 354 (1975).
6. D. B. Mountcastle, R. L. Biltonen and M. J. Halsey, Proc. Nat. Acad. Sci. USA, 75, 4906 (1978).

7. H. Hauser, S. A. Penkett and D. Chapman, Biochim. Biophys. Acta, 183, 466 (1969).
8. K. W. Butler, H. Schneider and I.C.P. Smith, Arch. Biochem. Biophys., 154, 548 (1973).
9. G. J. Giotta, D. S. Chan, and H. H. Wang, Arch. Biochem. Biophys., 163, 453 (1973).
10. M. S. Fernandez and J. Cerbón, Biochim. Biophys. Acta, 298, 8 (1973).
11. Y. Boulanger, S. Schreier, L. C. Leitch and I.C.P. Smith, Can. J. Biochem., 58, 986 (1980).
12. Y. Boulanger, S. Schreier and I.C.P. Smith, Biochemistry, 20, 6824 (1981).
13. M. F. Brown and J. Seelig, Nature, 269, 721 (1977).
14. A. P. Hornby and P. R. Cullis, Biochim. Biophys. Acta, 647, 285 (1981).
15. M. S. Fernandez, Biochim. Biophys. Acta, 597, 83 (1980).
16. M. S. Fernandez, Biochim. Biophys. Acta, 646, 27 (1981).
17. J. R. Nielsen, Lipids, 15, 481 (1980).
18. B. G. Covino and H. G. Vassalo, "Local Anesthetics: Mechanism of Action and Clinical Use", Grune and Stratton, New York, 1976.
19. J. R. Ernandes, S. Schreier and H. Chaimovich, Chem. Phys. Lipids, 16, 19 (1976).
20. S. Schreier, C. F. Polnaszek and I.C.P. Smith, Biochim. Biophys. Acta, 515, 395 (1978).
21. I. M. Cuccovia, P. S. Araujo, H. Chaimovich, W. A. Frezzatti, Jr. and S. Schreier, in preparation.
22. A. G. Lee, Biochim. Biophys. Acta, 514, 95 (1978).
23. M. J. Pringle, K. B. Brown and K. W. Miller, Mol. Pharmacol., 19, 49 (1981).
24. M. Menashe, C. F. Schmidt, T. G. Conley and R. L. Biltonen, Biophys. J., 27, 199a (1982).

ABOUT THE CONTRIBUTORS

Here are included biodata of only those authors who have
contributed to this volume. Biodata of contributors to Volumes
1 and 2 are included in those volumes.

T. Åkesson is with the Physical Chemistry 2, Chemical
Center, Lund, Sweden.

Pedro Soares de Araujo is Assistant Professor in the Department of Biochemistry, Institute of Chemistry, University of
Sao Paulo, Sao Paulo, Brazil. He received his Ph.D. degree in
Biochemistry and spent two years at the University of Utrecht,
The Netherlands, with Prof. G. de Haas investigating the association of phospholipase with micelles. His current research interests include analytical clinical chemistry and association of
proteins with interphases, and has published about 15 papers.

T. Assih is a graduate student in the Physics Department,
Universite des Sciences et Techniques du Languedoc, Montpellier,
France.

M. Barthe is Chemical Engineer of the Bordeaux Chemical
Ingeneer School. Her studies on microemulsions (at the Centre de
Recherche Paul Pascal in Bordeaux-Talence) are supported by the
French Oil Company Elf-Aquitaine in order to get the grade of
Doctor Ingineer.

Paul Becher is the founder of Paul Becher Associates Ltd.,
a consulting firm which he founded in 1981 after taking an early
retirement from ICI Americas Inc. He joined ICI Americas (former-
ly Atlas Powder Company) in 1957 and has had various research and
management positions. He received his Ph.D. degree in Chemistry
from Polytechnic Institute of Brooklyn in 1948. He has been very
active in the Division of Colloid and Surface Chemistry of the
American Chemical Society serving as Chairman in 1974 and was a
member of the Editorial Advisory Board of J. Colloid Interface
Science (1967-1969) and since 1969 has been the editor of Section
46, Surface Active Agents, of Chemical Abstracts. He has been

2157

very active in the Delaware Section of the ACS and the Society of
Cosmetic Chemists, and was chairman of the Gordon Research Conference on Chemistry at Interfaces. In 1965, he shared with Mr.
W. C. Griffin the Literature Award of the Society of Cosmetic
Chemists. He has authored or edited a number of books including
an ACS monograph Emulsions: Theory and Practice which has been
translated into Hungarian, Spanish and Chinese. He is the editor
of Encyclopedia of Emulsion Technology and coeditor of J.
Dispersion Sci. Technol., and has authored more than 40 papers.

A. M. Bellocq is Maitre de Recherches at CNRS at the Centre
de Recherche Paul Pascal in Bordeaux-Talence, France. She
received her Doctorat es Sciences in 1969 from the University of
Bordeaux. Her research activities are focused on conformational
studies of peptides, and microemulsions.

Guy C. Berry is Professor of Chemistry and Polymer Science
in the Department of Chemistry, Carnegie-Mellon University. He
received his B.S. Chemical Engineering, M.S. Polymer Science, and
Ph.D. degrees from the University of Michigan. He was Visiting
Professor at Colorado State University (1979) and the University
of Tokyo (1973), and is associated with the University of Pittsburgh as an Adjunct Professor. His research interests include the
physical chemistry of polymers in dilute solution and the rheo-
logy of polymers and their concentrated solutions. He has about
50 publications in these areas and serves as a consultant to a
number of industrial firms and government agencies.

J. Biais is Assistant Professor at the University of
Bordeaux, France. He received his Doctorat es Sciences in 1968.
He was the guest of the Chemical Institute for Surface Chemistry
in Stockholm for three months in 1980. He is a member of the
GRECO "Microemulsion" (French cooperative action on Microemulsion
CNRS) and is consultant of the French Oil Company Elf-Aquitaine.
His research (at the Centre de Recherche Paul Pascal in Bordeaux-
Talence) is focused on molecular interactions in liquids,
thermodynamical properties and model of microemulsions.

Christian Boned has since 1975 been involved in a research
program on microemulsion systems for enhanced oil recovery. He
received his D.Sc. degree in 1973 from the Universite de Pau et
des Pays de l'Adour, Pau, France and was appointed Assistant
Professor of Physics with the same institution in 1967. During
1967-1975 his main research interest has been the dielectric
behavior of pure or doped polycrystalline ice. He is the author
or coauthor of some 35 papers.

T. A. Bostock is in the Department of Chemistry, Sheffield
City Polytechnic, England and was SERC CASE research student.
Received Ph.D. (CNAA, Sheffield City Polytechnic).

P. Bothorel is Professor of Chemistry at the University of Bordeaux, France. He is also Assistant Director of the Centre de Recherche Paul Pascal, CNRS Laboratory in Talence. He is the Scientific Director of the GRECO "Microemulsions" (French cooperative action on microemulsions) and is consultant to the French oil company, Elf-Aquitaine. His research activities are focused on aliphatic chain conformations, biological membrane models, and microemulsions.

J. F. Boyce is a Ph.D. candidate in the Department of Biophysics, University of Western Ontario, London, Ontario, Canada.

Martine Buzier is Attachée de Recherches (CNRS) and is in the Laboratoire de Biophysique, University de Nancy, Nancy, France. Her major research interest is in the properties of nonionic surfactants.

Francoise Candau is "Maitre de Recherches" at the Centre de Recherches sur les Macromolecules, CNRS, Strasbourg, France. She received her D.Sci. degree in 1971 from the University of Strasbourg. Her current research interests are the amphiphile properties of copolymers and microemulsions.

S. J. Candau is "Directeur de Recherches" at the University of Louis Pasteur, Strasbourg, France. He received his D. Sc. degree in 1963 from the University of Strasbourg. He has published in the field of acoustic properties of liquid crystals and quasielastic light scattering studies of gels and micellar solutions.

J. Caspers is with the Laboratoire de Chimie Physique des Macromolecules aux Interfaces, Universite Libre de Bruxelles. Has a Ph.D. degree and has 22 publications.

A. M. Cazabat is with the Ecole Normale Superieure in Paris, France. She was a graduate student at the Ecole Normale Superieure de Jeune Filles. She received her Ph.D. in 1970 and Thesis in 1976.

D. Chatenay is with the Ecole Normale Superieure, Paris, France. Was a graduate student at Ecole Superieure d'Optique (1976-1979) and received a fellowship from the Institut Francais du Petrole.

Ying-Chech Chiu is currently Professor in the Department of Chemistry, Chung-Yuan Christian University, Chung-Li, Taiwan. Before her present position, she worked for 12 years for oil companies in the United States. She received her Ph.D. degree in Physical Chemistry from Baylor University in 1965 and carried out postdoctoral work with Prof. R. M. Fuoss at Yale University and Prof. J. O'M. Bockris at the University of Pennsylvania. Her research interests are in electrochemistry, surface and colloid chemistry, tertiary oil recovery, detergents and petrophysics. She is the author of a number of research publications and patents.

J. H. R. Clarke is a Lecturer in Chemistry at U.M.I.S.T., Manchester, England, and has been involved for the last ten years in research on the dynamical properties of liquids using polarized and depolarized light scattering, and molecular dynamics computer simulation.

Marc Clausse is currently at the Universite de Technologie de Compiegne, located near Paris, which he joined in October 1982. Before his current position, he was Professor at the Universite de Pau et des Pays de l'Adour, Pau, France where he received his D.Sc. degree in 1971. During 1975-1977 he was appointed a Charge de Mission with the French Ministry of Foreign Affairs where he managed the Middle East and Asia Section of the Cultural, Scientific and Technical Cooperation Division. During spring 1982, he was a NATO and CIES sponsored Visiting Professor at the University of Florida, Gainesville, and was a keynote speaker of the "International Energy Information Forum and Workshop for Educators" held June 9-12, 1982 in Gatlinburg, TN. He has recently been vice president of the Club Energetique-Thermodynamique, an association whose main objective is the development in France of energy oriented teaching and research programs. His recearch interests concern the electrical properties of microheterogeneous materials and the physicochemical properties of emulsions and microemulsions, and has published some 40 papers on these topics.

B. Clin is Charge de Recherche at CNRS at the Centre de Recherche Paul Pascal, Talence, France. He received his Doctorat es Sciences in 1972. His research activity is mainly focused on NMR, and microemulsions.

Iolanda Midea Cuccovia is Assistant Professor in the Department of Biochemistry, Institute of Chemistry, University of Sao Paulo, Sao Paulo, Brazil. She received her Ph.D. degree in Biochemistry. She has 12 publications in the field of micellar modified reactions and is currently interested in studying the effect of amphiphile aggregate structure on chemical reactivity.

P. Delord is Professor of Mineralogy and Crystallography in Montpellier. His initial research field was the study of thermotropic liquid crystals using x-ray diffraction, but for the last few years he has been working in micellar solutions and microemulsions by small angle x-ray scattering and neutron diffraction.

Eric Derouane joined the Facultes Universitaire de Namur as Professor and Director of the Laboratory of Catalysis in 1973. He is presently a Research Scientist in the central research division of Mobil Research and Development Corp. He received his Docteur en Sciences from the University of Liege. He has received several scientific distinctions and has edited two books and has authored/coauthored about 150 papers. Until Jan. 1982 he acted as President of the European Association on Catalysis, and is presently an editor of the Journal of Molecular Catalysis. His primary interests are catalysis, more specifically by zeolites, and the understanding and utilization of metal-support interactions.

Alexander Derzhanski is Professor in the Institute of Solid State Physics, Bulgarian Academy of Sciences, Sofia, Bulgaria. His Ph.D. dissertation was on the methodology of NMR and Doctor of Sciences dissertation on liquid crystals. He was a winner of the National "Obreshkov" Award in Physics in 1980, and has published a number of papers in the areas of electronics, NMR and molecular physics.

J. G. Dore is Lecturer in Physics at the University of Kent at Canterbury, U.K. Received Ph.D. degree from the University of Birmingham, U.K. and has published on theory of liquids, hydrogen bonding, and neutron diffraction.

V. Eck is a research student in the Department of Physical Chemistry, Fritz-Haber-Institut der Max-Planck-Gesellschaft, Berlin working for his Ph.D. degree.

Hans-Friedrich Eicke is Professor of Physical Chemistry at the University of Basel, Switzerland. He received his Ph.D. from the University of Gottingen (Max-Planck - Institute for Physical Chemistry) in 1961. He has published more than 50 papers dealing with surface active electrolytes in hydrocarbon solutions, inverted micelles, and microemulsions. He is coauthor (with B. Lindman and H. Wennerström) of a monograph on micelles in aqueous and nonpolar surfactant solutions published by Springer Verlag, Berlin.

James H. Fendler is Professor of Chemistry at Clarkson College of Technology which he joined in Jan. 1982. Prior to coming to Clarkson, he was Professor of Chemistry at Texas A & M University. He obtained his Ph.D. degree in Physical Organic Chemistry in 1964 from the University of London, England, and the same university recognized his authorative standing by awarding him a D.Sc. degree in 1978 for his work in membrane mimetic chemistry. More recently, he has been honored with the 1983 Kendall Award, the ACS Award in Colloid or Surface Chemistry. He has been active in a number of research areas, and during the past year his research has continued in areas of colloid chemistry with significant new results. In addition to continuously investigated problems in membrane mimetic chemistry, he has put a major effort into research on polymerizing surfactant aggregates and on the ulilization of polymeric surfactant vesicles in photochemical energy conversion. He has more than 140 publications to his credit and is the coauthor (with E. J. Fendler) of the book Micellar Catalysis published in 1975, and most recently has authored the book Membrane Mimetic Chemistry published in 1982.

J. Ferreira is a research student at the Laboratoire de Chimie Physique des Macromolecules aux Interfaces, Universite Libre de Bruxelles, and has 7 publications.

P. D. I. Fletcher is a Research Fellow in the Chemical Laboratory, University of Kent at Canterbury, U.K. where he received his Ph.D. degree. He has published on reactions in micellar and microemulsion systems, and fluorescence lifetime measurements.

Tomlinson Fort, Jr. is presently Vice President for Academic Affairs at California Polytechnic State University, San Louis Obispo, California which he joined in 1982. Prior to his current position, he was Provost at the University of Missouri-Rolla (1980-1982). His professional experience includes seven years with duPont, eight years as a faculty at Case Western Reserve Universty, and seven years at Carnegie-Mellon University. He received his Ph.D. degree in Physical Chemistry at the University of Tennessee and spent a year at the University of Sydney, Australia where he did research in colloid science with Prof. A. E. Alexander. He has published nearly 70 research papers on various aspects of surface and colloid science. For five years he was codirector of a research program on enhanced oil recovery begun at Carnegie-Mellon University.

Wilson A. Frezzatti, Jr. was last year a graduate student in the Faculty of Pharmacy and Biochemistry, University of Sao Paulo, Sao Paulo, Brazil. He has a Fellowship from the Foundation for Support of Research of the State of Sao Paulo for introductory work in scientific research.

Luz Alicia Fucugauchi is Head of the Department of Nuclear Techniques and Positronium Chemistry at the Instituto Nacional de Investigaciones Nucleares in Mexico City. Had education at the University of Mexico and received M.Sc. in Physics in 1980. Has been Visiting Scientist in the United States and France, and is a member of the Academy of Scientific Research (Mexico). The research interests are nuclear techniques, radiochemistry, nuclear chemistry, and physical-organic chemistry and is the author of a number of research papers including some dealing with micelles and microemulsions.

J. M. Furois is with the Laboratoire de Chimie-Physique, Universite P et M Curie, Paris, France.

Cecilia M. C. Gambi is a researcher in the Physics Institute of the University of Florence, Italy. She is the recipient of Fellowship of the CNR by the College de France (Paris). She is working in the field of microemulsions and lyotropic liquid crystals and has published some papers.

E. Goormaghtigh is a research student at the Laboratoire de Chimie Physique des Macromolecules aux Interfaces, Universite Libre de Bruxelles, and has 12 publications.

P. Guering is currently with the Ecole Normale Superieure, Paris, France. Was a graduate student at Ecole Superieure d'Optique (1978-1981) and received a fellowship from CNRS.

Rene Hasse is with the Institute for Applied Physics, University of Basel, Basel, Switzerland.

Jean Heil obtained his Ph.D. degree in 1981 from the Universite de Pau et des Pays de l'Adour, Pau, France for his research on the electrical properties of microemulsion-type media. In 1978, he obtained also a degree in Business Administration. During 1982, he served with the French Army as a scientist attached to the "Institut Franco-Allemand de Recherches", Saint-Louis, France.

Ulf Henriksson is Research Associate in the Department of Physical Chemistry, The Royal Institute of Technology, Stockholm, Sweden, where Doctor of Thechnology was received in 1975. Has published on the applications of NMR spectroscopy in amphiphilic systems.

B. Hickel is with the CEN-Saclay, Gif sur Yvette, France.

J. F. Holzwarth has been the Leader of a research group at the Fritz-Haber-Institut of the Max-Planck-Society in Berlin since 1972. He is also holding a Lecturership in Physical Chemistry at the Freie Universitat Berlin. His research interests encompass the field of very fast reactions in solution, in particular fast electron transfer reactions in self-aggregating systems like micelles and vesicles, as well as interactions between nucleic acids and dyes. He has successfully carried out the development of extremely fast experimental methods such as the integrated flow (time resolution 10^{-5}s) and the iodine-laser temperature jump (time resolution 10^{-9}s).

A. M. Howe is a SERC research student at the University of Kent at Canterbury, U.K.

Abul Hussam will be receiving his Ph.D. in Chemistry from the University of Pittsburgh in 1982. He obtained his B.S. (Honors) and M.S. degrees in Chemistry from the University of Dacca, Bangladesh. His current research interests include electrochemistry in non-aqueous media, solute-solvent interactions as studied by spectroscopic techniques, and microemulsions.

Jyi-feng Jeng is with the Department of Chemical Engineering, Rice University, Houston, TX.

Bo Jönsson is Assistant Professor in Physical Chemistry, University of Lund, Sweden. His research interests are in the field of computer simulations of liquids and solutions.

C. N. Joo is presently Chairman and Professor, Department of Biochemistry, Yonsei University, Seoul, Korea. He received his Ph.D. degree in Biochemistry in 1963 from the University of Liverpool and carried out postdoctoral work at NRC Canada. During 1972-1975, he was CIDA Fellow at NRC of Canada, and was President of Korean Biochemical Society during 1979-1980. Presently he is delegate to the International Union of Biochemistry General Assembly for the Korean Biochemical Society. He has published 40 papers about the effects of Korean ginseng saponins on biological systems.

R. K. Kubik is a Research Assistant (Ph.D. student) at the Institute for Physical Chemistry, University of Basel, Switzerland, and has authored/coauthored several research papers.

P. Lalanne is Assistant Professor at the University of Bordeaux (France). He received his Doctorat es Sciences in 1976. His research is focused on nuclear magnetic resonance, molecular dynamics in liquids, thermodynamical properties and model of microemulsions.

Jacques Lang is Maitre de Recherche at the Centre de Recherches sur les Macromolecules, CNRS, Strasbourg, France. He received his D.Sc. in 1968 from the University of Strasbourg. His research has concerned fast kinetics of proton transfer reactions and micellar systems, and more recently he has been involved in the study of microemulsions.

D. Langevin is Maitre de Recherche, CNRS, at the Laboratoire de Spectroscopie Hertzienne de l'ENS, Paris, France. She received her Doctor es Sciences in 1974 from Paris.

F. C. Larche is Lecturer in Materials Science at the Universite des Sciences et Techniques du Languedoc, Montpellier, France. He received his Ph.D. degree in Materials Science from

MIT. He has been working in the thermodynamics of multicomponent
stressed solids and mechanical properties of glasses, and he is
currently interested in the thermodynamics and phase transformation kinetics of surfactant solutions.

L. Leger is currently with the College de France, Paris,
and holds Ph.D. and Thesis degrees.

Y. S. Leong is presently a candidate for the D.Sci. degree
at the "Centre de Recherches sur les Macromolecules", Strasbourg,
France. His research concerns polymerization in microemulsions.

Panagiotis Lianos is Assistant Professor at the University
of Crete, Iraklion, Greece. He received his Ph.D. in 1978 from
the University of Tennessee, Knoxville. He specializes in photo-
physics and is currently interested in using fluoroscence probes
in the study of organized assemblies.

Bjorn Lindman is Professor of Physical Chemistry, University of Lund, Sweden. His research interests are in the fields
of micelles, microemulsions, and other surfactant systems, ion
binding to polyelectrolytes and biomacromolecules and applications of NMR spectroscopy to physico-chemical problems.

P. Linse is with the Division of Physical Chemistry I,
Chemical Center, Lund, Sweden.

Terrence D. Lomax is a graduate student working towards his
Ph.D. degree at the University of Auckland, New Zealand. He
received his M.Sc. in Chemistry in 1977. At present, he is studying the decomposition reactions of esters by aggregate systems in
nonpolar solvents.

Rafael Lopez is with the Instituto Nacional de Investigaciones Nucleares in Mexico City. He received his B.Sc. in Chemistry in 1980 from the University of Mexico and has coauthored
research papers. In 1981 he was Visiting Scientist at Texas A & M
University.

N. Lufimpadio is currently with the Laboratory of Catalysis
of the Facultes Universitaires de Namur in Belgium where he is
working toward his Ph.D. degree in Chemistry. He received his
"Licence en Sciences Chimiques" in 1972 from the University of
Zaire and was headmaster of Secondary School before coming to
Namur. He is the author of two books on chemistry accepted the
the Education Ministry of Zaire as official manuals for Chemistry
education in Zairian Secondary Schools. His current research
interest is in the field of structure and properties of reversed
micelles and the preparation of small metallic particles in
reversed micelles.

Raymond Mackay is Professor of Chemistry at Drexel University, Philadelphia. He received his Ph.D. degree in Chemistry in 1966 from the State University of New York, Stony Brook. His research work in recent years has been in the area of colloid and surface science. His research efforts involve the study of chemical reactions in microemulsions as well as physicochemical studies of these media. He is also investigating reactions of gases with submicron aerosols. He is an Associate Editor of the J. American Oil Chemists' Society.

M. P. McDonald is Principal Lecturer in Chemistry at the Sheffield City Polytechnic, England where Ph.D. degree was granted.

J. Meunier is Maitre de Recherche (CNRS) at the Ecole Normale Superieure in Paris, France. Was a graduate student at the Ecole Normale Superieure and holds Ph.D. (1965) and Thesis (1971) degrees.

Sonia Millán is Research Associate at the Instituto Nacional de Investigaciones Nucleares in Mexico City. She received Dr. de 3éme Cycle in 1981 from the Universite Louis Pasteur, Strasbourg, France, and has coauthored research publications.

Clarence A. Miller is Professor of Chemical Engineering at Rice University, Houston, TX. After receiving his Ph.D. Degree from the University of Minnesota, he spent twelve years on the faculty at Carnegie-Mellon University before assuming his present position in 1981. His research interests are in interfacial phenomena, with particular interest in interfacial stability, microemulsions, and enhanced oil recovery.

Kashmiri Lal Mittal is presently employed at the IBM Corporation in Hopewell Junction, NY. He received his M.Sc. (First Class First) in 1966 from Indian Institute of Technology, New Delhi, and Ph.D. in Colloid Chemistry in 1970 from the University of Southern California. In the last ten years, he has organized and chaired a number of very successful international symposia and in addition to this three-volume set, he has edited 14 more books as follows: Adsorption at Interfaces, and Colloidal Dispersions and Micellar Behavior (1975); Micellization, Solubilization, and Microemulsions, Volumes 1 & 2 (1977); Adhesion Measurement of Thin Films, Thick Films and Bulk Coatings (1978); Surface Contamination: Genesis, Detection, and Control, Volumes 1 & 2 (1979); Solution Chemistry of Surfactants, Volumes 1 & 2 (1979); Solutions Behavior of Surfactants - Theoretical and Applied Aspects, Volumes 1 & 2 (1982); Physicochemical Aspects of Polymer Surfaces, Volumes 1 & 2 (1983); and Adhesion Aspects of Polymeric Coatings (1983). In addition to these volumes he has published about 50 papers in the areas of surface and colloid chemistry,

adhesion, polymers, etc. He has given many invited talks on the
multifarious facets of surface science, particularly adhesion, on
the invitation of various societies and organizations in many
countries all over the world, and is always a sought-after
speaker. He is a Fellow of the American Institute of Chemists and
Indian Chemical Society, is listed in American Men and Women of
Science, Who's Who in the East, Men of Achievement and other
refrerence works. He is or has been a member of the Editorial
Boards of a number of scientific and technical journals. Presently, he is Vice President of the India Chemists and Chemical
Engineers Club.

Janos B. Nagy is Professor of Chemistry at Facultes Universitaires Notre-Dame de la Paix, Namur, Belgium. He received his
Ph.D. degree in 1970 from Catholic University of Louvain,
Belgium. He has published about 70 papers dealing with molecular
interactions in the liquid and adsorbed states, structure and
properties of reversed micelles, preparation of small metallic
catalysts and the characterization of solid catalysts by high
resolution solid state NMR techniques.

A. W. Neumann is Professor in the Department of Mechanical
Engineering, University of Toronto, Canada. He received Dr.
rer.-nat. from Mainz, West Germany. His research interests
include applications of surface thermodynamics to technological
and biological systems.

John David Nicholson is currently teaching at Giggleswick
School, N. Yorkshire, England. He received his Ph.D. degree in
Chemistry in Dec. 1982 from UMIST, Manchester, England and his
research has been concerned with study of microemulsions using
laser light scattering.

Luc Nicolas-Morgantini is currently preparing his Ph.D.
thesis on the subject of microemulsions at the Universite de Pau
et des Pays de l'Adour, Pau, France where he graduated in
Chemistry and Physics in 1981.

Charmian J. O'Connor is Associate Professor of Chemistry at
the University of Auckland, New Zealand, where she received her
Ph.D. and D.Sc. degrees. She has had visiting appointments at
University College and Imperial College, London; University of
California, Santa Barbara ; Texas A & M University; and Nagasaki
University. Her research interests include micellar catalysis and
high pressure kinetics of octahedral substitution reactions and
has over 90 publications.

Lars Ödberg is research director at the Institute for Surface Chemistry, Stockholm, Sweden. He obtained his Ph.D. from the
Royal Institute of Technology, Stockholm in 1970 doing research

in the NMR field. During the period 1971-1978 he was lecturer at
the University of Linkoping. He has published about 35 papers in
the field of NMR, light scattering and surface chemistry.

Nicole Ostrowsky has been with the Physique of Liquids
group at the University of Nice, France since 1976. Prior to her
current position, she was at the University of Paris where she
received her Ph.D. in 1970. She graduated from the Ecole Normale
Superieure in Paris in 1966 after passing the agregation de Physique and her 3rd cycle thesis on Optical Pumping. This was
followed (1967-1969) by work at Harvard in Prof. Bloembergen's
laboratory. Her major contributions have been in the light
scattering study of structural relaxation in viscous fluid (near
the glass transition) and more recently she has started a study
on the formation, the shapes and interactions of vesicles.

M. Paillette is Maitre de Recherche (CNRS) at the Groupe de
Physique des Solides de l'ENS, Paris, France. Received Ph.D.
degree in 1962 and Thesis in 1969.

N. M. Perrins is an SERC research student at the University
of Kent at Canterbury, U.K.

Jean Peyrelasse received his D.Sc. degree in 1974 from the
Universite de Pau et des Pays de l'Adour, Pau, France, and was
appointed Assistant Professor of Physics with the same institution in the same year. His main research interest is dielectric
methods as applied to the study of classical liquids, liquid mixtures and microemulsions, and has some 20 papers to his credit.

M. P. Pileni is with the Laboratoire de Chimie-Physique,
Universite P et M Curie, Paris, France.

Syed Qutubuddin is presently Assistant Professor in the
Chemical Engineering Department, Case Western Reserve Unversity,
Cleveland. He pursued his Ph.D. program at Carnegie-Mellon
University with thesis on microemulsions with ultralow interfacial tensions for enhanced oil recovery. Prior to coming to
USA, he was a Lecturer at Bangladesh University of Engineering
where he received his B.S. (Honors) in Chemical Engineering and
was awarded the University Gold Medal for Academic Excellence in
1975. His current research interests are in the areas of colloids
and interfacial phenomena, and has several publications.

Jean-Claude Ravey is Maitre de Recherches (CNRS) and is in
the Faculty of Sciences of Nancy I University, Nancy, France. He
received his D.Sc. in 1973 and his current research interests are
structural investigations of colloidal suspensions.

Rocio Reynoso is a Research Associate at the Instituto

Nacional de Investigaciones Nucleares in Mexico city. Received
Bs. in Chemical Engineering in 1978 from the Nacional Polytech-
nical Institute (Mexico) and has coauthored research publications.

B. H. Robinson is Lecturer in Chemistry at the University
of Kent at Canterbury, U.K. He received his D.Phil. degree from
the University of Oxford, U.K. He has published on fast reactions
in solution, proton transfer reactions, and reactions in micellar
and microemulsion systems.

Kenneth Rosenquist is Research Scientist in the Pharmaceutical and Cosmetic Section at the Institute for Surface
Chemistry in Stockholm, Sweden.

Y. Rotenberg is a Ph.D. candidate in the Department of
Mechanical Engineering, University of Toronto, Toronto, Canada.

D. Roux is Attache de Recherches at CNRS at the Centre de
Recherche Paul Pascal in Bordeaux-Talence, France. He is presently a graduate student and is studying microemulsions.

Eli Ruckenstein has been Distinguished Professor, Department of Chemical Engineering, SUNY Buffalo since 1981, and he
was Faculty Professor at the same institution, Faculty of Engineering and Applied Sciences from 1973-1981. He received his Ph.D.
degree from the Polytechnic Institute, Bucharest, Romania and was
Professor at this institute from 1949-1969. He has been the
recipient of the following awards: The George Spacu Award for
Research in Surface Phenomena of the Romanian Academy of Sciences
(1963); two research (1958 and 1964) and one teaching (1961)
awards from the Romanian Department of Education; and the Alpha
Chi Sigma Award (1977) of the American Institute of Chemical
Engineers. He has published extensively (over 350 papers) dealing
with many subjects. His research interests are transport
phenomena, catalysis, separation processes, colloids and interfaces, and biophysics. He has been visiting professor at the
Catholic University, Leuven, and Technion-Israel, and is listed
in World Who's Who in Science.

J. M. Ruysschaert is Associate Professor, Universite Libre
de Bruxelles, Belgium. Has a Ph.D. degree and about 100
publications.

Lisbeth Rydhag is Section Manager for the Pharmaceutical
and Cosmetic Research and the Food Research Section at the
Institute for Surface Chemistry in Stockholm, Sweden. She has a
Ph.D. degree.

S. A. Safran is a Research Physicist (Theoretical and

Condensed Matter Physics) at Exxon Research and Engineering Co., Linden, NJ which he joined in 1980. He received his Ph.D. degree in Physics in January 1978 from MIT and had a postdoctoral appointment at Bell Labs before moving to Exxon. He has about 30 publications in the areas of light scattering in magnetic semi-conductors, intercalation compounds, phase transitions, and statistical mechanics of microemulsions.

Zoltan A. Schelly is Professor of Physical Chemistry, The University of Texas at Arlington. He received his D.Sc. degree in 1976 from the Technical University of Vienna, Austria. He has had faculty and research appointments at various universities. In 1974, he was an Alexander von Humboldt Fellow at the Max-Planck-Institute for Biophysical Chemistry in Gottinigen, West Germany. Presently, he is on the editorial board of Advances in Molecular Relaxation and Interaction Processes. His main research interests are the relaxation kinetics of fast chemical and physical rate processes, including micelle formation, solubilization, and interfacial reactions; and chemical instabilities, bifurcations, and oscillating reactions.

Shirley Schreier is Associate Professor in the Department of Biochemistry, Institute of Chemistry, University of Sao Paulo, Sao Paulo, Brazil. She received her Ph.D. degree in Physical Biochemistry. She has 35 publications, mostly dealing with the use of the spin label method for the study of membranes, and has written several book chapters. She has been a Visiting Scientist at the NRC, Canada; National Biomedical ESR Center, Medical College of Wisconsin; and Max-Planck-Institut fur biophysikalische Chemie, Gottingen. Her other interest relate to the study of detergent aggregates and conformational properties of peptide hormones.

S. Schürch is Associate Professor of Biophysics and Medicine, University of Western Ontario, London, Ontario, Canada where he received his Ph.D. degree. His research interests include the applications of surface thermodynamics to biological systems.

Donatella Senatara is Associate Professor in the Physics Department at the University of Florence, Italy. She is a member of the Scientific Committee of the Italian Society of Pure and Applied Biophysics. Her primary research is in liquid state physics (water dispersed systems, lyotropic liquid crystals, phase transitions), but she has worked in a number of other areas including biological impedances determination, biorhythms, electromagnetic pollution (non-thermal effect) and biomembrane model systems. She has published several papers in these areas.

Juan Serrano is a Research Associate at the Instituto Nacional de Investigaciones Nucleares in Mexico City. He received his M.Sc. in 1982 in Nuclear Sciences from the University of Mexico and has coauthored research publications. During 1980-1981 he was Visiting Scientist in the United States.

E. Sjöblom is a Research Scientist and Section Manager for the Chemical-technical Research Section at the Institute for Surface Chemistry, Stockholm, Sweden. She received her B.S. degree in 1975 from the University of Stockholm. She has about fifteen publications in the area of microemulsions.

O. Sorba is a graduate student at the Ecole Normale Superieure, Paris, France.

Didier Sornette is currently working for his Ph.D. at the University of Nice on vesicles interactions and deformabilities in external constraints. He graduated from the Ecole Normale Superieure in Paris in 1981 after obtaining the agregation de Physique and his 3rd cycle thesis on the formation and stability of vesicles.

Per Stenius is Director of the Institute for Surface Chemistry, Stockholm, Sweden and Associate Professor of Colloid Chemistry at the Royal Institute of Technology, Stockholm. He received his Ph.D. from the University Åbo Akademi (Turku), Finland in 1973. He is a member of the Royal Academy of Engineering Sciences in Sweden. He serves on the editorial boards of Journal of Colloid and Interface Science, and Colloids and Surfaces and is an associate member of the IUPAC commission on Colloid Science. He has published about 90 papers on association colloids, polymer latexes, colloidal stability and surface characterization.

C. Stil is a research student at the Laboratorie de Chimie Physique des Macromolecules aux Interfaces, Universite Libre de Bruxelles, Bruxelles, Belgium.

Peter Stilbs is currently Associate Professor of Physical Chemistry at the University of Uppsala, Uppsala, Sweden. He received his Ph.D. from the Lund Institute of Technology in 1974. He has published about 60 papers dealing with the application of magnetic resonance methods for the study of physico-chemical phenomena.

Th. F. Tadros is Research Associate at ICI Plant Protection Division, Bracknell, England which he joined in 1969. He received his Ph.D. in 1962 from Alexandria University and was Lecturer until 1966. In 1969 he spent a sabbatical year at Wageningen with Prof. Lyklema. He has authored/coauthored 55 papers in the area of colloid and interface science and is one of the editors of Colloids and Surfaces.

G. J. T. Tiddy is Senior Scientist, Unilever Research Laboratory, Port Sunlight, England. He received his Ph.D. from East Anglia.

Christian Tondre is Maitre de Recherches at the CNRS (E.R.A. No. 222, University of Nancy I, France). He received his D.Sc. in 1969 from the University of Strasbourg. He has published in the area of relaxation methods and fast kinetics and his current research interests include micelles, microemulsions, polyelectrolyte catalysis and fast reaction mechanisms.

C. Toprakcioglu is a Research Fellow in the Chemistry and Physics Laboratory, University of Kent at Canterbury, U.K. Received Ph.D. degree from the University of Cambridge, U.K.

H. Wennerstrom has been Professor of Physical Chemistry at the University of Stockholm since 1980. He received his Ph.D. from the University of Lund in 1974. His research fields are NMR relaxation theory, electrostatic interactions in liquids, and thermodynamic properties of surfactant-water systems.

Aristotelis Xenakis is a graduate student in Molecular Chemistry at the University of Nancy I, Nancy France. He received his earlier degrees at the University of Athens, Greece.

Raoul Zana is Directeur de Recherche at the Centre de Recherches sur les Macromolecules, CNRS, Strasbourg, France. He received his D.Sc. degree in 1964 from the University of Strasbourg. His current research interests include equilibrium and dynamic aspects of micellar solutions, microemulsions, and polyelectrolyte solutions.

Antonia Zheliaskova is Senior Research Fellow in the Institute of Solid State Physics, Bulgarian Academy of Sciences, Sofia, Bulgaria. Her Ph.D. dissertation was on research into semiconductors using the NMR method. She is interested in biophysics, semiconductors, molecular physics and NMR and has published in these areas.

Abdallah Zradba is presently a Research Fellow at the Universite de Pau et des Pays de l'Adour, Pau, France and is preparing his Ph.D. thesis. He received his graduate degree in Physics in 1980 from the Facultes des Sciences, Rabat, Morocco. After completion of his thesis, he will return to Morocco as an Assistant Professor with the Ecole Normale Superieure de Takkadoum located near Rabat.

Iris Zschokke is with the Institute for Applied Physics, University of Basel, Basel, Switzerland.

INDEX